Student Study Guide

to accompany

Chemistry

The Molecular Nature of Matter and Change

Seventh Edition

Martin S. Silberberg
and
Patricia G. Amateis
Virginia Tech

Prepared by
Elizabeth Bent Weberg

STUDENT STUDY GUIDE TO ACCOMPANY
CHEMISTRY: THE MOLECULAR NATURE OF MATTER AND CHANGE, SEVENTH EDITION

Published by McGraw-Hill Education, 2 Penn Plaza, New York, NY 10121.

This book is printed on acid-free paper.

1 2 3 4 5 6 7 8 9 0 ROV/ROV 1 0 9 8 7 6 5 4

ISBN: 978-0-07-813161-5
MHID: 0-07-813161-8

www.mhhe.com

To the Students

This study guide highlights important ideas from each chapter of the text, and sometimes offers a different angle from which to view a concept. Seeing the same material from a different point of view can sometimes be the switch to the "light bulb effect".

How to Use This Study Guide

Study with a **pencil,** lots of **scrap paper,** and **actively work through the problems.** It's easy to fool yourself into thinking you understand a concept when simply reading through example problems.

Teach what you just learned to someone else.

Fiddle around. Don't expect to see answers immediately. Keep at it. Like learning to play the violin, learning chemistry takes lots of "fiddling."

Create mental pictures of *sticky molecules in motion.*
The molecular models in your textbook are extremely helpful visual aids.

Believe in atoms.

Nothing exists except atoms and empty space; everything else is opinion.
Democritus (460-370 BC)

Acknowledgments

Thanks to Lora Neyens at McGraw-Hill for her help and support of the study guide revisions.

I gratefully acknowledge the comments and suggestions from my father, Henry Bent, many of which have been incorporated throughout the study guide (for example, the treatment of temperature scales in Chapter 1, relating electron configurations to a periodic table in Chapter 8, bonding models in Chapter 9, and practice problems for the ends of Chapters 14 and 23).

Contents

Keys to the Study of Chemistry

Atoms exist! Richard Feynman (1918-1988), Nobel Prize Physicist and brilliant teacher, describes the fundamental importance of this fact to all scientific knowledge as:

"If, in some cataclysm, all of scientific knowledge were to be destroyed, and one sentence passed on to the next generations of creatures, what statement would contain the most information in the fewest words? I believe it is the *atomic hypothesis* (or the atomic *fact*, or whatever you wish to call it) that

all things are made of atoms—little particles that move around in perpetual motion, attracting each other when they are a little distance apart, but repelling upon being squeezed into one another.

In that one sentence, you will see, there is an enormous amount of information about the world, if just a little imagination and thinking are applied."

(Six Easy Pieces, Richard P. Feynman, Basic Books, 1995)

1.1 CHEMISTRY, MATTER, AND ENERGY

Chemistry is the science that deals with the specific properties, especially the composition and structure, of matter and its transformations. It isn't a chem-mystery! We are all familiar with matter—the stuff that is all around us. What you may not be as familiar with are the terms that chemists use to describe matter and the way in which chemists view matter as composed of tiny particles in random motion. Chemistry is a physical science because it deals with "physical quantities"—things that can be expressed by a number followed by a unit.

A chemist's concern

Which of the following would likely concern a chemist?

a) Water, like all matter, falls a distance of $\frac{1}{2}gt^2$ during a given time, *t*.
b) Water has a density of 1.0 g/mL at 25°C and 1 atm.
c) Water can be electrolytically decomposed into the elements hydrogen and oxygen.
d) The formula for water is H_2O.
e) The structure of a molecule of water is:

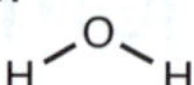

f) Water reacts violently with sodium metal to produce aqueous sodium hydroxide and hydrogen gas.
g) Water is one of the products when natural gas burns.
h) Water is the only product of combustion of hydrogen.
i) Salt water boils at a higher temperature than pure water.
j) A cupful of water molecules as a liquid occupies a volume of approximately 100 gallons at 100°C and atmospheric pressure, as a gas.
k) Water in rivers can erode land to form canyons.
l) Hard water contains calcium and magnesium ions.
m) Water is a good solvent for table salt and for table sugar.
n) Water and oil do not mix.
o) Water expands when it freezes.
p) Water is the chief constituent of our bodies.

answer: All of the above might concern a chemist except for "a", "k", and "p" which would be of more concern to a physicist, geologist, and biologist, respectively.

Properties of Matter

Matter is anything that has mass and occupies a volume. We describe matter by its **physical** and **chemical properties.** The first time you filled a car with gasoline, you most likely made observations about the properties of gasoline. First, as you began to pump the gasoline into the car, you may have noticed that gasoline has a distinct odor. When you overfilled the tank, and the gas spilled down the car, you probably noticed it was a clear liquid, but it (fortunately) did not dissolve the paint off your parents' car. As the gas landed in a puddle of rainwater, you might have noticed that it floated on top of the water. You noticed the sign that asked you to stop your engine and refrain from smoking, so you speculated that gasoline is flammable.

Gasoline's odor, color, and density are some of its **physical properties** because they do not involve interaction with other substances. The flammability of gasoline is a **chemical property** because it involves interactions with other substances. To react chemically, and rearrange their atoms, molecules must collide physically. When gasoline ignites, it reacts with oxygen to form new substances including carbon dioxide and water. This is an example of a **chemical change** or **chemical reaction.**

A **physical change** occurs when a substance changes state but not composition. When water boils or freezes, it undergoes a physical change because its composition remains H_2O whether it is ice, liquid water, or steam. A **chemical change** or **chemical reaction** occurs when a substance is converted into a different substance. When iron rusts, it reacts with oxygen to form iron oxide. A chemical change occurs; iron oxide is a substance with a set of properties different from iron.

The States of Matter

If we make the assumption that two atoms cannot be at the same place at the same time, broadly speaking, there are three ways molecules may be arranged:

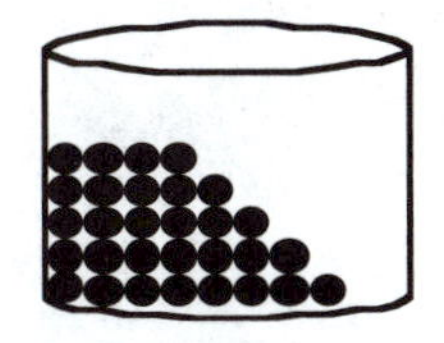
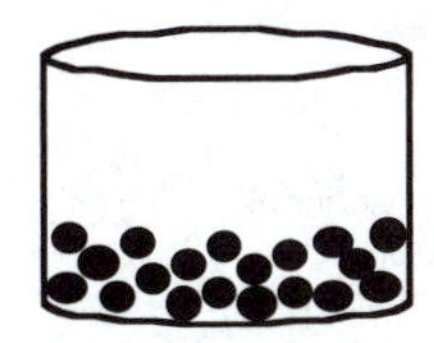
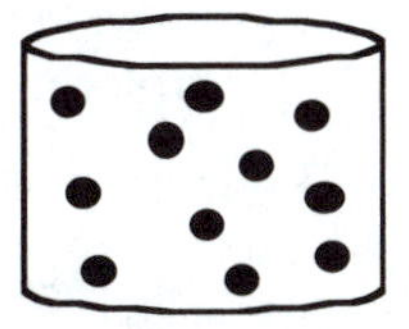
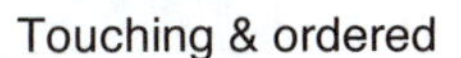

Touching & ordered — Touching & disordered — Not touching

States of matter exercise

1) The arrangements of molecules above represent the three phases of matter called:

____________	____________	____________
Fixed shape Fixed volume	Takes the shape of the container Fixed volume	Takes the shape of the container Takes the volume of the container

2) For water, these three phases of matter have the special names:

____________ ____________ ____________

3) The Greeks gave these three states of matter the names:

____________ ____________ ____________

answers: 1) solid, liquid, gas; 2) ice, water, steam; 3) earth, water, air.

(Note that the fourth type of matter that the Greeks classified, fire, is the thermal energy (heat) absorbed or liberated in passing from one state of matter to another.)

Energy

What is **energy?** When we say we feel energetic, or we want a candy bar for quick energy, we have an intuitive idea that energy is something that makes us move and get things done. In fact, we define mechanical energy as the capacity to do work. We define work as the distance an object moves times the force acting on the object $(W = f \times d)$.

Energy of mechanical systems (systems in which there are no changes in temperature or chemical composition) is divided into **potential energy** (energy of position) and **kinetic energy** (energy of motion). When we hold a ball above the ground, it contains potential energy that is converted to kinetic energy as the ball is dropped. The sum of the potential energy and kinetic energy for the ball as it falls is constant:

$$E_{\text{kinetic}} + E_{\text{potential}} = \text{constant}$$

As the ball comes to rest, the energy is not lost, it is converted to other forms through **work** and **heat.**

Thermochemistry, the study of heat transferred in chemical reactions, will be discussed in Chapter 6. Scientists know from experience that the energy of the universe is constant; energy cannot be created or destroyed. Mathematically we write:

$$E_{univ.} = \text{constant}$$

Therefore:

$$(E_{univ.})_{final} = (E_{univ.})_{initial}$$

$$(E_{univ.})_{final} - (E_{univ.})_{initial} = 0$$

$$\Delta \boldsymbol{E}_{\mathbf{univ.}} = \mathbf{0}$$

where $\Delta E_{univ.}$ means the difference between E_{final} and $E_{initial}$.

1.2 THE ORIGINS OF MODERN CHEMISTRY

As early as the 1st century AD, people studied chemistry primarily in an attempt to change metals into gold and to find remedies for diseases. In the 17th century, scientists proposed an incorrect theory to explain why substances change form (undergo a chemical transformation) as they burn. Scientists proposed that combustible materials released an undetectable substance, **phlogiston,** when they burned. These scientists would say that wood (a combustible material) contains phlogiston; as wood burns, ashes form as the phlogiston "burns out" of the wood. A French chemist named Antoine Lavoisier made careful quantitative measurements to show that oxygen, a component of air, is required for combustion and combines with a substance as it burns. Lavoisier's theory of combustion explained why substances require air for combustion and why metals gain mass as they burn. Lavoisier is considered to be the father of modern chemistry.

A theory of combustion

How did the phlogiston theory of combustion explain the following observations?

a) Metals undergo a chemical change as they are heated.

b) Substances require air for combustion.

c) Charcoal burns for only a short time in a closed vessel.

d) Metals gain mass as they burn.

answers:

a) A substance called phlogiston flows out of a combustible material as it burns.

b) Air is needed to attract the phlogiston out of a combustible substance.

c) Burning in a vessel stops when the air is saturated with phlogiston.

d) Some scientists suggested that phlogiston has a negative mass!

See page 9 in your textbook for a discussion of the phlogiston fiasco.

1.3 THE SCIENTIFIC APPROACH

The scientific method is a series of steps often used to solve problems. You have probably used the scientific method in daily life without analyzing the steps. Let's consider a problem you might face, and then notice the steps you go through as you solve the problem.

You commute to class on a bicycle and get a flat rear tire. An unobservant person might never notice the tire is flat and simply wonder why the hills seem so tough on the way home. However, you **observe** that the tire is flat and you remember seeing glass on the road. You **hypothesize** that the tire is flat because glass punctured the inner tube.

You place the inner tube in water **(experiment)** and observe bubbles, which support your hypothesis that glass punctured the inner tube. You patch the inner tube, and ride your bike to class the next day. Again, you obtain a flat rear tire. You hypothesize that your patch leaked so you check the inner tube again in water (experiment). Bubbles escape from the patch, so you hypothesize that perhaps the glue to your patch kit was old, and to avoid patching problems, you replace the inner tube with a new one.

After riding your bike, again the rear tire is flat. This time you notice that the tube always punctures in approximately the same place. A **model** or **theory** begins to emerge that something on the bike must be causing the flat tires. You hypothesize that there is a sharp object inside the tire, so you remove the tire and check it carefully for a foreign object. Finding nothing, you continue to check the bike and notice that there is no rim tape, and a spoke pokes through the rim in the same place you found the punctures in the inner tubes. You apply rim tape and ride to class the next day. After class, you return to your bike to find that the FRONT tire is flat. The next day you walk to class and order Kevlar® inner tubes.

You put the scientific method to work: you made **observations**, formed **hypotheses** via imagination, knowledge, and logic to explain the observations, ran **experiments** to support or disprove the hypotheses, and eventually formed a **model (theory)** based on experiments. Even though you solved the immediate problem, you determined that new technology was required to avoid future problems.

1.4 MEASUREMENT IN SCIENTIFIC STUDY

No measurement can be 100 percent accurate (according to the laws of physics), but scientists are getting incredibly close. Consider being able to detect or measure:

- The stomping of soccer fans after a goal is scored 30 miles away. (A seismologist at Washington University detected these tiny vibrations on an earthquake gauge.)
- A clock that doesn't gain or lose one second in a billion years. (Researchers at the National Institute of Standards and Technology built such a clock based on the "ticks" from a single atom of mercury.)
- The force needed to make one atom hop over another. (A physicist at IBM Research Division in San Jose, California used two powerful microscopes to measure this force—61 trillionths of an ounce.)
- The shrinking distance between Florida and Canada. (A geophysicist at Purdue University used satellite technology to measure minute changes in locations of GPS stations, and discovered that Florida moves three-hundredths of an inch closer to Canada each year as the continental plate recovers from the weight of ice during the last Ice Age.)
- A temperature of only 800 trillionths of a degree Fahrenheit above absolute zero. (Nobel Prize winner Wolfgang Ketterle cooled atoms to this temperature in his lab at the Massachusetts Institute of Technology.)
- The sideways motion of a star trillions of miles away, even though its speed is less than 10 inches per hour. (Astronomers from the Harvard-Smithsonian Center for Astrophysics plan to make these measurements using a laser device.)
- The weight of a single molecule with an uncertainty of less than a billionth of a trillionth of an ounce. (A physicist at the California Institute of Technology made this measurement.)

Obtaining these levels of precision and accuracy is more than an academic pursuit. The technology developed to make these measurements can be used to improve global positioning systems, space navigation, wireless communications, national security sensors, and biomedical techniques. Perhaps most incredible from a chemist's point of view, the ability to produce extremely minute intervals of time may soon allow scientists to observe the behavior of electrons inside an atom (Boyd, McClatchy Newspapers).

SI Units

The seven fundamental units of the SI system are based on the metric system. We use decimal prefixes and scientific notation to write quantities that are much larger or smaller than the base units. We will discuss four of the fundamental SI units that will be used in this study guide (mass, length, time, and temperature). We will also discuss density and volume whose units are derived from the fundamental units of length and mass. The SI unit for "amount," the mole, will be discussed in Chapter 3.

Common Prefixes Used in the SI System

Prefix	Symbol		Conventional notation		Exponential notation
mega	m	=	1,000,000	=	10^6
kilo	k	=	1000	=	10^3
hecto	h	=	100	=	10^2
deka	da	=	10	=	10^1
-	-	=	1	=	10^0
deci	d	=	0.1	=	10^{-1}
centi	c	=	0.01	=	10^{-2}
milli	m	=	0.001	=	10^{-3}
micro	μ	=	0.000001	=	10^{-6}
nano	n	=	0.000000001	=	10^{-9}

Examples: 1 kg = 1000 g 1 nm = 1×10^{-9} m, or 1×10^{9} nm = 1 m

Summary of SI Units

Physical quantity	Quantity's symbol	Name of SI unit	Abbreviation	Other common units
Mass	m	kilogram	kg	gram, pound, ton
Length	l	meter	m	foot, yard, mile, km
Time	t	second	s	minute, hour
Temperature	T	Kelvin	K	mK (milliKelvin)
Amount	n	mole	mol	dozen, gross

Length

The SI unit for length is the meter (m). A meter is a little longer than a yard. A centimeter (1/100th of a meter) is a little smaller than half an inch.

$$1 \text{ in} = 2.54 \text{ cm}$$

Conversion between yards and meters

How many yards are in 1 meter?

solution:

$$1.00 \text{ m} \times \frac{100.\text{ cm}}{1.00 \text{ m}} \times \frac{1.00 \text{ in}}{2.54 \text{ cm}} \times \frac{1.00 \text{ yd}}{36.0 \text{ in}} = 1.09 \text{ yd}$$

A meter is slightly longer than a yard.

Conversion between angstroms and nanometers

Two units frequently used in chemistry to express dimensions useful on an atomic scale are the angstrom (1 Å = 10^{-8} cm or 1 Å = 10^{-10} m) and the nanometer (1 nm = 10^{-7} cm or 1 nm = 10^{-9} m). How many angstroms are in 1 nanometer?

solution:

$$1 \text{ nm} \times \frac{10^{-7} \text{ cm}}{1 \text{ nm}} \times \frac{1 \text{ Å}}{10^{-8} \text{ cm}} = 10 \text{ Å}$$

An angstrom is 1/10th the size of a nanometer.

Volume

Volume is derived from the SI unit for length, the meter. A cube that measures one meter on each edge has a volume of $(1\text{m})^3 = 1 \text{ m}^3$. A cubic meter is pretty large. In chemistry we generally base volume on a cubic decimeter, which is also called a liter (L). A liter is slightly larger than a quart. We often do not use entire liters of solutions in the laboratory, so in these cases we measure volume in milliliters (mL). A mL is one thousandth the size of a liter. One milliliter equals one cubic centimeter (cm^3).

The Equivalence of a mL and a cm^3

$$1 \text{ cm}^3 \text{ means: } (1 \text{ cm})^3 = (1)^3 (\text{cm})^3 = 1 (\text{cm})^3$$

$$\mathbf{1 (cm)^3} = (10^{-2}\text{m})^3 = (10^{-2})^3\text{m}^3 = \mathbf{10^{-6}m^3}$$

$$\mathbf{1 \text{ mL}} = 10^{-3} (\text{dm})^3 = 10^{-3}(10^{-1}\text{m})^3 = 10^{-3}(10^{-1})^3\text{m}^3 = 10^{-3} \times 10^{-3}\text{m}^3 = \mathbf{10^{-6}m^3}$$

$$\mathbf{1 \text{ mL} = 1 \text{ cm}^3}$$

Volume determination by water displacement

The volume of irregularly shaped solids is often determined from the volume of water they displace. A graduated cylinder contains 15.50 mL of water. When a nickel is added, the volume increases to 16.05 mL. What is the volume of the nickel in cm^3 and in^3?

solution:

The change in volume of the water in the graduated cylinder must be due to the nickel's volume. The volume change is:

$$\Delta V = \text{volume after} - \text{volume before}$$

$$= 16.05 \text{ mL} - 15.50 \text{ mL}$$

$$= 0.55 \text{ mL}$$

Since a mL = cm^3, the volume of the nickel in cubic centimeters is 0.55 cm^3. We will use the relationship 1 in = 2.54 cm to find the volume of the nickel in inches:

$$0.55 \text{ cm}^3 \times \left(\frac{1 \text{ in}}{2.54 \text{ cm}}\right)^3 = 0.034 \text{ in}^3$$

You measure the diameter of a nickel to be 2.10 cm and you measure its thickness to be 0.18 cm. Calculate the volume of the nickel from these measurements.

$$V = \text{area} \times \text{thickness}$$

The area of a circle is πr^2, so we can write:

$$V = \pi r^2 \times \text{thickness} = \pi \times (2.10 \text{ cm}/2)^2 \times 0.18 \text{ cm} = 0.62 \text{ cm}^3$$

Compare this volume to the volume obtained by the water displacement method (0.55 cm^3). Which is likely to be more accurate? (The accuracy of each measurement depends upon the accuracy of your measuring devices, but water displacement is the potentially more accurate method since the nickel could have rounded edges and other deviations from a perfect cylinder.)

Mass

The SI unit for mass is the kilogram (kg). A kilogram is a little more than twice the size of a pound.

$$1 \text{ kg} = 2.205 \text{ lb}$$

The kilogram is the only remaining international standard in the metric system that is still a man-made object rather than being defined by a physical property of nature that does not vary. The meter, for example, is now defined in terms of the speed of light (the length of the path traveled by light in a vacuum during a time interval of 1/299,792,458 of a second), and a second itself is based on atomic clocks. The official kilogram, by contrast, is a lump of metal (a golf ball-sized cylinder of platinum and iridium) more than 130 years old stored under glass domes in a vault in a basement near Paris, accessed with three independent keys. The problem is it has been slowly losing mass compared to its copies—about 50 micrograms (around the weight of a grain of sand)—which obviously defeats its one purpose: constancy.

Experts have been wrestling for over 30 years on how to redefine a kilogram in terms of a fundamental constant of nature. Researchers have been using two different types of experiments to work towards a redefinition. The first uses a sophisticated scale, a “watt balance” to define the kilogram in terms of Planck's constant (a number used in quantum mechanics and which you will encounter in Chapter 7), and the second method involves counting the atoms in a sphere of crystalline silicon to define the kilogram in terms of Avogadro's constant (Chapter 3). So far the experimental results diverge by a small amount, and averaging incompatible results doesn't sit well with many scientists. The hope is that the existing measurements can be reconciled by 2015 when the International General Conference on Weights and Measures will meet. One insider admits, "The present situation is pretty flaky."

Conversion between grams and pounds

How many grams are in 1 pound?

solution:

$$1.00 \text{ lb} \times \frac{1.00 \text{ kg}}{2.20 \text{ lb}} \times \frac{1000 \text{ g}}{1.00 \text{ kg}} = 454 \text{ g}$$

Mass and weight are not synonyms. Mass refers to a quantity of matter; it does not change depending on where it is measured. Weight is dependent on the quantity of matter and the gravitational field pulling on it. Your weight is different on the moon than on earth. Your mass is the same on any planetary body.

Density

Density, derived from the SI units for mass and volume, is the mass per unit volume of a substance.

$$d = \frac{m}{V}$$

The SI unit for density is therefore kg/m^3. However, in chemistry, we generally use g/L or g/mL.

Density is a characteristic physical property of a substance (for a given temperature and pressure). Density does not depend on the amount of a substance present, so we call it an **intensive property.** Volume and mass do depend on the amount of a substance present, so we call them **extensive properties.** The density of water at ordinary pressure and room temperature is 1.00 g/mL. Substances that float in water will have densities less than 1.00 g/mL. Objects that sink in water (such as a nickel) will have densities greater than 1.00 g/mL. We say that muscle tissue is more dense than fat. For a given volume, muscle tissue will weigh more than fat.

Density of a nickel

A nickel weighs approximately 5 grams. Calculate the density of the nickel discussed on the previous page using the volume obtained by the water displacement method.

solution:

$$d = \frac{m}{V} = \frac{5 \text{ g}}{0.55 \text{ cm}^3} = 9 \text{ g/cm}^3$$

Let's compare this number to the density of elemental nickel. The Handbook of Chemistry and Physics lists the density as 8.90 g/mL. Our calculated number of 9 g/mL suggests that a nickel could be elemental nickel. A nickel is actually made from a mixture of 25% nickel and 75% copper. Copper has a density of 8.92 g/mL which is close enough to the density of nickel that our density determination is not accurate enough to distinguish pure nickel from a nickel/copper mixture.

Density of water

Most cooks have used the relationship "a pint's a pound the world around" to convert between volume and mass. How close is 1 lb/pt to the density of water (1 g/mL)?

solution: We will use the conversions:

$$1 \text{ L} = 1.06 \text{ qt}, \ 1 \text{ lb} = 454 \text{ g}$$

$$\frac{1 \text{ lb}}{1 \text{ pt}} \times \frac{454 \text{ g}}{1 \text{ lb}} \times \frac{2 \text{ pt}}{1 \text{ qt}} \times \frac{1.06 \text{ qt}}{1 \text{ L}} \times \frac{1 \text{ L}}{1000 \text{ mL}} = 0.962 \text{ g/mL}$$

The statement that "a pint's a pound" is a good approximation for the density of water (1.00 g/mL).

Density example problem

Which has the larger mass: 0.500 lb of gold, or 25.0 mL of lead (d = 11.3 g/mL)?

solution: Convert the mass of gold to grams:

$$0.500 \text{ lb} \times \frac{454 \text{ g}}{1 \text{ lb}} = 227 \text{ g}$$

Convert the volume of lead to mass in grams using density:

$$d = \frac{m}{V} \qquad 11.3 \text{ g/mL} = \frac{m}{25.0 \text{ mL}} \qquad m = 282 \text{ g}$$

25.0 mL of lead has a larger mass than 0.500 lb of gold.

Temperature

Temperature is an important physical quantity in chemistry because it affects chemical reaction rates and chemical stability. The three temperature scales you will encounter and their relationships to one another are sketched below.

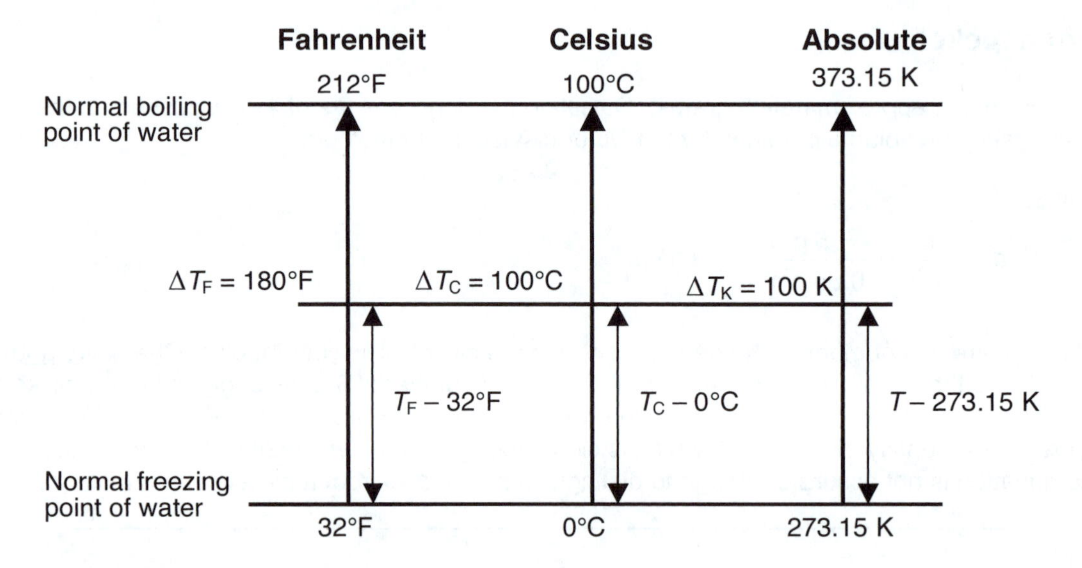

Notice there are 100 degrees between the freezing and boiling points of water on the absolute and Celsius temperature scales, so a Kelvin is the same size as a degree Celsius. The freezing point of water is defined to be 0° on the Celsius scale and 273.15 on the Kelvin scale, so to convert from Celsius to Kelvin, you add 273.15:

$$T(\text{in K}) = T(\text{in °C}) + 273.15$$

On the Fahrenheit scale, water freezes at 32° and boils at 212°, so there are 180 degrees between the freezing and boiling points of water. A Fahrenheit degree is almost one half (100/180 = 5/9) the size of a Celsius or Kelvin degree.

$$T(\text{in °F}) = \frac{9}{5}T(\text{in °C}) + 32$$

$$T(\text{in °C}) = \frac{5}{9}(T(\text{in °F}) - 32)$$

Comments about the Temperature Scales

- The Kelvin scale is also known as the **absolute scale** because 0 K is based on absolute zero, the hypothetical temperature at which all molecular motion ceases.
- A Kelvin is the same size as a degree Celsius; a Celsius or Kelvin degree is 1.8 (or 180/100 = 9/5) times larger than a Fahrenheit degree.
- Absolute temperature does not require the degree sign: the freezing point of water on the absolute scale is $T = 273.15$ K.

Warm-up problem

What is the temperature on the Celsius scale of a thermal state halfway between the normal boiling point and the normal freezing point of water ($T_{1/2}$)? On the Fahrenheit scale? On the Kelvin scale?

solutions:

Celsius: $T_{1/2} = 50°C$

Fahrenheit: $T_{1/2} = 122°F$

Kelvin: $T_{1/2} = 323.15$ K

How did you get 122°F? Probably you took the difference between the freezing and boiling points (212 – 32 = 180) and divided by 2 (180/2 = 90), and added it to 32 (90 + 32 = 122).

Or, you might have taken the average of 212 and 32: $\left(\frac{212 + 32}{2}\right) = 122$

How can we show algebraically that $T_{1/2}$ is half-way up the temperature scales? We take the ratio of the difference in degrees between $T_{1/2}$ and T_{nfp} to the difference in degrees between T_{nbp} and T_{nfp}:

(continued on next page)

Fahrenheit: $$\frac{T_{1/2}\,°F - 32°F}{212°F - 32°F} = \frac{122°F - 32°F}{180°F} = \frac{1}{2}$$

Celsius: $$\frac{T_{1/2}\,°C - 0°C}{100°C - 0°C} = \frac{50°C - 0°C}{100°C} = \frac{1}{2}$$

Kelvin: $$\frac{T_{1/2}\text{ K} - 273.15\text{ K}}{373.15\text{ K} - 273.15\text{ K}} = \frac{323.15\text{ K} - 273.15\text{ K}}{100\text{ K}} = \frac{1}{2}$$

We can generalize this formula for any temperature:

$$\boldsymbol{\frac{T_F - 32°F}{180°F} = \frac{T_C}{100°C} = \frac{T - 273.15\text{ K}}{100\text{ K}}}$$

Temperature conversion practice

Rearrange the equations in bold on page 14 to obtain equations that will 1) convert from Fahrenheit to Celsius, 2) convert from Celsius to Fahrenheit, 3) convert from Celsius to Kelvin.

solutions:

1) We will use the equations containing T_C and T_F, and solve for T_C:

$$\frac{T_F - 32°F}{180°F} = \frac{T_C}{100°C}$$

$$T_C = \frac{100°C}{180°F} \times \left(T_F - 32°F\right) = \frac{5°C}{9°F} \times \left(T_F - 32°F\right)$$

Check: If $T_F = 122°F$, what is T_C?

$$T_C = \frac{5°C}{9°F} \times \left(122°F - 32°F\right) = 50°C$$

2) We will use the same equations as part "1," and solve for T_F:

$$T_F = \left(\frac{180°F}{100°C} \times T_C\right) + 32°F = \left(\frac{9°F}{5°C} \times T_C\right) + 32°F$$

Check: If $T_C = 50°C$, what is T_F?

$$T_F = \left(\frac{9°F}{5°C} \times 50°C\right) + 32°F = 122°F$$

3) We will use the equations containing T_C and T, and solve for T:

(continued on next page)

$$\frac{T_C}{100°C} = \frac{T - 273.15\ K}{100\ K}$$

$$\mathbf{T = (T_C/°C + 273.15)\ K}$$

Check: If $T_C = 50°C$, what is T?

$$T = (50°C/°C + 273.15)\ K = 323.15\ K$$

Additional checks for temperature conversions

1) If $\mathbf{T_C = 0°C}$ (the freezing point of water), what is T_F?

$$T_F = \left(\frac{9°F}{5°C} \times 0°C\right) + 32°F = 32°F$$

2) If $\mathbf{T_C = 100°C}$ (the boiling point of water), what is T_F?

$$T_F = \left(\frac{9°F}{5°C} \times 100°C\right) + 32°F = 212°F$$

3) If $\mathbf{T_F = 32°F}$, what is T_C?

$$T_C = \frac{5°C}{9°F} \times (32°F - 32°F) = 0°C$$

4) If $\mathbf{T_F = 212°F}$, what is T_C?

$$T_C = \frac{5°C}{9°F} \times (212°F - 32°F) = \frac{5°C}{9°F} \times 180°F = 100°C$$

5) If $\mathbf{T_C = 0°C}$, what is T?

$$T = (0°C/°C + 273.15)\ K = \mathbf{273.15\ K}$$

6) If $\mathbf{T_C = 100°C}$, what is T?

$$T = (100°C/°C + 273.15)\ K = \mathbf{373.15\ K}$$

Some common temperatures

1) What is 98.6°F (body temperature) on the Celsius scale?

$$T_C = \frac{5°C}{9°F} \times (T_F - 32°F) = \frac{5°C}{9°F} \times (98.6°F - 32°F) = 37.0°C$$

2) What is 25°C (room temperature) on the Fahrenheit scale? On the absolute scale?

$$T_F = \left(\frac{9°F}{5°C} \times T_C\right) + 32°F = \left(\frac{9°F}{5°C} \times 25°C\right) + 32°F = 77°F$$

(continued on next page)

$$T = \left(\frac{T_C}{°C} + 273.15\right) K = (25 + 273.15)\ K = 298.15\ K$$

3) When is $T_F = T_C$?

Let $T_F = T_C = T$ and substitute into the equation for T_F:

$$T = \left(\frac{9}{5} \times T\right) + 32$$

Solving for T gives:

$$T = \left(-\frac{5}{4}\right) \times 32 = -40 = T_F = T_C$$

So, –40°F = –40°C.

4) What is absolute zero ($T = 0$) on the Celsius and Fahrenheit scales?

$T = (T_C/°C + 273.15)$ K, therefore, $T_C = (T/K - 273.15)°C$

$T_C = (0/K - 273.15)°C = -273.15°C$

$$T_F = \left(\frac{9°F}{5°C} \times T_C\right) + 32°F = \left(\frac{9°F}{5°C}\right) \times (-273.15°C) + 32°F = -459.67\ °F$$

Time

The SI base unit of time is the **second (s).** The length of a second was originally based on regular events in the heavens, but in 1964 scientists adopted the vibration rate of cesium atoms as the standard for measuring time. One second is defined, by international agreement, as 9,192,631,770 vibrations of cesium-133. The cesium atomic clock is accurate to within 1 second in 20 million years.

In Chapter 16 we will study the rates of chemical reactions and factors that influence those rates. As you will see, reactions may take years, or they may be over in less than a nanosecond (10^{-9} s).

Units and Conversion Factors

When we measure a physical quantity, we must always express a number and a unit for it to be meaningful. "I was doing 60" may be understood to mean 60 miles per hour in America, but in Germany, "doing 60" would be poking along at about 37 mph, since Germans measure speed in kilometers per hour. A physical quantity, is therefore, a quantity that can be expressed by the product of a pure number and a physical unit:

$$\text{physical quantity} = \text{pure number} \times \text{physical unit}$$

For example, if the mass of an average college student is 140 pounds, we write:

$m_{\text{avg. college student}} = 140 \times \text{lb}$, or leaving out the multiplication sign,

$m_{\text{avg. college student}} = 140\ \text{lb}$

Conversion factors convert one unit to another but do not change the physical quantity because a conversion factor equals 1. If we take an equation such as: 1 lb = 454 grams, we can obtain two conversion factors by:

a) dividing both sides of the above equation by 454 g:

$$\frac{1\ \text{lb}}{454\ \text{g}} = \frac{454\ \text{g}}{454\ \text{g}}$$

The conversion factor is: $\frac{1\ \text{lb}}{454\ \text{g}}$ which equals 1.

b) dividing both sides of the first equation by 1 lb:

$$\frac{1\ \text{lb}}{1\ \text{lb}} = \frac{454\ \text{g}}{1\ \text{lb}}$$

The conversion factor is: $\frac{454\ \text{g}}{1\ \text{lb}}$ which equals 1.

If we multiply a physical quantity by a "conversion factor," the pure number and physical unit will change, but the physical quantity will not change since we multiply by 1. The use of conversion factors in calculations is often referred to as the **unit-factor method** or **dimensional analysis.**

How to solve chemistry problems

If the average college student weighs 140 pounds, how much does the average college student weigh in kilograms?

1) **Know the problem.**
Read the problem carefully before you begin to write.

2) **Plan the problem-solving approach and estimate the answer.**
Often you will not know how to solve a problem at the outset—in fact, for real problems, we never do. That's OK! Instead of concentrating on what you don't know, concentrate on what you do know. Start by thinking about what information you have, and what you are trying to find. Guess what the answer will be, even if it is as general as: a really big number, a really small number, or something in between.

   a) **Write down what you know as mathematical equations.**
   This exercise will serve a couple of purposes. It gets you started (often the hardest part), and it will give you a chance to think about tools you may need to solve the problem.

   We know that "the average college student weighs 140 pounds," or:

$$m_{\text{avg. college student}} = 140\ \text{lb}$$

   b) **Write down what you want to know.**
   This will remind you of where the problem eventually needs to end up.

$$m_{\text{avg. college student}}\ \text{in kg} = ?$$

   c) **Write down equations you know that relate to the given information.**
   Even if you do not yet have a specific road map in mind, writing down conversions or equations you might use will eventually show you how to link the known to the unknown.

(continued on next page)

Equalities we could use to convert from lb to kg are:

$$1 \text{ lb} = 454 \text{ g}$$

$$1 \text{ kg} = 1000 \text{ g}$$

Now we can see the conversion factors to use to take us from pounds to grams to kilograms.

3) Solution: perform calculations.
We convert 140 pounds first to grams and then to kilograms. The conversion factors must contain the unit you wish to convert to in the numerator so that the unit you wish to convert from will be in the denominator and cancel out:

$$140 \text{ lb} \times \frac{454 \text{ g}}{1 \text{ lb}} \times \frac{1 \text{ kg}}{1000 \text{ g}} = 64 \text{ kg}$$

4) Check the answer to see if it makes sense.
Are the units correct? Is the answer approximately the right size? A kilogram is larger than a pound, so our average college student should weigh fewer kilograms than pounds: 64 is smaller than 140, so the answer appears to be correct.

A Few Hints:

- **Include units when working problems.**
  Take the time to write them out so you can easily cancel units that appear in both the numerator and denominator.

- **Play around with the information.**
  If you get through the first two steps and cannot see a route to the answer, play around with the information given, and it will often become clear how to obtain the answer.

- **Teach the material to someone else** (or explain it to yourself out loud).
  An often-heard comment is, "I didn't really understand the material until I had to teach it."

- **Work problems more than one way.**
  Don't work problems as quickly as you can, but try working fewer problems more slowly.

- **Translate the short-hand (math) into long-hand (English) in your mind.**
  Remember that mathematical equations, which may look "foreign," are just a short-hand way to say something that most likely takes up lots of space if written with words.

- **Review!**
  If you can work problems again the next day with no notes, chances are you understand the material and will be able to apply it for an exam.

Follow-up problem

What other information could we obtain from the previous problem?
We could determine the number of lb in a kg by using our relationship that:

$$140 \text{ lb} = 64 \text{ kg}$$

(continued on next page)

Dividing each side of the above equation by 64 kg gives:

$$\frac{140 \text{ lb}}{64 \text{ kg}} = \frac{64 \text{ kg}}{64 \text{ kg}} = 2.2 \frac{\text{lb}}{\text{kg}}$$

Alternatively, we could start with a kilogram, and use the knowledge that there are 1000 g in a kg, and 454 g in a pound to obtain:

$$1.00 \text{ kg} \times \frac{1000 \text{ g}}{1.00 \text{ kg}} \times \frac{1.00 \text{ lb}}{454 \text{ g}} = 2.20 \text{ lb}$$

Conversion problem

1 gram of water at 25°C and atmospheric pressure occupies a volume of 1 cm^3 (about a thimble full). How much volume does it occupy in cubic inches? (Use the conversion factor 1 in = 2.54 cm.)

solution:

1 in = 2.54 cm

$(1)^3 \text{ in}^3 = (2.54)^3 \text{ cm}^3$ (Cube each number/unit combination)

$$1 \text{ cm}^3 \times \left(\frac{1 \text{ in}}{2.54 \text{ cm}}\right)^3 = 0.06 \text{ in}^3$$

Note: Be sure to cube the entire conversion factor.

1.5 UNCERTAINTY IN MEASUREMENT

Every measurement we make includes some uncertainty. As we discussed at the beginning of section 1.4, scientists are increasingly able to make measurements with almost ridiculous levels of accuracy, but these measurements are dependent on sophisticated technology. When we weigh ourselves on a regular bathroom scale, we might tell someone our weight to within a pound, or perhaps, half a pound, but it would not be too believable to claim our weight was 145.259 pounds. An analytical balance in the lab is accurate to .0001 grams. It would be just as unbelievable to claim a reagent measured on this balance weighed 0.1357964 grams. We always estimate the rightmost digit when reading measuring devices, and we assume an uncertainty in this digit.

Significant digits are ones that we record. The exception to this is zeros that are used only to position the decimal point.

Significant digit examples

How many significant digits are in each of the following?

1) 323 and 3.23

   Both numbers have 3 significant digits because all non-zero digits are significant.

2) 3003 and 3.003

(continued on next page)

Both numbers have 4 significant digits because zeros between significant digits are significant.

3) 0.00102

This number has 3 significant digits because zeros used to position the decimal point (those to the left of the first non-zero digit) are not significant because they are not measured.

4) 3.30 and 3.00

Both numbers have 3 significant digits because a zero that ends a number and is to the right of the decimal point is significant.

5) 500

Since there is no decimal point, we assume the zeros are not significant, and the number has only one significant digit. A zero that ends a number and is to the left of the decimal point is significant only if it was actually measured. We can write the number in scientific notation to clearly show the number of significant digits: 5×10^2 has 1 significant digit, 5.0×10^2 has 2 significant digits, and 5.00×10^2 (or 500.) has 3 significant digits.

Exact numbers are ones which have no uncertainty because they are not measured, but refer to a number of items, or are part of a definition: 4 oranges, 12 molecules, 0.9144 meters in a yard. Exact numbers never limit the number of significant figures in a calculation because they can be assumed to have an infinite number of significant figures.

Calculators are useful devices for plugging out answers, but too often, it is tempting to write down the whole string of numbers the calculator spews out without evaluating if all the digits are meaningful. Suppose you want to divide the weight of the gear for a camping trip equally among three people, and the total weight of the gear is 64 pounds according to your bathroom scale. You divide 64 by 3 on your calculator and announce that each person will carry 21.333333 pounds. Obviously this answer has exceeded a meaningful number of significant digits. The following rules tell how many significant figures to show in the final answer of a calculation.

Significant Digits in Calculations

1) For multiplication and division: The answer contains the same number of significant figures as there are in the measurement with the fewest significant figures.
2) For addition and subtraction: The answer has the same number of decimal places as there are in the measurement with the fewest decimal places.

Rules for Rounding Off

1) If the digit removed is more than 5, the preceding number increases by 1. (2.46 becomes 2.5)
2) If the digit removed is less than 5, the preceding number is unchanged. (2.43 becomes 2.4)

(continued on next page)

3) If the digit removed is 5, the preceding number increases by 1 if it is odd and remains unchanged if it is even. (2.45 becomes 2.4; 2.35 becomes 2.4)

4) Round off only the final answer in a calculation.

Significant digits example problem

Let's work through the significant digits in the calculation of the density of a nickel described earlier.

1) To find the volume of the nickel, you measure the volume of water it displaces in a graduated cylinder (initial volume = 15.50 mL, final volume = 16.05 mL).

$$\begin{aligned}\Delta V &= \text{volume after} - \text{volume before}\\ &= 16.05 \text{ mL} - 15.50 \text{ mL}\\ &= 0.55 \text{ mL}\end{aligned}$$

Both numbers in the calculation contain 2 places past the decimal, so your answer contains two decimal places past the zero.

2) You estimate the mass of a nickel to be 5 g. What is the density of the nickel?

$$d = \frac{m}{V} = \frac{5 \text{ g}}{0.55 \text{ mL}} = 9 \text{ g/mL}$$

Since you only know the mass of the nickel to one significant digit, this value limits the number of significant digits in your answer to one significant digit.

3) If you were to weigh the nickel on an analytical balance and obtain its mass to be 5.0012 g, your answer becomes:

$$d = \frac{m}{V} = \frac{5.0012 \text{ g}}{0.55 \text{ mL}} = 9.1 \text{ g/mL}$$

In this case, the fewest number of significant digits in the density calculation is 2 (the volume measurement), so your answer contains two significant digits.

It is important to be aware of the level of accuracy of measuring devices you use in the lab. What would be the problem with weighing an open beaker of water on an analytical balance? Analytical balances are accurate to 0.0001 grams, and are sensitive enough to measure weight loss due to water evaporation. They are also sensitive enough to measure weight gain due to fingerprints on glassware. Don't hope the weight gains and weight losses will cancel out! It is good practice to use covered flasks and a paper towel or tongs when handling glassware that will be weighed on an analytical balance.

Precision, Accuracy, and Instrument Calibration

Precision and accuracy are not synonyms. **Precision** refers to how close measurements in a series are to each other. **Accuracy** refers to how close a measurement is to the real value. You can have extremely precise measurements that are inaccurate. This situation usually occurs if there is a **systematic error** in the measurements that causes them to all read higher or lower than they should. Systematic error can be minimized through calibration of the measuring device.

Random error produces measurements that are both higher and lower than the average value. Random error always occurs, but its size depends on the instrument's precision and the skill of the measurer. Precise measurements have low random error. Accurate measurements generally have low random error as well. In some cases, even when there is high random error, the average value of the measurements may still be accurate.

Precision and accuracy

A nickel weighs 5.0012 g on a calibrated balance. Two students weigh the nickel five times and record the weights shown below:

Student #1		Student #2	
#	weight	#	weight
1	4.8002	1	5.1020
2	4.8004	2	4.8882
3	4.8000	3	5.1095
4	4.8002	4	4.9055
5	4.8003	5	5.0012
Avg.	4.8002	Avg.	5.0013

Student #1 produced measurements that were precise, but not very accurate. A systematic error appears to have occurred, such as using a poorly calibrated balance. Student #2 did not have a good day. The measurements were not precise, and although reading #5 and the average were accurate, the size of the random error indicates a need for more work on weighing technique by the measurer.

CHAPTER CHECK

Make sure you can...

- Distinguish between physical and chemical properties and changes.
- Form a mental picture of the molecular arrangements for the three states of matter.
- Define kinetic and potential energy and describe their relationship.
- Explain the phlogiston theory of combustion.
- Summarize the scientific approach.
- List the common units of mass, length, temperature, volume, and time.
- Find the density from mass and volume.
- Convert between units and between temperature scales.
- Determine the number of significant digits required in answers to calculations.

Chapter 1 Exercises

1.1) Nail polish remover, which is usually acetone, sometimes ethyl acetate (or a mixture of both), often with glycerol and various fragrances, has some characteristic properties. List some of the properties of nail polish remover and characterize them as chemical or physical properties. For the physical properties, state whether they are intensive or extensive.

1.2) Indicate whether the following processes involve a chemical change or a physical change.

a. The melting of an ice cube on a hot griddle.

b. The discharge of a dry cell battery as it runs down.

c. The formation of carbon on adding sulfuric acid to table sugar.

d. The cooling of skin as rubbing alcohol evaporates from it.

e. The rising of bread dough.

1.3) a. Picture a pendulum swinging. At what point on the swing is the potential energy of the pendulum the greatest? At what point is the potential energy the lowest? At what point is the kinetic energy the greatest?

b. When a glass vase is dropped to the floor, the glass breaks. Describe what happened to the potential energy of the vase.

1.4) The scientific method is an iterative approach to problem solving involving observation, hypothesis, prediction, experiment, and observation. Explain how you could use the scientific method to determine why the electricity in your home has suddenly cut off.

1.5) a. If there are 5280 feet in one mile, 0.6214 mile in one kilometer, and 12 inches in one foot, how many inches are there in 10 km? How many strides, a stride being, e.g., 36"?

b. Calculate the area in square meters of a swimming pool 150 feet long and 45.0 feet wide. (1 foot = 0.3048 meter exactly).

1.6) Calculate the volume in cubic centimeters of a piece of aluminum wire with a diameter of 3.00 mm and a length of 6.25 m.

1.7) How many milliliters are in a 12.0 oz can of soda?

1.8) The largest super tankers have the capacity to carry approximately 6.7×10^5 m^3 of oil. How many barrels of oil will the tanker carry? (Hint: 42 gallons = 1 barrel)

1.9) An empty bottle has a mass of 100 grams. When the bottle is filled with pure water (density 1.00 g/cm^3), the mass of the full bottle is 250 grams. When filled with an unknown liquid, the mass of the full bottle is 320 grams. What is the density of the unknown liquid?

1.10) a. The number presented on an electronic calculator as the answer to the problem 7.23×5.19 is 37.5237. How should this number be reported?

b. There are exactly 12 inches in one foot and exactly 2.54 cm in one inch. How many significant figures should be written in the answer to the following problem:

$$0.2252 \text{ ft} \times 12 \text{ in/ft} \times 2.54 \text{ cm/in}$$

1.11) The development of new materials exhibit superconductivity at temperatures that can be maintained with liquid nitrogen. The boiling point of liquid nitrogen is 77.0 K. What is the boiling point in Fahrenheit?

1.12) Daniel Fahrenheit chose the lowest temperature he could achieve with a mixture of snow, salt, and water as the zero point of his scale. He chose body temperature as another point on his scale and gave it a value of 100 (Linus Pauling suggested, "perhaps he had a slight fever"). As a result, the freezing point of water on the Fahrenheit scale is 32°F, and the boiling point of water is 212°F. The difference between the freezing point and boiling points of water is 180 Fahrenheit degrees.

Suppose you developed a new temperature scale (°N), with a value of 0°N at the boiling point of liquid nitrogen (–196°C), and a value of 100 for the freezing point of water. What would normal body temperature (37°C) be on your new scale?

Chapter 1 Answers

1.1) A chemical property is a property that involves a chemical change.
An extensive property is a property that depends upon the amount of substance.
An intensive property is independent of the amount of substance.

Some of the properties that you might have thought of are:

Smell or odor	Physical	Intensive
Temperature	Physical	Intensive
Ability to dissolve nail polish	Physical	Intensive
Ability to dissolve varnishes and paints	Physical	Intensive
High volatility, evaporates easily	Physical	Intensive
Mass	Physical	Extensive
Volume	Physical	Extensive
Liquid state	Physical	Intensive
Translucent	Physical	Intensive
Density	Physical	Intensive
Ability to mix with water	Physical	Intensive
Flammable	Chemical	

1.2)
a. physical: a phase change from solid to liquid
b. chemical; a chemical reaction occurs in the battery
c. chemical; a different substance is formed
d. physical; a phase change from liquid alcohol to vapor
e. chemical: a chemical reaction producing carbon dioxide gas causes the bread to rise

1.3)
a. The potential energy is the greatest at the top of the swing when the pendulum is momentarily stationary. The potential energy is the lowest at the bottom of the swing when the pendulum is moving fastest—this is where the kinetic energy is the greatest.

b. The potential energy is converted to kinetic energy as the vase falls. When the vase hits the ground, some of the kinetic energy is absorbed by the glass—so much so that some of the chemical bonds within the glass break and the vase breaks. Energy is also absorbed by the ground; the ground gets slightly warmer.

1.4) The scientific method involves a cyclic reasoning process that might have proceeded as follows:

Observation:	lights go out; appliances don't work
Hypothesis:	electricity blackout in entire neighborhood
Prediction:	lights out in neighbors' houses
Experiment:	look at neighbors' houses
Observation:	lights are on
Hypothesis:	problem is in your own house
Prediction:	main fuse blown
Experiment:	replace main fuse
Observation:	lights still don't work; appliances don't work
Hypothesis:	electric line down outside house
Prediction:	...and so on

1.5) a. $10.\ \text{km} \times \frac{0.6214\ \text{mi}}{1\ \text{km}} \times \frac{5280\ \text{ft}}{1\ \text{mi}} \times \frac{12\ \text{in}}{1\ \text{ft}} = 3.9 \times 10^5\ \text{in}$

$$3.9 \times 10^5\ \text{in} \times \frac{1\ \text{stride}}{36\ \text{in}} = 1.1 \times 10^4\ \text{strides}$$

b. $A = \text{length} \times \text{width};\ A = 150\ \text{ft} \times 45.0\ \text{ft} = 6750\ \text{ft}^2$

Convert units of ft^2 to m^2: $6750\ \text{ft}^2 \times \frac{(0.3048\ \text{m})^2}{(1\ \text{ft})^2} = 627\ \text{m}^2$

1.6) diameter = 3.00 mm; radius = 3.00 mm/2 = 1.50 mm or 0.150 cm

length = 6.25 m or 625 cm

$V = \text{area} \times \text{length} = \pi r^2 \times \text{length} = \pi(0.150\ \text{cm})^2 \times 625\ \text{cm} = 44.2\ \text{cm}^3$

1.7) We have to make an assumption about the density of the liquid in the soda can; we'll assume its density is close to water (1 g/mL):

$$12.0\ \text{oz} \times \frac{1\ \text{lb}}{16\ \text{oz}} \times \frac{454\ \text{g}}{\text{lb}} \times \frac{1\ \text{mL}}{\text{g}} = 340\ \text{mL}$$

1.8) $6.7 \times 10^5\ \text{m}^3 \times \left(\frac{10\ \text{dm}}{\text{m}}\right)^3 \times \frac{1\ \text{L}}{1\ \text{dm}^3} \times \frac{1.06\ \text{quart}}{1\ \text{liter}} \times \frac{1\ \text{gallon}}{4\ \text{quarts}} \times \frac{1\ \text{barrel}}{42\ \text{gallons}} = 4.2 \times 10^6\ \text{barrels}$

1.9) First find the volume of the bottle:

$$V = \frac{m}{d} = \frac{(250\ \text{g} - 100\text{g})}{1\ \text{g/mL}} = 150\ \text{mL}$$

Now use the volume of the bottle to determine the density of the unknown liquid:

$$d = \frac{m}{V} = \frac{(320 - 100\ \text{g})}{150\ \text{mL}} = 1.47\ \text{g/mL}$$

1.10) a. Electronic calculators often display as many digits in an answer to a problem as the display will accommodate. The calculator has no knowledge of the uncertainty in the data given to it. Each number in the product has 3 significant figures and the answer should be written with the same number of significant figures: 37.5.

b. In this calculation the conversion factors are infinitely precise—there are *exactly* 12 inches in one foot, and *exactly* 2.54 cm in one inch. The answer should have the same number of significant figures as the least precise number involved in the calculation (0.2252)—i.e., four.

1.11) –196°C, –321°F

1.12) Any two points, experimentally reproducible physical phenomena, can be used to define a temperature scale. The two points can be given any convenient numerical values. On your new scale (°N), the two points are: boiling point of liquid N_2 = 0°N (corresponds to –196°C), and freezing point of water = 100°N (corresponds to 0°C). The conversion factor for °C to °N is (100°N/196°C).

An interval of 37°C gives 37°C × (100°N/196°C) = 19°N, so body temperature on the new scale is: 100°N + 19°N = 119°N.

2

The Components of Matter

2.1 ELEMENTS, COMPOUNDS, AND MIXTURES

A pure substance is a type of matter whose composition is fixed. Elements and compounds are pure substances. Elements consist of only one type of atom—elements may exist as individual atoms or as molecules (two or more atoms chemically bonded together). Compounds have two or more different elements that are chemically bound together.

Mixtures are impure substances: they consist of pure substances physically mixed together in no specific ratio. Mixtures may be either homogeneous (no visible boundaries) or heterogeneous (boundaries are visible).

Pure substances

Choose words from the list below to complete the following paragraph:

(a) compounds	(c) elements	(e) silver (Ag)
(b) copper (Cu)	(d) gold (Au)	(f) water (H_2O)

There are two kinds of pure substances. In one kind, the atoms are all the same. Such substances are called _______________(1). In the other kind of pure substance, all the molecules are the

(continued on next page)

same, but not all the atoms in each molecule are the same. They are called _______________(2). Among the commonest examples of naturally occurring elements are the three "coinage metals": _______________(3) (the most common coinage metal), _______________(4) (the metal used in the manufacture of tableware, mirrors, and jewelry), and _______________(5) (one of the more precious metals, famous for not dissolving in nitric acid). The most common compound, _______________(6), is the major component of most living things.

answers: 1-c, 2-a, 3-b, 4-e, 5-d, 6-f.

It is easy to confuse the terms molecule and compound. Molecule is the more general term that refers to the smallest particle of a substance that retains the properties of the substance, and is composed of two or more atoms. A compound is a special kind of molecule in which all of its atoms are not the same.

Pure substances (molecules, compounds, and elements)

Classify the following as: a) molecules, b) compounds, or c) elements.

1) H_2O	3) CO_2	5) Au	7) N_2
2) O_2	4) S_8	6) $C_{12}H_{22}O_{11}$	8) N_2O_4

answers: 1-a,b; 2-a,c; 3-a,b; 4-a,c; 5-c; 6-a,b; 7-a,c; 8-a,b.

Impure substances (mixtures)

Choose words from the list below to complete the following paragraph:

(a) aqueous
(b) gaseous
(c) heterogeneous
(d) homogeneous
(e) mixture
(f) pure substances
(g) solution

A _______________(1) is a combination of pure substances. There are two kinds of mixtures. A _______________(2) mixture has visible boundaries; a _______________(3) mixture has no visible boundaries. A homogeneous mixture is also called a _______________(4). A substance dissolved in water is an _______________(5) solution. Air is a _______________(6) solution consisting of mainly oxygen and nitrogen molecules. We live in a world of predominately mixtures. Air, the oceans, soil, rocks, and living things all consist of _______________(7) mixed together.

answers: 1-e, 2-c, 3-d, 4-g, 5-a, 6-b, 7-f.

2.2/2.3 THE ATOMIC VIEW OF MATTER and DALTON'S ATOMIC THEORY

Evidence for Atoms (Dalton's Induction)

In 460 BC, a Greek philosopher, Democritus, postulated that all things are composed of atoms (fr. *atomos*, indivisible)—tiny particles, imperceptible to the senses that are indivisible and indestructible. His theory that everything, including human existence, was a product of random atomic collisions was not popular among his contemporaries and successors, who were comfortable with their belief that the gods of the Olympus steered their fate. It would take more than 2000 years before experimental results on the masses of compounds and chemical reactions led to the laws of mass conservation and constant composition, which in turn led John Dalton (1766-1844) to postulate the existence of atoms.

1) Law of Mass Conservation

Magnesium burns in oxygen to produce a white powder, magnesium oxide:

$$2Mg(s) + O_2(g) \rightarrow 2MgO(s)$$

Lavoisier found that the mass of the reactants (magnesium and oxygen) used up, equals the mass of the product (magnesium oxide) formed. Early scientists made many measurements on the masses of reactants and products of various reactions and found that mass is always conserved in chemical transformations. John Dalton postulated that this would be the case if all matter were composed of indestructible particles he called atoms, which in chemical reactions merely rearrange.

2) Law of Constant Composition

Early scientists discovered that pure water is always 11.2 percent hydrogen and 88.8 percent oxygen by mass. Dalton postulated that this was further evidence for the existence of atoms, since the result could be explained if all molecules of water have the same chemical formula and if each kind of atom has on the average the same mass. The mass of an element in a compound divided by the total mass of the compound is that element's **mass fraction.** For water, the mass fraction of hydrogen is 0.112 and the mass fraction of oxygen is 0.888. The sum of the mass fractions equals 1.000.

Law of constant composition

1) What mass of hydrogen can be obtained from:

a) 18.0 g of water; b) 1.00 g of water; c) 10.0 g water?

solutions:

Water is always 11.2% hydrogen by mass so:

a) mass of hydrogen in 18.0 g water = 0.112 × 18.0 g = 2.02 g

b) mass of hydrogen in 1 g water = 0.112 × 1.00 g = 0.112 g

c) mass of hydrogen in 10 g water = 0.112 × 10.0 g = 1.12 g

2) What mass of hydrogen can be obtained from:

a) 18.0 lb of water; b) 1.00 lb of water; c) 10.0 lb of water?

solutions:

a) mass of hydrogen in 18.0 lb water = 0.112 × 18.0 lb = 2.02 lb

b) mass of hydrogen in 1.00 lb water = 0.112 × 1.00 lb = 0.112 lb

c) mass of hydrogen in 10.0 lb water = 0.112 × 10.0 lb = 1.12 lb

A Test of the Atomic Model of Matter (Dalton's Deduction—the Law of Multiple Proportions)

If atoms exist, what could we predict about compounds?

Let: M_C = mass of 1 carbon atom

M_O = mass of 1 oxygen atom

Suppose carbon and oxygen form two compounds, as, in fact, they do. One compound might have the formula CO, e.g., and the other CO_2. If so, one would predict that:

$$\frac{\text{mass of oxygen in } CO_2 / \text{mass of carbon in } CO_2}{\text{mass of oxygen in } CO / \text{mass of carbon in } CO} = \frac{2M_O/M_C}{M_O/M_C} = \frac{[(16+16)\div 12]}{(16\div 12)} = \frac{2}{1}$$

which, in fact, is the case. This law is known as the **law of multiple proportions** stated below:

When two elements form a series of compounds, the ratios of the masses of the second element that combine with a fixed mass of the first element can always be reduced to small whole numbers.

The law of multiple proportions

1) Hydrogen and oxygen form two compounds: water, H_2O, and hydrogen peroxide, H_2O_2. Evaluate the following ratio of ratios:

$$\frac{\text{mass of oxygen in } H_2O_2 / \text{mass of hydrogen in } H_2O_2}{\text{mass of oxygen in } H_2O / \text{mass of hydrogen in } H_2O}$$

solution:

$$\frac{2M_O/2M_H}{M_O/2M_H} = \frac{[(16+16)\div(1+1)]}{[16\div(1+1)]} = \frac{2}{1}$$

2) Given that:

$$\frac{\text{mass of H in hydrocarbon I} / \text{mass of C in hydrocarbon I}}{\text{mass of H in hydrocarbon II} / \text{mass of C in hydrocarbon II}} = \frac{2}{1}$$

Which of the following atomic models are consistent with the cited data:

	A	B	C	D	E
Hydrocarbon I	CH_2	CH_2	C_2H_4	CH_4	CH_4
Hydrocarbon II	CH	C_2H_2	C_2H_2	CH_2	C_2H_4

solution:

All 5 cases are consistent with the data. The case that Dalton discovered was E, CH_4 (methane, natural gas, or marsh gas), and C_2H_4 (ethylene).

Dalton's Theory Stated:

1. Each element is made up of tiny particles called atoms.
2. Atoms of one element cannot be converted into atoms of another element. Chemical reactions involve reorganization of the atoms, not changes in the atoms themselves (law of mass conservation).
3. Atoms of an element are identical, and are different from atoms of other elements.
4. Compounds form when atoms combine with each other. A given compound always has the same relative numbers and types of atoms (law of constant composition).

After Dalton published his theory, he prepared the first table of atomic masses. Because the masses of individual atoms are so small, he could only determine the relative masses of the elements. Dalton assigned an atomic mass of 1 to hydrogen, the lightest known substance.

Relative atomic masses

It was known in Dalton's day that water was composed of hydrogen and oxygen and that every gram of hydrogen required 8 grams of oxygen to form water. Based on this knowledge, Dalton assigned the atomic mass of oxygen to be 8. What did Dalton assume was the molecular formula for water?

answer:

Dalton assumed that the molecular formula for water was HO. We know now that the molecular formula for water is H_2O, and the atomic mass of oxygen is 16. Dalton's billiard ball view of the atom was extremely useful for explaining many scientific observations, but it could not explain why elements combine in specific ratios of atoms to form compounds.

Avogadro's Hypothesis: Evidence for Molecules of Elements

Experiments by Gay-Lussac on the volumes of gaseous reactants and products showed:

hydrogen(*g*)	+	chlorine(*g*)	→	hydrogen chloride(*g*)
1 vol.		1 vol.		**2 vol.**
□	+	□	=	□□

One might have expected that the reaction of 1 volume of hydrogen gas with 1 volume of chlorine gas would yield 1 volume of hydrogen chloride gas:

hydrogen(*g*)	+	chlorine(*g*)	→	hydrogen chloride(*g*)
1 vol.		1 vol.		**1 vol.**
□	+	□	=	□

To explain the confusing experimental results, Amadeo Avogadro made two proposals.

Avogadro's Proposals

- Equal volumes of gases contain equal number of gas particles. This equal volume—equal number hypothesis makes sense if the distances between gas particles is very great compared to the sizes of the particles so the size of the particles is insignificant, and the volume is determined by the number of molecules present.
- Hydrogen and chlorine exist as **polyatomic molecules** that split apart and recombine to form gaseous hydrogen chloride molecules.

Avogadro's concept of polyatomic gaseous molecules preserved the idea that atoms are not created or destroyed in chemical reactions. Instead, molecules are created and destroyed.

The evidence for diatomic hydrogen molecules

Let's assume, as Avogadro did, that:

a) equal number of gas particles take up equal volume
b) atoms cannot be split

Which of the following possibilities for reaction of hydrogen with chlorine obey both of Avogadro's assumptions and yield the experimental result found above (1 vol. hydrogen + 1 vol. chlorine = 2 vol. hydrogen chloride)?

	Volumes of gases:
$H + Cl \rightarrow HCl$	1 vol. + 1 vol. = 1 vol.
$H_2 + Cl_2 \rightarrow 2HCl$	**1 vol. + 1 vol. = 2 vol.**
$H_3 + Cl_3 \rightarrow 3HCl$	1 vol. + 1 vol. = 3 vol.
$H_3 + Cl_3 \rightarrow H_3Cl_3$	1 vol. + 1 vol. = 1 vol.
$H_4 + Cl_4 \rightarrow 4HCl$	1 vol. + 1 vol. = 4 vol.
$H_4 + Cl_4 \rightarrow 2H_2Cl_2$	**1 vol. + 1 vol. = 2 vol.**
$H_4 + Cl_4 \rightarrow H_4Cl_4$	1 vol. + 1 vol. = 1 vol.
$H_6 + Cl_6 \rightarrow 2H_3Cl_3$	**1 vol. + 1 vol. = 2 vol.**

answer:

The possibilities that give the experimentally found volumes of reactants and products are written in bold in the right column. Notice that the hydrogen and chlorine molecules that could produce 2 volumes of HCl all contain an **even number** of atoms.

One volume of hydrogen gas has never produced more than 2 volumes of gaseous product. Therefore, it is not necessary to assume more than 2 atoms per molecule of hydrogen. Chemists use the symbol H_2 for the hydrogen molecule.

In summary:

Avogadro's concept of diatomic gaseous molecules preserved the idea that:

_______________(1) are not created or destroyed in chemical reactions.

Instead, _______________(2) are created and destroyed.

answers: (1) atoms, (2) molecules.

2.4 OBSERVATIONS THAT LED TO THE NUCLEAR ATOM MODEL

Research into the nature of electricity led to the discovery of electrons, the negatively charged particles in atoms. Soon after this discovery, experiments revealed the presence of an atom's nucleus—a tiny, positively charged central core with mass. These discoveries are summarized below.

The Discovery of Electrons

The first convincing evidence for subatomic particles came when experimenters observed rays of light when conducting electricity through gases at low pressures. **J.J. Thomson** (1856-1940) concluded that the rays appeared when negatively charged particles collided with the gas molecules in the tube. He named the particles **electrons.** Thomson determined the mass-to-charge ratio of the electron was the same regardless of what gas was in the tube, so he concluded that the electron was a fundamental particle found in all atoms. Since atoms were known to be electrically neutral, Thomson proposed that the negatively charged electrons were distributed randomly in a diffuse cloud of positive charge like raisins in a plum pudding.

Robert Millikan measured the charge of the electron by measuring the effect of an electrical field on the rate at which charged oil drops fell under the influence of gravity. He used this value and Thomson's mass-to-charge ratio to calculate an electron's mass to be 9.11×10^{-28} g. This mass is approximately 1/2000 the mass of hydrogen, the lightest element!

The Evidence for a Nucleus Containing Positively Charged Particles

Ernest Rutherford (1871-1937) established that an atom contains a tiny, positively charged center called an **atomic nucleus.** He bombarded a piece of thin gold foil with alpha particles (a helium nucleus, ^{4}He, which is emitted from certain radioactive materials). Since the mass of an alpha particle is 7300 times that of an electron, Rutherford expected all the alpha particles to smash right through the gold foil with, at most, very minor deflections in their paths. Unexpectedly, a tiny percentage of the alpha particles reflected back from the foil! The plum pudding model could not explain these results. Rutherford postulated that the alpha particles bounced off a tiny center of positive charge. He presented a new picture of an atom as mostly space occupied by electrons, but which contained a tiny region of concentrated positive charge which he called the **nucleus.** He postulated that the nucleus contained positively charged particles he called **protons,** which accounted for most of the mass of the atom.

The Discovery of Neutrons

Rutherford's work did not account for all the mass of an atom. Twenty years later, **James Chadwick** discovered the **neutron.** This uncharged particle is present in the nucleus with the protons and accounts for the remaining mass of the atom.

2.5 A MODERN (CA. YEAR 2014) VIEW OF THE ATOM

Today we picture an atom as a positively charged nucleus surrounded by a negatively charged electron cloud. The electrons whiz around the nucleus, held there by the attraction of the nucleus. The nucleus is tiny and heavy—it is extremely dense! It contains protons (p^+) with a positive charge, and neutrons (n^0) with no charge. The protons and neutrons account for essentially all the mass of the atom. An atom is electrically neutral, so the number of protons and electrons in an atom are equal.

The Atomic Theory Today

1. All matter is composed of atoms. Although atoms are composed of smaller particles (electrons, protons, and neutrons), an atom is the smallest body that retains the chemical and physical properties of the element.
2. Atoms of one element can be converted into atoms of another element only in nuclear processes. They are never transformed by a chemical reaction.
3. All atoms of an element have the same number of protons and electrons, which determines the chemical behavior of the element. Isotopes of an element differ in their number of neutrons and hence differ in their mass number. The mass of a sample of an element is the average of the masses of its naturally occurring isotopes.
4. Compounds are formed by the chemical combination of two or more elements in specific ratios as originally stated by Dalton.

Terminology for the Atom

Atomic Number (*Z*): The number of protons in the nucleus of an atom.

$$Z = \#\text{protons}$$

Mass Number (*A*): The total number of protons and neutrons in an atom.

$$A = \#\text{protons} + \#\text{neutrons}$$

Each proton and neutron has one unit of mass. Notice that the mass number (A) minus the atomic number (Z) = the number of neutrons (N):

$$A - Z = (\#\text{protons} + \#\text{neutrons}) - \#\text{protons} = \#\text{neutrons} = N$$

Atomic Symbol: A symbol for each element based on its English, Latin, or Greek name. The convention is to write the atomic symbol with the atomic number as a left subscript, and the mass number as a left superscript.
For atom X, we write:

$$^{A}_{Z}\mathbf{X}$$

Isotopes: Atoms with the same number of protons (the same element), but different numbers of neutrons and therefore different mass numbers.

Atomic Mass Unit (amu): 1/12 the mass of a carbon-12 atom. The mass of a carbon-12 atom is defined as exactly 12 atomic mass units.

Atomic Mass: The average of the masses of an element's naturally occurring isotopes weighted according to their abundances.

Terminology for the atom (exercise)

Choose words from the list below to complete the following paragraphs:

(a) atomic mass unit
(b) atomic number
(c) isotope(s)
(d) mass number
(e) neutrons
(f) protons
(g) seven
(h) six
(i) twelve

Elements are defined by the number of _____________(1) in their nucleus, which is the element's _____________(2). Every carbon atom contains _____________(3) protons in its nucleus and therefore has an atomic number of _____________(4). Essentially all the mass of an atom is contributed by its __________ and __________(5 & 6), and the sum of these two gives the _____________(7). The mass number varies depending on the number of _____________(8) an atom contains in its nucleus (its _____________(9)). A ^{13}C atom contains _____________(10) protons and _____________(11) neutrons in its nucleus. The chemical properties of an element are primarily determined by the number of electrons, so _____________(12) have similar chemical behavior.

The mass unit which relates the relative masses of atoms is the _____________(13). The mass of a ^{12}C atom is defined to be exactly _____________(14) amu. The nucleus of a ^{12}C contains _____________(15) protons and _____________(16) neutrons. The atomic mass unit is based on this particular _____________(17) of carbon, not on naturally occurring carbon which is a mixture of _____________(18).

answers:
1-f, 2-b, 3-h, 4-h, 5-f, 6-e, 7-d, 8-e, 9-c, 10-h, 11-g, 12-c, 13-a, 14-i, 15-h, 16-h, 17-c, 18-c.

Atomic mass exercise

Consider the lightest element, hydrogen. Hydrogen has an atomic mass of 1.00794 amu. Why isn't it exactly 1.000?

solution:

Atomic mass is based on the carbon-12 isotope and is defined to be exactly equal to 12 amu. However, the mass of 6 protons and 6 neutrons does not add to 12. The mass of an ordinary hydrogen atom (a proton) is 1.007825, and the mass of a neutron is 1.008665. The loss of mass accompanying the formation of a heavier atom from hydrogen atoms and neutrons is due to the very large amount of energy that is evolved in the formation of heavier atoms, an amount given by Einstein's equation, $E = mc^2$. This decrease in mass is described by a quantity called the *packing fraction.*

Most naturally occurring hydrogen atoms (99.985%) contain no neutrons in their nucleus (atomic mass of ^{1}H = 1.007825), but a small number of naturally occurring hydrogen atoms (0.015%) contain 1 neutron in their nucleus ($A = 2$, atomic mass of ^{2}H = 2.0140 amu). The small number of hydrogen atoms containing a neutron causes the average mass of naturally occurring hydrogen atoms to be slightly greater than 1.007825, or 1.00794 amu.

2.6 THE PERIODIC TABLE

In 1871, Dmitri Mendeleev organized the elements by increasing atomic mass and arranged them so that elements with similar chemical properties would lie in the same column. The modern periodic table of the elements is based on Mendeleev's earlier version.

Periodic table exercise

Choose words from the list below to complete the following paragraphs:

(a) atomic number
(b) group(s)
(c) inner transition
(d) letters
(e) main-group
(f) period(s)
(g) proton
(h) transition

The elements are arranged in order of increasing _______________(1); each element takes the previous element and adds one _______________(2) to its nucleus. The vertical columns are called _______________(3) and are numbered 1 to 8 with either the letter A or B. The horizontal rows are called _______________(4) and are numbered 1 to 7. A newer system numbers the groups from 1 to 18 without _______________(5). Elements in a _______________(6) have similar chemical properties; elements in a _______________(7) have different chemical properties.

The eight A-groups contain the _______________(8) or **representative** elements. The 10 B-groups contain the _______________(9) elements. The _______________(10) elements are usually placed below the main body of the table, but actually fit between Ba and Hf (lanthanides), and Ra and Rf (actinides).

answers: 1-a, 2-g, 3-b, 4-f, 5-d, 6-b, 7-f, 8-e, 9-h, 10-c.

Classification of the elements

Elements are classified as **metals, nonmetals,** or **metalloids** depending on their physical and chemical properties. Match the characteristics (a-n) to the classification (metal, nonmetal, metalloid) that it best describes:

a) shiny solids at room temperature
b) group 8A elements, the relatively unreactive noble (or inert) gases
c) properties between those of metals and nonmetals
d) conduct heat and electricity poorly
e) group 7A elements, the highly reactive halogens
f) good conductors of heat and electricity
g) easily tooled into sheets and wires
h) group 1A elements, the highly reactive alkali metals
i) transition elements
j) dull, brittle solids or gases at room temperature

(continued on next page)

k) inner transition elements

l) elements in the upper right portion of the periodic table

m) elements that lie along the staircase line on the periodic table

n) group 2A elements, the reactive alkaline earth metals

answers:

Metals (a,f,g,h,i,k,n); **Nonmetals** (b,d,e,j,l); **Metalloids** (c,m).

2.7 IONIC AND COVALENT BONDING

Most elements don't occur naturally in elemental form, but in combination with other elements as compounds. Elements combine in two general ways, both of which involve the elements' electrons. In one case, electrons are transferred between atoms (ionic bonds); in the other case, the electrons are shared between atoms (covalent bonds).

Bonding Terminology

Ionic compounds: form when an atom of one element transfers electrons to atoms of another element

Covalent compounds: form when atoms share electrons

Ion: a charged particle

Cation: a positively charged particle

Anion: a negatively charged particle

Monoatomic ion: an ion derived from a single atom

Polyatomic ions: two or more atoms bonded covalently and having a net positive or negative charge, i.e.: NH_4^+, SO_4^{2-}

Coulomb's Law

An **ionic bond** forms when two atoms approach, and one (or more) electron(s) move from one atom to the other, creating a positive/negative cation/anion pair. Objects of opposite charge (the cation and anion) attract each other through an attractive Coulomb force. For chemistry, we use the form of Coulomb's law:

$$E = k\frac{Q_1Q_2}{r}$$

Where k is a constant (2.31×10^{-19} J∗nm), Q_1 and Q_2 are the charges of the ions, and r is the distance between the nuclear centers.

The strength of the ionic bonding will be greater the higher the charges on the ions, and smaller the size of the ions. Although no bond is 100% ionic (there is always a little sharing of electrons), many bonds come close. If we wanted to find the energy of attraction of a single NaCl ion pair, we would need to know the ionic radii of Na^+ and Cl^- (0.095 nm and 0.181 nm respectively) and their ionic charges. Then, to find the ionic energy, we simply substitute into Coulomb's Law:

$$E = 2.13\times10^{-19}\,\text{J}\cdot\text{nm}\frac{(-1)(+1)}{(0.095+0.181)\text{nm}} = -8.37\times10^{-19}\,\text{J}$$

Sodium chloride, written NaCl(*s*), is not a collection of pairs of sodium and chloride ions as the formula seems to imply, but is a three-dimensional array of alternating Na^+ and Cl^- ions. The number of sodium ions must equal the number of chloride ions because ionic compounds have no charge. Although both sodium and chlorine are only one electron away from having the same number of electrons as a noble gas, some elements must lose or gain two or three electrons to obtain a noble gas electron configuration.

Ionic bonding exercise

Choose words from the list below to complete the following paragraphs:

(a) anions	(e) eleven	(h) lose	(k) noble gases
(b) argon	(f) gain	(i) NH_4^+	(l) polyatomic
(c) cations	(g) ion	(j) neon	(m) seventeen
(d) Cl^-			

When an atom, or small group of atoms, gains or loses electrons, it becomes a charged particle or an ______________(1). Negatively charged particles called ______________(2) attract positively charged particles called ______________(3) to form ionic compounds. We can predict how many electrons an element is likely to gain or lose by examining its position in the periodic table. The elements in group 8A, the ______________(4), are particularly stable elements. <u>Elements strive to obtain the same number of electrons as the nearest noble gas</u>. Elements on the left hand side of the periodic table, the metals, ______________(5) electrons to become cations. Elements on the right hand side of the periodic table, the nonmetals, ______________(6) electrons to become anions.

Table salt (sodium chloride) is a familiar ionic compound. We would predict that a sodium atom with ______________(7) electrons would ______________(8) an electron to obtain the same number of electrons as the noble gas ______________(9), and that chlorine with ______________(10) electrons would ______________(11) an electron to obtain the same number of electrons as the noble gas ______________(12).

Some ionic compounds are formed with ______________(13) ions, ions consisting of two or more atoms covalently bonded which stay together as a unit during reactions. An example of an ionic compound formed with a polyatomic ion is ammonium chloride, NH_4Cl. ______________(14) is the cation and ______________(15) is the anion.

<u>answers</u>: 1-g, 2-a, 3-c, 4-k, 5-h, 6-f, 7-e, 8-h, 9-j, 10-m, 11-f, 12-b, 13-l, 14-i, 15-d.

When elements share electrons, they form **covalent bonds**. Covalent bonds usually occur between nonmetals, and result in the formation of molecules. Pure covalent bonds (where electrons are shared equally) occur between diatomic molecules such as H_2, F_2, and Cl_2. Figure 2.14 in your text shows how a covalent bond forms between two hydrogen atoms. As two H atoms approach, the positive nucleus of one atom attracts the negative electron of the other atom. Although the two nuclei (and the two electrons) repulse each other, these repulsions are relatively weak. When the attractive and repulsive forces balance, an H_2 molecule forms with the hydrogen atoms separated by a distance of 0.74 Å.

Covalent bonding exercise

Choose words from the list below to complete the following paragraphs:

(a) anions	(c) H_2O	(e) O_2	(g) separate
(b) cations	(d) nonmetals	(f) same	(h) share

Covalent compounds form when atoms ____________(1) electrons, which usually occurs between elements found in the upper right portion of the periodic table, the ____________(2). In the case of an oxygen molecule, two oxygen atoms share electrons to form the molecule ____________(3). In the case of water, two hydrogen atoms and an oxygen atom share electrons to form the molecule ____________(4).

Many compounds are not purely ionic, or purely covalent; their bonds possess some of each character. The purest type of covalent bond is one between two of the ____________(5) atoms since their nuclei attract the electrons to the same extent.

A covalent compound such as methanol, CH_3OH, consists of molecules that loosely associate with each other, but remain ____________(6). In contrast, an ionic compound such as magnesium chloride ($MgCl_2$), consists of a continuous array of ions which alternates positively charged ____________(7) and negatively charged ____________(8).

answers: 1-h, 2-d, 3-e, 4-c, 5-f, 6-g, 7-b, 8-a.

2.8 NAMING COMPOUNDS

Ionic Compounds

Let's consider the name for NaCl, sodium chloride. Notice that:

- The cation is named first followed by the anion.
- The name of the cation (sodium) is the same as the name of the neutral metal.
- The name of the anion (chloride) takes the root of the nonmetal name (chlorine) and adds the suffix "-ide."

Naming ionic compounds

What are the names for the following ionic compounds:

a) $MgCl_2$ b) CaO c) CuCl d) $CuCl_2$

(continued on next page)

<u>answers</u>:

a) magnesium chloride; b) calcium oxide; c) copper(I) chloride; d) copper(II) chloride.

Notice that <u>two different compounds cannot have the same name</u>, so for metals (such as copper and iron) that can form cations with different charges, we indicate the metal's ionic charge by writing it in Roman numerals (in parentheses) after the metal ion's name. You will often see the common names for these compounds: the Latin root metal is followed by the suffix "-ous" for the ion with the lower charge and "-ic" for the ion with the higher charge:

$CuCl$	=	copper(I) chloride	=	cupr**ous** chloride
$CuCl_2$	=	copper(II) chloride	=	cupr**ic** chloride
$FeCl_2$	=	iron(II) chloride	=	ferr**ous** chloride
$FeCl_3$	=	iron(III) chloride	=	ferr**ic** chloride

(Elements that form only one cation do not need to be identified by a Roman numeral: group 1A elements form only 1^+ ions, group 2A elements form only 2^+ ions, and aluminum forms only Al^{3+}.)

Ionic Compounds Containing Polyatomic Ions

Polyatomic ions stay together as a charged unit. When two or more of the same polyatomic ion are present in the formula, the ion appears in parentheses with the subscript written outside.

Examples of ionic compounds containing polyatomic ions:

a) silver nitrate = $AgNO_3$
b) silver nitrite = $AgNO_2$
c) calcium nitrate = $Ca(NO_3)_2$
d) ammonium chloride = NH_4Cl
e) calcium phosphate = $Ca_3(PO_4)_2$

The Naming Convention for Oxoanions

When a polyatomic ion contains one or more oxygen atoms, we say it is an **oxoanion.** Often, there are families of oxoanions that differ only in the number of oxygen atoms.

a) With two oxoanions in the family, we use the suffixes:

"-ate" for the ion with more O atoms

"-ite" for the ion with fewer O atoms.

SO_4^{2-} is sulf**ate**

SO_3^{2-} is sulf**ite**

b) With four oxoanions in the family, in addition to the suffixes "-ate" and "-ite", we use the prefixes "per-" (more than) and "hypo-" (less than) to name the members of the series with the most and fewest oxygen atoms, respectively:

ClO_4^- is **per**chlor**ate**

ClO_3^- is chlor**ate**

ClO_2^- is chlor**ite**

ClO^- is **hypo**chlor**ite**

Naming Acids

An acid is a molecule with one or more H^+ ions associated to an anion. Acids dissolve in water to produce a solution containing free H^+ ions (protons). There are two classes of acids in terms of naming them:

a) If the anion does not contain oxygen, the acid is named with the prefix "hydro-" and the suffix "-ic." When gaseous HCl (hydrogen chloride) dissolves in water, it forms **hydrochloric acid.**

b) If the anion contains oxygen, the names are similar to those of the oxoanions, except:

the anion "-ate" suffix becomes an "-ic" suffix in the acid:

H_2SO_4 = sulfuric acid

the anion "-ite" suffix becomes an "-ous" suffix in the acid:

H_2SO_3 = sulfurous acid

Binary Covalent Compounds

Let's consider the name for CO_2, carbon dioxide. Notice that:

- The first element named (carbon) is the one in the lower group number in the periodic table. (An exception to this rule is when the compound contains oxygen and a halogen. In this case, the halogen is named first.) If both elements are in the same group, the one with the higher period number is named first.
- The second element (oxygen) is named with the suffix "-ide".
- Greek numerical prefixes indicate the number of atoms of each element in the compound. The first word has a prefix only when more than one atom of the element is present. The second word always has a numerical prefix.

Naming binary covalent compounds

Give names for the following covalent compounds:

a) PCl_3
b) NO_2
c) Cl_2O_7
d) CO
e) H_2O
f) NH_3

answers:

a) phosphorus trichloride
b) nitrogen dioxide
c) dichlorine heptaoxide
d) carbon monoxide
e) dihydrogen monoxide
f) nitrogen trihydride

The last two molecules are almost always referred to by their common names, water and ammonia.

Alkanes

Alkanes are a special type of binary covalent compound that contain only carbon and hydrogen. Each carbon atom in an alkane forms four covalent bonds. Straight chain alkanes contain chains of carbon atoms with no branches. Alkanes are named with a prefix plus the suffix "*-ane.*" The first 10 straight-chain alkanes are listed in Table 2.7 on p. 72 of the textbook. In Chapter 15, you will get a chance to practice naming more complex organic compounds.

Naming alkanes

Name the following straight-chain alkanes: (Keep in mind the lightest alkanes exist in ambient conditions as gases; linear alkanes with five or more carbon atoms are liquids at room temperature.)

a) the lightest alkane; a colorless, odorless, flammable gas that is a product of decomposition of organic matter in marshes, giving it the nickname "swamp gas"

b) a colorless, odorless gas, heavier than air, and used from tanks for fuel

c) a volatile liquid containing six carbon atoms

d a volatile, flammable liquid, the primary component in gasoline

answers:

a) methane, CH_4; b) propane, C_3H_8; c) hexane, C_6H_{14}; d) octane, C_8H_{18}.

Molecular Masses from Chemical Formulas

We calculate the molecular mass of a compound by adding together its atomic masses:

Molecular mass = sum of atomic masses

Calculating molecular mass

1) What is the molecular mass of an ammonia molecule?

solution:

We calculate the molecular mass by adding together the atomic masses:

Molecular mass of NH_3 = (1 × atomic mass of N) + (3 × atomic mass of H)

= (14.01 amu) + (3 × 1.008 amu)

= 17.03 amu

2) What is the molecular mass of magnesium nitrate?

solution:

Again, we add together the atomic masses to calculate the molecular mass:

(continued on next page)

Molecular mass $Mg(NO_3)_2$

$= (1 \times \text{atomic mass of Mg}) + 2 \times [(1 \times \text{atomic mass of N}) + (3 \times \text{atomic mass of O})]$

$= 24.31\ \text{amu} + 2 \times [(14.01\ \text{amu}) + (3 \times 16.00\ \text{amu})]$

$= 148.47\ \text{amu}$

Notice that the atomic mass of Mg is used for the cation Mg^{2+} even though a magnesium ion is two electrons lighter than a magnesium atom. The two electrons lost by Mg are gained by the anion, so electron masses are balanced in neutral compounds.

3) What is the molecular mass of sucrose, $C_{12}H_{22}O_{11}$?

solution:

Molecular mass $C_{12}H_{22}O_{11}$

$= (12 \times 12.01\ \text{amu}) + (22 \times 1.008\ \text{amu}) + (11 \times 16.00\ \text{amu})$

$= 342.3\ \text{amu}$

2.9 CLASSIFICATION OF MIXTURES

In the natural world, matter usually occurs as mixtures. An often-challenging aspect of lab work involves separating mixtures so pure substances can be identified. The two broad classes of mixtures are **heterogeneous mixtures,** which have one or more visible boundaries between the components, and **homogeneous mixtures,** which have no visible boundaries because the components are mixed as individual atoms, ions, and molecules. A homogeneous mixture is also called a **solution.** Solutions may exist as solids, liquids, or gases. Solutions in water are called **aqueous solutions.**

Physical processes including filtration, crystallization, extraction, distillation, and chromatography can separate mixtures. Page 78, "Tools of the Laboratory" in your textbook provides good descriptions and drawings of basic separation techniques.

Separation techniques

You are familiar with mixture separations even if you have never performed them in a chemistry lab. Determine the type of separation technique used for each of the following processes:

a) Straining orange juice to remove the pulp

b) Pouring hot water over coffee grounds to make a cup of coffee

c) Pouring noodles and hot water in a colander

d) Producing moonshine

e) Making rock candy

answers:

a) filtration; b) extraction; c) filtration; d) distillation; e) crystallization.

Use the exercise on the following page on the classification of matter to check your understanding of pure and impure substances.

Classification of matter

Use the appropriate classifications of matter listed below (lettered a-g) to describe the substances (numbered 1-8) that follow:

a) element
b) molecule
c) compound
d) homogeneous mixture (solution)
e) heterogeneous mixture
f) pure substance
g) impure substance

1) Pure air

(d,g): Pure air is a homogeneous mixture or solution of primarily nitrogen and oxygen. (Air that is around us typically contains dust and other particles that cause air to be a heterogeneous mixture.)

2) Stainless steel

(d,g): Stainless steel is a solid solution of iron with chromium and sometimes nickel or manganese.

3) Glass

(d,g): Glass is a homogeneous mixture (solution) of silica, or of oxides of boron or phosphorus and a stabilizer.

4) Wood

(e,g): Wood is a heterogeneous mixture of many substances; the proportions of these substances vary depending on the type of wood and where it grows.

5) Iron

(a,f): Iron is an element and therefore is also a pure substance. It has the symbol Fe.

6) Water

(b,c,f): Pure water is a compound made up of hydrogen and oxygen in a fixed ratio (H_2O). (Drinking water is a solution containing dissolved minerals.)

7) Milk

(e,g): Milk is a heterogeneous mixture or a colloidal suspension of many elements and compounds.

8) F_2, Cl_2, Br_2, I_2, H_2, N_2, O_2, P_4, S_8

(a,b,f): All of the above are elements which exist in nature as molecules. The halogens (F_2, Cl_2, Br_2, I_2), hydrogen (H_2), nitrogen (N_2), and oxygen (O_2) exist as diatomic molecules. Some elements, such as phosphorus (P_4) and sulfur (S_8), exist as larger molecules.

CHAPTER CHECK

Make sure you can...

- Classify pure substances as elements or compounds.
- Classify impure substances (mixtures) as homogeneous or heterogeneous.
- Explain how the laws of mass conservation and constant composition led to Dalton's postulate that all matter is composed of atoms and to his deduction of the law of multiple proportions.
- State the postulates of Dalton's atomic theory.
- Describe experiments that led to Avogadro's concept of diatomic gaseous molecules.
- Describe Thomson's and Millikan's contributions to the discovery of electrons.
- Explain why the "plum pudding model" of an atom could not explain results of Rutherford's experiments.
- State the postulates of the atomic theory today.
- Define: atomic number (Z), mass number (A), atomic symbol, amu, atomic mass, isotopes.
- Calculate atomic mass from relative abundances of isotopes.
- Identify regions of the periodic table.
- List characteristics of metals, nonmetals, and metalloids.
- Define: ionic, covalent, ion, cation, anion, monoatomic ion, polyatomic ion.
- Explain how Coulomb's Law leads to ionic bonding.
- Name ionic and covalent compounds.
- Find molecular mass from a chemical formula.

Chapter 2 Exercises

2.1) Which of the following is a pure substance?

seawater
air
methane gas
orange juice
homogenized milk
a cup of coffee

2.2) An atom, ion, or molecule of a particular element X has the general symbol shown below. In this symbol various information is included at one or more of the four corners. What information is usually located at each corner?

$$^{0}_{0}X^{0}_{0}$$

2.3)
a. The nucleus of a neutral atom contains 42 protons and 54 neutrons. What is the element?
b. A neutral atom contains 110 neutrons and 74 electrons. What is the element?
c. If a chlorine atom has a mass number of 37, how many neutrons are there in its nucleus?
d. One isotope of nitrogen has the symbol ^{14}N. How many protons, neutrons, and electrons are there in one ^{14}N atom?

2.4) Copper exists in nature as a mixture of two isotopes. One isotope contains 34 neutrons and has a relative abundance of 69.77%. The average atomic mass of copper is 63.55. How many neutrons must there be in the second isotope?

2.5) Alloys of zirconium, Zr, are used in the fabrication of nuclear reactors. The element has five naturally occurring isotopes:

^{90}Zr, 89.9043 amu (51.46%)

^{91}Zr, 90.9053 amu (11.23%)

^{92}Zr, 91.9046 amu (17.11%)

^{94}Zr, 93.9061 amu (17.40%)

^{96}Zr, 95.9082 amu (2.80%)

What is the average atomic mass of Zr?

2.6) Iridium has two naturally occurring isotopes. ^{191}Ir and ^{193}Ir have atomic masses of 190.9609 amu and 192.9633 amu, respectively. The average atomic mass for iridium is 192.22 amu. What is the percent natural abundance for each isotope?

2.7) a. Which one of the following elements exists as a diatomic gas under normal conditions?

bromine	calcium	chlorine
sodium	argon	gallium

b. Which one of the following elements exists as a monatomic gas under normal conditions?

sulfur	oxygen	bromine
potassium	krypton	silicon

c. Atoms of which of the following elements form covalent bonds with themselves in molecules?

sodium	neon	cadmium
nitrogen	phosphorus	iodine

2.8) What is the formula of the ionic compound:

a. that contains only potassium and carbonate ions?
b. that contains only calcium and carbonate ions?
c. that contains only copper(II) and bisulfate ions?
d. that contains only ammonium and phosphate ions?

2.9) a. What is wrong with the following formulas?

$SrCl$ $Na_2(PO_4)_3$ $Ba(SO_4)$ $MgNO_3$ Li_2Br_2

b. What is wrong with the following names?

$NaCl$	sodium monochloride
KNO_3	potassium nitrite
$CaBr_2$	dibromine calcium
$NaHCO_3$	sodium carbonate
$(NH_4)_2HPO_4$	diammonium hydrogen phosphate

2.10) Name the following compounds:

a. NaI
b. NH_4Cl
e. Fe_2S_3
f. $PbCrO_4$

c. ZnO
d. $Al(OH)_3$
g. $Mg(ClO_3)_2$
h. H_3PO_3

2.11) Give the formula for each of the following compounds:

a. nitrogen triiodide
b. copper(II) sulfate
c. diphosphorous pentoxide
d. cobalt(II) nitrite
e. sulfurous acid
f. sodium carbonate
g. hydrobromic acid
h. sodium hydride

2.12) Which compounds contain only ionic bonds, only covalent bonds, and both covalent and ionic bonds?

NaCl NH_4Br H_2O KNO_3 $(NH_4)_2HPO_4$ NH_3

2.13) Calculate the molecular mass of:

nitric acid, HNO_3
carbon tetrachloride, CCl_4
acetic acid, CH_3CO_2H
toluene, $C_6H_5CH_3$

2.14) Cinnabar is reddish ore from which mercury is derived. 8.62 g of Hg are in each 10.0 g of cinnabar. What is the mass fraction of mercury in cinnabar? What mass of cinnabar is necessary to yield 1.00 kg of Hg?

2.15) a. What ending do the names of monatomic negative ions have?

b. What ending do the names of nonmetal positive ions such as H_3O^+ or NH_4^+ have?

c. If there are two possible polyatomic oxyanions of an element, what ending does the anion having the fewer oxygens have?

d. In accordance with the rules for assigning names to oxyanions, what would the name for the XeO_6^{4-} ion be?

Chapter 2 Answers

2.1) methane gas (CH_4); seawater, air, orange juice, milk, and a cup of coffee are all mixtures

2.2) upper left corner: the mass number of the atom; example ^4He
lower left corner: the atomic number; example $_{27}Co$
upper right corner: the charge, if any; example Zn^{2+}
lower right corner: number of atoms in the molecule; example S_8

The atomic number, *Z*, is the number of protons in the nucleus. The mass number, *A*, is the protons and neutrons in the nucleus.

2.3) a. Mo; indicated by the number of protons, regardless of the number of neutrons.
b. W; the number of electrons equals the number of protons.
c. 20; the atomic number of chlorine is 17, therefore the number of neutrons is 37 – 17 = 20.
d. 7 protons, 7 neutrons, 7 electrons; the atomic number of nitrogen is 7, therefore there are 7 protons. There must then be 7 electrons. Since the mass number is 14, there must also be 7 neutrons.

2.4) For the isotope with 34 neutrons, $A = 63$ (# protons + # neutrons = 29 + 34 = 63). Since the average atomic mass of copper (63.55) is greater than 63, our second isotope must contain more than 34 neutrons. Let Y be the atomic mass of the second isotope. We know that the average atomic mass of copper is the sum of the masses of its isotopes weighted by their relative abundance:

$$(0.6977 \times 63) + (0.3023 \times Y) = 63.55$$

$$Y \sim 65$$

The number of neutrons in the second isotope is therefore:

$$\# \text{ neutrons} = \text{mass number} - \# \text{ protons} = 65 - 29 = 36$$

Check: 36 is greater than 34; the second isotope contains more neutrons as we predicted.

2.5) 91.22 amu

2.6) The atomic mass of ^{193}Ir is closer to the average atomic mass for the iridium (192.22), so it should have the higher natural abundance.

Let Y = the fraction of ^{193}Ir, and $(1 - Y)$ = the fraction of ^{191}Ir, then:

$$[Y \times (192.9633)] + [(1 - Y) \times (190.9609)] = 192.22$$

$$192.9633Y + 190.9609 - 190.9609Y = 192.22$$

$$Y = 0.6288, \text{ or } 62.9\% \ ^{193}Ir$$

$$(1 - Y) = 0.3712, \text{ or } 37.1\% \ ^{191}Ir$$

Check: ^{193}Ir has the greater abundance, as we predicted.

2.7) a. Chlorine. Argon is the only other gas in the list and is monatomic. Bromine is a liquid.
b. Krypton. Oxygen, the other gas in the list, is diatomic O_2.
c. Phosphorus forms P_4 molecules (and other allotropes).
Iodine and nitrogen form diatomic molecules I_2 and N_2.
Neon, the noble gas in the list, is monatomic. It does not form bonds.
Sodium and cadmium are metals and form metallic solids.

2.8) a. K_2CO_3
b. $CaCO_3$
c. $Cu(HSO_4)_2$
d. $(NH_4)_3PO_4$

2.9) a. Strontium is in group 2 and invariably forms the ion Sr^{2+}; there should be two chloride ions to match the strontium ion: $SrCl_2$.

The phosphate ion has a –3 charge, so the ratio of sodium ions to phosphate ions should be 3:1, not 2:3: Na_3PO_4.

The ratio of atoms is correct in this formula but no parentheses are required around the sulfate ion since there's only one of them: $BaSO_4$.

Magnesium is in group 2 and invariably forms the ion Mg^{2+}; there should be two nitrate ions for one magnesium ion: $Mg(NO_3)_2$.

The ratio of atoms is correct in this formula but the ratio should be expressed as simply as possible: LiBr.

b. sodium chloride; prefix "mono-" not required
potassium nitrate, not nitrite
calcium bromide; the metal is placed first in name; prefix "di-" not required
sodium bicarbonate; not carbonate

ammonium hydrogen phosphate; the number of hydrogen atoms still attached equals one and is specified; the number of ammonium ions follows automatically

2.10)
a. sodium iodide
b. ammonium chloride
c. zinc oxide
d. aluminum hydroxide
e. iron(III) sulfide
f. lead(II) chromate
g. magnesium chlorate
h. phosphorous acid

2.11)
a. NI_3
b. $CuSO_4$
c. P_2O_5
d. $Co(NO_2)_2$
e. H_2SO_3
f. Na_2CO_3
g. HBr
h. NaH

2.12) NaCl: ionic bond

NH_4Br: the ammonium ion contains covalent bonds; the bond between the NH_4^+ and the Br^- is ionic.

H_2O: covalent bonds

KNO_3: the anion NO_3^- contains covalent bonds; the bond between the potassium and the nitrate ion is ionic.

NH_4HSO_4: both the cation NH_4^+ and the anion HSO_4^- contain covalent bonds, and together form an ionic bond.

NH_3: covalent bonds

2.13) Sum the atomic masses to obtain the molecular mass:

HNO_3	$1.008 + 14.01 + (3 \times 16.00) = 63.02$ amu
CCl_4	$12.011 + (4 \times 35.45) = 153.81$ amu
CH_3CO_2H	$(2 \times 12.011) + (4 \times 1.008) + (2 \times 16.00) = 60.05$ amu
$C_6H_5CH_3$	$(7 \times 12.011) + (8 \times 1.008) = 92.141$ amu

The units of these molecular masses are amu (dalton) for the mass of an individual molecule, or grams for the mass of one mole of the molecules.

2.14) $\frac{8.62}{10.00} = 0.862$

$(0.862)Y = 1.00$ kg; therefore $Y = 1.00/0.862 = 1.16$ kg of cinnabar

2.15)
a. Monatomic negative ions have the ending "-ide."
b. Nonmetal positive ions have the ending "-ium" (hydronium, ammonium).
c. The ending of the anion having the fewer oxygens has the ending "-ite."
d. The name for the XeO_6^{4-} ion is ***per**xen**ate***.

3

Stoichiometry

3.1 THE MOLE

Review of Atomic Mass

We learned in Chapter 2 that **atomic mass** is the average of the masses of an element's naturally occurring isotopes weighted according to their abundances. Atomic masses are given in **atomic mass units** (amu) arbitrarily defined as exactly 1/12 the weight of a ^{12}C atom:

$$\text{mass of one } ^{12}\text{C atom} = 12.000 \text{ amu}$$

Carbon found on the earth (natural carbon) does not have an atomic mass of 12.000 amu, but 12.011 amu because it is a mixture of the isotopes ^{12}C and ^{13}C. Both isotopes contain 6 protons, but while ^{12}C also contains 6 neutrons, ^{13}C contains 7 neutrons giving it an atomic mass of 13.003. The average mass of these isotopes (weighted for abundance) gives the atomic mass of naturally occurring carbon (12.011 amu). No single carbon atom has an atomic mass of 12.011, however for stoichiometric purposes, we consider carbon to be composed of only one type of atom with a mass of 12.011.

Atomic mass units allow us to know only the relative atomic masses of the elements. The periodic table lists these relative masses: a hydrogen atom (atomic weight = 1.008) is 1/12 as heavy as a carbon atom; a helium atom (atomic weight = 4.003) is 1/3 as heavy as a carbon atom (or 4 times as heavy as a hydrogen atom).

Relative atomic masses

1) How many grams of iron would contain the same number of iron atoms as the number of carbon atoms contained in 1.000 g of carbon?

(continued on next page)

solution:

Iron has an atomic weight of 55.85 meaning that it is 55.85/12.01 times as heavy as carbon. If you have 1 gram of carbon, you would need to weigh out:

$$\frac{55.85 \text{ g Fe}}{12.01 \text{ g C}} \times 1.000 \text{ g C} = 4.650 \text{ g Fe}$$

Does our answer make sense? Think of the iron atoms as red marbles that weigh more than yellow marbles (carbon atoms). If we have the same number of red and yellow marbles, the red marbles (iron atoms) will weigh more than the yellow marbles. Our answer indicates that more than 1 g of iron is needed to contain the same number of atoms as a gram of carbon. Since an iron atom weighs between 4 and 5 times that of a carbon atom, our answer appears to be correct.

2) How many grams of iron would contain the same number of iron atoms as the number of carbon atoms contained in 12.01 g of carbon?

solution:

$$\frac{55.85 \text{ g Fe}}{12.01 \text{ g C}} \times 12.01 \text{ g C} = 55.85 \text{ g Fe}$$

Check: The iron sample should be 4.65 times as heavy as the carbon sample (see part 1). Our answer appears to be correct.

The above example illustrates that you can weigh out the atomic weight of any element (in grams) to give you the same number of atoms of that element:

12.000 g carbon-12	=	N carbon-12 atoms
1.008 g hydrogen	=	N hydrogen atoms
55.85 g iron	=	N iron atoms
107.868 g silver	=	N silver atoms

What is N? N has been found to be **6.022×10^{23}.** This number (six hundred and two sextillion) is called **Avogadro's number.** It is the number equal to the number of carbon atoms in exactly 12 grams of pure carbon-12.

The SI unit for amount of substance is a **mole.** One mole (abbreviated 1 mol) contains 6.022×10^{23} entities. We work with moles of atoms or molecules because we need a huge number of these minuscule particles in order to measure them conveniently in the laboratory. In this chapter we will see why the mole is an extremely useful counting unit.

1 mole = **602,200,000,000,000,000,000,000 entities**

602 sextillion entities

6.02×10^{23} entities

the amount of substance that contains the same number of entities as there are atoms in 12 grams of carbon-12

Avogadro's number of entities

Bridging between amu's and grams

Avogadro's number tells us the number of atoms of an element it takes to weigh the element's atomic mass in grams. It relates a number of atoms to a mass:

12.000 g ^{12}C	≈	**6.022×10^{23}** ^{12}C atoms	=	1 mole ^{12}C atoms
1.008 g H	≈	**6.022×10^{23}** H atoms	=	1 mole H atoms
55.85 g Fe	≈	**6.022×10^{23}** Fe atoms	=	1 mole Fe atoms
107.868 g Ag	≈	**6.022×10^{23}** Ag atoms	=	1 mole Ag atoms
16.00 g O	≈	**6.022×10^{23}** O atoms	=	1 mole O atoms

Figure 3.1 on page 92 in your text shows a mole of some familiar substances. As you can see, 602 sextillion atoms or molecules of a substance (copper, water, salt, sugar, aluminum), is an amount of material that you could weigh in grams or measure in a graduated cylinder. A mole therefore allows us to bridge between the atomic world of atoms and molecules to the macroscopic world that we observe and can easily measure.

Three physical quantities that we use to tell us how much of a substance there is are mass, volume, and amount:

physical quantity	some common physical units
mass (*m*)	gram, lb, ton
volume (*V*)	liter, mL, quart
amount (*n*)	mole, dozen, trio

Using the term "amount" as a physical quantity can be confusing because we usually think of mass and volume as amounts, too. It may be clearer to think of the physical quantity "amount" (*n*) as being a "population of entities." The difference between a mole and the pure number 6.022×10^{23} is that a mole (or a dozen, or a trio) is a unit of population defined in terms of the basic unit of population—things, items, or **entities.** A mole is a chemist's dozen*, and is used in the same way as a dozen. For example, if we have 12 eggs, we can say:

mass of eggs = 500 **grams**

amount, or population of eggs = 12 **entities,** or 1 **dozen entities**

If we have 6.022×10^{23} carbon atoms, we can say:

mass of atoms = 12.01 **grams**

amount, or population of atoms = 6.022×10^{23} **entities,** or 1 **mole entities**

Problems are less likely to become confused if you include the entity that the mole is referring to: 1 mole H_2, 12.01 g C/mole C, 5 mole electrons, etc.

*Keep in mind that, as your textbook discusses on page 92, the difference between a dozen and a mole is that we use the word dozen to refer to objects that do not have a fixed mass. A dozen eggs may weigh 500 grams, or a dozen eggs may weigh 450 grams. Likewise, a dozen doughnuts you buy would not weigh the same as a dozen doughnuts bought by a friend, even if they were the same type of doughnut. Whereas a mole in chemistry is used to refer to atoms or molecules that have fixed masses. A mole of carbon atoms always weighs 12.01 grams.

Ratios of mass, volume, and amount

Chemists often refer to ratios of these three units of measure (mass, volume and amount). Although you have not encountered all of these ratios yet, you can match the ratio to the name given for it if you keep in mind that **molar** means **"divide by *n*,"** and **specific** means **"divide by *m*."**

Names:	Ratios:
a) density (*d*)	1) m/V
b) molar mass (**M**)	2) V/n
c) concentration (*c*)	3) n/V
d) molar volume	4) m/n
e) specific volume	5) V/m
f) specific amount	6) n/m

answers:

a) 1, b) 4, c) 3, d) 2, e) 5, f) 6.

We discussed density (mass/volume) in Chapter 1. In this chapter, we will discuss **molar mass,** the mass per unit amount, and **concentration,** the amount per unit volume.

The Atomic Point of View of Mass and Molar Mass

Chemists make atomic models of everything. We saw in Chapter 1 atomic models for solids, liquids, and gases, and we discussed in Chapter 2 models for molecules, elements, and compounds. Now we consider the atomic model chemists have of mass.

Chemists think of the total mass (*m*) being equal to the mass per unit population times the population:

Total mass = mass per unit population × population

Total mass = molar mass × amount

$$m = \mathbf{M} \times n$$

Molar mass, M is therefore:

$$\mathbf{M} = m/n$$

The Common Units for M, Molar Mass

$$\frac{\mathbf{g}}{\textbf{mole atoms}}$$ (often written g/mol)

Although "molar" simply means divide by *n*, and does not require that *n* be in units of moles, the units used for molar mass in this course will be g/mol.

Molar mass

Choose words from the list below to complete the following paragraphs:

(a) amu/atom (c) average (e) mole
(b) atom (d) g/mole

12.000 grams of carbon-12 contains one ______________(1) of carbon-12 atoms. Therefore, the molar mass of carbon-12 equals 12 ______________(2). 12.000 amu of carbon-12 contains one ______________(3) of carbon-12. Therefore, the atomic mass of carbon-12 equals 12 ___________(4). The mass (in amu) of one atom of an element is numerically the same as the mass (in grams) of one mole of atoms of an element. We work with grams and moles of atoms because we can measure these amounts in the laboratory.

Naturally occurring carbon is a mixture of predominately carbon-12 and carbon-13. The molar mass of naturally occurring carbon is 12.011 g/mole. No single carbon atom has an atomic mass of 12.011 amu/atom, but the ________________(5) mass of the carbon atoms in a sample will equal 12.011 amu.

answers: 1-e, 2-d, 3-b, 4-a, 5-c.

Conversion between grams and amu's

1) What is the mass of 6.022×10^{23} hydrogen atoms, H, in amu? In grams?

solution:

The mass of 1 H atom = 1.00 amu, therefore:

mass of 6.022×10^{23} H atoms = 6.022×10^{23} amu

The mass of 6.022×10^{23} H atoms = 1.00 g, hydrogen's atomic mass in grams.

The conversion between grams and atomic mass units is therefore:

$$\mathbf{6.022 \times 10^{23}\ amu = 1.00\ g}$$

We can use this equality to derive two useful conversion factors:

$$\frac{1.00\ \text{g}}{6.022 \times 10^{23}\ \text{amu}} = 1 \quad \text{and} \quad \frac{6.022 \times 10^{23}\ \text{amu}}{1.00\ \text{g}} = 1$$

2) What is the mass of a hydrogen atom in grams?

solution:

We expect the mass of one hydrogen atom in grams to be a very small number:

$$1\ \text{H atom} \times \frac{1.00\ \text{g H}}{6.022 \times 10^{23}\ \text{atoms}} = 1.66 \times 10^{-24}\ \text{g}$$

Since one hydrogen atom weighs one atomic mass unit, we can write:

$$\mathbf{1.00\ amu = 1.66 \times 10^{-24} g}$$

Additional exercises

1) What mass of hydrogen can be obtained from:

a) 18.0 g of water, b) 1.00 g of water, c) 10.0 g of water?

solutions:

The atomic mass of a H_2O molecule is:

$$2 \times \frac{1.008 \text{ amu}}{\text{H atom}} + \frac{16.00 \text{ amu}}{\text{O atom}} = \frac{18.02 \text{ amu}}{\text{water molecule}}$$

The percent hydrogen by mass is:

$$\frac{2 \times 1.008 \text{ amu}}{\text{H atom}} \times \frac{\text{water molecule}}{18.02 \text{ amu}} \times 100\% = 11.2\%$$

Water is always 11.2% hydrogen by mass so:

a) mass of hydrogen in 18 g water = 0.112×18.0 g = 2.02 g

b) mass of hydrogen in 1.00 g water = 0.112×1.00 g = 0.112 g

c) mass of hydrogen in 10.0 g water = 0.112×10.0 g = 1.12 g

2) What mass of hydrogen can be obtained from:

a) 18.0 lb of water, b) 1.00 lb of water, c) 10.0 lb of water?

answers: a) 2.02 lb; b) 0.112 lb; c) 1.12 lb.

3) How many molecules of water are there in:

a) 18 amu of water, b) 18.0 g of water, c) 18.0 lb of water?

solutions:

a) The atomic mass of a H_2O molecule is 18.02 amu (from part 1).

Therefore, **1 molecule** of water weighs 18 amu.

b) The molar mass of a water molecule is:

$$2 \times \frac{1.008 \text{ g}}{\text{mole H atoms}} + \frac{16.00 \text{ g}}{\text{mole O atoms}} = \frac{18.02 \text{ g}}{\text{mole water molecules}}$$

Therefore, 1 mole **(6.022×10^{23} molecules)** of water weighs 18.0 g.

c) We know from part "b" that 6.022×10^{23} water molecules weigh 18.0 g.

There are 454 g in 1 lb, therefore:

$$18.0 \text{ lb} \times \frac{454 \text{ g}}{1 \text{ lb}} \times \frac{6.022 \times 10^{23} \text{ } H_2O \text{ molecules}}{18.0 \text{ g}} = 2.73 \times 10^{26} \text{ } H_2O \text{ molecules}$$

(continued on next page)

Let's check our answers to see if they make sense. The 18 amu of water in part "a" is an atomic size sample of water, so our number of molecules should be small. Parts "b" and "c" ask for the number of molecules in a sample of water we can gulp or guzzle. There should be trillions of molecules in these samples. In fact there are over a sextillion; our answers appear plausible.

4) You may have heard the saying that "a pint of liquid water is a pound the world around." How many water molecules are there in a cup of water?

solution:

We know that:

1 pint water ≈ 1 lb water

1/2 pint water = 1/2 lb water = 1 cup water (1 pint = 2 cups)

18.0 lb water = 2.73×10^{26} water molecules (from exercise 3c)

Therefore:

$$\tfrac{1}{2} \text{ lb water} = \frac{18.0 \text{ lb water}}{36} = \frac{2.73 \times 10^{26}\ H_2O \text{ molecules}}{36}$$

$$= \mathbf{8 \times 10^{24} \text{ water molecules}}$$

We could also work this problem by the factor label method:

$$\text{\# molecules} = 1 \text{ cup } H_2O \times \frac{1 \text{ pint}}{2 \text{ cups}} \times \frac{1 \text{ lb } H_2O}{1 \text{ pint } H_2O} \times \frac{454 \text{ g}}{1 \text{ lb}} \times \frac{6.022 \times 10^{23} \text{ amu}}{1 \text{ g}} \times \frac{1\ H_2O \text{ molecule}}{18.0 \text{ amu}}$$

$$= 8 \times 10^{24}\ H_2O \text{ molecules}$$

Notice that if we substitute variables for the numbers in the above calculation, we obtain the equation:

$$n = \left(V_{H_2O}\right) \times 1 \times \left(d_{H_2O}\right) \times 1 \times 1 \times \left(\frac{1}{\mathbf{M}_{H_2O}}\right)$$

$$n = \left(V_{H_2O}\right) \times \left(d_{H_2O}\right) \times \left(\frac{1}{\mathbf{M}_{H_2O}}\right)$$

$$n = \frac{Vd}{\mathbf{M}}$$

We can arrive at this same equation using the equations for molar mass and density:

$$\mathbf{M} = m/n \rightarrow n = m/\mathbf{M}$$

$$d = m/V \rightarrow m = Vd$$

Substituting $m = Vd$ into the equation for n gives:

$$n = \frac{Vd}{\mathbf{M}}$$

Converting from mass to atoms

You weigh a nickel and find that it weighs 5.0012 grams. How many nickel atoms are in the nickel? A nickel is 25.00% elemental nickel and 75.00% copper.

solution:

First calculate the mass of elemental nickel in a nickel: $0.2500 \times 5.0012\text{ g} = 1.250\text{ g}$

Next convert the mass of the nickel to an amount (moles of nickel):

$$\frac{1.250\text{ g}}{58.69\text{ g/mole}} = 0.02130\text{ moles}$$

Finally, convert moles of nickel to atoms of nickel:

$$0.02130\text{ moles Ni} \times 6.022 \times 10^{23}\text{ atoms/mole} = 1.283 \times 10^{22}\text{ atoms}$$

A nickel weighs less than 1/10th of a mole of nickel atoms (58.69 g), and our answer is a less than 1/10th of Avogadro's number, so it appears to be the right magnitude. We could also work this problem by first converting grams to amu's, and then converting amu's to atoms:

$$1.250\text{ g} \times \frac{1\text{ amu}}{1.661 \times 10^{-24}\text{ g}} = 7.527 \times 10^{23}\text{ amu}$$

$$7.527 \times 10^{23}\text{ amu} \times \frac{1\text{ Ni atom}}{58.69\text{ amu}} = 1.283 \times 10^{22}\text{ atoms}$$

This problem illustrates how molar mass (**M,** g/mol) relates a mass (in grams) to an amount (mols). Avogadro's number (6.022×10^{23} atoms/mol) relates an amount (moles) to a population of atoms.

Converting from mass to moles and molecules

A typical baking powder contains 28.0% sodium bicarbonate, $NaHCO_3$. How many moles and molecules of sodium bicarbonate are in 1.00 gram of this baking powder?

solution:

First we calculate the molar mass of $NaHCO_3$:

M ($NaHCO_3$) = molar mass Na + molar mass H + molar mass C + (3 × molar mass O)

= 22.99 g/mol + 1.01 g/mol + 12.01 g/mol + 3(16.00 g/mol) = 84.01 g/mole

We use the molar mass to convert from grams of baking powder to moles:

1.00 g of the baking powder contains: $0.280 \times 1.00\text{ g} = 0.280\text{ g } NaHCO_3$

$$\frac{0.280\text{ g}}{84.01\text{ g/mole}} = 3.33 \times 10^{-3}\text{ moles}$$

Finally, we convert from moles to molecules:

$$3.33 \times 10^{-3}\text{ moles} \times \frac{6.022 \times 10^{23}\text{ molecules}}{\text{mole}} = 2.01 \times 10^{21}\text{ molecules}$$

3.2 DETERMINING THE FORMULA OF AN UNKNOWN COMPOUND

Empirical and Molecular Formulas

Empirical formulas show the relative number of atoms of each element in a compound using the smallest whole numbers possible. **Molecular formulas** show the actual number of atoms in a molecule. Water has a molecular formula of H_2O; its empirical formula is also H_2O. Hydrogen peroxide has a molecular formula H_2O_2. Its empirical formula is HO. Many different compounds may have the same empirical formula. Table 3.2 on page 106 in your textbook lists six compounds that all have the empirical formula CH_2O. Different compounds may have the same molecular formula because the same atoms may bond together in different arrangements. We need **structural formulas** to distinguish between these structural isomers. Table 3.3 on page 106 in your text lists the very different properties of two pairs of structural isomers: ethanol and dimethyl ether, both of which have the molecular formula C_2H_6O, and butane and 2-methylpropane, both of which have the molecular formula C_4H_{10}.

Types of Formulas

Chemical formulas use element symbols and numerical subscripts to show the type and number of each atom present in a compound. There are three types of chemical formulas:

1) **Empirical formula:** shows the simplest whole number ratio of atoms of each element in a compound.

2) **Molecular formula:** shows the actual number of atoms of each element in the molecule.

3) **Structural formula:** shows the actual number of atoms of each element in a molecule and the bonds between them.

Ethene (or ethylene) has a molecular formula of C_2H_4. Its empirical formula is CH_2, and its structural formula shows the bonds between the atoms:

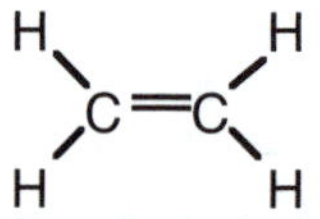

```
H       H
 \     /
  C = C
 /     \
H       H
```

Methane has a molecular formula of CH_4. Its empirical formula is the same as its molecular formula, CH_4, and its structural formula is:

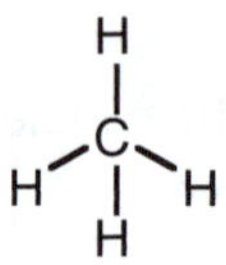

```
    H
    |
    C
  / | \
 H  |  H
    H
```

Ionic compounds such as sodium chloride generally use empirical formulas. NaCl is an empirical formula that tells us there is one sodium ion for every chloride ion.

Determining empirical and molecular formulas

Ethanol contains the elements carbon, hydrogen, and oxygen. The weight percentages of each element was measured and found to be:

$$\%C = 52.1;\ \%H = 13.1;\ \%O = 34.7$$

Determine the empirical formula. The molar mass of ethanol is found to be about 46.1 g/mole. Determine its molecular formula.

solution:

We need to determine the relative numbers of C, H, and O atoms. We'll assume, for convenience, we have a 100.0 gram sample of ethanol. In 100.0 grams of ethanol, there are 52.1 g C, 13.1 g H, and 34.7 g O. We use the molar masses to find the moles of each element:

$$\frac{52.1 \text{ g C}}{12.01 \text{ g/mole}} = 4.34 \text{ mole C}$$

$$\frac{13.1 \text{ g H}}{1.008 \text{ g/mole}} = 13.0 \text{ mole H}$$

$$\frac{34.7 \text{ g O}}{16.00 \text{ g/mole}} = 2.17 \text{ mole O}$$

At this point in the problem we can write a preliminary equation: $C_{4.34}H_{13.0}O_{2.17}$

However, the empirical formula is written with the simplest whole numbers, so we need to divide each number of moles by the smallest number (2.17):

$$C: \frac{4.34}{2.17} = 2.00;\quad H: \frac{13.0}{2.17} = 5.99;\quad O: \frac{2.17}{2.17} = 1.00$$

The empirical formula for ethanol is: $\mathbf{C_2H_6O}$

The molar mass of the empirical formula is:

$$\mathbf{M} = (2 \text{ mol C} \times 12 \text{ g/mol}) + (6 \text{ mol H} \times 1 \text{ g/mol}) + (1 \text{ mol O} \times 16 \text{ g/mol}) = 46 \text{ g/mol}$$

Since the molar mass of ethanol equals the molar mass of the empirical formula, the molecular formula is the same as the empirical formula: molecular formula = $\mathbf{C_2H_6O}$.

Combustion Analysis

When a substance containing carbon and hydrogen (an organic compound) is burned in pure oxygen, all the carbon in the combusted material converts to its oxide, CO_2, and all the hydrogen converts to its oxide, H_2O. If we collect the carbon dioxide and water vapor in separate containers, we can determine their masses, and calculate the masses of carbon and hydrogen in the combusted material. If the organic compound also contains oxygen, nitrogen, or a halogen, we calculate its mass by subtracting the masses of C and H from the original mass of the compound. This technique, called "combustion analysis," uses a combustion apparatus to selectively absorb carbon dioxide and water in separate chambers. A drawing of this apparatus is shown in Figure 3.4 on page 104 in your text.

Chemical composition from combustion analysis

A 105.5 mg sample of a white substance, suspected of being cocaine ($C_{17}H_{21}NO_4$), forms 279.3 mg of CO_2 and 66.46 mg of H_2O on combustion. Chemical analysis shows that the compound contains 4.68% N by mass. Would you conclude that the white solid is cocaine?

solution:

First calculate the mmol of CO_2 and H_2O formed:

$$\frac{279.3 \text{ mg } CO_2}{44.01 \text{ mg/mmole}} = 6.346 \text{ mmol } CO_2$$

$$\frac{66.46 \text{ mg } H_2O}{18.02 \text{ mg/mmole}} = 3.688 \text{ mmol } H_2O$$

Since all the carbon in the unknown converts to CO_2, we know that:

mmol CO_2 = mmol C = 6.346 mmol C,

and we can use the molar mass of carbon to calculate mg of carbon in the unknown:

g C = 6.346 mmol × 12.01 mg/mmol = **76.22 mg C**

Since all the hydrogen in the unknown converts to H_2O, we know that:

mmol H_2O = 1/2 × mmol H in unknown = 7.376 mmol H,

and we can use the molar mass of hydrogen to calculate mg of H in the unknown:

g H = 7.376 mmol × 1.008 mg/mmol = **7.435 mg H**

The compound contains 4.68% N by weight, so in 105.5 mg, there is:

105.5 mg × 0.0468 = **4.94 mg N**

The mass in the unknown sample from carbon, hydrogen, and nitrogen is:

76.22 mg C + **7.435** mg H + **4.94** mg N = 88.59 mg

The difference between the sum of the masses of C, H, and N and the total mass of the unknown must be due to the mass of oxygen: 105.5 mg – 88.59 = **16.9 mg O**

Calculate the percent of each element by mass:

$$C: \frac{76.22 \text{ mg}}{105.5 \text{ mg}} \times 100 = 72.25\% \qquad N: \frac{4.937 \text{ mg}}{105.5 \text{ mg}} \times 100 = 4.680\%$$

$$H: \frac{7.435 \text{ mg}}{105.5 \text{ mg}} \times 100 = 7.047\% \qquad O: \frac{16.9 \text{ mg}}{105.5 \text{ mg}} \times 100 = 16.0\%$$

Total = 100%

Our last step is to compare this composition to that of cocaine ($C_{17}H_{21}NO_4$):

M (cocaine) = (17 × 12.01 g/mol) + (21 × 1.008 g/mol) + 14.01 g/mol + (4 × 16.00 g/mol)
= 303.4 g/mol

(continued on next page)

One mole of cocaine weighs 303.4 g and contains:

(17 mol × 12.01 g/mol) = 204.2 g C (1 mol × 14.01 g/mol) = 14.01 g N

(21 mol × 1.008 g/mol) = 21.17 g H (4 mol × 16.00 g/mol) = 64.00 g O

The percent of each element by mass is:

$$\text{C:}\ \frac{204.2\ \text{mg}}{303.4\ \text{mg}} \times 100 = 67.30\% \qquad \text{N:}\ \frac{14.01\ \text{mg}}{303.4\ \text{mg}} \times 100 = 4.618\%$$

$$\text{H:}\ \frac{21.17\ \text{mg}}{303.4\ \text{mg}} \times 100 = 6.978\% \qquad \text{O:}\ \frac{64.00\ \text{mg}}{303.4\ \text{mg}} \times 100 = 21.09\%$$

Total = 99.99%

Our results indicate that the unknown is not cocaine because the composition by mass is not the same. (The unknown is $C_{18}H_{21}NO_3$, codeine.)

3.3 BALANCING CHEMICAL EQUATIONS

Chemical reactions occur when atoms recombine to form new molecules or compounds. Let's imagine hydrogen and fluorine reacting to form hydrogen fluoride gas: diatomic H_2 and F_2 molecules break apart, and the atoms rearrange to form hydrogen fluoride (HF). The chemical equation is:

$$H_2(g) + F_2(g) \longrightarrow 2HF(g)$$

A balanced equation has the same number of each type of atom on both sides of the equation. This equation tells us that 1 hydrogen molecule reacts with 1 fluorine molecule to produce 2 hydrogen fluoride molecules, or 1 million H_2 molecules react with 1 million F_2 molecules to produce 2 million HF molecules, or 1 mole of H_2 molecules react with 1 mole of F_2 molecules to produce 2 moles of HF molecules.

Steps for Balancing an Equation

1) Write chemical formulas for reactants to the left of an arrow, and the chemical formulas for the products to the right of the arrow. Put a blank in front of each substance to remind us to account for its atoms.

2) Balance the number of each atom on the two sides of the equation by placing coefficients in front of chemical formulas; begin with the most complex substance, the one with the largest number of atoms or different types of atoms.

3) Adjust the coefficients to the smallest whole number coefficients. In the final form, a coefficient of 1 is implied, so we don't need to write it.

4) Check to make sure the equation is balanced.

5) Specify the states of matter: solid (*s*), liquid (*l*), gaseous (*g*), aqueous solution (*aq*).

Balancing Reactions

Many reactions in our daily life, such as burning gasoline to run our cars, burning natural gas to heat our houses, and digesting food to provide our energy are combustion reactions. The combustion of a hydrocarbon (a compound containing carbon and hydrogen), or a carbohydrate (a compound containing carbon, hydrogen, and oxygen) in the presence of O_2, produces carbon dioxide (CO_2) and water (H_2O). The easiest way to balance a combustion reaction is to balance the atoms in the order: carbon, hydrogen, and then oxygen. The oxygen is easily balanced at the end because the coefficients of the oxygen molecule can be changed without affecting the number of any other atoms.

Balancing a combustion reaction

Balance the equation for the combustion of ethanol, C_2H_6O.

solution:

First write the reactants and products:

$$C_2H_6O + O_2 \rightarrow CO_2 + H_2O$$

Balance the carbon atoms by placing a 2 in front of CO_2:

$$C_2H_6O + O_2 \rightarrow 2CO_2 + H_2O$$

Balance the hydrogen atoms by placing a 3 in front of H_2O:

$$C_2H_6O + O_2 \rightarrow 2CO_2 + 3H_2O$$

Balance the oxygen atoms by placing a 3 in front of O_2:

$$C_2H_6O + 3O_2 \rightarrow 2CO_2 + 3H_2O$$

Specify states of matter:

$$C_2H_6O(l) + 3O_2(g) \rightarrow 2CO_2(g) + 3H_2O(g)$$

Check: 2 carbon atoms on both sides, 6 hydrogen atoms on both sides, and 7 oxygen atoms on each side.

Remember that a coefficient operates on all the atoms in the formula. Add only coefficients to balance equations, never add or change subscripts (the chemical formulas themselves cannot be changed).

3.4 CALCULATING AMOUNTS OF REACTANT AND PRODUCT

Our balanced equation for the combustion of ethanol:

$$C_2H_6O(l) + 3O_2(g) \rightarrow 2CO_2(g) + 3H_2O(g)$$

tells us that one molecule of ethanol reacts with three molecules of oxygen to produce two molecules of carbon dioxide and three molecules of water, or 1 mole of ethanol molecules reacts with 3 moles of oxygen molecules to produce 2 moles of carbon dioxide molecules and 3 moles of water molecules. We say that 1 mol C_2H_6O is stoichiometrically equivalent to 3 mol O_2. Which of the statements on the following page about the above balanced equation is not true?

3 mol O_2 is stoichiometrically equivalent to 2 mol CO_2

2 mol CO_2 is stoichiometrically equivalent to 3 mol H_2O

3 mol O_2 is stoichiometrically equivalent to 3 mol H_2O

1 mol C_2H_6O is stoichiometrically equivalent to 2 mol CO_2

2 mol C_2H_6O is stoichiometrically equivalent to 6 mol O_2

3 grams O_2 is stoichiometrically equivalent to 2 grams CO_2

The last statement is false! The coefficients in the balanced equation do not directly relate masses of molecules, only numbers of molecules. Let's find out how many grams of CO_2 form when 3 grams of O_2 are consumed in the combustion of ethanol. We'll use steps 1-4 in your textbook on page 91 to solve this problem.

1. The balanced equation is:

$$C_2H_6O + 3O_2 \rightarrow 2CO_2 + 3H_2O$$

2. Convert the given mass (3.0 g O_2) to moles:

$$\text{Moles } O_2 \text{ consumed} = 3.0 \text{ g } O_2 \times \frac{1 \text{ mol } O_2}{32 \text{ g } O_2} = .094 \text{ mol } O_2$$

3. The molar ratio from the balanced equation tells us how many moles of CO_2 form:

$$0.094 \text{ mol } O_2 \times \frac{2 \text{ mol } CO_2}{3 \text{ mol } O_2} = .063 \text{ mol } CO_2$$

4. Convert moles of CO_2 to grams of CO_2:

$$0.063 \text{ mol } CO_2 \times \frac{44 \text{ g } CO_2}{1 \text{ mol } CO_2} = 2.8 \text{ g } CO_2$$

5. Shown in one equation, steps 2-4 are:

$$3.0 \text{ g } O_2 \times \frac{1 \text{ mol } O_2}{32 \text{ g } O_2} \times \frac{2 \text{ mol } CO_2}{3 \text{ mol } O_2} \times \frac{44 \text{ g } CO_2}{1 \text{ mol } CO_2} = 2.8 \text{ g } CO_2$$

The combustion of C_2H_6O with 3 moles of oxygen would produce 2 moles of carbon dioxide. However, the combustion of C_2H_6O with 3.0 **grams** of oxygen does not produce 2.0 grams of carbon dioxide, but produces 2.8 grams.

Reactions that Occur in Sequence

Often a product of one reaction becomes a reactant for the next in a sequence of reactions. When the same substance that forms in one reaction reacts in the next, we eliminate it in an overall (net) equation. Make sure the equations are adjusted arithmetically to cancel the common substance(s) before adding the equations together to obtain the overall balanced equation. On the next page, we will work through problem 3.77 in your text on the preparation of lead from lead(II) sulfide.

Preparation of lead from lead(II)sulfide

Lead can be prepared from galena (lead(II)sulfide) by first roasting the galena in oxygen gas to form lead(II)oxide and sulfur dioxide. Heating the metal oxide with more galena forms the molten metal and more sulfur dioxide. Write a balanced equation for each step, and write an overall balanced equation for the process. How many metric tons of sulfur dioxide form for every metric ton of lead obtained?

solution:

First write a balanced reaction for lead(II)sulfide with oxygen:

$$2PbS(s) + 3O_2(g) \rightarrow 2PbO(s) + 2SO_2(g)$$

Then write the reaction for the metal oxide with more galena:

$$2PbO(s) + PbS(s) \rightarrow 3Pb(l) + SO_2(g)$$

Write an overall balanced equation for the process (canceling the lead oxide):

$$2PbS(s) + 3O_2(g) \rightarrow \cancel{2PbO}(s) + 2SO_2(g)$$

$$\cancel{2PbO}(s) + PbS(s) \rightarrow 3Pb(l) + SO_2(g)$$

$$3PbS(s) + 3O_2(g) \rightarrow 3Pb(l) + 3SO_2(g) \quad \text{or,} \quad PbS(s) + O_2(g) \rightarrow Pb(l) + SO_2(g)$$

Calculate metric tons of sulfur dioxide formed for every metric ton of lead obtained:

$$1 \text{ metric ton Pb} \times \frac{\text{mol Pb}}{207.21 \text{ g Pb}} \times \frac{1 \text{ mol } SO_2}{1 \text{ mol Pb}} \times \frac{64.06 \text{ g } SO_2}{\text{mol } SO_2} = 0.309 \text{ metric ton } SO_2$$

Limiting Reactant

In a chemical reaction, when one reactant is used up, the reaction stops. The **limiting reactant** is simply the reactant that runs out first. If the reactants react in a 1:1 mole ratio, the limiting reactant is the reactant with the fewest moles of material. Otherwise, we use the balanced equation to determine which reactant disappears first.

Limiting reactant example

Ammonia (17.03 g/mol) reacts with copper(II) oxide (79.55 g/mol) to form nitrogen gas (28.02 g/mol), copper (63.55 g/mol), and water (18.02 g/mol). If we react 40.00 g of ammonia with 50.00 g of copper(II) oxide, which is the limiting reactant? How much copper can be produced in this reaction?

solution:

Write the balanced equation:

$$2NH_3(g) + 3CuO(s) \rightarrow N_2(g) + 3Cu(s) + 3H_2O(g)$$

Calculate moles of each reactant:

(continued on next page)

$$NH_3: \frac{40.00 \text{ g}}{17.03 \text{ g/mole}} = 2.349 \text{ mole } NH_3$$

$$CuO: \frac{50.00 \text{ g}}{79.55 \text{ g/mole}} = 0.6285 \text{ mole CuO}$$

Using the coefficients from the balanced equation, determine how many moles of CuO would theoretically react with 1.468 moles of NH_3:

$$2.349 \text{ mole } NH_3 \times \frac{3 \text{ mole CuO}}{2 \text{ mole } NH_3} = 3.523 \text{ mole CuO}$$

There are only 0.6286 moles of CuO available, so it is the limiting reactant. We will base the amount of Cu formed on the limiting reactant, CuO:

$$\text{g Cu produced} = 0.6285 \text{ mole CuO} \times \frac{3 \text{ mole Cu}}{3 \text{ mole CuO}} \times 63.55 \text{ g/mole} = 39.94 \text{ g Cu}$$

The amount of product formed based on the limiting reactant is the **theoretical yield,** the yield that could be obtained if 100% of the limiting reactant becomes product.

Real Life: Theoretical, Actual, and Percent Yields

The actual yield that you can obtain from a chemical reaction is never as high as the theoretical yield. Some loss of product always occurs either due to physical loss of material to laboratory equipment, or to competing side reactions. The amount of product that is actually obtained is the **actual yield.** The **percent yield** (% yield) is the ratio of actual to theoretical yield:

$$\textbf{\% yield} = \frac{\textbf{actual yield}}{\textbf{theoretical yield}} \times \textbf{100\%}$$

Percent yield

If 5.00 g each of ammonia, oxygen, and methane react, how much hydrogen cyanide can be produced? If 2.10 g of HCN is produced, what is the percent yield?

solution:

The balanced equation is:

$$2NH_3(g) + 3O_2(g) + 2CH_4(g) \rightarrow 2HCN(g) + 6H_2O$$

Determine how much hydrogen cyanide could be produced if each reactant is completely used up:

$$NH_3: \frac{5.00 \text{ g}}{17.03 \text{ g/mole}} \times \frac{2 \text{ mole HCN}}{2 \text{ mole } NH_3} = 0.294 \text{ mole HCN}$$

(continued on next page)

$$O_2: \frac{5.00\text{ g}}{32.00\text{ g/mole}} \times \frac{2\text{ mole HCN}}{3\text{ mole }O_2} = 0.104\text{ mole HCN}$$

$$CH_4: \frac{5.00\text{ g}}{16.04\text{ g/mole}} \times \frac{2\text{ mole HCN}}{2\text{ mole }CH_4} = 0.312\text{ mole HCN}$$

Oxygen produces the least amount of hydrogen cyanide, so it is the limiting reactant. The amount of HCN produced if all the oxygen reacts is (the theoretical yield):

$$0.104\text{ mol HCN} \times 27.03\text{ g/mol} = 2.81\text{ g HCN}$$

If 2.10 g of HCN is actually produced, the percent yield is:

$$\%\text{ yield} = \frac{\text{actual}}{\text{theoretical}} \times 100 = \frac{2.10\text{ g}}{2.81\text{ g}} \times 100 = 74.7\%$$

When might an "actual" yield appear higher than the theoretical yield?

Atom Economy

A high percent yield in chemical reactions is desirable, but it does not indicate how efficiently the reactants have been used in generating the desired product. “Atom economy” is a method of expressing how efficiently a particular reaction makes use of the reactant atoms:

$$\text{atom economy} = \frac{\text{mass of atoms in desired product}}{\text{mass of atoms in reactants}} \times 100\%$$

This approach helps in comparing different pathways to a desired product (see the comparison in your text of two synthetic routes for the production of maleic anhydride, an important intermediate chemical).

CHAPTER 3 SUMMARY

Units of Measure

SI unit for amount (n) = **mole**

mass (m), volume (V), number (n):

density (d) = mass/volume = m/V

concentration = amount/volume = n/V (special case: **molarity (M)** = moles/L)

molar mass (M) = mass/amount = m/n

1 mole = Avogadro's number of entities

= 6.022×10^{23} **entities**

= the amount of substance that contains the same number of entities as there are atoms in 12.00 g of carbon-12.

Balanced Equations

Equations are balanced when the same number of each atom type appears on both sides of the equation.

Combustion reactions of hydrocarbons or carbohydrates in the presence of oxygen always produce $\mathbf{CO_2}$ and $\mathbf{H_2O}$. The amounts of CO_2 and H_2O formed can be used to determine the composition of the combusted substance.

The atoms in combustion reactions should be balanced in the order: C, H, and then O.

The **limiting reactant** is the one that is used up first in a reaction and causes the reaction to stop. The amount of the limiting reactant determines the **theoretical yield.**

$$\%\text{ yield} = \frac{\text{actual yield}}{\text{theoretical yield}} \times 100\%$$

$$\text{atom economy} = \frac{\text{mass of atoms in desired product}}{\text{mass of atoms in reactants}} \times 100\%$$

CHAPTER CHECK

Make sure you can...

- Define a mole.
- Know Avogadro's number.
- Define molar mass, **M,** and give its common units.
- Write a mathematical relationship between volume, density, molar mass, and amount.
- Convert between mass and amount (atoms or moles).
- Define empirical, molecular, and structural formulas.
- Determine empirical formulas from the percentages, by mass, of elements in a compound.
- Write a balanced equation for the combustion of a hydrocarbon.
- Determine a molecular formula from combustion analysis.
- Know how to write an overall equation for reactions that occur in sequence.
- Determine which reactant is the limiting reagent.
- Calculate % yield from actual and theoretical yields.
- Understand the concept of atom economy and how it differs from percent yield.

Chapter 3 Exercises

3.1) a. How many moles of carbon atoms are there in 24 grams of carbon?
b. How many moles of oxygen atoms are there in 64 grams of oxygen?
c. How many moles of hydrogen molecules are there in 20 grams of hydrogen?

3.2) a. What is the molar mass of methane CH_4 in g/mol?
b. What is the molar mass of potassium sulfide K_2S in g/mol?
c. What is the molar mass of neon gas in g/mol?

3.3) a. How many g of nitrogen contain the same number of moles of atoms as 24.00 g of hydrogen?
b. Which sample contains the greatest number of atoms?

One mole of hydrogen molecules H_2
One mole of carbon atoms
Two moles of helium atoms
One mole of uranium atoms
One-half mole of methane molecules CH_4

c. Which sample has the greatest mass?
One mole of iron atoms
One mole of sulfur atoms
One mole of nickel atoms
One mole of potassium atoms
One mole of tin atoms

3.4) a. How many moles of nitrogen are there in 12 moles of TNT (trinitrotoluene; $CH_3C_6H_2(NO_2)_3$)?
b. How many moles of nitrogen are there in 3 moles of ammonium nitrate (NH_4NO_3)?

3.5) What is the percentage by mass of:
a. sulfur in sulfur dioxide SO_2
b. carbon in ethanol C_2H_5OH
c. iron in limonite $2Fe_2O_3 \cdot 3H_2O$

3.6) a. Malonic acid has the composition: carbon 34.6%; hydrogen 3.9%; oxygen 61.5%
What is its empirical formula?

b. 10.00 g of an oxide of uranium is treated with fluorine to produce oxygen and 13.45 grams of uranium hexafluoride UF_6. What is the empirical formula of the oxide?

3.7) The elemental analysis of estrone, one of the female sex hormones, reported the compound to be 80.00% C, 8.15% H, and 11.85% O by mass. The mass spectrum shows the molar mass to be 270 g/mole. What are the empirical and molecular formulas for estrone?

3.8) Talc, the principle ingredient of baby powder, has the following elemental composition:

19.24% Mg, 0.53% H, 29.56% Si, and 50.67% O by mass

Further analysis shows that each formula contains two hydroxide groups. What is the empirical formula for talc?

3.9) Ibuprofen, an alternative analgesic to aspirin, contains only C, H, and O. It is sold under trade names such as Motrin and Advil. The combustion of 24.72 mg of Ibuprofen produced 68.64 mg of CO_2 and 12.96 mg of H_2O. What is the empirical formula for Ibuprofen?

3.10) a. What coefficient is necessary for nitric acid to balance the equation:

$$3Cu + HNO_3 \rightarrow 3Cu(NO_3)_2 + 2NO + 4H_2O$$

b. What coefficients are necessary to balance the equation:

$$Pb(NO_3)_2 \rightarrow PbO + NO_2 + O_2$$

3.11) a. When benzene C_6H_6 burns in air, carbon dioxide and water are produced. What is the ratio of the moles of carbon dioxide produced to the moles of water produced?

b. Aluminum carbide reacts with water as follows:

$$Al_4C_3 + 12H_2O \rightarrow 3CH_4 + 4Al(OH)_3$$

How many grams of methane will be produced if 12 g of aluminum carbide are used?

3.12) Sodium azide, NaN_3, is used to generate the gas that fills automobile air bags during a collision. What mass of N_2 gas is produced by the decomposition of 715 grams of NaN_3?

$$2NaN_3 \rightarrow 2Na + 3N_2$$

3.13) a. If 7.00 grams of ammonia and 14.0 grams of oxygen react to form nitric oxide and water as shown in the equation below, which one of the two reactants limits the amount of product formed?

$$4NH_3 + 5O_2 \rightarrow 4NO + 6H_2O$$

b. How much of the non-limiting reactant remains after all the limiting reactant has been used up?

c. How much NO is formed?

3.14) Phosphoric acid H_3PO_4 is manufactured by treating mineral phosphates with concentrated sulfuric acid H_2SO_4. A typical phosphate mineral, $Ca_5(PO_4)_3F$, reacts with sulfuric acid as shown by the equation:

$$Ca_5(PO_4)_3F + 5H_2SO_4 \rightarrow 3H_3PO_4 + 5CaSO_4 \cdot 2H_2O + HF$$

If 5.00 tons of this mineral is treated with 6.00 tons of sulfuric acid to produce 2.70 tons of phosphoric acid, what is the percent yield of phosphoric acid?

3.15) Liquid propane (LP gas) is the fuel used in outdoor gas grills and is commonly used for heating in rural areas. The use of liquid propane is a safe fuel as long as an abundant source of oxygen is available:

$$C_3H_8 + 5O_2 \rightarrow 3CO_2 + 4H_2O$$

If 55.0 g of propane burns in the presence of 192 g of oxygen, what is the limiting reactant? How much of the excess reactant remains?

3.16) 2.50 g of elemental phosphorous, P_4, was burned in an excess of oxygen to produce a compound with the elemental composition of 43.7% P and 56.3% O by mass. The mass spectrum of the product indicates it has a molar mass between 280 and 285 g/mol:

a. What is the molecular formula of the product?

b. Give the name of the product.

c. Write a balanced chemical equation for the formation of the product.

d. What mass of product was produced?

Chapter 3 Answers

3.1)
a. $24\text{ g C} \times 1\text{ mol}/12.01\text{g} = 2.0\text{ mol C}$
b. 4.0 mol oxygen atoms
c. 10. mol hydrogen molecules

3.2) a. Molar mass $CH_4 = (1 \times 12.01) + (4 \times 1.008) = 16.04$ g/mol; b. 110.27 g/mol; c. 20.18 g/mol

3.3)
a. $24.00\text{ g H} \times 14.01\text{ g N/mol}/1.008\text{ g H/mol} = 333.6\text{ g}$
b. One-half mole of methane (contains 2.5 mol atoms)
c. Each sample contains one mole of atoms. The sample with the greatest mass is the substance with the greatest molar mass: one mole of tin atoms has a mass = 118.7 g.

3.4)
a. Every mole of TNT contains 3 moles of N, so 12 moles of TNT contains $(3 \times 12) = 36$ mol N.
b. 6 mol

3.5) a. The molar mass of SO_2 = 32.066 + 2(15.9994) = 64.0648 g/mol

$$\% \text{ of sulfur} = \frac{32.066}{64.0648} \times 100 = 50.05\ \%$$

b. 52.14%; c. 59.82%

3.6) a. $C_3H_4O_4$

b. All the uranium from the reactant ends up in the product, UF_6. We can use the amount of the product formed and its molecular formula to figure out the moles of uranium:

Molar mass of UF_6 = (238.0 g U/mol) + (6 × 19.00 g F/mol) = 352.0 g/mol

$$\text{Grams of uranium} = \frac{238.0}{352.0} \times 13.45 \text{ g} = 9.09 \text{ g}$$

9.09 g of uranium came from the reactant. The rest of the reactant mass (10.00 – 9.09 = 0.91 g) comes from oxygen. Converting from grams to moles we obtain:

$$9.09 \text{ g} \times \frac{1 \text{ mol}}{238.0 \text{ g}} = .0382 \text{ mol U}; \quad 0.91 \text{ g} \times \frac{1 \text{ mol}}{16.00 \text{ g}} = .057 \text{ mol O}$$

The empirical formula is: $U_{0.038}O_{.057} = U_1O_{1.5} = U_2O_3$

3.7) empirical formula: $C_9H_{11}O$; molecular formula: $C_{18}H_{22}O_2$

3.8) $Mg_3H_2Si_4O_{12}$

3.9) $C_{11}H_{10}O_2$

3.10) a. 8

b. $2Pb(NO_3)_2 \rightarrow 2PbO + 4NO_2 + O_2$

3.11) a. One mole of benzene will produce 6 moles of carbon dioxide on combustion, and 3 moles of water. The ratio between the compounds, water and carbon dioxide, is:

6 moles CO_2 : 3 moles H_2O, or 2:1

b. 4.0 grams CH_4

3.12) Change grams of NaN_3 to moles, use the balanced equation to determine moles of N_2, and convert mass:

$$715 \text{ g NaN}_3 \times \frac{1 \text{ mol NaN}_3}{65.02 \text{ g NaN}_3} \times \frac{3 \text{ mol N}_2}{2 \text{ mol NaN}_3} \times \frac{28.02 \text{ g N}_2}{\text{mol N}_2} = 462 \text{ g N}_2$$

3.13) a. Determine how much NO forms from the ammonia and the oxygen if both were completely used:

$$7.00 \text{ g NH}_3 \times \frac{1 \text{ mol NH}_3}{17.03 \text{ g NH}_3} \times \frac{4 \text{ mol NO}}{4 \text{ mol NH}_3} = 0.411 \text{ mol NO}$$

$$14.0 \text{ g } O_2 \times \frac{1 \text{ mol } O_2}{32.00 \text{ g } O_2} \times \frac{4 \text{ mol NO}}{5 \text{ mol } O_2} = 0.350 \text{ mol NO}$$

Oxygen limits the amount of NO formed.

b. $0.350 \text{ mol NO} \times \frac{4 \text{ mol } NH_3}{4 \text{ mol NO}} = 0.350 \text{ mol } NH_3$ reacted; 0.062 mol left

$0.062 \text{ mol } NH_3 \times 17.03 \text{ g/mol} = 1.0 \text{ g } NH_3$ left over

c. $0.350 \text{ mol NO} \times 30.01 \text{ g/mol} = 10.5 \text{ g NO}$

3.14) 100% yield = 2.91 tons; actual yield is 2.70 tons, or 92.8%

3.15) Limiting reactant is O_2; 2.1 g C_3H_8 remains

3.16) a. We'll use a molar mass of 283:

$0.437 \times 283 = 123.67$ g P; 123.67 g / 30.97 g/mol = 3.99 (~4) mol P

$0.563 \times 283 = 159.329$ g O; 159.329 g / 16.00 g/mol = 9.96 (~10) mol O

The molecular formula is P_4O_{10}.

b. tetraphosphorous decoxide

c. $P_4 + 5O_2 \rightarrow P_4O_{10}$

d. 2.50 g of P_4 / 123.9 g/mol = .0202 mol P_4 = mol of P_4O_{10}

0.0202 mol P_4O_{10} × 283.88 g/mol = 5.73 g P_4O_{10}

Check: The molar mass of P_4O_{10} is more than twice that of P_4. Since one mole of P_4 produces one mole of P_4O_{10}, we would expect the mass of P_4O_{10} that forms to be more than twice the mass of P_4 that reacts. 5.73 g is more than two times 2.50 g, so our answer appears reasonable.

Three Major Classes of Chemical Reactions

CHAPTER OVERVIEW

Atoms and molecules are in constant motion. We can picture them as "sticky" electron clouds that bounce into each other and may "stick" together, or fall apart and stick to new atoms. These atomic dances are called chemical reactions. The historical classification of reactions as combination, decomposition, and displacement reactions does not require much knowledge of chemistry (chemists didn't know much chemistry at first). As scientists learned more about the chemistry involved in reactions, they were able to classify reactions according to how they behave. We will discuss precipitation, acid-base, and redox reactions in this chapter. We also discuss the structure of water, and how its bent geometry determines the types of compounds that it can dissolve, and we end with a discussion on the reversibility of reactions: the extent to which the atomic dances favor the reactants or the products.

4.1 WATER AS A SOLVENT AND SOLUTION CONCENTRATION

We have already encountered the idea that a solution is a homogenous mixture with uniform composition. A solution contains a **solute**, which is dissolved in a **solvent.** In an aqueous solution, the solvent is water, and as we will see, because of an uneven charge distribution in water molecules, the molecules interact strongly with some substances.

Let's consider two types of solids, both of which dissolve in water. An ionic solid breaks apart into charged ions as it dissolves. A covalent compound stays intact as uncharged molecules. These two types of solutions are depicted on the following page:

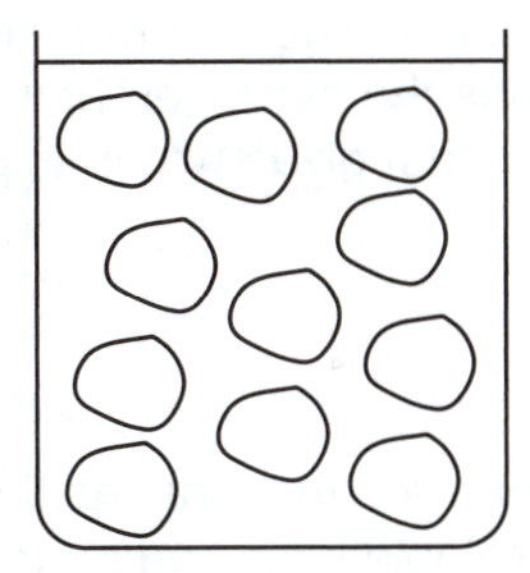

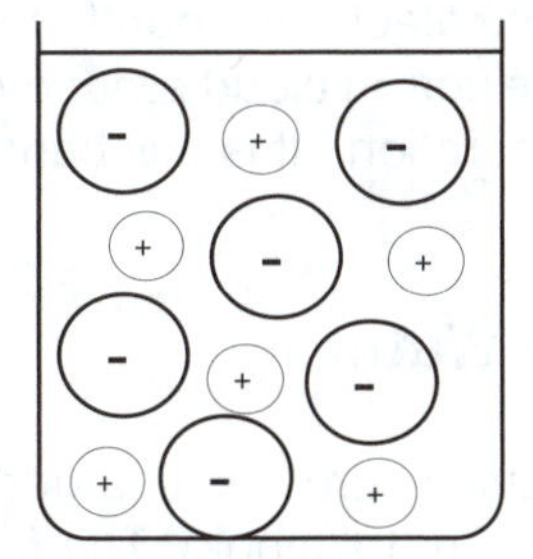

Which of the drawings above represents a table salt (NaCl) solution and which represents a sugar solution? (Answer: The drawing on the right represents a sodium chloride solution with charged ions (Na^+ and Cl^-); the drawing on the left represents a sugar solution with intact $C_{12}H_{22}O_{11}$ molecules.) We show that ions or molecules are in solution (or **solvated)** with the symbol "*aq*" which is the symbol for aqueous, in the chemical equation:

$$NaCl(s) \xrightarrow{H_2O} Na^+(aq) + Cl^-(aq)$$

$$C_{12}H_{22}O_{11}(s) \xrightarrow{H_2O} C_{12}H_{22}O_{11}(aq)$$

An ionic compound that is insoluble contains electrostatic attraction between its ions that are too strong for the water molecules to overcome. We'll see in the next section that silver bromide (AgBr) precipitates out in the double displacement reaction of silver nitrate ($AgNO_3$) with potassium bromide (KBr). Silver bromide is an insoluble compound often used in photographic films to record an image when exposed to light. If silver bromide were soluble, the silver bromide crystals, and hence the recorded image, would dissolve as soon as the film was immersed in aqueous developer solution. However, as we will see later, even compounds that are considered insoluble dissolve to a very small extent.

Ionic vs. Molecular

How could you tell a sugar and a salt solution apart? One way would be to taste them, but tasting is not recommended for safety reasons. (Older chemists may attest that tasting used to be a common method for identification; we won't comment on the effects this method had on the chemists who used it.) A safer method to tell a sugar and salt solution apart would be to immerse electrodes into the two solutions. The salt solution would conduct a current as its charged particles (ions) move towards oppositely charged electrodes (see Figure 4.3 on page 141 in your text). The sugar solution would not conduct a current because its molecules have no charge. A sodium chloride solution is called an **electrolytic solution** because it conducts a current, a sugar solution is called a **nonelectrolytic solution.**

The Polar Nature of Water

Why is water attracted to sodium and chloride ions? Water is a **polar** molecule with an uneven charge distribuion; it has a partially negative pole that attracts cations, and a partially positive pole that attracts anions:

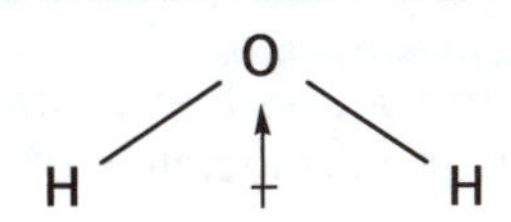

The oxygen atom attracts electrons more strongly than hydrogen, so each O–H bond in the water molecule is a **polar covalent bond;** the electrons spend more time near the oxygen atom than around the hydrogen atoms. The shape of the electron cloud around oxygen pushes the hydrogen atoms away from a linear orientation to a bent configuration. It is this bent configuration that gives water its positive and negative poles.

Ionic Compounds in Water

Polar water molecules surround charged ions and separate the ions, as they become **solvated** (see Figure 4.2 on page 140 in your textbook). The formula for a soluble ionic compound lets us know how many ions will be in solution when it is dissolved. If the attractions between ions are greater than the attraction between the ions and water, the ionic compound will only dissolve to a small extent.

Calculating the Number of Moles of Ions in Solution

If 5.0 g of the soluble salt, lithium sulfate, is dissolved in water, how many ions will be in solution once the solid has dissolved?

solution:

The formula for lithium sulfate is Li_2SO_4, so on dissolving in water, we write:

$$Li_2SO_4(s) \xrightarrow{H_2O} 2Li^+(aq) + SO_4^{2-}(aq)$$

Next we need to determine moles of lithium sulfate in 5.0 g of lithium sulfate:

$$5.0 \text{ g } Li_2SO_4 \times \frac{\text{mole } Li_2SO_4}{109.94 \text{ g } Li_2SO_4} = 0.045 \text{ moles } Li_2SO_4$$

Each mole of lithium sulfate produces 2 lithium ions and 1 sulfate ion, so moles of ions equals 3 x the moles of lithium sulfate, or 0.14 moles of ions. The number of ions in solution is therefore:

$$0.14 \text{ mol ions} \times \frac{6.02 \times 10^{23} \text{ ions}}{\text{mol ions}} = 8.4 \times 10^{22} \text{ mol ions}$$

Covalent Compounds in Water

The polar nature of a water molecule explains why some covalent compounds such as table sugar dissolve in water, and others such as octane, the major component in gasoline, do not dissolve in water. Sucrose contains polar covalent bonds that interact with those of water. Hydrocarbons such as octane do not contain polar bonds, and therefore do not dissolve in water. The general rule is that "like dissolves like."

Acids are a special case of covalent molecules that contain hydrogen and interact so strongly with water that they dissociate into ions. We show the dissociation of a fully dissociated "strong" acid such as nitric acid as:

$$HNO_3(aq) \xrightarrow{H_2O} H^+(aq) + NO_3^-(aq)$$

The H^+ ion is simply a bare proton. It is an extremely small volume of positive charge and associates strongly to the negative pole of water molecules. Often an aqueous H^+ ion is written as H_3O^+ to indicate this strong association:

$$HNO_3(aq) \xrightarrow{H_2O} H_3O^+(aq) + NO_3^-(aq)$$

Solution Stoichiometry

Concentration is the amount per unit volume (n/V). For the specific case when n = moles, and V = liters, concentration is given the special name **molarity (M).**

$$\textbf{Molarity} = \frac{\textbf{moles of solute}}{\textbf{liters of solution}}$$

Molarity

What is the molarity of a 750 mL solution containing 14.8 g of silver nitrate?

solution:

The molar mass of silver nitrate, $AgNO_3$ = 169.9 g/mole, so moles of silver nitrate in 14.8 g is:

$$\frac{14.8 \text{ g}}{169.9 \text{ g/mole}} = 0.08711 \text{ mole}$$

The molarity is:

$$\frac{0.08711 \text{ mole}}{0.750 \text{ liter}} = 0.116\ M$$

Remember that molarity refers to moles of a substance in a volume of solution, not a volume of solvent. If you were asked to make up a 0.116 M silver nitrate solution, you would add 14.8 g of silver nitrate to a volumetric flask and dilute to 750 mL with water. You would not add 750 mL of water because the total volume would be greater than 750 mL, and the solution would be too dilute.

Molarity exercise

How many grams of $ZnCl_2$ (136.29 g/mol) are present in 600 mL of a 0.500 M $ZnCl_2$ solution?

solution:

This problem is easily solved if you write down what is given, and watch units so they cancel to give your answer in grams of zinc chloride:

$$\text{g } ZnCl_2 = (0.600 \text{ L}) \times (0.500 \text{ mol/L}) \times (136.29 \text{ g/mol})$$

$$= 40.9 \text{ g } ZnCl_2$$

Check:

(continued on next page)

600 mL is close to a half liter; the molarity of the solution is half a mole per liter, so there should be approximately a quarter moles of $ZnCl_2$ present; one quarter of 140 is 35, not too far from our answer, so our answer appears reasonable.

Limiting reactant for a reaction in solution

The reaction of silver nitrate with potassium bromide is:

$$AgNO_3(aq) + KBr(aq) \longrightarrow AgBr(s) + KNO_3(aq)$$

If you react 45.0 mL of a 1.25 *M* $AgNO_3$ solution with 75.0 mL of a 0.800 *M* KBr solution, what is the limiting reactant, and how much AgBr is produced?

solution:

First calculate the moles of reactants:

$AgNO_3$: (0.0450L) × (1.25 mol/L) = 0.0562 mol

KBr: (0.0750L) × (0.800 mol/L) = 0.0600 mol

Since the reactants react in a 1:1 molar ratio, the reactant with the fewer number of moles ($AgNO_3$) is the limiting reactant.

We use the amount of the limiting reactant, $AgNO_3$, to determine the theoretical yield of silver bromide:

$$0.0562 \text{ mol } AgNO_3 \times \frac{1 \text{ mol AgBr}}{1 \text{ mol } AgNO_3} \times \frac{187.8 \text{ g AgBr}}{1 \text{ mol AgBr}} = 10.6 \text{ g AgBr}$$

4.2 WRITING EQUATIONS FOR AQUEOUS IONIC REACTIONS

When two aqueous ionic compounds are mixed, a precipitate, (*s*), may form. If we introduce silver and bromide ions into the same flask, a pale yellow solid (silver bromide) precipitates out of solution. Silver halide crystals used in photographic film are commonly produced by introducing silver nitrate and potassium bromide into an aqueous solution. Silver nitrate and potassium bromide are soluble in water, which means they dissociate into solvated ions:

$$AgNO_3(s) \xrightarrow{H_2O} Ag^+(aq) + NO_3^-(aq)$$

$$KBr(s) \xrightarrow{H_2O} K^+(aq) + Br^-(aq)$$

The four ions are the potential reactants. The potential products are AgBr and KNO_3. It turns out that potassium nitrate (KNO_3) is soluble, and silver bromide (AgBr) is the insoluble precipitate that forms. There are three ways you may see this reaction written:

1) Total ionic equation

$$Ag^+(aq) + NO_3^-(aq) + K^+(aq) + Br^-(aq) \longrightarrow AgBr(s) + NO_3^-(aq) + K^+(aq)$$

Total ionic equations show <u>all the ions, even if they are not involved in a reaction</u>. In this equation, potassium and nitrate ions are not involved in the reaction (potassium nitrate is soluble), and they are referred to as **spectator ions.**

2) Net ionic equation

$$Ag^+(aq) + Br^-(aq) \longrightarrow AgBr(s)$$

Net ionic equations eliminate spectator ions and record only ions involved in precipitation.

3) Molecular equation

$$AgNO_3(aq) + KBr(aq) \longrightarrow AgBr(s) + KNO_3(aq)$$

Molecular equations show all reactants and products as undissociated compounds. Historically, this equation would have been called a <u>double-displacement reaction</u>.

4.3 WRITING EQUATIONS FOR AQUEOUS IONIC REACTIONS

We can predict which combinations of ions will precipitate out from aqueous solutions by using <u>solubility rules</u> summarized below. Table 4.1 on page 153 in your textbook also summarizes these rules.

Solubility Rules for Ionic Compounds in Water

1. **Most salts of group 1A ions (Li^+, Na^+, K^+, etc.) and NH_4^+ are soluble.**
2. **Most nitrates (NO_3^-), perchlorates (ClO_4^-), and acetates (CH_3COO^-) are soluble.**
3. **Most chloride, bromide and iodide salts are soluble;** notable exceptions are $AgCl$, $PbCl_2$, $CuCl$, and Hg_2Cl_2, which are insoluble.
4. **Most common sulfates (SO_4^{2-}) are soluble;** exceptions are $CaSO_4$, $SrSO_4$, $BaSO_4$, Ag_2SO_4, and $PbSO_4$, which are insoluble.

   <u>Note</u>: NO_3^-, SO_4^{2-}, and ClO^{4-} are all anions of strong acids that lose H^+ readily to water (see the list of strong acids in the next section on acids and bases). These anions are able to spread their negative charge out over several oxygen atoms, and therefore have low charge densities that do not attract other cations strongly, so they tend to form soluble salts, especially with cations of low charge.
5. **Most hydroxide salts are only slightly soluble;** notable exceptions are $NaOH$, KOH, and $Ca(OH)_2$, which are soluble.

   <u>Note</u>: OH^- has only a –1 charge, but it is localized on one oxygen atom, and therefore attracts cations strongly, and tends to form insoluble salts.
6. **Most sulfides (S^{2-}), carbonates (CO_3^{2-}), and phosphates (PO_4^{3-}) are only slightly soluble;** exceptions are those of group 1A and NH_4^+, and the group 2A sulfides.

   <u>Note</u>: S^{2-}, CO_3^{2-}, and PO_4^{3-} are anions of weak acids. These ions have higher charges spread over fewer oxygen atoms than the anions of strong acids. The higher charge densities attract cations, so they tend to form insoluble salts.

Predicting Whether a Precipitation Will Form

When two aqueous ionic solutions are mixed, look at all the reactant ions, consider all possible cation/anion combinations, and then decide whether any combination is insoluble using the above solubility rules. For example, if silver nitrate and potassium chloride are mixed, the reactant ions are Ag^+, NO_3^-, K^+, and Cl^-. The cation/anion combinations are: AgCl and KNO_3. Since most salts of group 1A ions are soluble, we would expect KNO_3 to be soluble. Most chloride salts are soluble, but AgCl is an exception, and we would predict that a precipitate (AgCl(*s*)) would form.

4.4 ACID-BASE REACTIONS

For now, we define an acid as a substance that produces protons (H^+ ions) when dissolved in water, and a base as a substance that produces hydroxide (OH^-) ions when dissolved in water.

Acids and bases

Hydrochloric acid (a) and sodium hydroxide (b) are a common acid and base:

a) $HCl(aq) \rightarrow H^+(aq) + Cl^-(aq)$

b) $NaOH(s) \rightarrow Na^+(aq) + OH^-(aq)$

Choose words from the list below to complete the following paragraphs:

(a) dissociates	(d) strong	(g) Mg^{2+}
(b) low	(e) weak	(h) OH^-
(c) partly	(f) HF	

Hydrochloric acid is a ______________(1) acid because it ______________(2) completely into ions. Likewise, sodium hydroxide is a ______________(3) base because it ______________(4) completely into ions. Most acids and bases are not strong, they dissociate only ______________(5). Hydrogen fluoride (HF) is a weak acid; when it dissolves in water, it only partially ionizes; the concentration of H^+ and F^- ions is ______________(6) compared to the concentration of undissociated ______________(7) molecules.

Solubility is different than extent of dissociation. Magnesium hydroxide does not dissolve readily in water, but the molecules that do go into solution dissociate completely to ______________(8) and ______________(9) ions. In contrast, HF, a ______________(10) acid, is completely soluble in water, but does not dissociate to a very great extent.

answers: 1-d, 2-a, 3-d, 4-a, 5-c, 6-b, 7-f, 8-g, 9-h, 10-e.

Not all bases contain OH^- as part of their structure. Bases may remove a proton from water producing hydroxide ion. Ammonia (NH_3) is a base because it obtains a proton from water leaving OH^- ions in solution:

$$NH_3(g) + H_2O(l) \rightarrow NH_4^+(aq) + \mathbf{OH^-}(aq)$$

Soluble metal oxides act as strong bases in solution because the oxide ion reacts quickly with water to form hydroxide:

$$O^{2-}(aq) + H_2O(l) \rightarrow 2\mathbf{OH}^-(aq)$$

The Strong Acids and Bases

ACIDS		BASES
Hydrochloric acid	HCl	Hydroxides of the 1A and 2A metals:
Hydrobromic acid	HBr	$LiOH$, $NaOH$*, KOH*, $RbOH$, $CsOH$
Hydroiodic acid	HI	$Mg(OH)_2$, $Ca(OH)_2$, $Sr(OH)_2$, $Ba(OH)_2$
Nitric acid	HNO_3	Metal oxides
Perchloric acid	$HClO_4$	
Sulfuric acid	H_2SO_4	* most commonly used in the lab

Neutralization Reaction

Reaction of an acid and a base produces water and a salt (on evaporation) in a **neutralization** reaction. Analogous to precipitation reactions discussed on page 74 and 75, we can write these neutralization reactions as total ionic equations showing all the ions involved, net ionic equations, which do not show the spectator ions, or as molecular equations, which do not indicate the presence of ions. Historically, acid-base reactions were written as molecular equations and would have been called double displacement reactions.

The net ionic equation shows that the key event in an acid-base reaction is a proton transfer process with the formation of water molecules: a proton transfers from the acidic species, H_3O^+ to the basic species, OH^- producing two water molecules.

Neutralization reaction

Write the 1) total ionic equation, 2) net ionic equation, and 3) molecular equation for the reaction of hydrochloric acid with sodium hydroxide.

answers:

1) total ionic equation:

$$H^+(aq) + Cl^-(aq) + Na^+(aq) + OH^-(aq) \rightarrow Na^+(aq) + Cl^-(aq) + H_2O(l)$$

2) net ionic equation (the key event):
(Na$^+$ and Cl$^-$ are the spectator ions and therefore not included.)

$$H^+(aq) + OH^-(aq) \rightarrow 2H_2O(l)$$

3) molecular equation:

$$NaOH(aq) + HCl(aq) \rightarrow NaCl(aq) + H_2O(l)$$

Historically, this acid-base reaction would have been written as the molecular equation above and would have been called a double-displacement reaction.

Notice that it is the net ionic equation in part 2 on the previous page that clearly shows that an acid-base reaction is a proton transfer process producing two water molecules. Figure 4.13 on page 163 in your textbook shows how to picture this process on an atomic level. Remember that although the net ionic equation does not include the spectator ions, Na^+ and Cl^-, they are in solution, and would crystallize as table salt (NaCl) if the water were evaporated. The **Brønsted-Lowry** definition of acids and bases recognizes the proton transfer nature of acid-base reactions. They defined an acid as a molecule (or ion) that donates a proton, and a base as a molecule (or ion) that accepts a proton.

Reaction of a Weak Acid and Strong Base

Acid-base reactions may form a gaseous product. A common example, especially to kitchen chemists, is the formation of carbon dioxide gas (CO_2) from reaction of acetic acid (a weak acid and a component of vinegar), with baking soda, $NaHCO_3$. In water, baking soda dissociates into Na^+ and HCO_3^-. HCO_3^- acts as a base and accepts a proton from acetic acid (CH_3COOH) to form H_2CO_3, which immediately decomposes into water and carbon dioxide. The net ionic equation is:

$$HCO_3^-(aq) + CH_3COOH(aq) \rightarrow CH_3COO^-(aq) + CO_2(g) + H_2O(l)$$

Since acetic acid is a weak acid, it exists mainly as the molecule CH_3COOH in solution. Only a small number of molecules of acetic acid dissociate into CH_3COO^- and H^+ ions until it is in the presence of a base. When a base is present (such as HCO_3^-), the base "steals" the acidic proton from acetic acid to produce the anion CH_3COO^-.

We can determine the concentration of an unknown acid (or base) by titration with a base (or acid) of known concentration. Titrations depend on **indicators** to indicate the point at which neutralization of the acid and base occurs. Indicators are themselves weak acids and are useful in that they are one color in their acidic form, and a different color in their basic form.

Let's suppose we want to know the concentration of acetic acid in vinegar. We would place a measured volume of the vinegar into a flask, and add a few drops of indicator solution that is clear in its acidic form. We would then add a standardized basic solution drop-wise into the acid, swirling the flask. The hydroxide ions from the base immediately react with protons from the acid to form water molecules. When all the moles of H^+ ion present in the vinegar react with an equivalent number of moles of OH^- ion added drop-wise, we are at the **equivalence point.** A tiny amount of excess OH^- ions causes the indicator to become deprotonated to its basic form, and its color changes to pink indicating to us that we have reached the end point of the titration.

Acid-base titration

We find that 22.3 mL of 0.240 *M* NaOH is required to react with a 50.0 mL sample of vinegar. Calculate the concentration of acetic acid in the vinegar.

First write down the equation for the reaction of OH^- with acetic acid:

$$CH_3COOH(aq) + OH^-(aq) \rightarrow CH_3COO^-(aq) + H_2O(l)$$

Notice that the reactants are consumed in a 1:1 mole ratio. Even though acetic acid contains more than one hydrogen atom, it is only the hydrogen bonded to the oxygen which is acidic.

Now we will determine how many moles of hydroxide ions were required to reach the end point:

(continued on next page)

$$22.3 \text{ mL} \times \frac{1 \text{ L}}{1000 \text{ mL}} \times 0.240 \text{ mole/L} = 0.00535 \text{ mole}$$

Since moles OH^- = moles HC_2H_3O, we can calculate the concentration of acetic acid:

$$\frac{0.00535 \text{ mole acetic acid}}{0.0500 \text{ L}} = 0.107\ M$$

4.5 OXIDATION-REDUCTION (REDOX) REACTIONS

Oxidation-reduction reactions involve the net movement of electrons from one reactant to another. When two atoms stick together to form a chemical bond, one atom may attract the electrons more strongly than the other atom. The electrons spend more time around the atom with the stronger attraction, and if this change in the distribution of electrons occurs to a great enough extent, we can say electrons transfer from one atom to another. Reactions in which these electron transfers occur are called **oxidation-reduction** (or redox) reactions. **Oxidation** is the loss of electrons, **reduction** is the gain of electrons. **"Oil rig"** is a good mnemonic device for remembering this.

Oxidation and reduction

$$2Na(s) + F_2(g) \rightarrow 2NaF(s)$$

Consider the above reaction, then choose words from the list below to complete the following paragraph:

(a) gains	(c) oxidized	(e) reduced
(b) loses	(d) oxidizing agent	(f) reducing agent

When sodium and fluorine react, sodium ______________(1) an electron to become Na^+ and we say it's ______________(2). Fluorine ______________(3) an electron to become F^- and we say it's ______________(4). Alternatively, we can say that sodium is the ______________(5) because it causes fluorine to be reduced and fluorine is the ______________(6).

answers: 1-b, 2-c, 3-a, 4-e, 5-f, 6-d.

Let's compare the reaction between hydrogen and fluorine and the reaction between sodium and fluorine. The reactions are:

$$1/2H_2(g) + 1/2F_2(g) \rightarrow HF(g)$$

$$Na(s) + 1/2F_2(g) \rightarrow NaF(s)$$

In both reactions, there is an exchange of electrons between atoms, however, there is a difference in the extent to which the electrons are transferred. In the case of sodium fluoride formation, we think of the electron as being completely transferred from sodium to fluorine to produce ions that are electrostatically attracted to each other to form an ionic solid. In the case of hydrogen fluoride formation, a hydrogen electron is displaced towards the fluorine atom but is shared with the hydrogen atom to form a covalent bond. Since the electrons in the H–F covalent bond are displaced strongly toward fluorine, for bookkeeping purposes, we think of fluorine as having gained an electron.

It is convenient to think of redox reactions as occurring in two steps, although it is important to remember that the oxidation and reduction occur simultaneously. In the above case with NaF formation, we can write:

Oxidation: $Na \rightarrow Na^+ + e^-$ (electron lost by sodium)

Reduction: $1/2F_2 + e^- \rightarrow F^-$ (electron gained by F)

We can keep track of which atoms lose electrons and which gain them with **oxidation numbers.** The oxidation number refers to the charge an atom would have if the electrons that bond two atoms together are assigned to the element which attracts the electrons more strongly. For covalent compounds, the oxidation number is not an accurate description of the electron distribution because the electrons are only partially transferred from one atom to another and so do not carry a full positive or negative charge. However, for bookkeeping purposes, we assign oxidation numbers as whole numbers and keep in mind that it is only the atoms in ionic compounds that come close to completely transferring electrons between atoms.

Rules for Assigning Oxidation Numbers

1) An atom in its elemental form has an oxidation number of 0:

   O.N. of N_2, H_2, and Na = 0.

2) A monatomic ion has an oxidation number equal to the ion charge:

   O.N. of $Na^+ = +1$; of $Mg^{2+} = +2$.

3) The sum of oxidation numbers for the atoms in a compound equals 0. The sum of oxidation numbers for the atoms in a polyatomic ion equals the ion charge.

   For the compound NH_3 (ammonia), the oxidation numbers for nitrogen (–3) and 3 hydrogens (+1 each) add up to zero:

   $(-3) + (3 \times 1) = 0.$

   For the polyatomic ion, NH_4^+, the oxidation numbers for nitrogen (–3) and 4 hydrogens (+1 each) add up to +1, the charge on the ammonium ion:

   $(-3) + (4 \times 1) = +1.$

4) Some elements (listed on the next page) have the same oxidation number in almost all their compounds:

   Group 1A elements = **+1**

   Group 2A elements = **+2**

   Fluorine = **–1**

   Hydrogen usually has an oxidation number of **+1;** when it is in combination with metals, it has an oxidation number of –1.

   O.N. of hydrogen in:

   $H_2O = +1$
   NaH = –1

   Oxygen usually has an oxidation number of **–2;** it has an oxidation number of –1 in peroxides, and –1/2 in superoxides.

   O.N. of oxygen in:

   $H_2O = -2$
   H_2O_2 (hydrogen peroxide) = –1
   KO_2 (a superoxide) = –1/2

(continued on next page)

Group 7A elements usually have an oxidation number of **–1;** notable exceptions are when these elements are in combination with oxygen.

O.N. of Cl in:

$HCl = -1$
$HAuCl_4 = -1$
$HClO_4 = +7$

Most elements are more versatile than the ones listed above. Transition metals may have as many as five possible oxidation numbers. Metals tend to lose electrons, so their oxidation numbers are positive. With a few exceptions, an element's group number in the periodic table is its highest possible oxidation number. Once we understand oxidation numbers, the words oxidation and reduction make more sense:

an atom which is **oxidized** undergoes an **increase** in its oxidation number;
an atom which is **reduced** undergoes a **decrease** (or **reduction**) in its oxidation number.

Determining oxidation numbers

Determine the oxidation number of each element in the following compounds:

a) phosphorus trichloride b) sulfuric acid c) perchloric acid

answers:

a) PCl_3: Halides usually have an oxidation number of –1, so we assign a –1 oxidation number to chlorine. Since there are three chloride ions, phosphorus must have a +3 oxidation number to make the compound neutral.

b) H_2SO_4: Hydrogen in combination with nonmetals has an oxidation number of +1. Oxygen has an oxidation number of –2 in all compounds except peroxides, superoxides and with fluorine. The four oxygen atoms give a –8 charge, and the two hydrogen atoms give a +2 charge, so to make the compound neutral, we assign an oxidation number of +6 to sulfur.

c) $HClO_4$: First we'll assign hydrogen a +1 oxidation number, and oxygen a –2 oxidation number. Cl must have a +7 oxidation number to make the compound neutral.

In redox equations, all electrons must be accounted for. Redox reactions can be balanced using the method summarized below.

Rules for Balancing Redox Equations

1) Assign oxidation numbers to all atoms.

2) Identify the oxidized and reduced atoms (atoms whose oxidation numbers change).

3) Make the number of electrons lost equal the number of electrons gained by multiplying by appropriate factors; use the factors as balancing coefficients.

4) Complete the balancing by inspection.

Balancing a redox equation

Balance the equation for the reaction that occurs between chloride and permanganate ions in acidic solution. The equation is:

$$MnO_4^-(aq) + H^+(aq) + Cl^-(aq) \rightarrow Mn^{2+}(aq) + Cl_2(g) + H_2O(l)$$

solution:

First we assign oxidation numbers to all atoms using the rules listed. In the permanganate ion, MnO_4^-, oxygen is assigned an oxidation number of –2 giving a negative charge of 4 × –2 = –8. The charge on the polyanion is –1, so we know that manganese must be contributing a +7 charge.

Reactants		Products	
Atom	Oxidation number	Atom	Oxidation number
Mn	+7	Mn	+2 (reduced)
O	–2	O	–2
H	+1	H	+1
Cl	–1	Cl	0 (oxidized)

Manganese is reduced since its oxidation number decreases from +7 to +2; chlorine is oxidized since its oxidation number increases from –1 to 0.

The number of electrons lost by chlorine = 1; the number of electrons gained by manganese = 5. To make electrons lost = electrons gained, we multiply electrons lost by chlorine by 5 and use this factor as a balancing coefficient:

$$MnO_4^-(aq) + H^+(aq) + 5Cl^-(aq) \rightarrow Mn^{2+}(aq) + 5/2Cl_2(g) + H_2O(l)$$

Multiply through by two to eliminate fractional coefficients:

$$2MnO_4^-(aq) + 2H^+(aq) + 10Cl^-(aq) \rightarrow 2Mn^{2+}(aq) + 5Cl_2(g) + 2H_2O(l)$$

Now we balance the remaining elements. We balance oxygen atoms with water molecules, and hydrogen atoms with H^+ ions. The oxygen is balanced first. Since $2MnO_4^-$ contains 8 oxygen atoms, there must be eight water molecules each containing one oxygen atom. Then to balance the hydrogen, 16 H^+ must be present on the left:

$$2MnO_4^-(aq) + 16H^+(aq) + 10Cl^-(aq) \rightarrow 2Mn^{2+}(aq) + 5Cl_2(g) + 8H_2O(l)$$

Redox Titrations

Analogous to acid-base titrations described earlier, we can determine an unknown concentration of a reducing agent by titrating with a known concentration of an oxidizing agent (or vice versa). As your textbook discusses on page 164, permanganate ion, MnO_4^- often serves as an indicator in these reactions because in its oxidized form (as the MnO_4^- ion) it is a deep purple color but once it is reduced to Mn^{2+}, it is nearly colorless. Let's work through follow-up problem 4.13 from page 162 in your textbook.

Follow-up problem 4.13

A 2.50 mL sample of low-fat milk was treated with sodium oxalate, and the precipitate was dissolved in H_2SO_4. This solution required 6.53 mL of 4.56×10^{-3} *M* $KMnO_4$ to reach the end point.

(continued on next page)

a) Calculate the molarity of Ca^{2+} in the milk.

b) What is the concentration of Ca^{2+} in g/L? Is this value consistent with the typical value in milk of about 1.2 g Ca^{2+}/L?

solution:

We have no good way to measure directly the concentration of calcium ions in our milk solution, so we pull them out of solution (precipitate them) with an anion whose concentration we can measure.

$$Ca^{2+}(aq) + C_2O_4^{2-}(aq) \rightarrow CaC_2O_4(s)$$

The oxalate $C_2O_4^{2-}$ ion concentration can be measured because of its redox chemistry with permanganate ion, a good indicator. We've made progress, but how do we titrate a solid (CaC_2O_4)? We don't! We dissolve the precipitate with dilute sulfuric acid to make a solution that contains equal concentrations of oxalate ions and calcium ions. We'll measure the concentration of oxalate ions by titrating with a permanganate solution of known concentration. The net ionic equation (unbalanced) is:

$$MnO_4^-(aq) + C_2O_4^{2-}(aq) + H^+(aq) \rightarrow Mn^{2+}(aq) + CO_2(g) + H_2O(l)$$

MnO_4^-: we titrate with this; $C_2O_4^{2-}$: what we're measuring; H^+: from the sulfuric acid

First we assign oxidation numbers to all atoms using the rules listed earlier.

Reactants		Products	
Atom	Oxidation number	Atom	Oxidation number
Mn	+7	Mn	+2 (reduced)
O	–2	O	–2
H	+1	H	+1
C	+3	C	+4 (oxidized)

The number of electrons lost by carbon = 1; the number of electrons gained by manganese = 5. To make electrons lost = electrons gained, we multiply electrons lost by carbon by 5 and use this factor as a balancing coefficient:

$$MnO_4^-(aq) + 5/2C_2O_4^{2-}(aq) + H^+(aq) \rightarrow Mn^{2+}(aq) + 5CO_2(g) + H_2O(l)$$

Multiply through by two to eliminate fractional coefficients:

$$2MnO_4^-(aq) + 5C_2O_4^{2-}(aq) + 2H^+(aq) \rightarrow 2Mn^{2+}(aq) + 10CO_2(g) + 2H_2O(l)$$

Now we balance the remaining elements. We balance oxygen atoms with water molecules, and hydrogen atoms with H^+ ions.

$$2MnO_4^-(aq) + 5C_2O_4^{2-}(aq) + 16H^+(aq) \rightarrow 2Mn^{2+}(aq) + 10CO_2(g) + 8H_2O(l)$$

Now our problem is simply a matter of finding the moles of permanganate ion and using the stoichiometry of our balanced equation to determine the moles of oxalate ion which equals our moles of calcium ion.

$$\text{Moles } MnO_4^- = 6.53 \text{ mL soln.} \times \frac{1 \text{ L}}{1000 \text{ mL}} \times \frac{4.56 \times 10^{-3} \text{ mol } MnO_4^-}{1 \text{ L soln.}}$$

$$= 2.98 \times 10^{-5}$$

(continued on next page)

$$\text{Moles } C_2O_4^{2-} \text{ titrated} = 2.98 \times 10^{-5} \text{ mol } MnO_4^- \times \frac{5 \text{ mol } C_2O_4^{2-}}{2 \text{ mol } MnO_4^-}$$

$$= 7.44 \times 10^{-5} = \text{moles } Ca^{2+}$$

Finally, the molarity of the calcium ions in the milk is:

$$\frac{7.44 \times 10^{-5} \text{ mol } Ca^{2+}}{2.50 \text{ mL milk}} \times \frac{1000 \text{ mL milk}}{1 \text{ L milk}} = 2.98 \times 10^{-2} M$$

For part (b) we need to convert our molarity (mol/L) to g/L and compare to a typical value in milk:

$$\frac{2.98 \times 10^{-2} \text{ mol } Ca^{2+}}{1 \text{ L milk}} \times \frac{40.08 \text{ g } Ca^{2+}}{\text{mol } Ca^{2+}} = 1.19 \text{ g } Ca^{2+}/\text{ L}$$

Comment: Our calcium ion concentration is typical of that found in milk (1.2 g Ca^{2+}/L).

4.6 ELEMENTS IN REDOX REACTIONS

As we stated at the beginning of this chapter, the historical classification of reactions as combination, decomposition, and displacement reactions does not require much knowledge of chemistry. To classify the reactions, we simply look to see if reactants combine, decompose, or change partners.

Historical classification of reactions

Match the following reactions with their historical classification:

1) X + Y → Z
2) Z → X + Y
3) X + YZ → XZ + Y
4) WX + YZ → WZ + XY

a) Single displacement
b) Combination
c) Decomposition
d) Double displacement or metathesis

answers: 1-b, 2-c, 3-a, 4-d.

Types of chemical reactions

Choose words from the list below to complete the following paragraph:

(a) atoms
(b) combination
(c) compound(s)
(d) decomposition
(e) displacement
(f) electricity
(g) elements
(h) heat
(i) molecules

Just as the name implies, a _______________(1) reaction is one in which two (or more) _____________ or _______________(2 & 3) stick together to form one substance. Combination reactions can involve two elements, an element and a compound, or two compounds. When a reactant absorbs

(continued on next page)

enough energy to break apart into two or more products, a ______________(4) reaction occurs. Usually the starting material is a ______________(5), and the products are smaller ______________(6) and/or ______________(7). Compounds generally decompose by ______________(8) (thermal decomposition) or ______________(9) (electrolytic decomposition). You can think of ______________(10) reactions as molecules and atoms changing partners. They are often referred to as single-displacement and double-displacement (metathesis) reactions.

answers: 1-b, 2-a, 3-i, 4-d, 5-c, 6-c, 7-g, 8-h, 9-f, 10-e.

Combining Two Elements

As you look at a periodic table, it might appear that there are quite a few elements that could potentially combine together to form a compound. However, not all elements have the right arrangement of electrons to be reactive. Nonmetals, on the right hand side of the table (except for the noble gases in the last column), tend to be reactive. The reactions in the next exercise will show combination reactions of chlorine, as well as elements that combine with oxygen to form oxides. Oxides themselves tend to be reactive since they may react with more oxygen, with each other, or with water to form larger compounds.

Other reactive species are the alkali metals and the alkaline earth metals found on the far left hand side of the periodic table. Some of these metals must be stored under oil because they are so reactive with water. They exist in nature most often in the form of salts.

Metals react with nonmetals to form ionic compounds, such as the reaction between an alkali metal and a halogen, or the reaction of a metal with oxygen to form the ionic oxide. Figure 4.19 in your textbook on page 175 shows the reaction of potassium metal with the nonmetal chlorine to form potassium chloride. Two nonmetals react to form covalent compounds with shared electrons, such as the reaction between nitrogen and hydrogen to form ammonia, or reaction between nitrogen and oxygen to form nitrogen monoxide.

Combination of elements

Reactions of the nonmetal, chlorine

$$Cl_2(g) + 2Na(s) \rightarrow 2NaCl(s)$$

$$Cl_2(g) + Mg(s) \rightarrow MgCl_2(s)$$

$$3Cl_2(g) + 2Fe(s) \rightarrow 2FeCl_3(s)$$

$$Cl_2(g) + 3F_2(g) \rightarrow 2ClF_3(g)$$

Fill in the blanks based on the above reactions and the list of words below:

(a) alkali metal
(b) alkaline earth metal
(c) nonmetal
(d) **reactive**
(e) transition metal

Chlorine, a nonmetal, reacts with sodium, an ______________(1); chlorine reacts with magnesium, an ______________(2); chlorine reacts with iron, a ______________(3); and chlorine reacts with fluorine, another ______________(4). **Conclusion: chlorine is** ______________(5).

answers: 1-a, 2-b, 3-e, 4-c, 5-d.

(continued on next page)

Reactions of the nonmetal, oxygen

Almost every element combines with oxygen to form an oxide.

A) Name the following oxides:

a) CO_2
b) H_2O
c) NO
d) CO
e) CaO
f) Fe_2O_3

answers:

a) carbon dioxide
b) water
c) nitrogen monoxide
d) carbon monoxide
e) calcium oxide
f) iron(III) oxide

B) Fill in the appropriate oxides from the list above:

Oxides surround us:

1) we exhale _______________ and _______________.
2) our cars exhale ____________, ____________, ____________, and ____________.
3) rust is _______________.
4) lime is _______________.

Conclusion: oxygen is _______________(5).

answers: 1-a,b; 2-a,b,c,d; 3-f; 4-e; 5-reactive.

Combining an Element and a Compound

As mentioned above, many oxides are reactive and may react with more oxygen to form "higher" oxides, such as the reaction of nitrogen monoxide (NO) with oxygen to form nitrogen dioxide (NO_2). Similarly, many nonmetal halides combine with additional halogen, such as the reaction of phosphorus trichloride (PCl_3) with chlorine to form phosphorus pentachloride (PCl_5).

Decomposition

Give products for the reactions a-c below. When compounds decompose thermally, they often release familiar gaseous products such as carbon dioxide (equation "a" below), oxygen (equation "b"), or water (equation "c"):

a) $CaCO_3(s) \rightarrow$
b) $2HgO(s) \rightarrow$
c) $Mg(OH)_2(s) \rightarrow$

answers:

a) $CaO(s)$ + $\mathbf{CO_2(g)}$
b) $2Hg(l)$ + $\mathbf{O_2(g)}$
c) $MgO(s)$ + $\mathbf{H_2O(g)}$

Electrical energy can decompose compounds through the process of electrolysis. Water can be decomposed into the elements hydrogen and oxygen:

$$2H_2O(l) \xrightarrow{\text{electricity}} 2H_2(g) + O_2(g)$$

This reaction is useful for generating a potential fuel, H_2.

Displacement Reactions

In single-displacement reactions, metals may displace hydrogen gas from water, steam, or an acid. Metals may also displace other metal ions from solution. Metals can be ranked by their ability to displace H_2 from various sources or to displace one another from solution. Figure 4.16 on page 139 in your textbook shows the activity series of the metals with the most active metal (Li) at the top, and the least active metal (Au) at the bottom. Notice that the most reactive metals (Li, K, Ba, Ca, Na) displace H_2 from water, while slightly less reactive metals (Mg, Al, Mn, Zn, Cr, Fe, Cd) displace H_2 from steam. Still less reactive metals (Co, Ni, Sn, Pb) do not react with water but react with acids, from which H_2 is displaced more easily. The stable and unreactive coinage metals (Cu, Ag, Au), and Hg cannot displace H_2 from any source.

Halogens may displace another halide ion from solution. Reactivity decreases down group 7A, so the activity series for halogens is:

$$F_2 > Cl_2 > Br_2 > I_2$$

Displacement reactions

1) Single-Displacement Reactions

Using the activity series of the metals (shown on page 139 in your textbook), predict the products for the following reactions:

a) $Ba(s) + 2H_2O(l) \rightarrow$

b) $Cu(NO_3)_2(aq) + Zn(s) \rightarrow$

c) $2KBr(aq) + Cl_2(aq) \rightarrow$

answers:

a) $Ba(OH)_2(aq) + H_2(g)$

b) $Cu(s) + Zn(NO_3)_2(aq)$

c) $Br_2(aq) + 2KCl(aq)$

Notice that single-displacement reactions are all oxidation-reduction processes. In the examples above, barium reduces H^+ to H_2, zinc reduces Cu^{2+} to Cu^0, and Br^- reduces Cl_2 to Cl^-. Barium, zinc, and Br^- are oxidized.

2) Double-Displacement Reactions

Three common double-displacement reactions are:

a) precipitation of an insoluble ionic compound

b) neutralization of an acid and a base to form water

c) reaction of a carbonate, nitrite, or sulfite with an acid to form a gas

Predict the products for the following reactions:

a) precipitation of the solid ionic compound, $AgBr(s)$:

$AgNO_3(aq) + KBr(aq) \rightarrow$

b) neutralization of an acid (HCl) and base (NaOH) to form water:

$NaOH(s) + HCl(aq) \rightarrow$

c) reaction of a carbonate with an acid to form a gas (CO_2):

$MgCO_3(s) + 2HCl(aq) \rightarrow$

(continued on next page)

answers:

a) $AgBr(s)$ + $KNO_3(aq)$
b) $NaCl(aq)$ + $H_2O(l)$
c) $MgCl_2(aq)$ + $H_2O(l)$ + $CO_2(g)$

Combustion Reactions

All combustion reactions are redox processes because elemental oxygen is a reactant. When we burn propane (C_3H_8) to grill food, the carbon-carbon and carbon-hydrogen bonds break, and if the combustion is complete, the C and H atoms all combine with oxygen to produce carbon dioxide and water:

$$C_3H_8(g) + 5O_2(g) \rightarrow 3CO_2(g) + 4H_2O(g)$$

The carbon atom in propane, with a negative oxidation number, is oxidized to a +4 oxidation number in carbon dioxide. We burn mixtures of hydrocarbons in the form of coal, gasoline, and natural gas to run our cars and heat our homes.

4.7 THE REVERSIBILITY OF REACTIONS AND THE EQUILIBRIUM STATE

If X reacts with Y to produce Z, we can write this as:

$$X + Y \rightarrow Z$$

The above reaction suggests that the reaction continues until either X or Y is consumed. Many reactions stop before either X or Y is consumed. The reason for this is that at the same time X and Y are combining to form Z, a certain percent of Z is decomposing to reform reactants X and Y. When the forward reaction occurs at the same rate as the reverse reaction, there is no further change in the amounts of the reactants or products and we say that the system is in **equilibrium.** We use double arrows to indicate a reaction at equilibrium:

$$X + Y \rightleftharpoons Z$$

If the product, Z, is removed from the system, because, for example, it is a gas that escapes, or it is a precipitate, the reaction will continue to completion. Equilibrium can only be established when all the reactants and products are kept in contact with each other.

Weak acids and bases dissociate to only a small extent because dissociation quickly becomes balanced by reassociation. HF molecules dissociate in water to H^+ and F^- ions, but as these ions form, many of them react together to reform HF molecules. At any given time, only a small percentage of the HF molecules are dissociated, and so we say that the equilibrium lies to the left:

$$HF(aq) \leftharpoondown\!\!\!\rightharpoonup H^+(aq) + F^-(aq)$$

CHAPTER CHECK

Make sure you can...

- Describe how ionic and polar compounds dissolve in water.
- Write total ionic, net ionic, and molecular equations for precipitation reactions.
- List the important solubility rules for ionic compounds in water.

- Identify strong acids and bases, and know the distinction between strong and weak acids and bases.
- Calculate a concentration from an acid-base titration.
- Identify oxidized and reduced species in redox reactions.
- Assign oxidation numbers to atoms in compounds.
- Balance a redox equation.
- Identify reactions as combination, decomposition, or displacement.
- Explain how combustion reactions are redox processes
- Describe a weak acid or base as a reversible reaction in equilibrium.

Chapter 4 Exercises

4.1) Which solution has the greatest number of ions or molecules?

a. 400 mL of 1.20 M $Ni(NO_3)_2$

b. 700 mL of 2.00 M $C_{12}H_{22}O_{11}$

c. 200 mL of 1.50 M $Fe_2(SO_4)_3$

4.2) a. When 8.0 grams of sodium hydroxide is dissolved in sufficient water to make 400. mL of solution, what is the molarity of the solution?

b. If this solution is then poured into a volumetric flask and made up to 1.0 L in volume, what would the molarity of the solution become?

4.3) a. How many moles of nitric acid are there in a 75 mL sample of a 0.60 M solution of nitric acid?

b. If this sample is made up to 2.0 L in volume, what would the molarity of the solution be?

4.4) A laboratory technician has been asked to prepare 750 mL of a 0.200 M $H_2SO_4(aq)$ solution. The only sulfuric acid solution the technician has available in the stockroom is 6.00 M $H_2SO_4(aq)$. Describe how to prepare the 0.200 M $H_2SO_4(aq)$ solution.

4.5) Mylanta, a commonly used over-the-counter antacid, is used to relieve heartburn. Each gelcap contains 311 mg of calcium carbonate and 232 mg of magnesium carbonate:

$$CaCO_3 + 2HCl(aq) \rightarrow CaCl_2(aq) + CO_2(g) + H_2O$$

$$MgCO_3 + 2HCl(aq) \rightarrow MgCl_2(aq) + CO_2(g) + H_2O$$

How many milliliters of stomach acid, 0.030 M HCl, will two Mylanta gelcaps neutralize?

4.6) Which of the following substances would you expect to be insoluble in water?

barium hydroxide
hydrochloric acid
lithium sulfate
ammonium nitrate
silver chloride
lithium carbonate
calcium carbonate
barium sulfate
lead iodide
ammonium acetate
potassium cyanide
lead acetate
strontium hydroxide
ammonium nitrate
silver nitrate
cadmium acetate

4.7) Write net ionic equations for:

a. the precipitation reaction $Pb(NO_3)_2 + 2KI \rightarrow PbI_2(s) + 2KNO_3$

b. the neutralization of a strong acid by a strong base

c. the evolution of sulfur dioxide when sodium sulfite is treated with a dilute solution of HCl

4.8) a. If the concentration of sodium ions in an aqueous solution of sodium sulfate is 2.4×10^{-2} mol/L, how many moles of sodium sulfate must have been dissolved per liter to make the solution?

b. How many moles of ammonium ions are there in a 200 mL sample of 0.10 *M* ammonium phosphate?

4.9) a. If it requires 50. mL of 0.030 *M* HCl to titrate a 100. mL sample of potassium hydroxide to the equivalence point, how many moles of potassium hydroxide are in the sample?

b. What was the original concentration of the potassium hydroxide?

c. What is the concentration of the potassium chloride salt at the equivalence point?

4.10) 37.12 mL of 0.1151 *M* NaOH is required to neutralize a 20.00 mL sample of citric acid ($H_3C_6H_5O_7$).

$$H_3C_6H_5O_7 + 3NaOH \rightarrow 3H_2O + Na_3C_6H_5O_7$$

What is the molarity of the citric acid solution?

4.11) Lactic acid, $HC_3H_5O_3$, is responsible for the sour taste in out-of-date milk. 57.04 mL of 0.09827 *M* NaOH is required to neutralize a 50.00 mL lactic acid solution.

$$HC_3H_5O_3 + NaOH \rightarrow H_2O + NaC_3H_5O_3$$

What is the molarity of the lactic acid solution?

4.12) In the following reactions, identify the element oxidized, the element reduced, the oxidizing agent, and the reducing agent:

a. $Sn + 2NaOH \rightarrow Na_2SnO_2 + H_2$

b. $2KCl + MnO_2 + 2H_2SO_4 \rightarrow K_2SO_4 + MnSO_4 + Cl_2 + 2H_2O$

4.13) When heated in the presence of base, an aqueous solution of chlorine disproportionates to chloride ion and chlorate ion. What ratio of chlorate ion to chloride ion is produced in this reaction? The unbalanced redox reaction is:

$$Cl_2 \rightarrow Cl^- + ClO_3^-$$

4.14) Copper metal reacts with concentrated nitric acid to produce nitrogen monoxide and copper nitrate. Calculate the mass of nitrogen monoxide produced when 5.00 g of copper is consumed. The unbalanced reaction is:

$$Cu(s) + HNO_3(aq) \rightarrow Cu^{2+}(aq) + NO_3^-(aq) + NO(g) + H_2O$$

4.15) How many milliliters of a 0.10 *M* potassium dichromate solution are required to titrate 50. mL of an acidic 0.10 *M* iron(II) chloride solution? The unbalanced chemical reaction is:

$$Fe^{2+} + H^+ + Cr_2O_7^{2-} \rightarrow Cr^{3+} + Fe^{3+} + H_2O$$

4.16) Solutions of hydrogen peroxide are sold over-the-counter as topical antiseptics. 30.0 mL of H_2O_2 is first acidified and then titrated to a pale pink with 26.5 mL of 0.200 *M* $KMnO_4$ solution.

$$H_2O_2 + MnO_4^- + H^+ \rightarrow O_2 + Mn^{2+} + H_2O$$

a. Balance the net oxidation-reduction reaction above.

b. What is the molarity of the H_2O_2 solution?

c. If the density of the solution is 1.00 g/mL, what is the percent H_2O_2 by mass of the hydrogen peroxide solution?

4.17) Liquid bleach is normally sold as 5% solutions of sodium hypochlorite, NaOCl. Quality control analysis of the NaOCl content of bleach solutions can be carried out using an iodometric titration. The hypochlorite solution is treated with an excess of iodide solution:

$$OCl^- + I^- + H^+ \rightarrow Cl^- + I_2 + H_2O$$

The resulting iodine is titrated with a standard sodium thiosulfite, $Na_2S_2O_3$ solution:

$$S_2O_3^{2-} + I_2 + H_2O \rightarrow I^- + SO_4^{2-}$$

a. Balance each of the oxidation-reduction reactions above.

b. A 100 mL bleach sample is treated with an iodide solution and then titrated with 47.24 mL of a 0.347 M $Na_2S_2O_3$ solution. What is the molarity of the bleach solution?

c. Assume the density of the solution is 1.00 g/mL, what is the percent NaOCl by mass in the bleach solution?

4.18) A metallurgist determines the amount of manganese in stainless steel by using the method of Volhard oxidation-reduction titration. A 0.5229 g sample of stainless steel is dissolved in nitric acid and titrated to a pale pink end point with 33.81 mL of 1.296×10^{-3} *M* $KMnO_4$ solution:

$$Mn^{2+} + MnO_4^- + H_2O \rightarrow MnO_2 + H^+$$

a. Balance the net oxidation-reduction reaction above.

b. What is the mass of manganese in the stainless steel sample?

c. What is the percent Mn by weight in the steel sample?

4.19) How many milliliters of a 0.10 *M* potassium dichromate solution are required to titrate 50. mL of an acidic 0.10 *M* iron(II) chloride solution? The unbalanced chemical reaction is:

$$Fe^{2+} + H^+ + Cr_2O_7^{2-} \rightarrow Cr^{3+} + Fe^{3+} + H_2O$$

Chapter 4 Answers

4.1) a. $0.400\ L \times 1.20\ mol/L = 0.480\ mol\ Ni(NO_3)_2$; there are 3 mol of ions ($Ni^{2+} + 2NO_3^-$) for each mol of $Ni(NO_3)_2$, so moles of ions = $3 \times 0.480 = 1.44$ mol

b. $0.700\ L \times 2.00\ mol/L = 1.40\ mol\ C_{12}H_{22}O_{11}$

c. $0.200\ L \times 1.50\ mol/L = 0.300\ mol\ Fe_2(SO_4)_3$; there are 5 ions ($2Fe^{3+} + 3SO_4^{2-}$) for each mol of $Fe_2(SO_4)_3$, so moles of ions = $5 \times 0.300 = 1.50$ mol

4.2) a. We need to know moles of NaOH per liter of solution:

$$M = \frac{8.0\ g \times 1\ mol/40.0\ g}{0.400\ L} = 0.50\ mol/L$$

b. The moles of NaOH remain the same (0.20 mol); the solution volume increases to 1 L, so molarity should decrease: $M = 0.20$ mol/L

4.3) a. $0.075\ L \times 0.60\ mol/L = .045$ moles HNO_3

b. The moles of nitric acid remain the same; the solution volume increases to 2.0 L so the molarity should decrease. The new solution contains 0.045 moles in 2.0 L. The molarity is: 0.045/2 = 0.023 *M*

4.4) First determine how many moles of H_2SO_4 are required in the final solution:

$$0.200\ mol/L \times 0.750\ L = 0.150\ moles$$

Volume of the more concentrated 6.00 *M* solution that contains 0.150 moles of H_2SO_4:

0.150 mol/6.00 mol/L = 0.025 L, or 25 mL

Use 25.0 mL of 6.00 *M* H_2SO_4 diluted to 750 mL, and mix.

4.5) Each mole of $CaCO_3$ and $MgCO_3$ will neutralize 2 moles of HCl.

0.311 g $CaCO_3$ / 100.087 g/mol = 0.003107 mol $CaCO_3$

0.282 g $MgCO_3$ / 84.3142 g/mol = 0.002752 mol $MgCO_3$

Total moles = 0.003107 + 0.002752 = .005859

Moles HCl neutralized = 2 × .005859 = .01172

The molarity of the HCl is 0.030 *M*, so 0.01172 mol / 0.030 mol/L = .390 L, or 390 mL

Two gelcaps will neutralize 390 × 2 = 780 mL of 0.030 *M* HCl

4.6) All the substances are soluble <u>except</u> lead iodide, silver chloride, calcium carbonate, and barium sulfate.

4.7) a. First, write all components in the form in which they predominantly exist under the conditions of the reaction. Strong electrolytes in solution are in ionic form, weak electrolytes in solution are written in molecular form, evolved gases and precipitates are written in molecular (or formula unit form):

$$Pb^{2+}(aq) + 2NO_3^-(aq) + 2K^+(aq) + 2I^-(aq) \rightarrow PbI_2(s) + 2K^+(aq) + 2NO_3^-(aq)$$

Eliminate spectator ions to obtain the net ionic equation:

$$Pb^{2+}(aq) + 2I^-(aq) \rightarrow PbI_2(s)$$

b. The net ionic equation for the neutralization of any strong acid by any strong base is:

$$H_3O^+(aq) + OH^-(aq) \rightarrow 2H_2O(l)$$

The cation of the strong base, and the anion of the strong acid are both spectator ions of the neutralization reaction and do not appear in the net ionic equation.

c. $SO_3^{2-}(aq) + 2H^+(aq) \rightarrow H_2O(l) + SO_2(g)$

4.8) a. Each mol of Na_2SO_4 produces 2 mol of Na^+ ions:
In one liter, there are 2.4×10^{-2} mol Na^+ ions so half this number of moles of sodium sulfate must have been dissolved, or 1.2×10^{-2} mol.

b. Moles of ammonium ions = 3 × moles of $(NH_4)_3PO_4$ = 3 × 0.10 mol/L × 0.200 L = 0.060 mol

4.9) a. Mol KOH = mol HCl = 0.050 L × 0.030 mol/L = 0.0015 mol;
b. Molarity = 0.0015 mol/0.100 L = 0.015 *M*
c. Molarity = 0.0015 mol ÷ (0.100 L + 0.050 L) = 0.010 *M*

4.10) Citric acid contains 3 acidic protons, so we need 3 mol of NaOH to neutralize one mol of citric acid.

Moles of NaOH = 0.03712 L × 0.1151 mol/L = 0.0042725 mol

Moles of citric acid = 1/3 × 0.0042725 = 0.001424167

Molarity of the citric acid solution = 0.001424167 ÷ 0.02000 L = 0.07121 *M*

4.11) 0.1121 *M*

4.12) a. Tin is oxidized (from 0 to +2); Hydrogen is reduced (from +1 to 0); Hydrogen of the hydroxide is the oxidizing agent, it oxidizes the tin; Tin is the reducing agent—it reduces the hydrogen.

b. Chlorine is oxidized (from –1 to 0); Manganese is reduced (from +4 to +2); Manganese of the manganese dioxide is the oxidizing agent, it oxidizes the chlorine; Chlorine (as chloride) is the reducing agent, it reduces the manganese.

4.13) A disproportionation reaction is a redox reaction in which the same species is both oxidized and reduced. In this reaction, chlorine is oxidized and reduced. In any redox reaction, the oxidation and reduction must be balanced. Compare the decrease in oxidation number of the chlorine that is reduced to the increase in the oxidation number of the chlorine that is oxidized:

Reduction: $Cl_2 \rightarrow Cl^-$ (change of –1)
Oxidation: $Cl_2 \rightarrow ClO_3^-$ (change of +5)

In order to balance the two processes, five times as much chloride ion as chlorate ion must be produced.

4.14) The balanced redox reaction is:

$$3Cu(s) + 8HNO_3(aq) \rightarrow 3Cu^{2+}(aq) + 6NO_3^-(aq) + 2NO(g) + 4H_2O$$

The stoichiometry indicates that 2 moles of NO are produced for every 3 moles of copper consumed. This same information could have been derived from the fact that the copper metal is oxidized to Cu^{2+} (a change of +2), and the nitrate ion is reduced to nitrogen oxide (a change from +5 to +2, or a change of –3). Mass of NO produced = 1.57 grams.

4.15) Iron(II), Fe^{2+}, is oxidized to iron(III), Fe^{3+}: a change of +1
The dichromate ion, $Cr_2O_7^{2-}$, is reduced to chromium(III), Cr^{3+}: a total change of +6 (+3 for each Cr in the dichromate ion).

One mole of dichromate can oxidize six moles of iron(II). Alternatively, the equation can be balanced and the mole ratio derived directly from the stoichiometry:

$$6Fe^{2+} + 14H^+ + Cr_2O_7^{2-} \rightarrow 2Cr^{3+} + 6Fe^{3+} + 7H_2O$$

Since the concentrations of the two solutions are equal (both 0.10 *M*), 50 mL of dichromate solution requires only 50 × 1/6 mL of iron(II) solution (or 8.3 mL).

4.16) a. $5H_2O_2 + 2MnO_4^- + 6H^+ \rightarrow 5O_2 + 2Mn^{2+} + 8H_2O$; b. 0.442 *M*; c. 1.50%

4.17) a. $OCl^- + 2I^- + 2H^+ \rightarrow Cl^- + I_2 + H_2O$

$S_2O_3^{2-} + 4I_2 + 5H_2O \rightarrow 8I^- + 2SO_4^{2-} + 10H^+$

b. Mol bleach = mol OCl^- = mol I_2 = 4 × mol $S_2O_3^{2-}$ = mol sodium thiosulfite
Mol $Na_2S_2O_3$ = 0.04724 L × 0.347 mol/L = 0.01639 mol
Mol I_2 = mol OCl^- = 4 × 0.01639 mol = .06557 mol; Molarity = 0.06557mol ÷ 0.100 L = 0.656 *M*

c. We use moles of NaOCl from part b and change moles to grams:
0.0656 mol × 74.44 g/mol = 4.88 g
100 mL bleach solution has a mass of 100 g (density = 1.00 g/mL), therefore the percent NaOCl by mass is: 4.88 g ÷ 100 g (× 100%) = 4.88%

4.18) a. $3Mn^{2+} + 2MnO_4^- + 2H_2O \rightarrow 5MnO_2 + 4H^+$; b. 0.003611 g; c. 0.6906%

4.19) Iron(II), Fe^{2+}, is oxidized to iron(III), Fe^{3+}: a change of +1
The dichromate ion, $Cr_2O_7^{2-}$, is reduced to chromium(III), Cr^{3+}: a total change of +6 (+3 for each Cr in the dichromate ion).

One mole of dichromate can oxidize six moles of iron(II). Alternatively, the equation can be balanced and the mole ratio derived directly from the stoichiometry:

$$6Fe^{2+} + 14H^+ + Cr_2O_7^{2-} \rightarrow 2Cr^{3+} + 6Fe^{3+} + 7H_2O$$

Since the concentrations of the two solutions are equal (both 0.10 *M*), 50 mL of dichromate solution requires only 50 × 1/6 mL of iron(II) solution (or 8.3 mL).

5

Gases and the Kinetic-Molecular Theory

"Molecules in Motion"

INTRODUCTION

Hydrogen and helium are both gaseous elements that are less dense than air, but hydrogen balloons aren't found at typical birthday parties. Hydrogen is extremely flammable when activated, while helium is inert at all temperatures and, in fact, is the most inert substance in the universe. In spite of their chemical differences, the physical behavior of hydrogen and helium is similar. For example, balloons filled to the same volume with hydrogen and helium at the same temperature and pressure, contain the same number of hydrogen molecules as helium atoms, even though a helium atom is about twice the mass of a hydrogen molecule. Chapter 5 explores the similar **physical** behavior of gases.

5.1 THE PHYSICAL STATES OF MATTER

In Chapter 1 we saw that, generally, there are three ways molecules may be arranged:

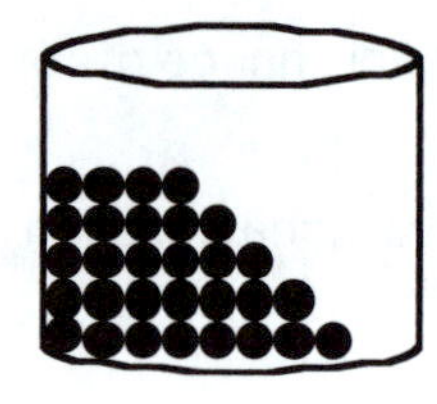

Touching & ordered
(solid)

Touching & disordered
(liquid)

Not touching
(gas)

Elemental solids, liquids, and gases

Choose words from the list below to complete the following paragraph:

(a) bromine
(b) chlorine
(c) helium
(d) hydrogen
(e) mercury
(f) neon
(g) nitrogen
(h) oxygen
(i) solids

Most elements are ____________(1) at ambient temperature and pressure. The only two elements that are liquids at ambient conditions are: ____________(2), a metal, and ____________(3), a nonmetal. Some familiar elemental gases are: ____________(4), the lightest; ____________(5), an inert, nonflammable gas used for balloons and airships; ____________(6), the major component of air; ____________(7), the most abundant element on earth which supports combustion; ____________(8), a greenish-yellow poisonous gas; and ____________(9), used in tubes to produce colored light.

answers: 1-i, 2-e, 3-a, 4-d, 5-c, 6-g, 7-h, 8-b, 9-f.

The drawing on the previous page representing gas molecules indicates that gas molecules are not touching. With that drawing in mind, two important points to remember about gases are:

- At normal conditions, gases are chiefly empty space.
- Mechanically and thermally, all gases behave the same way: they exhibit the same compressibility, and the same thermal expansion.

The same volume of all gases at the same temperature and pressure contain the same number of gas molecules. By contrast, the volume of solids and liquids, where molecules are touching, depend on the size of the molecules, their interactions, and, in the case of solids, the type of crystal lattice they form.

Comparison of liquid and gaseous volumes

1) How much volume does a mole of water molecules in the gaseous state at 273 K and 1 atm occupy?

2) How much volume does this same amount of liquid water occupy (where the molecules are touching one another)?

answers:

1) One mole of any gas (including water) at 273 K and 1 atm takes up **22.4 L.**

2) 6.02×10^{23} water molecules (1 mole) weigh 18.0 grams; the density of water is 1.00 g/mL. A mole of water molecules in the liquid state occupies:

$$6.02 \times 10^{23} \text{ H}_2\text{O molecules} \times \frac{18.0 \text{ amu}}{\text{H}_2\text{O molecule}} \times \frac{1.00 \text{ g}}{6.02 \times 10^{23} \text{ amu}} \times \frac{1.00 \text{ mL}}{1 \text{ g}} = 18.0 \text{ mL}$$

A drawing illustrates the difference in volumes of a mole of gaseous water and a mole of liquid water:

(continued on next page)

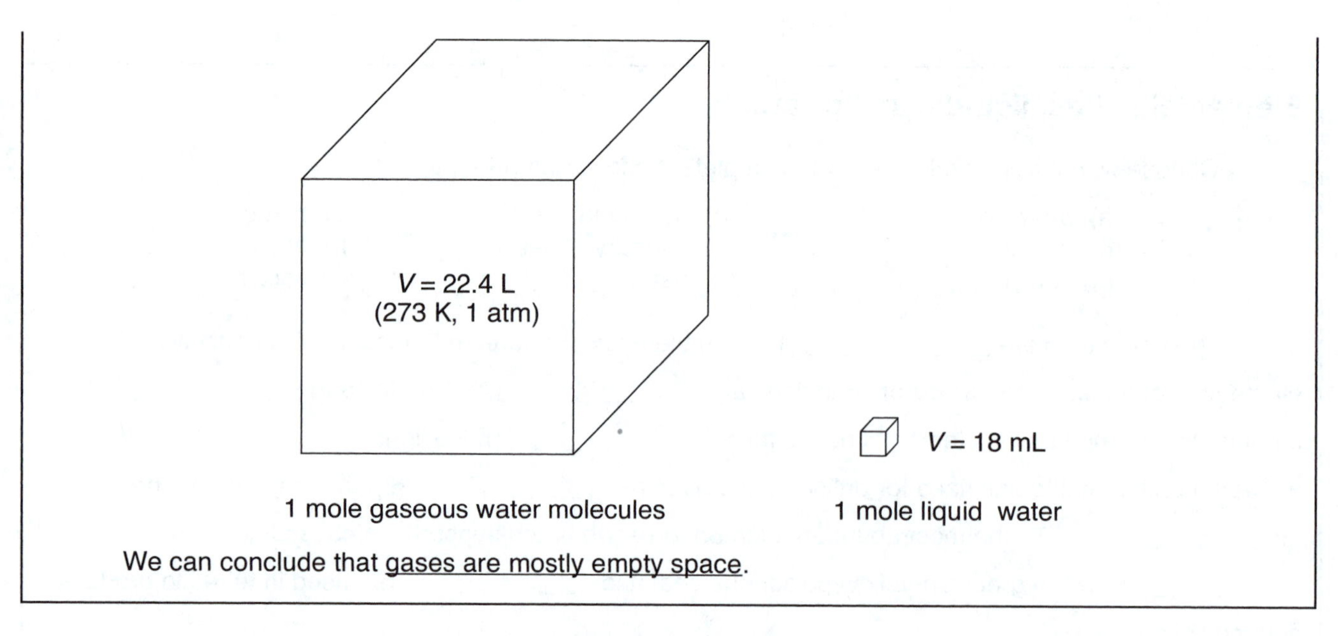

Since gases are mostly space, they are easily compressed. They also have a low viscosity (they flow freely), and they typically have low densities. Densities of gases are often given in grams per liter instead of grams per milliliter because gases contain so much empty space and occupy such a large volume compared to solids and liquids.

5.2 GAS PRESSURE

Gases expand to fill a container. As gas particles collide with the walls of a container, they exert some force on the walls. The force per unit area that a gas exerts on its container is pressure (P):

$$P = \frac{\text{force}}{\text{area}}$$

You are probably already aware of the definition of pressure. Think, for example, why you might choose to lie down on thin ice instead of standing on it. The gravitational force exerted by your body to the ice is the same whether you are standing or lying, but by increasing the area of your body in contact with the ice when you lie down, you decrease the pressure at any point. The same concept applies to wearing snowshoes, or being able to lie (somewhat comfortably) on a bed of nails.

Atmospheric Pressure

Air is a gaseous mixture containing mainly nitrogen and oxygen, but also a small amount of other gases. Gravity holds these gases around the earth in a layer about 50 miles thick. Our atmosphere exerts a pressure of about 15 pounds per square inch at the earth's surface. The atmospheric pressure varies depending on the altitude and weather conditions. Sometimes it is useful to picture the atmosphere as a weight exerting force on a solution surface or on a container:

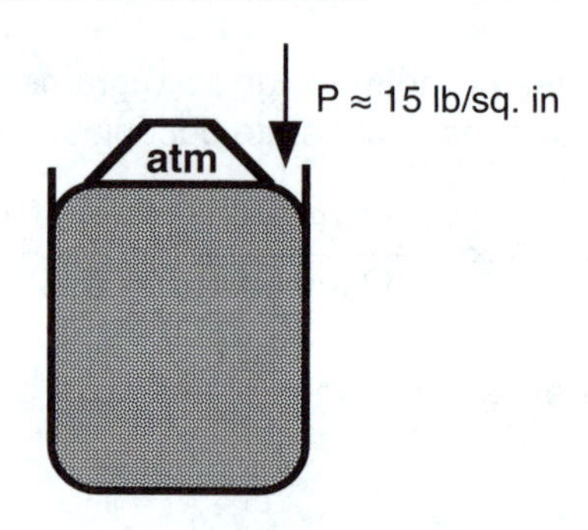

It is easy to forget about atmospheric pressure since its source, air, is invisible, but it is critical to remember its presence when working problems that involve gases. Expanding gases do work as they push back the weight of the atmosphere. In order to exhale, you must overcome atmospheric pressure.

What would happen if you filled a tube (closed at one end) with mercury and inverted it into a dish of mercury?

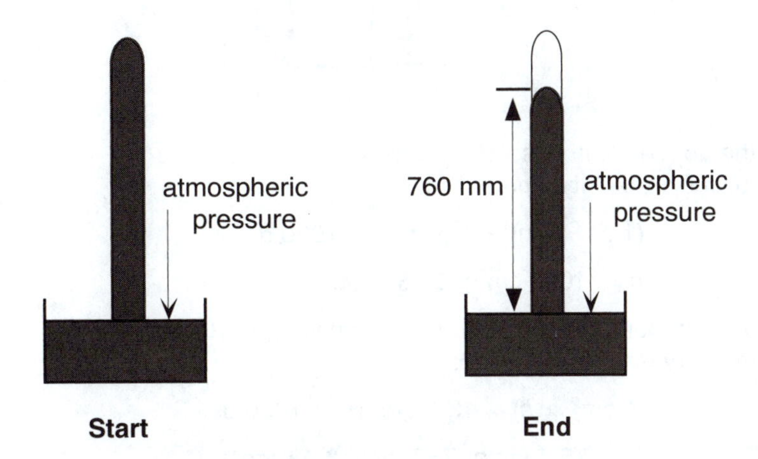

As the diagram shows, mercury flows out of the tube until the pressure of the column of mercury on the surface of the mercury in the dish is equal to the pressure the atmosphere exerts on the dish of mercury. Typically, the atmosphere exerts enough pressure to hold up a column of mercury about 760 mm high.

A **barometer** is a device, such as the one pictured above, used to measure atmospheric pressure. If a student in Denver (altitude 1524 m or 5000 ft.) made a barometer, the mercury in the column would be only about 700 mm high. Denver is almost a mile above sea level and therefore the blanket of gases forming the atmosphere is thinner and exerts less pressure than is typical at sea level. The atmospheric pressure might become that low at sea level during a hurricane.

A **manometer** measures the pressure exerted by a gas in a closed container. Manometers work by comparing the pressure of a gas in a flask to a known pressure, either the atmosphere (barometric pressure) or a vacuum (0 pressure).

Choice of liquid for a barometer

Why do barometers use the poisonous liquid mercury? Let's suppose we wanted to make a barometer containing water. How high would the column need to be to hold a column of water supported by atmospheric pressure? Hint: Use the pressure of the atmosphere as 14.7 lb/in^2.

solution:

What do we know about water? Its density is 1.00 g/mL; does this information, along with the information given about atmospheric pressure, suggest a way to obtain a unit of length for the column height?

Let's rewrite the density of water as 1.00 g/cm^3, and convert units of pressure to similar units (g/cm^2) using the equalities 1 lb = 454 g and 1 in = 2.54 cm:

$$P_{atm} = 14.7\frac{\text{lb}}{\text{in}^2} \times 454\frac{\text{g}}{\text{lb}} \times \left(\frac{1 \text{ in}}{2.54 \text{ cm}}\right)^2 = 1030\frac{\text{g}}{\text{cm}^2}$$

(continued on next page)

Let's consider a column of water which is a cm^2 in cross section. How high a column (h), of water in this column weighs 1030 g?

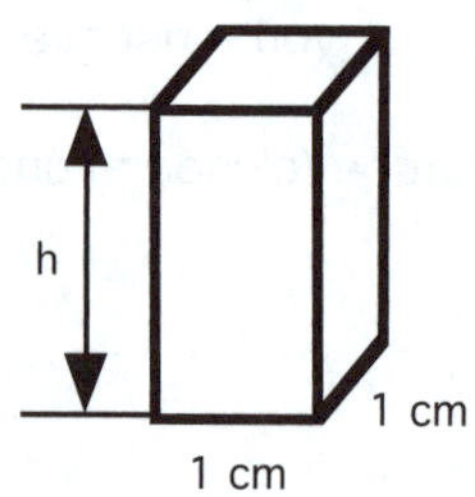

The volume of the above column is: $V = 1\ cm^2 \times h$
We know that volume × density = mass, so

$$(1\ cm^2 \times h) \times 1\ g/cm^3 = 1030\ g$$

h = 1030 cm or **33.8 feet!**

How high a column of mercury in the above column weighs 1030 g?
The density of mercury is 13.6 g/cm^3, so:

$$(1\ cm^2 \times h) \times 13.6\ g/cm^3 = 1030\ g$$

h = 75.7 cm or **757 mm (2.48 feet)**

Two and a half feet is a more convenient height for a barometer column than 34 feet. Mercury's high density makes it a useful liquid for barometers.

The ratio of heights of the liquid columns in a barometer is inversely related to the densities of the liquids for a given pressure. For the above example we could write:

$$\frac{h\,(H_2O)}{h\,(Hg)} = \frac{d\,(Hg)}{d\,(H_2O)}$$

$$\frac{1030\ cm}{75.7\ cm} = 13.6 = \frac{13.6\ g/cm^3}{1.00\ g/cm^3}$$

People in the 1600's knew that they couldn't lift water more than about 33 feet with a suction pump, but they didn't know why at first. An Italian scientist, Torricelli, recognized that the maximum height of the water column was a measure of the atmospheric pressure, and that once a pump lowered the gas pressure at the top of the pipe to zero, the water in the pipe would stop rising, no matter how fast the pump worked.

Some Units of Pressure

Most chemists express pressure in **mm Hg** or **atmospheres (atm).** A mm Hg is sometimes called a **torr** in honor of the scientist Torricelli who invented the barometer:

1 atmosphere = 760 mm Hg = 760 torr (Hg at 0°C)

The SI unit for pressure uses the SI unit for force (Newton, N) and area (m^2) and is called a **pascal** (Pa). A pascal is small compared to an atmosphere or a torr and is not used often in the United States:

(continued on next page)

1 Pa = 1 N/m²

1 atm = 101,325 Pa = 760 mm Hg

Pressure of a Gas in a Container

The pressure of a gas inside a container (which is not pressurized or evacuated) equals the pressure of the atmosphere pressing against the container; if this were not the case, the container would collapse or explode.

5.3 THE GAS LAWS

Consider a cylinder of gas equipped with a piston initially under atmospheric pressure, P_{atm}.

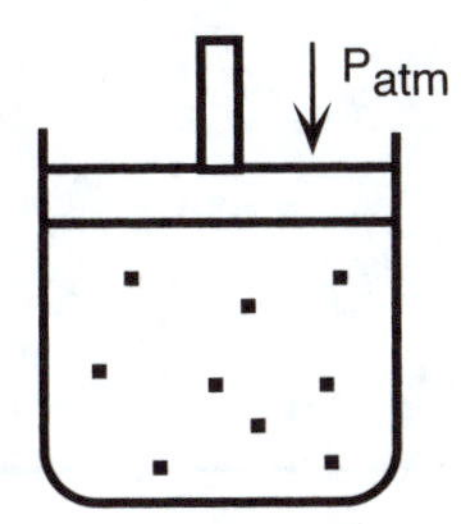

What could we do to change the volume of the gas? If we increase the pressure on the piston, we would compress the gas and decrease its volume. We could also raise the temperature, which increases the molecular motion of the gas molecules, and therefore increases the volume that the gas molecules occupy. Likewise, the volume increases if we increase the number of molecules. These relationships are summarized in the **gas laws.**

The Gas Laws

1) **Boyle's Law:** At constant temperature, the volume of a fixed amount of gas is inversely proportional to the applied pressure. As the piston goes down, pressure increases as volume decreases:

$$V = \frac{k}{P} \quad (T \text{ and } n \text{ fixed})$$

where V = volume, P = pressure, T = temperature, n = moles, and k = a constant. We can use Boyle's law to solve problems that compare two sets of volume and pressure values:

$$V_1 = \frac{k}{P_1} \quad \text{and} \quad V_2 = \frac{k}{P_2}$$

$$k = P_1V_1 = P_2V_2$$

$$\text{or } V_2 = V_1 \times \frac{P_1}{P_2}$$

We can see from this last equation that when $P_2 > P_1$, $V_2 < V_1$, or, as pressure increases (P_2 increases), volume decreases (V_2 decreases).

(continued on next page)

2) **Charles's Law:** At constant pressure, the volume of a fixed amount of gas is directly proportional to its absolute temperature. As a gas is heated and molecular motion increases, the piston goes up as the volume increases.

$$V = kT \quad (P \text{ and } n \text{ fixed})$$

To compare two sets of volume and temperature values, we can use the relationship:

$$\frac{V_1}{T_1} = \frac{V_2}{T_2} \quad \text{or,} \quad V_2 = V_1 \times \frac{T_2}{T_1}$$

3) **Avogadro's Law:** At fixed temperature and pressure, equal volumes of any ideal gas contain equal numbers of particles.

$$V \propto n \quad (T \text{ and } P \text{ fixed})$$

To compare two sets of volume and number of moles, we can use the relationship:

$$\frac{V_1}{n_1} = \frac{V_2}{n_2}$$

$$\text{or} \quad V_2 = V_1 \times \frac{n_2}{n_1}$$

Gases at Standard Conditions

As we saw in section 5.1, the same volume of all gases at the same temperature and pressure contain the same number of gas molecules. Experiments have shown that one mole of molecules of any gas at 273 K and 1 atmosphere occupies 22.4 L (the standard molar volume, V_m).

V_m (1 atm, 273 K) = 22.4 L/mol molecules or atoms (all gases)

We define the conditions, $T = 0°C$ (273.15 K) and $P = 760.$ torr as **standard temperature and pressure (STP).** Notice that STP conditions involve the usual pressure, but not the usual ambient temperature in laboratories.

Combining the gas laws

1) What is the volume (V) of 1 mole of a gas at 1 atm and 273 K?

$n = 1$ mole $\quad T = 273$ K $\quad P = 1$ atm

$V = 22.4$ L (see above, "Gases at Standard Conditions")

2) What is the volume of 2 moles of a gas at 1 atm and 273 K?

$n = 2$ moles $\quad T = 273$ K $\quad P = 1$ atm

$$V = 22.4\text{ L} \times \frac{2\text{ mol}}{1\text{ mol}} = 44.8\text{ L}$$

(Common sense: twice the number of molecules takes up twice the volume)

(continued on next page)

3) What is the volume of n moles of a gas at 1 atm and 273 K?

$n = n$ $T = 273$ K $P = 1$ atm

$$V = 22.4\,\text{L} \times \frac{n}{1\text{ mol}} \quad \text{(general case of exercise \#2)}$$

4) What is the volume of 1 mole of a gas at 2 atm and 273 K?

$n = 1$ mole $T = 273$ K $P = 2$ atm

$$V = 22.4\,\text{L} \times \frac{n}{1\text{ mol}} \times \frac{1\text{ atm}}{2\text{ atm}} = 11.2\,\text{L}$$

(Boyle's law: twice the pressure decreases volume to 1/2)

5) What is the volume of n moles of a gas at pressure P and 273 K?

$n = n$ $T = 273$ K $P = P$

$$V = 22.4\,\text{L} \times \frac{n}{1\text{ mol}} \times \frac{1\text{ atm}}{P} \quad \text{(general case of exercise \#4)}$$

6) What is the volume of 1 mole of a gas at 1 atm and 546 K?

$n = 1$ mole $T = 546$ K $P = 1$ atm

$$V = 22.4\,\text{L} \times \frac{n}{1\text{ mol}} \times \frac{1\text{ atm}}{P} \times \frac{546\text{ K}}{273\text{ K}} = 44.8\,\text{L}$$

(Charles's law; twice the temperature doubles the volume)

7) What is the volume of n moles of a gas at pressure P and temperature T?

$n = n$ $T = T$ $P = P$

$$V = 22.4\,L \times \frac{n}{1\text{ mol}} \times \frac{1\text{ atm}}{P} \times \frac{T}{273\text{ K}} \quad \text{(general case of exercise \#6)}$$

8) Rewrite the expression for volume from exercise 7 with numbers (constants) and variables gathered together:

$$V = \underbrace{\frac{22.4\text{ L} \times 1\text{ atm}}{1\text{ mol} \times 273\text{ K}}}_{R} \times \frac{nT}{P}$$

The constants in this equation are called "R," the gas law constant. What is the value for the constant, R, in this equation (note this R value contains temperature in the Kelvin scale)?

$$R = \frac{PV}{nT} = 22.4\text{ L} \times \frac{1\text{ atm}}{1\text{ mol}} \times 273\text{ K} = 0.0821\,\frac{\text{L}\cdot\text{atm}}{\text{mol}\cdot\text{K}}$$

We can rewrite the equation for R above as: $\boldsymbol{PV = nRT}$

This equation, which contains the gas laws of Avogadro, Boyle, and Charles, is called the

Ideal Gas Law.

Remember that a law generalizes behavior observed in many experiments, and may not apply to every situation. In fact, real gases always break the gas laws! The gas laws apply strictly only to ideal gases. We assume ideal gas particles have no volume and no interactions. Real gases, at low pressure and high temperature, behave much like ideal gases, and in these cases, the gas laws provide a close approximation for the behavior of real gases.

Gas law problem

Suppose we fill an uninflated balloon with 15.0 g of dry ice (solid CO_2). The balloon's capacity is 7.5 L. Is the balloon likely to pop as the dry ice vaporizes? Assume the temperature is 298 K.

solution:

To work this problem, we need to first calculate the volume of 15.0 g of gaseous CO_2 ($M = 44.0$ g/mol) and compare the value to 7.5 L (the balloon capacity).

What do we know?

$$T = 298 \text{ K} \qquad P = 1.0 \text{ atm}$$

$$n = \frac{15.0 \text{ g}}{44.0 \text{ g/mol}} = 0.341 \text{ mol} \qquad R = 0.0821 \text{ L·atm/mole·K}$$

Solve for volume, V:

$$V = \frac{nRT}{P} = \frac{0.341 \text{ mol} \times 0.0821 \frac{\text{L·atm}}{\text{mol·K}} \times 298 \text{ K}}{1.0 \text{ atm}} = 8.3 \text{ L}$$

8.3 L is larger than the balloon's capacity, so BLAM!...the balloon will pop.

5.4 REARRANGEMENTS OF THE IDEAL GAS LAW

In this section, we will see how to use the ideal gas law to find gas density, molar mass, and the partial pressure of gases in a mixture.

Density of an ideal gas

Use the molar mass of helium and its standard molar volume to determine the density of helium at STP. Derive an equation for the density of a gas under any conditions.

solution:

The molar volume of any gas at STP, $V_m = 22.4$ L/mole.
The molar mass of helium is $\mathbf{M} = 4.003$ g/mole.

Therefore:

$$d_{STP} = \frac{m}{V_m} = \frac{m/n}{V_m/n} = \frac{4.003 \text{ g/mol}}{22.4 \text{ L/mol}} = 0.179 \text{ g/L}$$

We know from the ideal gas law that $V = nRT/P$. Substituting this expression into the equation for density above gives:

$$d_{STP} = \frac{m}{V_m} = \frac{m}{nRT/P} = \frac{\mathbf{M}\,P}{R\,T} \qquad \text{where } \mathbf{M} = \text{molar mass, } m/n.$$

At constant *T* and *P*, the density of a gas is directly proportional to its molar mass. The heavier the atom or molecule, the denser it will be. Xenon has a greater mass than argon, so xenon is denser than argon. This relationship does not hold with solids and liquids. Compare carbon (**M** = 12.0 g/mole) in its diamond form to sodium (**M** = 23.0 g/mole). A sodium atom is heavier than a carbon atom, but it is less dense. Diamonds sink in water (density = 3.51 g/mL) while sodium floats on water (as it reacts vigorously, density = 0.97 g/mL). The densities of solids and liquids depend both on their molar mass and on their volume, which in turn depends on interactions between molecules.

Density of gases

We know from experience that helium balloons rise in air. Will a nitrogen (N_2) filled balloon float or sink? What about balloons filled with oxygen (O_2), carbon dioxide (CO_2), methane (CH_4), and propane (C_3H_8)? (Air is approximately 78% nitrogen and 21% oxygen.)

solution:

Since the density of a gas is directly proportional to molar mass, we will compare molar masses of the gases in the balloons to that of air to determine if the balloon will sink or float.

The molar mass of air (78% nitrogen and 21% oxygen) is approximately:

$$\mathbf{M} = (.78 \times 28) + (.21 \times 32) = 28.6 \text{ g/mole}$$

The molar mass of nitrogen is 28 g/mole, which is less than the molar mass of air, so it would seem the balloon should rise slightly in air. However, the weight of the balloon may prevent the balloon from rising.

The molar mass of oxygen (32 g/mole), carbon dioxide (44 g/mole) and propane (44 g/mole) are all larger than the molar mass of air, so a balloon filled with those gases would sink.

The molar mass of methane is 16 g/mole. A large balloon filled with methane will rise. A small one might not rise since the weight of the balloon might prevent it from doing so.

Density and the ideal gas law

What is the density in g/L of radon at 25°C and 1.00 atm?

answer:

We can substitute *m*/**M** for *n* in the ideal gas law, and then we solve for *m*/*V* (density):

$$PV = \left(\frac{m}{\mathbf{M}}\right)RT$$

$$d = \frac{m}{V} = \frac{P\mathbf{M}}{RT} = \frac{1.00 \text{ atm} \times 222.02 \text{ g/mol}}{0.0821\frac{\text{L·atm}}{\text{mol·K}} \times 298 \text{ K}} = 9.07 \text{ g/L}$$

Radon is a very dense gas. Radon detectors should be placed in the basement of a house to determine if house levels are high. The equation used to work this problem indicates that the density of a gas is inversely proportional to the temperature. As temperature increases, the space between molecules increases, volume increases, and density decreases.

Mixtures of Gases: Dalton's Law of Partial Pressures

In a mixture of unreacting gases, the total pressure is the sum of the partial pressures of the individual gases:

$$P_{total} = P_1 + P_2 + P_3 \ldots$$

Dalton's law is useful when a gas is collected by displacement of water because the collected gas is a mixture of the gas produced by a chemical reaction and water vapor. Water molecules escape from the surface of the water and collect in the space above the liquid. The pressure exerted by these water molecules on the container is called the **vapor pressure.** When the rate of escape of the water molecules from the water surface equals the rate of return to the water, the vapor pressure remains constant.

Partial pressures

When aluminum reacts with concentrated hydrochloric acid, hydrogen gas evolves. The hydrogen is collected by displacement of water at 22°C at a total pressure of 750. torr. If the volume of the collected gas is 365 mL, how many moles of hydrogen were collected? The vapor pressure of water at 22°C is 21 torr.

solution:

The collected gas is a mixture of hydrogen and water vapor. By Dalton's law:

$P_{tot} = P_{H_2} + P_{H_2O}$ The partial pressure of water in the sample is its vapor pressure.

The partial pressure of hydrogen gas is: $P_{H_2} = P_{tot} - P_{H_2O}$

$$= 750.\text{ torr} - 21\text{ torr} = 729\text{ torr}$$

$$= 729\text{ torr} \times \left(\frac{1\text{ atm}}{760.\text{ torr}}\right) = 0.959\text{ atm}$$

Use the ideal gas law to calculate moles of hydrogen:

$$n_{H_2} = \frac{P_{H_2}V}{RT} = \frac{0.959\text{ atm} \times 0.365\text{ L}}{0.0821\frac{\text{L}\cdot\text{atm}}{\text{mol}\cdot\text{K}} \times 295\text{ K}} = 0.0144\text{ mol } H_2\text{ gas}$$

THE IDEAL GAS LAW AND REACTION STOICHIOMETRY

In the laboratory, you are more likely to measure the volume, temperature, and pressure of a gas than its mass. The ideal gas law gives you the means to convert these measurements (P, V, and T) to moles of gas for use in reaction stoichiometry. Figure 5.13 on page. 222 in your textbook shows how you can combine a gas law problem with a stoichiometry problem to convert between gas variables (pressure, temperature, and volume) and moles of gaseous reactants and products. It is useful to check your answers in these problems by comparing your answer with the molar volume of an ideal gas at STP (22.4 L at 273.15 K and 1 atm). Use what you know about the effects of temperature and pressure and number of molecules on the volume of a gas (review the beginning of section 5.3) to determine if your answer seems to make sense.

Reaction stoichiometry

We can use the ideal gas law to determine the volumes of gases involved in chemical reactions under given conditions. For example, how many liters of pure oxygen, measured at 740 mm Hg and 24°C, would be required to burn 1.00 g of octane, $C_8H_{18}(l)$ to carbon dioxide and water?

solution:

First, we write the balanced equation for the reaction:

$$C_8H_{18}(l) + \tfrac{25}{2}O_2(g) \rightarrow 8CO_2(g) + 9H_2O(l)$$

Using the reaction stoichiometry, determine how many moles of O_2 are required to burn 1.00 g of octane:

$$\frac{1.00 \text{ g } C_8H_{18}}{114 \text{ g/mol}} \times \frac{\frac{25}{2} \text{ mol } O_2}{1 \text{ mol } C_8H_{18}} = 0.110 \text{ mol } O_2$$

Now use the gas law to calculate the volume of O_2: (first convert pressure to atm, and temperature to Kelvin, since R is in units of L·atm/mole·K):

$$P = 740 \text{ mm Hg} \times \frac{1 \text{ atm}}{760 \text{ mm Hg}} = 0.974 \text{ atm}; \quad T = 24 + 273 = 297 \text{ K}$$

$$V = \frac{nRT}{P} = \frac{0.110 \text{ mol} \times 0.0821\frac{\text{L·atm}}{\text{mol·K}} \times 297 \text{ K}}{0.974 \text{ atm}} = 2.75 \text{ L}$$

This reaction is analogous to the one that occurs in an automobile engine. A vehicle that gets 25 miles per gallon at 60 miles per hour would use 1 gram of gasoline almost every half second. Air is only 21% oxygen by volume, so the volume of air that your car engine would need to completely combust the octane would be approximately 13 L every half second! Hence the need for air filters.

5.5 THE KINETIC-MOLECULAR THEORY

Chemists theorize about what happens at the molecular level to explain observed macroscopic behavior. The kinetic-molecular theory proposes a model for molecular behavior that explains the observed macroscopic behavior of ideal gases. This theory is widely accepted because it can be used with the principles of physics to derive an expression for pressure that agrees with the ideal gas law. In addition, it correctly predicts the relative effusion rates of gases summarized by Graham's law. This theory can be stated in terms of three postulates:

Postulates of the Kinetic-Molecular Theory

1) **Particle size:** Gas molecules are so small compared to the space between molecules that their volume is assumed to be zero.
2) **Particle interaction:** Gas molecules exert no forces on each other; they are assumed neither to attract nor to repel each other.
3) **Particle motion:** Gas molecules are in constant random motion. It is the collisions of the particles with the walls of the container that are the cause of the pressure exerted by the gas. The collisions are elastic, which means that no energy is lost to frictional forces (their total kinetic energy remains constant).

The Kinetic-Molecular Theory's Explanation for Ideal Gas Behavior

The kinetic-molecular theory states that collisions of gas particles with the walls of its container cause gas pressure. The kinetic-molecular theory can be combined with the laws of mechanics to predict that the pressure of a gas in a container will depend on the speed and mass of the particles.

Pressure:

$$P = \frac{\frac{1}{3}mu^2}{V} = \frac{nRT}{V}$$

kinetic-molecular theory (molecular behavior) — experimental results (macroscopic behavior)

where u is the average molecular speed of the gas particles.

We will use the above equations to obtain equations for a) the average kinetic energy of gas particles and, b) the average speed of the gas particles.

Kinetic energy

Use the equations above, and the equation for kinetic energy (K.E. = $1/2mu^2$), to obtain an equation for the average kinetic energy of gas molecules moving through space.

solution:

We know:

$$P = \frac{\frac{1}{3}mu^2}{V} \text{ (kinetic-molecular theory) and, } \text{K.E.} = \tfrac{1}{2}mu^2$$

We can multiply each side of the K.E. equation by 2/3 to obtain:

$$\tfrac{2}{3}\text{K.E.} = \tfrac{1}{3}mu^2$$

and substitute this into the equation for pressure to give:

$$P = \frac{\frac{2}{3}\text{K.E.}_{avg}}{V} \quad \text{or} \quad \text{K.E.}_{avg} = \tfrac{3}{2}PV$$

Using the ideal gas law:

$$\mathbf{K.E.}_{avg} = \tfrac{3}{2}nRT$$

The kinetic energy of a mole of gas is directly proportional to temperature and has the same value for all gases.

- Heavier molecules do not bombard container walls with more energy than lighter gas molecules because the kinetic energy of an object is related to its mass and its speed.
- For two gases at the same temperature, the gas molecule with the higher mass will move more slowly than the lighter gas molecule.

Average speed of gas molecules

Rearrange the equations for pressure of gas particles to calculate the average speed of gas molecules.

solution:

$$P = \frac{\frac{1}{3}mu^2}{V} = \frac{nRT}{V} \quad \text{(from the kinetic-molecular theory and the ideal gas law)}$$

therefore,

$$\tfrac{1}{3}mu^2 = nRT$$

Solving for u gives:

$$u = \left(\frac{3nRT}{m}\right)^{\frac{1}{2}} \quad \text{so,} \quad \boldsymbol{u} = \left(\frac{\mathbf{3}\boldsymbol{RT}}{\mathbf{M}}\right)^{\frac{1}{2}}$$

The gas constant R is not useful with the units liter·atm/mole·K for problems such as these. R can be expressed as 8.31 J/mole·K (1 J = 1 kg·m^2/s^2) which will provide speed in cm/s. The equation above shows that at a given temperature, lighter molecules move faster than heavier molecules. Figure 5.20 on page 228 in your textbook shows the relative speed distributions of hydrogen, helium, water, nitrogen, and oxygen molecules.

Average speed of nitrogen

Find the average speed of a nitrogen molecule in air at room temperature (25°C).

solution:

$$u = \left(\frac{3RT}{\mathbf{M}}\right)^{\frac{1}{2}}$$

$$u = \left(\frac{3 \times (8.31 \times 10^7 \text{ erg/mol}\cdot\text{K}) \times (25 + 273)\text{ K}}{28.0 \text{ g/mol}}\right)^{\frac{1}{2}}$$

$$= 5.15 \times 10^4 \text{ cm/s} = 515 \text{ m/s}$$

If we convert our answer to miles per hour (1 m/s = 2.24 miles/h, a slow walk), we can see that the average speed of a nitrogen molecule at room temperature is about 1150 miles per hour!

In the above problem, we solved for u, the average speed of a nitrogen molecule. In fact, the many collisions among gas particles produces a wide distribution of speeds, and most nitrogen molecules will not be moving at the average speed, but will be moving either faster or slower than the average speed. As temperature increases, average velocity increases and the range of velocities becomes much larger. Figure 5.14 on page 225 in your textbook shows the distribution of molecular speeds at three temperatures for nitrogen molecules. Does our answer of 515 m/s for the average speed of a nitrogen molecule at room temperature (298 K) seem reasonable based on Figure 5.14 in your text?

We can compare this very high average speed of molecules to the speed of sound in air. Molecular motion propagates sound waves, so the speed of sound and the speed of the gas molecules through which it passes should be approximately equal. The speed of sound in air is about 800 miles an hour, which is close to the average speed of an oxygen or nitrogen molecule at room temperature.

Molar energy of gases

What is the energy of one mole of a gas at room temperature? Use R with units of energy, 1.987 cal/mole K.

solution:

$$K.E._{avg} = \tfrac{3}{2} nRT$$

$$K.E._{avg} = \tfrac{3}{2} \times 1\text{ mol} \times 1.987 \tfrac{\text{cal}}{\text{mol·K}} \times 298\,K = 888\text{ cal/mol}$$

$$K.E._{avg} = \tfrac{3}{2} \times 1\text{ mol} \times 8.314 \tfrac{\text{J}}{\text{mol·K}} \times 298\,K = 2480\text{ J/mol}$$

Note: The kinetic energy of a mole of molecules of any gas at room temperature is 888 calories or 2480 J.

Values for *R*

0.0821 L·atm/mole·K	(useful in gas law problems)
8.314 J/mole·K	(useful for determining speed and energy in J)
1.987 cal/mole·K	(useful for determining energy in calories)

Effusion and Diffusion

We saw in the previous section that the kinetic-molecular theory could be used to derive an expression for pressure that agrees with the ideal gas law. The kinetic-molecular theory also satisfactorily explains the observed behavior of diffusion and effusion. You are probably familiar with the word diffusion, but may not be as familiar with the word effusion—but no need for confusion!

Diffusion is the movement of one gas through another.

Effusion is the movement of a gas into a vacuum.

Effusion is a fast process, diffusion happens more slowly. When we open a perfume bottle, it takes time for everyone in the room to smell it. Why is this the case if, as we have seen, gas particles at room temperature move at very high speeds (nitrogen has an average velocity of 1150 miles per hour or 515 m/s)? The answer is that the path of a gas particle is extremely convoluted because of collisions with nitrogen and oxygen in the air, so most likely, a gas molecule travels a long distance before it reaches the opposite side of the room. Effusion happens quickly because there are no molecules in a vacuum for the gas molecules to collide with.

The average distance a molecule travels between collisions is called its **mean free path.** For a nitrogen molecule in air at room temperature, it travels, on average, 6.6×10^{-8} m before colliding with another molecule or atom. This distance is equal to about 180 molecular diameters between collisions. You can use the average speed nitrogen molecules travel and their mean free path to calculate their number of collisions per second or their **collision frequency.** Predict an order of magnitude for the collision frequency of nitrogen molecules at room temperature, and then work through the problem on the following page to calculate the answer.

Collision frequency

Use the average speed (515 m/s) of a nitrogen molecule in air at room temperature and its mean free path (6.6×10^{-8} m) to calculate its collision frequency (number of collisions per second).

Collision frequency can be calculated by taking the ratio of the average speed (distance per second) and the mean free path (distance per collision):

$$\text{collision frequency} = \frac{\text{distance/second}}{\text{distance/collision}} = \frac{\text{collisions}}{\text{second}}$$

$$= \frac{515 \text{ m/s}}{6.6 \times 10^{-8} \text{ m/collision}} = 7.7 \times 10^{9} \text{ collisions/s}$$

Graham's Law of Effusion: The rate of effusion of a gas is inversely proportional to the square root of its molar mass:

$$\text{rate} = \frac{c}{\sqrt{\mathbf{M}}}$$

The ratio of the effusion rates of two gas molecules such as helium and argon is:

$$\frac{\text{rate}_{Ar}}{\text{rate}_{He}} = \frac{\sqrt{\mathbf{M}_{He}}}{\sqrt{\mathbf{M}_{Ar}}}$$

The lighter gas (He) effuses faster than the heavier gas (Ar).

Graham's law also applies to diffusion. The ratio of the diffusion of two gases moving through another gas or a mixture of gases such as air is:

$$\frac{\text{time}_A}{\text{time}_B} = \frac{\text{rate}_B}{\text{rate}_A}$$

Graham's law and molar mass

Graham's law is generally used to determine the molar mass of an unknown gas. In an effusion experiment, it requires 68 seconds for a sample of N_2 to effuse down a capillary, and 85 seconds for a different gas to effuse under the same conditions of temperature and pressure. Was the gas ethane (C_2H_6) or propane (C_3H_8)?

solution:

The measured times vary inversely with the rates of effusion (a high rate of effusion = short time):

$$\frac{\text{time}_A}{\text{time}_B} = \frac{\text{rate}_B}{\text{rate}_A}$$

Substituting into Graham's law gives: $\frac{\text{time}_{unk}}{\text{time}_{N_2}} = \frac{\sqrt{\mathbf{M}_{unk}}}{\sqrt{\mathbf{M}_{N_2}}}$

$$\mathbf{M}_{unk} = \mathbf{M}_{N_2} \times \left(\frac{\text{time}_{N_2}}{\text{time}_{unk}}\right)^2$$

$$\mathbf{M}_{unk} = 28 \text{ g/mol} \times \left(\frac{85 \text{ s}}{68 \text{ s}}\right)^2 = 44 \text{ g/mol}$$

The unknown gas is propane, which has a molar mass of 44 g/mole.

5.6 REAL GASES: DEVIATIONS FROM IDEAL BEHAVIOR

Real gases

Choose words from the list below to complete the following paragraphs describing the behavior of real gases:

(a) container	(e) ideal	(i) **pressures**	(m) volume
(b) distance	(f) interact	(j) Real	(n) **volumes**
(c) free	(g) low	(k) speed	
(d) high	(h) molecules	(l) temperature	

No gas exactly follows the ______________(1) gas law. Ideal gas behavior is approached by real gases under certain conditions. The kinetic-molecular theory successfully explains ideal behavior by making the assumptions that gas particles do not ______________(2) and have no ______________(3). ______________(4) gases have a finite volume and exhibit interparticle attractive forces. This last section discusses how the ideal gas law can be changed to more closely reflect the behavior of real gases.

At ______________(5) temperatures (close to a gas's condensation point), and ______________(6) pressures (above 10 atm), gas particles begin to ______________(7) appreciably. As applied pressure increases, the average ______________(8) between molecules becomes smaller, and attraction to nearby molecules decreases the ______________(9) of a gas particle, and therefore the force with which it hits the container walls. Lowering the ______________(10) has the same effect because it slows the molecules and attractive forces become more significant. **Real gases exhibit lower ______________(11) than the ideal gas law would predict.**

At low pressures, we assume that the gas particles themselves do not contribute significantly to the volume and therefore we assume that the ______________(12) volume inside the container equals the ______________(13) volume. At ______________(14) pressures, this assumption no longer holds. As the pressure increases and the free volume decreases, eventually, the volume of the ______________(15) themselves takes up a significant proportion of the container volume, and the container volume is significantly larger than the free volume. **At very high pressures, real gases exhibit larger ______________(16) than the ideal gas law would predict based on free volume inside the container.**

answers: 1-e, 2-f, 3-m, 4-j, 5-g, 6-d, 7-f, 8-b, 9-k, 10-l, 11-i, 12-c, 13-a, 14-d, 15-h, 16-n.

Initially, as pressure increases, interactive forces become important and PV/RT decreases below the ideal value. As the pressure increases further, however, the volume of the molecules becomes significant, and outweighs the importance of the intermolecular attractions so PV/RT increases above the ideal value.

An equation for real gases was developed in 1873 by Johannes van der Waals which takes into account molecular interactions and molecule volume:

$$\left(P + n^2\frac{a}{V^2}\right)(V - nb) = nRT$$

adjusts P up   adjusts V down

where a and b are van der Waals constants, experimentally determined positive numbers that depend on the particular gas. Your textbook lists van der Waals constants for some common gases in Table 5.4 (p. 216).

SUMMARY

The Physical Attributes of Gases

Gases are mostly empty space at normal conditions.

- easily compressed
- low viscosities
- low densities

Mechanically, all gases behave the same way.

1 mole of any gas at STP = 22.4 L (standard molar volume)

Standard temperature and pressure (STP) conditions: $T = 0°C$ (273 K), $P = 1$ atm.

The density of a gas is directly proportional to its molar mass.

Pressure

Definition: $P = \frac{\text{force}}{\text{area}}$

A barometer measures atmospheric pressure (typically 760 mm Hg).

Units of pressure: 1 atm = 760 mm Hg = 760 torr

The Gas Laws

Boyle's Law: $V = k/P$ (T and n fixed) $P_1V_1 = P_2V_2$

Charles's Law: $V = bT$ (P and n fixed) $\frac{V_1}{T_1} = \frac{V_2}{T_2}$

Avogadro's Law: $V \propto n$ (T and P fixed) $\frac{V_1}{n_1} = \frac{V_2}{n_2}$

Ideal Gas Law: $PV = nRT$ ($R = 0.0821$ L·atm/mole·K)

The ideal gas law may also be written in the form: $m/V = d = P\mathbf{M}/RT$.

Dalton's Law of Partial Pressures

For a mixture of gases, $P_{total} = P_1 + P_2 + P_3$ ...

For a gas, X, collected over water, $P_X = P_{tot} - P_{water}$

The Kinetic-Molecular Theory

Assumptions:

- Gas particles have no volume.
- Gas particles do not interact.
- Gas particles are in constant motion.
- The average kinetic energy of a gas molecule is directly proportional to the absolute temperature of the gas.

Average speed of gas molecules: $u = \left(\frac{3RT}{\mathbf{M}}\right)^{\frac{1}{2}}$

u = 1150 mi/hr for a nitrogen molecule at room temperature

Molecular energy:

K.E.$_{(avg)}$ = 3/2 RT

K.E.$_{(avg)}$ = 2480 J/mole for any gas at room temperature

Graham's Law of Effusion and Diffusion: $\frac{\text{rate}_A}{\text{rate}_B} = \frac{\sqrt{\mathbf{M}_B}}{\sqrt{\mathbf{M}_A}}$

Real Gases: van der Waals Equation: $\left(P + n^2\frac{a}{V^2}\right)(V - nb) = nRT$

CHAPTER CHECK

Make sure you can...

- Describe the physical properties of gases.
- Know the standard molar volume of a gas at STP.
- Relate the density of a gas to its molar mass.
- Define pressure and understand how a barometer works.
- Use Boyle's, Charles's, and Avogadro's gas laws.
- Write the ideal gas law in two useful forms.
- Explain Dalton's law of partial pressures;
  know how to find the pressure of a gas collected over water.
- List the assumptions of the kinetic-molecular theory.
- Find the average speed of gas molecules at a given temperature.
- Find the average molecular energy of gases at a given temperature.
- Recall and be able to use Graham's law of effusion and diffusion.
- Understand the corrections van der Waal's equation makes for real gases.

Chapter 5 Exercises

5.1) A U-tube manometer is open to the atmosphere on the right and connected to a sealed glass vessel on the left. If the atmospheric pressure is 760 torr, and the difference in the levels of mercury in the two sides is 15 cm, so that the level on the side open to the atmosphere is the higher, calculate the pressure of the gas in the sealed vessel on the left.

5.2) a. If a mixture of 4.0 mol of nitrogen and 1.0 mol of oxygen is placed in a 5.0 liter container at 27°C, what pressure is exerted by the gas mixture?

b. At a constant temperature, how many liters will a gas occupy at a pressure of 3.0 atm if it occupies 8.0 liters at a pressure of 4.5 atm?

5.3) Scuba divers use compressed air tanks for under water air sources. A scuba tank contains 12.00 L of compressed air at 200. atm. How much volume would the air in the tank occupy at 1.25 atm?

5.4) A bicycle pump contains a column of air 40.0 cm long with a 5.00 cm diameter at 1.00 atm pressure. How far must the handle be compressed to create a pressure of 3.00 atm within the bicycle pump?

5.5) Expandable balloons are used in weather and atmospheric research. Such an instrument balloon, at take-off, has a volume of 14.0 cubic meters at 748 mm Hg and 18°C. As the balloon rises through the atmosphere, it expands to 900. cubic meters when the pressure drops to 8.72 mm Hg. What is the temperature at this point?

5.6) Methane gas, CH_4, contributes to the process of global warming. An average dairy cow produces 24.0 lb of methane gas each year. What volume of CH_4 is produced at 25°C and 750. mm Hg?

5.7) A 10 liter vessel of nitrogen gas at 1 atm pressure is heated from a temperature of 200 K to a temperature of 300 K. The volume of the container does not change. How much of the nitrogen gas would have to be released from the container in order to reduce the pressure back to 1 atm while maintaining the temperature at 300 K?

5.8) a. A 7.20 liter flask contains 20. grams of an unknown gas at a pressure of 1.02 atm and a temperature of 87°C. What is the molar mass of the gas?

b. What is the density of ammonia gas at 2.00 atm pressure and 25.0°C?

5.9) The molar volume of any gas at standard temperature and pressure is 22.4 L/mol.

a. What is the density of PF_3 at STP?
b. If the pressure is held constant at 1.00 atm, what is the density of PF_3 at 25°C? at 40.°C?
c. How does temperature affect the density of a gas?
d. If the temperature is kept constant at 273 K, what is the density of PF_3 at 2.00 atm? At 5 atm?
e. How does pressure affect the density of a gas?

5.10) a. What is the partial pressure of helium gas in a mixture of 24 grams of oxygen and 24 grams of helium held in a vessel at a pressure of 1350 torr?

b. A 2.0 liter vessel of hydrogen gas at 1.0 atm and a 3.0 liter vessel of nitrogen gas at 1.5 atm are connected by a valve. If the valve between the flasks is opened and the gases are allowed to mix at constant temperature, what is the final pressure in the apparatus?

5.11) Halothane, C_2HF_3BrCl, is a commonly used gaseous anesthesia. One suggested dosage is 3.0% halothane by volume. What is the partial pressure of halothane if air administered to the patient is at a pressure of 775 mm Hg?

5.12) 834 mL of O_2 was collected over water at 24°C and 726 mm Hg. What is the mass of the dry oxygen gas collected?

5.13) If an unknown gas, which creates copious white fumes in the presence of hydrochloric acid, diffuses 3 times faster than carbon tetrachloride CCl_4, what might the gas be?

5.14) The early enrichment of uranium-235 involved the slightly different rates of diffusion between UF_6 prepared with uranium-238 and UF_6 prepared with uranium-235. What is the relative rate of diffusion for $^{235}UF_6$ to $^{238}UF_6$?

5.15) 2.50 L of CO_2 and 4.50 L of NH_3 are injected into a reaction chamber at 11.0 atm pressure and heated to 200.°C to produce urea:

$$2NH_3 + CO_2 \rightarrow CO(NH_2)_2 + H_2O$$

What mass of urea can be produced from this reaction?

5.16) A hot air balloon is filled with 1500 L air at 51°C and 785 mm Hg. What is the density of the air inside the balloon? Assume the air is 80% N_2 and 20% O_2 by mass.

5.17) 21.0 L of N_2 gas and 50.0 L of H_2 gas is mixed and pumped at 5.00 atm into a reaction column containing a catalyst which is heated to 500°C. Ammonia was produced with a 87.0% yield. The unreacted N_2 and H_2 are recycled back to the reaction column:

$$N_2 + 3H_2 \rightarrow 2NH_3$$

What is the volume of NH_3 produced during the first pass through the reaction column?

5.18) Which of the following gases would you expect to be most ideal in its behavior? Which one would you expect to deviate the most from ideal behavior?

nitrogen	water	methane
hydrogen sulfide	hydrogen chloride	hydrogen
carbon dioxide	ammonia	argon
nitric oxide		

5.19) If the temperature of an ideal gas is increased while keeping the volume constant, the mean free path of the gas molecules remains unchanged. Explain, on a qualitative basis, how this can be.

Chapter 5 Answers

5.1) If the level of mercury on the atmosphere side is higher, then the gas inside the glass vessel is pushing down with greater pressure than the atmosphere; the pressure in the glass vessel must be higher than atmospheric pressure.

Pressure inside vessel = atmospheric pressure + difference in the two levels in mm
= 760 torr + 150 mm Hg
= 910 torr

5.2) a. You assume the gases behave ideally and use the ideal gas law, $PV = nRT$

$P = 5.0 \text{ mol} \times 0.08206 \text{ L·atm K}^{-1}\text{mol}^{-1} \times 300.15 \text{ K}/5.09 \text{ L} = 25 \text{ atm}$

b. Use Boyle's law: $P_1V_1 = P_2V_2$; $V_1 = 4.5 \text{ atm} \times 8.0 \text{ L}/3.0 \text{ atm} = 12 \text{ L}$

5.3) There is a large difference in pressure between 200. atm. and 1.25 atm, so we expect a large difference in the volume of the air. Use Boyle's law: $P_1V_1 = P_2V_2$

$12.00 \text{ L} \times 200. \text{ atm} = V_2 \times 1.25 \text{ atm}$
$V_2 = 1920 \text{ L air}$

5.4) Use Boyle's law: $P_1V_1 = P_2V_2$; First find the volume of the pump column:

$V = \text{area} \times \text{length} = \pi(2.50 \text{ cm})^2 \times 40.0 \text{ cm} = 785 \text{ cm}^3$
$1 \text{ atm} \times 785 \text{ cm}^3 = 3.00 \text{ atm} \times V_2$
$V_2 = 262 \text{ cm}^3$; length of bicycle pump $= V \div \text{area} = 262 \text{ cm}^3 \div 19.63 \text{ cm}^2 = 13.3 \text{ cm}$

5.5) 218 K

5.6) Use the ideal gas law (make sure to check all units!):

Moles of CH_4 produced = 24.0 lbs × 454 g/lb × 1 mol/16 g = 681 mol
$V = nRT/P = 681 \text{ mol} \times .0821 \text{ L·atm/mol·K} \times 298 \text{ K} / .987 \text{ atm} = 1.68 \times 10^4 \text{ L}$

Check: At STP, 1 mol of a gas occupies 22.4 L, so 681 mol of a gas would occupy:
$681 \text{ mol} \times 22.4 \text{ L/mol} = 1.52 \times 10^4 \text{ L}$

We calculated a higher volume for the methane gas than it would occupy at STP. We would predict this result because our cows are at higher temperature (298 K vs. 273 K at STP) and lower pressure (0.987 atm vs. 1 atm at STP) than STP conditions. Both of these differences should lead to a larger volume of methane gas. Our answer seems reasonable.

5.7) You can use the relationship $n_1T_1 = n_2T_2$, so that $n_2 = n_1 \times T_1/T_2 = n_1 \times 200/300$
One-third of the nitrogen gas must be removed to leave 2/3 behind.

5.8) a. density = $P\mathbf{M}/RT$; **M,** molar mass = 80. g/mol; b. density = 1.39 g/L

5.9) a. 3.92 g/L at STP; b. 3.60 g/L at 25°C, and 3.42 g/L at 40°C; c. An increase in temperature decreases the density; d. 7.84 g/L at 2.0 atm, and 19.6 g/L at 5 atm; e. An increase in pressure will increase the density.

5.10) Both problems involve Dalton's law of partial pressure. The partial pressure of a gas in a mixture of gases is the contribution that gas makes to the total pressure exerted by the mixture. The partial pressure is a molar quantity; it depends upon the mole fraction of the gas in the mixture.

a. mole fraction of He = 6.0/6.75 = 0.89; partial pressure of He = 0.89 × 1350 torr = 1200 torr

b. A useful way to express Dalton's law is to say that the partial pressure of a gas is the pressure that the gas would exert if it alone occupied the vessel.
The partial pressure of hydrogen gas = 0.40 atm; partial pressure of nitrogen gas = 0.90 atm.
The total pressure is the sum of the partial pressures = 0.40 + 0.90 = 1.30 atm.

5.11) 775 mm Hg × 0.03 = 23 mm Hg halothane

5.12) 1.02 g O_2

5.13) The molar mass of carbon tetrachloride is 153.8 g/mol. Graham's law of diffusion states that the relative rates of diffusion of two gases equal the square root of the inverse ratio of their molar masses. If one gas diffuses three times faster than another, then its molar mass must be nine times (3^2) less than the molar mass of the other gas. The molar mass of the unknown gas must be about 153.8 g/mol / 9 = 17.09 g/mol. This value suggests the gas is ammonia NH_3.

5.14) Lighter gases diffuse more quickly, so we expect the relative rate of diffusion of $^{235}UF_6$ to $^{238}UF_6$ to be greater than 1:

$$\frac{\text{rate } ^{235}UF_6}{\text{rate } ^{238}UF_6} = \sqrt{\frac{352}{349}} = 1.0043$$

5.15) Find out if NH_3 or CO_2 is the limiting reactant:

Moles NH_3: $n = PV/RT$ = 11.0 atm × 4.50 L / 0.0821 × 473 K = 1.275 mol

1.275 mol NH_3 would produce 0.6375 mol urea (from the balanced equation)

Moles CO_2: n = 11.0 atm × 2.50 L / 0.0821 × 473 K = 0.7082 mol

0.7082 mol CO_2 would produce 0.7082 mol urea

NH_3 is the limiting reactant; mass of urea = 0.6375 mol × 60.05 g/mol = 38.3 g urea

5.16) Use the ideal gas law written in terms of density and molar mass:

$d = P\mathbf{M} / RT$; **M** for air = 0.80 × 28 + 0.20 × 32 = 28.8 g/mol

d = 1.03 atm × 28.8 g/mol / 0.0821 × 324 K = 1.1 g/L

Remember the density of gases is often reported in g/L instead of g/mL because of the large volume gases typically occupy.

5.17) 29.0 L NH_3

5.18) The predominant reason for deviation from ideal behavior is intermolecular attraction. The greater the attraction between molecules, the greater is the deviation. The secondary reason for deviation is the size of the molecules. In Chapter 12 you will learn that hydrogen bonding, which occurs in water and ammonia, is the strongest intermolecular force. The weakest intermolecular forces occur between symmetrical molecules (and between atoms) such as carbon dioxide, methane, hydrogen, argon, and nitrogen.

Most ideal gas: hydrogen (very weak intermolecular attraction and low molar mass).

Least ideal gas: water (strong hydrogen bonding; in fact it's a liquid at room temperature and 1 atm pressure).

5.19) If the temperature of an ideal gas is increased, then the kinetic energy of the gas molecules increase, i.e., the molecules move faster. This suggests that the distance traveled between collisions should increase. However, increasing the temperature at constant volume increases the pressure of the gas and the collision frequency increases.

The velocity, or distance/time, increases

The collision frequency, or collisions/time increases

Dividing one by the other, distance/collision, or mean free path, remains constant

6

Thermochemistry

6.1 ENERGY, HEAT, AND WORK

Energy

All mechanical energy can be classified as kinetic energy (energy of motion), or potential energy (energy of position). When we hold a ball above the ground, it contains potential energy, which is converted into kinetic energy as the ball falls to the ground. At any given time, the sum of the potential and kinetic energies are a constant:

$$E_{\text{kinetic}} + E_{\text{potential}} = \text{constant} \quad \text{(purely mechanical system)}$$

where $E_{\text{kinetic}} = \frac{1}{2}mv^2$ and $E_{\text{potential}} = mgh$.

Units of energy

The SI unit for energy is the joule (J),

$$1\ \text{J} = \frac{1\ \text{kg m}^2}{\text{s}^2}$$

Units for a joule

Use (a) the potential energy equation, and (b) the kinetic energy equation to obtain the units for a Joule (kg m^2/s^2).

solution:

If we substitute kg for mass, m/s^2 for *g*, m for height, and m/s for velocity we obtain:

(continued on next page)

(a) $E_{pot} = mgh = \text{kg} \times \frac{\text{m}}{\text{s}^2} \times \text{m} = \frac{\text{kg m}^2}{\text{s}^2}$

(b) $E_{kin} = \frac{1}{2}mv^2 = \frac{1}{2} \times \text{kg} \times \text{m} \times \frac{\text{m}}{\text{s}^2} = \frac{1}{2}\frac{\text{kg m}^2}{\text{s}^2}$

You may be more familiar with the term calorie which is about 4 times larger than a joule:

1 cal = 4.184 J

A calorie is the amount of energy needed to raise 1 gram of water 1°C. A nutritional Calorie (note the capital C) is equivalent to 1000 calories, or 1 kilocalorie:

1 Cal = 1 kcal = 1000 cal

Demonstration that $E_{kin} + E_{pot}$ = constant

1) If a ball (mass 20 g) is dropped and falls for 1.5 seconds before it hits the ground, what is its potential energy before it falls, and what is its kinetic energy when it reaches the ground? The distance the ball drops from its initial height (h_i) to its final height (h_f) is $(h_i - h_f) = \Delta h = \frac{1}{2}gt^2$.

   The value for g is 9.8 m/s^2.

   solution:

   In order to solve for potential energy, ($E_{pot} = mgh$), we need to know the ball's initial height:

   $h = h_i$

   $\Delta h = h_i - h_f$

   $h = h_f = 0$

   We know from physics that the distance the ball will fall in 1.5 seconds is:

$$\Delta h = h_i - h_f = \frac{1}{2} \times 9.8\frac{\text{m}}{\text{s}^2} \times (1.5\ \text{s})^2 = 11.0\ \text{m}$$

$$E_{pot} = mgh = 0.020\ \text{kg} \times 9.8\frac{\text{m}}{\text{s}^2} \times 11.0\ \text{m} = 2.2\frac{\text{kg m}^2}{\text{s}^2} = 2.2\ \text{J}$$

   In order to solve for kinetic energy ($E_{kin} = \frac{1}{2}mv^2$), we need to know the ball's velocity as it reaches the ground:

$$\text{velocity,}\quad v_f = gt = 9.8\ \text{m/s}^2 \times 1.5\ \text{s} = 14.7\ \text{m/s}$$

$$E_{kin} = \frac{1}{2}mv^2 = \frac{1}{2}\left[0.020\ \text{kg} \times (14.7\ \text{m/s})^2\right] = 2.2\frac{\text{kg m}^2}{\text{s}^2} = 2.2\ \text{J}$$

   Notice that $E_{kin} = E_{pot} = 2.2$ J

2) Use algebra to show: $E_{kin} + E_{pot}$ = constant

   solution:

   We start with the equation for the distance an object falls:

(continued on next page)

$$\Delta h = h_i - h_f = \tfrac{1}{2}gt^2$$, and substitute for

t, $t = \frac{v_f}{g}$ $(v_i = 0)$:

$$\Delta h = h_i - h_f = \tfrac{1}{2}g\left(\frac{v_f}{g}\right)^2 = \tfrac{1}{2}\frac{v_f^2}{g}$$

Multiplying through by g gives:

$$gh_i - gh_f = \tfrac{1}{2}v_f^2$$

Multiplying both sides of the equation by m gives:

$$mgh_i - mgh_f = \tfrac{1}{2}mv_f^2$$

$$-\Delta(E_{pot}) = \Delta(E_{kin})$$

$$\Delta(E_{pot}) + \Delta(E_{kin}) = 0, \text{ or } \Delta(E_{pot} + E_{kin}) = 0$$

$$\boldsymbol{E_{pot} + E_{kin} = \text{constant}}$$

Conservation of Energy

Let's go back to the ball in free-fall. We know from experience that eventually the ball will come to rest. Does this mean that the kinetic and potential energy are lost? No. Experience teaches us that energy is not lost; it is converted to other forms. As the ball bounces on the ground, frictional forces cause the mechanical energy to be converted to thermal energy. When one form of energy disappears, equivalent amounts of other types of energy always appear. Experience tells us:

Energy cannot be created or destroyed,

only transformed.

This observation is the law of conservation of energy or the **first law of thermodynamics.**

We cannot in practice measure absolute energies, but we can measure energy changes. To measure a change in energy (ΔE), we measure the difference between a system's internal energy before some change, and its energy after the change:

$$\Delta E = E_{final} - E_{initial}$$

A system consists of the components whose change we wish to measure. How do we measure the energy change of a system? By looking at the surroundings. We can measure the energy change of the surroundings to infer the energy change of the system. We divide the surroundings into two types of energy changes: mechanical and thermal. We measure ΔE of the mechanical surroundings with meter sticks and stopwatches, and we use a thermometer to determine ΔE of the thermal surroundings. We obtain ΔE of the system from ΔE of the thermal plus mechanical surroundings.

In the case of the ball in free-fall, the ball is the system. In a chemical reaction, the system is often the reactants. The surroundings are everything else that is affected by the change. The system can gain energy from the surroundings or lose energy to the surroundings. Since we know the total amount of energy cannot change, we can write:

$$\Delta E_{total} = \Delta E_{sys} + \Delta E_{surr} = 0$$

$$E_{total} = E_{system} + E_{surr} = \text{constant}$$

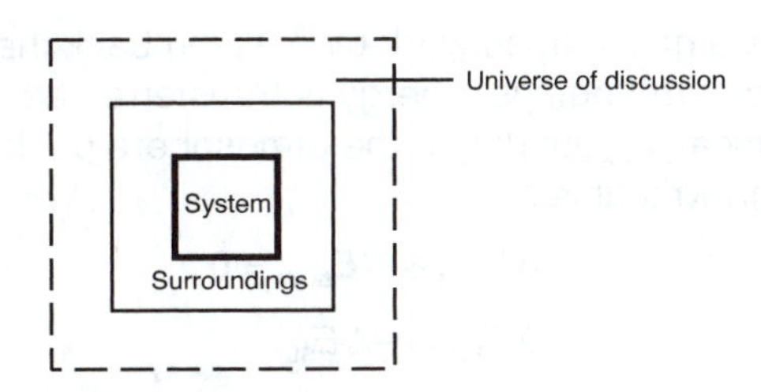

The figure shows the universe of discussion enclosed with a dotted line. The central box represents the system, which contains the components we wish to study. The outer box represents the surroundings, which contains everything else that is affected by the change. In a chemical reaction, if a gas is produced, it does work by pushing back the atmosphere (mechanical change of the surroundings). If it absorbs, or gives off heat, we can measure the thermal changes of the surroundings to determine the energy changes of the system.

Calorimeters are useful devices because they enclose the system in an insulated container. We use a thermometer to measure the change in thermal energy of the surroundings (often water) to determine the energy changes from a chemical reaction (the system).

Heat and Work

Heat and work defined

Choose words from the list below to complete the following paragraph:

(a) energy
(b) heat
(c) temperature
(d) work

Energy can be transferred from one system to another by either heat flow or by work. Heat is energy transferred due to differences in ______________(1) between the system and the surroundings. Energy transferred when an object is moved by a force is ______________(2). We saw that a bouncing ball causes its surroundings to ______________(3) up. It could also have done ______________(4) by causing some of the particles on the ground to move. When we eat food, we input calories (______________(5)) into our bodies which is converted to ______________(6) (to keep our body temperature at approx. 98.6°F), and into ______________(7) (to propel our bodies around if we choose to do so).

answers: 1-c, 2-d, 3-b, 4-d, 5-a, 6-b, 7-d.

Let's consider a reaction run in a flask exposed to the constant pressure of the atmosphere. We can picture the system and surroundings as follows:

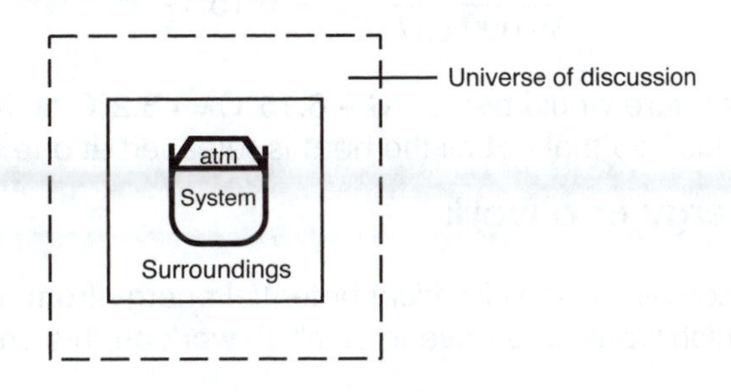

If a reaction produces a gas, it does work by pushing back the atmosphere. If the system is at a different temperature from the surroundings, energy is transferred as heat. We must consider both the energy change of the mechanical surroundings (the atmosphere pushing down on the reaction), and the thermal surroundings. We can write this as:

$$\mathbf{\Delta E_{sys} + \Delta E_{surr} = 0}$$

$$\Delta E_{sys} = -\Delta E_{surr}$$

$$\Delta E_{sys} = -(\Delta E_{thermal\ surr} + \Delta E_{mech\ surr})$$

$$\mathbf{\Delta E_{sys} = q + w}$$

where:

$q = -\Delta E_{therm\ surr}$ and is defined as energy transferred between the system and surroundings due to a temperature change.

$w = -\Delta E_{mech\ surr}$ and is defined as energy transferred by other mechanisms. Many different kinds of work are possible, but we will be concerned only with mechanical work that is due to volume changes against an external pressure ($P\Delta V$ work).

The equation is written from the system's perspective so q is positive when the system gains heat from the surroundings, and w is positive when work is done on the system by the surroundings.

Notice that the equations in bold above are simply different ways to express that the total amount of energy in the universe is constant:

$$\mathbf{\Delta E_{sys} = q + w}$$

$$\mathbf{\Delta E_{sys} - q - w = 0 = \Delta E_{total}}$$

$$\mathbf{\Delta E_{sys} + \Delta E_{therm\ surr} + \Delta E_{mech\ surr} = 0 = \Delta E_{total}}$$

$$\mathbf{E_{sys} + E_{therm\ surr} + E_{mech\ surr} = constant = E_{total}}$$

Conversion of energy into heat

If you weigh 65.0 kg (143 lb), and eat a 400. Calorie breakfast, what would be your body temperature if the energy were all converted to heat? (Assume a starting body temp. of 98.6°F or 37.0°C, and that, for simplicity's sake, you are 100% water.)

solution:

400 nutritional Calories = 400 kcal or 400,000 cal
1 gram of water requires 1 calorie to raise its temperature 1°C. It would therefore take 65,000 calories to raise 65,000 g (65 kg) of water 1°C or 65,000 cal/°C.

Our breakfast supplied us with 400,000 calories of energy, so 400,000 calories would raise 65,000 g by:

$$\frac{400{,}000\text{ cal}}{65{,}000\text{ cal/°C}} = 6.15\text{°C}$$

Our body temperature would be: 37°C + 6.15°C = 43.2°C or 109.6°F (probably fatal). We metabolize our food so that not all the heat is released at one time.

Conversion of energy into work

Assuming you convert all the chemical potential energy from a 400 Calorie breakfast into work with 100% efficiency, how high would you have to climb to work off this energy?

(continued on next page)

$E_{potential} = mgh$ (m = 65 kg, g = 9.8 m/s^2)

Solving for h gives: $h = E_{pot}/mg$

$E_{potential}$ from breakfast = 400 Cal = 400,000 cal

400,000 cal × 4.184 J/cal = 1,673,600 J (Remember: $J = \frac{kg\,m^2}{s^2}$)

Substituting into the equation for h gives:

$$h = \frac{1{,}673{,}600 \text{ kg m}^2/\text{s}^2}{65 \text{ kg} \times 9.8 \text{ m/s}^2} = 2600 \text{ m} \text{ (about 1.6 miles)}$$

Note: We cannot convert much of our food Calories into work because most of the energy is used to keep our body temperatures around 98°F.

A Note about *q*

Heat and temperature are not synonyms. Heat transfer is measured by a change in temperature. Unlike temperature, heat depends upon the amount of material involved. A five gallon container of gasoline sitting in a garage has the same temperature as a cupful of gasoline from the container, but the "heat content" of the five gallon container is much larger than the cupful, as one could easily demonstrate by lighting the two with a match.

A Note about *w*

In this chapter we are concerned with work done by expanding gases. The useful equation for pressure-volume work which can be derived from the definition of work is:

$$w = -P\Delta V$$

- The system loses energy when it does work ($\Delta V > 0$), (indicated by the negative sign).
- This equation assumes constant pressure conditions, in this case the constant pressure of the atmosphere.

Using $w = -P\Delta V$, we can write:

$$\Delta E_{sys} = q + w$$

$$\mathbf{\Delta E_{sys} = q_P - P\Delta V} \text{ (constant } P\text{, only } P\Delta V \text{ work done)}$$

The subscript "P" of q_P reminds us that this equation is written for constant pressure conditions. Note: when the volume is constant, $\Delta V = 0$, so $P\Delta V = 0$, and

$$\mathbf{\Delta E = q_V} \text{ (constant volume)}$$

State Functions

A property that is independent of the pathway is a state function. E is a state function because the difference in E_{final} and $E_{initial}$ is independent of the path taken between them. Another way to think about this is that the same ΔE can occur through any combinations of q and w. After we eat a 400 Calorie breakfast, we can either use the energy by taking a nap, in which case much of the energy is slowly(!) converted to heat, or we can climb 8000 ft to the top of a mountain, in which case some of the energy (≈ 25%) is converted into our ability to do work.

Another example of a state function is the altitude gain on a mountain climbing expedition. If you and a friend climb Long's Peak in Colorado, and one decides to scale the vertical east face, while the other takes the more circuitous, but decidedly safer "keyhole" route, the change in altitude from the trail head to the peak is a state function: it is the same regardless of the route taken.

6.2 ENTHALPY: CONSTANT PRESSURE CONDITIONS

For a universe containing a system with its mechanical and thermal surroundings we have seen:

$$\Delta E = q + w$$

When the system is at constant pressure and only $P\Delta V$ work is possible, we can write:

$$\Delta E_{sys} = q_P - P\Delta V_{sys}$$

We can solve this equation for q_P to give:

$$q_P = \Delta E + P\Delta V$$

We can substitute $\Delta E = E_{final} - E_{initial}$ and $\Delta V = V_{final} - V_{initial}$:

$$q_P = (E_{final} - E_{initial}) + P(V_{final} - V_{initial})$$

and rearrange the equation to give:

$$q_P = (E_{final} + PV_{final}) - (E_{initial} + PV_{initial})$$

We can simplify the above equation by defining:

$$H = E + PV$$

We call H **enthalpy.** Enthalpy simplifies the writing of equations when working under constant pressure conditions, the conditions you normally encounter in the laboratory. Enthalpy, like energy, is a state function and so a change in enthalpy, ΔH, is dependent only on the initial and final states of the system. Substituting H into the equation for q_P above gives:

$$q_P = H_{final} - H_{initial} = \Delta H$$

We can substitute ΔH into our equation for ΔE ($\Delta E = q_P - P\Delta V$) to give:

$$\Delta E = \Delta H - P\Delta V \text{ (constant } P\text{)}$$

ΔH and ΔE differ only by the $P\Delta V$ term, which for most chemical reactions is insignificant, so **usually**, $\Delta H \approx \Delta E$. $P\Delta V$ may become significant when:

1) There is a change in the number of moles of gases between reactants and products (the change in volume term, ΔV, may become significant). The volume change when dealing with liquids and solids is too small to cause the ΔV term to be significant.

2) Pressures are approximately those reached in the interior of the earth (the pressure term, P, becomes significant)—not likely a concern with reactions carried out in a laboratory!

Determining when $P\Delta V$ is significant

1) When 1 mole of water is vaporized at 1 atm pressure at 100°C, ΔH = **40.6 kJ.** Would you expect ΔE to be larger, smaller, or approximately the same as ΔH?

(continued on next page)

solution:

Because there is a large increase in volume between a mole of water and a mole of steam, the work term ($P\Delta V$) is significant (3.1 kJ). Since the system does work against the atmosphere as water is vaporized, it loses energy and ΔE is smaller than ΔH:

$$\Delta E = 40.6 \text{ kJ} - 3.1 \text{ kJ} = \mathbf{37.5 \text{ kJ}}$$

The amount of energy required to vaporize 1 mole of water (100°C, 1 atm), 37,500 J, or a little above 10^3 J, lies below the amount of heat released from combustion of 1 mole of glucose, and above 1 calorie. The order of magnitude for our value seems reasonable.

2) When 1 mole of water freezes, would you expect ΔE to be larger, smaller, or about equal to ΔH?

solution:

The volume change between water and ice is small, so the work term ($P\Delta V$) is insignificant. In fact, when a mole of water freezes, ΔE and ΔH differ by only 0.00017 kJ and:

$$\Delta H \approx \Delta E$$

Note on enthalpy: When a system is at constant pressure, and only $P\Delta V$ work is possible, enthalpy represents the heat liberated. It is merely a useful way to simplify the expression for q_P.

H, like E, is a state function; it is independent of the route used to reach the products.

$$\Delta H = H_{\text{products}} - H_{\text{reactants}}$$

- If $H_{\text{products}} > H_{\text{reactants}}$, $\Delta H > 0$ and the reaction is endothermic (absorbs heat from the surroundings).

- If $H_{\text{reactants}} > H_{\text{products}}$, $\Delta H < 0$, and the reaction is exothermic (gives off heat to the surroundings).

Some important types of enthalpy change are discussed later in section 6.4.

6.3 CALORIMETRY

How do we measure ΔH or ΔE in the laboratory? We measure a change in enthalpy or energy by measuring heat released or absorbed by the system from the thermal surroundings. We can't directly measure heat flow, but we can easily measure temperature changes that are proportional to the amount of heat transferred. We use a calorimeter to measure these temperature changes. The amount the temperature changes will depend on the material's heat capacity.

Heat Capacity

Heat capacity is the amount of heat required to raise the temperature of a substance 1 K.

Two specific types of heat capacity are:

Specific heat capacity (*c*): the amount of heat required to raise the temperature of **1 gram** of a substance by 1 K.

(continued on next page)

Molar heat capacity (*C*): the amount of heat required to raise the temperature of **1 mole** of a substance by 1 K.

Notice that specific heat capacity and molar heat capacity specify an amount of material; heat capacity is a term which does not specify an amount of material (its units are J/T). You cannot look up heat capacity in a table since a number would be meaningful only if it were defined for a specific amount of a system.

Water has a specific heat capacity of 4.184 J/g K or 1 calorie/g K (remember a calorie is the amount of heat required to raise 1 gram of water 1 K). Water has a high heat capacity! For comparison, metals typically have specific heat capacities of less than 1 J/g K.

Specific heat capacity

What is the amount of energy (in joules) required to bring a cup of water to boiling (100°C) from room temperature (25.0°C)?

solution:

Specific heat capacity of water = 4.184 J/g K. The amount of water is:

$$1 \text{ cup} \times \frac{1 \text{ pint}}{2 \text{ cups}} \times \frac{1.04 \text{ lb}}{1 \text{ pint}} \times \frac{454 \text{ g}}{1 \text{ lb}} = 236 \text{ g}$$

The temperature change is: $\Delta T = 100°C - 25.0°C = 75.0°C$, which also equals the ΔT in K.

Therefore, the amount of energy required is:

$$4.184 \text{ J/g K} \times 236 \text{ g} \times 75.0°C = 74{,}100 \text{ J or } 74.1 \text{ kJ}$$

Mathematically we write: $\boldsymbol{q = c \times m \times \Delta T}$

As long as we remember the definition for specific heat capacity (c), we can remember this equation by using unit analysis. The specific heat capacity must be multiplied by terms that provide q with energy units (typically joules):

$$\text{J} = \frac{\text{J}}{\text{g°C}} \times \text{g} \times \text{°C}$$

It can be confusing to keep heat capacity, specific heat capacity, and molar heat capacity straight unless you always include units in your equations. It is also helpful to notate in the algebraic equations what type of heat capacity you are using. We use a capital C to notate molar heat capacity and a small c for specific heat capacity. Your textbook does not use a symbol for heat capacity.

Note also that J/g K = J/g°C since the magnitude of a degree Celsius and a Kelvin are equivalent.

The amount of heat evolved or absorbed in a process can be measured using a **calorimeter.** A **coffee cup calorimeter** is a simple apparatus that constructs surroundings so that water captures and retains the heat released or absorbed by the system. The water in a Styrofoam cup is defined to be the thermal surroundings. A thermometer, lowered into the water through the cover of the calorimeter, monitors the temperature change of the surroundings (ΔT). We can weigh the amount of water in the coffee cup, and we know that the specific heat capacity of water is 4.184 J/g K. We can then calculate q_{surr}:

$$q_{surr} = c \times m \times \Delta T$$

Since heat gained by the surroundings = heat lost by the system (or vice versa), we can calculate q_{sys} by:

$$q_{surr} = -q_{sys}$$

Coffee cup calorimeter (determination of ΔH)

A 45.0 g spoon is heated to 100°C in boiling water and is dropped into a coffee cup calorimeter containing 50.0 g of water at 25.0°C. If the final water temperature is 28.6°C, calculate the specific heat capacity of the spoon. Is the spoon stainless steel or silver?

solution:

First we calculate the heat absorbed by the water. Next we set up an expression for the heat released by the spoon leaving the specific heat capacity of the spoon (c_{spoon}) as the variable we want to solve for. Finally, we use the relationship:

$$q(\text{spoon}) = -q(\text{water})$$

to solve for c_{spoon} and compare the value with the known specific heat capacities of steel and silver.

Heat absorbed by the water:

$$q(\text{water}) = c \times m \times (T_{final} - T_{initial})$$

$$= 4.184\ \text{J/g°C} \times 50.0\ \text{g} \times (28.6\text{°C} - 25.0\text{°C})$$

$$= 753.1\ \text{J}$$

Heat released by the spoon:

$$q(\text{spoon}) = c_{spoon} \times m \times (T_{final} - T_{initial})$$

$$= c_{spoon} \times 45.0\ \text{g} \times (28.6\text{°C} - 100\text{°C})$$

$$= -3210\ \text{g°C} \times c_{spoon}$$

To solve for c_{spoon} we use the relationship:

$$\Delta E_{sys} = -\Delta E_{surr} \text{ or, } q(\text{spoon}) = -q(\text{water})$$

Substituting into this equation we obtain:

$$-3210\ \text{g°C} \times c_{spoon} = -753.1\ \text{J}$$

$$c_{spoon} = 0.234\ \text{J/g°C}$$

The specific heat capacity for steel is 0.45 J/g°C and for silver is 0.237 J/g°C; therefore, the spoon is most likely silver. A note about heat capacity calculations: the problems are not difficult to set up algebraically, but they must be solved carefully so that minus signs are not misplaced. Remember that ΔT is always defined as $T_{final} - T_{initial}$.

Comments about the Coffee Cup Calorimeter

- We measure ΔT of the surroundings to calculate q_{surr}. Then we can solve for q_{sys} since we know:

$$q_{sys} = -q_{surr}$$

- We assume that the water in the coffee cup is the only element of the surroundings that changes temperature.

- It is a constant pressure system since the contents of the coffee cup are exposed to constant atmospheric pressure, so $q_p = \Delta H$ (by definition).

A coffee cup calorimeter is a simple calorimeter that you can easily make, but it isn't effective for reactions that involve gases, or ones that produce high temperatures. A **bomb calorimeter** can obtain precise heat flow measurements for these reactions. Your textbook shows a drawing of a bomb calorimeter in Figure 6.10 on page 265.

In a bomb calorimeter, weighed reactants are sealed in a rigid steel container called the "bomb" (and they wonder why people can be nervous in chemistry labs!). The bomb is immersed in a water-filled container equipped with a stirrer and thermometer. Electrical ignition starts the reaction in the bomb, and if the reaction is exothermic, it gives off heat. The volume inside the bomb does not change ($\Delta V = 0$). The surroundings that absorb the heat are not simply the water, but the water, the bomb, and the other parts of the calorimeter. Before a bomb calorimeter is useful for calculating energy changes, therefore, we must know the heat capacity of the entire calorimeter. We measure the heat capacity of the calorimeter by running a reaction in which a known amount of heat will be released.

Calibration of a bomb calorimeter

It is known that 1 gram of benzoic acid evolves 26.42 kJ of heat when combusted. If one gram of benzoic acid is burned in a bomb calorimeter containing 2.95 kg of water, and the temperature rises from 24.33°C to 26.67°C, what is the heat capacity of the bomb (including the water) (in kJ/°C)?

solution:

We know q and ΔT:

$$q = 26.42 \text{ kJ/g} \times 1 \text{ g} = 26.42 \text{ kJ}$$

$$\Delta T = T_{final} - T_{initial} = 26.67°\text{C} - 24.33°\text{C} = 2.34°\text{C}$$

We can solve for heat capacity using the relationship:

$$q = \text{heat capacity} \times \Delta T$$

$$\text{heat capacity} = \frac{q}{\Delta T} = \frac{26.42 \text{ kJ}}{2.34°\text{C}} = 11.3 \text{ kJ/°C}$$

Note that the heat capacity value for the bomb calorimeter doesn't specify an amount of material, so it is only useful if the calorimeter is set up with the same amount of water each time.

Now that we know the heat capacity of the bomb calorimeter, we can run a combustion reaction and determine the heat evolved.

Bomb calorimeter experiment

When hydrogen (H_2) burns in oxygen, it produces only water as a product. Hydrocarbons that contain carbon and hydrogen produce water and carbon dioxide. CO_2 is an undesirable product since it is one of the "greenhouse" gases that contributes to global warming. We can compare the heats of combustion of H_2 and methane (CH_4) using a bomb calorimeter.

When a 1.5 g sample of methane gas is burned with excess oxygen in the calorimeter, the temperature increases by 7.3°C. When a 1.15 g sample of hydrogen gas is burned, the temperature increase is 14.3°C. Calculate and compare the heat evolved when 1 gram of hydrogen is burned, and when 1 gram of methane is burned.

solution:

We know from our experiment with benzoic acid (previous page) that the heat capacity of our bomb calorimeter is 11.3 kJ/°C.

For methane, we calculate q:

$$q = \text{heat capacity} \times \Delta T = 11.3\ \text{kJ/°C} \times 7.3\text{°C} = 82\ \text{kJ for 1.5 g } CH_4$$

For 1 gram of CH_4,

$$q = \frac{82\ \text{kJ}}{1.5\ \text{g}} = 55\ \text{kJ/g}$$

For hydrogen we calculate q:

$$q = \text{heat capacity} \times \Delta T = 11.3\ \text{kJ/°C} \times 14.3\text{°C} = 162\ \text{kJ for 1.15 g } H_2$$

For 1 gram of H_2, $q = \dfrac{162\ \text{kJ}}{1.15\ \text{g}} = 141\ \text{kJ/g}$

More energy is released when 1 gram of hydrogen is combusted than when 1 gram of methane is combusted. Hydrogen is a potentially useful and clean fuel. It is not a practical fuel yet because it is difficult to store (see the discussion on fuel cells on page 410 of Chapter 21 of the study guide).

Comments about the Bomb Calorimeter

- Like the coffee cup calorimeter, we measure the temperature change of the surroundings to calculate q_{surr} and q_{sys}:

$$q_{surr} = \text{heat capacity} \times \Delta T$$

$$q_{sys} = -q_{surr}$$

- Unlike the coffee cup calorimeter, we do NOT assume that water is the only element of the surroundings that changes temperature. We take into account the temperature change of the entire calorimeter. The heat capacity of the calorimeter must be calculated in a separate experiment. In the coffee cup calorimeter, the temperature change of the Styrofoam cup is assumed to be negligible.
- A bomb calorimeter is a constant volume system: $\Delta V = 0$ and $P\Delta V = 0$. Since there is no work term, $q_V = \Delta E$.

In Summary:

Energy

Heat flow at constant volume is an <u>energy</u> change: $\Delta E = q_V$

A bomb calorimeter measures ΔE.

Enthalpy

Heat flow at constant pressure is an <u>enthalpy</u> change: $\Delta H = q_P$

A coffee cup calorimeter measures ΔH.

6.4 STOICHIOMETRY OF THERMOCHEMICAL EQUATIONS

A thermochemical equation is a balanced equation that includes the enthalpy change of reaction (ΔH). The heat of reaction, ΔH is dependent on the amount of substances reacting. ΔH for a reaction is equal in magnitude but opposite in sign for ΔH of the reverse reaction. For endothermic reactions, ΔH is positive; for exothermic reactions, ΔH is negative. For example, as your textbook describes on page 266, the decomposition of 2 mol of water to its elements, hydrogen and oxygen, is $\Delta H = 572$ kJ. Inversely, the formation of 2 mol of water from hydrogen and oxygen is exothermic and $\Delta H = -572$. ΔH values for a reaction refer to the amounts of substances (and their states of matter) for a specific equation. For the example above, the enthalpy of reaction of the decomposition of 1 mol of water to its elements is ½ that for 2 mol, or $\Delta H = 286$ kJ.

6.5 HESS'S LAW OF HEAT SUMMATION

Sometimes it may be difficult to measure ΔH for a chemical reaction directly because, for example, the change is very slow. Since H is a state function (it does not depend on the path), we can add together ΔH's for individual steps to obtain the ΔH in question:

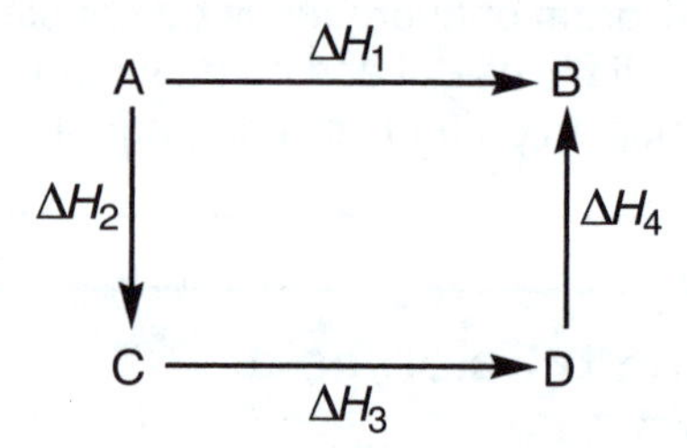

We use Hess's law of heat summation to write: $\Delta H_1 = \Delta H_2 + \Delta H_3 + \Delta H_4$

Hess's law

The standard enthalpy of combustion of elemental carbon to form carbon dioxide is –393.7 kJ/mol C. The standard enthalpy of combustion of carbon monoxide to form carbon dioxide is –283.3 kJ/mol CO. Use these data to calculate ΔH° for the reaction: $2C(s) + O_2(g) \rightarrow 2CO(g)$

<u>solution</u>:

The balanced equation for the combustion of carbon is:

$$C(s) + O_2(g) \rightarrow CO_2(g) \qquad \Delta H^\circ = -393.7 \text{ kJ/mol carbon}$$

(continued on next page)

The balanced equation for the combustion of carbon monoxide is:

$$CO(g) + \tfrac{1}{2}O_2(g) \rightarrow CO_2(g) \qquad \Delta H^\circ = -283.3 \text{ kJ/mol CO}$$

Manipulation of equations allows us to obtain the equation in question. We want 2 moles of carbon to appear on the left hand side of the final equation, so we multiply the equation for the combustion of C by 2:

$$2C(s) + 2O_2(g) \rightarrow 2CO_2(g) \qquad \Delta H^\circ = -787.4 \text{ kJ}$$

We need 2 moles of CO to be a product, so we write the reverse of the equation for the combustion reaction for CO and multiply it by 2. We inverse the sign of ΔH°, and multiply it by 2:

$$2CO_2(g) \rightarrow 2CO(g) + O_2(g) \qquad \Delta H^\circ = 566.6 \text{ kJ}$$

Adding the two equations and corresponding ΔH° values gives:

$$2C(s) + 2O_2(g) \rightarrow 2CO_2(g) \qquad \Delta H^\circ = -787.4 \text{ kJ}$$

$$2CO_2(g) \rightarrow 2CO(g) + O_2(g) \qquad \Delta H^\circ = 566.6 \text{ kJ}$$

$$2C(s) + O_2(g) \rightarrow 2CO(g) \qquad \Delta H^\circ = -220.8 \text{ kJ}$$

Most compounds have a negative heat of formation, and we see in this example that the heat of formation of carbon monoxide from the elements carbon and oxygen is negative.

Another example where Hess's law is frequently used is in using bond energy data to calculate heats of reaction when ΔH_f° data is unavailable. We can think of a reaction as a process in which reactant bonds are broken and product bonds are formed. Breaking bonds requires energy, and forming bonds releases energy. We can calculate ΔH_{rxn} by determining the energy required to break bonds in the reactants, and the energy released as bonds are formed in the products.

For the following reaction, we will use bond energy data to obtain ΔH_{rxn}:

$$CH_4 + Cl\text{—}Cl \longrightarrow CH_3Cl + H\text{—}Cl$$

			ΔH_{sys} (kJ/mol bonds)
1.	One H–C bond is broken	414.2	ΔH_1
2.	One Cl–Cl bond is broken	242.7	ΔH_2
3.	One C–Cl bond is formed	–330.5	ΔH_3
4.	One H–Cl bond is formed	–431.0	ΔH_4

Using Hess's law we write:

$$\Delta H_{rxn} = \Delta H_1 + \Delta H_2 + \Delta H_3 + \Delta H_4$$

$$\Delta H_{rxn} = 414.2 + 242.7 - 330.5 - 431.0 = -104.6 \text{ kJ}$$

The reaction is exothermic because more energy is released forming the C–Cl and H–Cl bonds than is needed to break the H–C and Cl–Cl bonds.

6.6 STANDARD ENTHALPIES OF REACTION

Determining the Standard Enthalpies of Reaction from the Standard Enthalpies of Formation

Heat of Formation (ΔH_f)

Consider the equation for magnesium burning in oxygen:

$$2Mg(s) + O_2(g) \rightarrow 2MgO(s); \quad \Delta H^o_{rxn} = -1203 \text{ kJ}$$

The negative ΔH tells us it's an exothermic reaction; that heat is given off. The superscript zero on the ΔH tells us the reactants and products are in their standard states. The amount of heat given off is related to the quantity of material. If we want to find the heat of formation of MgO from the elements, we would need to write the equation as:

$$Mg(s) + \tfrac{1}{2}O_2(g) \rightarrow MgO(s); \quad \Delta H^o_f = \frac{-1203 \text{ kJ}}{2} = -601.6 \text{ kJ}$$

We must divide the value for ΔH^o_{rxn} by two because ΔH^o_f is defined as formation of 1 mole of the compound.

Two Comments about Enthalpies of Formation

- An element in its standard state is assigned a $\Delta H^o_f = 0$.
- Most compounds have a negative ΔH^o_f.

We can view any reaction as a two-step process in which the first step is breaking the reactants down into their elements ($-\Delta H^o_f$), and the second step is combining the elements to form the products (ΔH^o_f). Using Hess's law, we find that the standard heat of reaction for any reaction is the sum of the standard heats of formation of the products minus the sum of the standard heats of formation of the reactants. This is written mathematically as:

$$\Delta H^o_{rxn} = \sum m\Delta H^o_{f\,(products)} - \sum n\Delta H^o_{f\,(reactants)}$$

where $\sum$ means "the sum of," and m and n are the coefficients of the reactants and products from the balanced equation.

Standard enthalpy for combustion reactions

Ethanol (C_2H_5OH) is sometimes used in gasoline to decrease the amount of undesirable combustion products. The standard heat of formation of ethanol is –279 kJ/mol. Compare the standard enthalpy of combustion per gram of ethanol to that per gram of octane (C_8H_{18}), a major component of gasoline. ΔH^o_f for octane is –269 kJ/mol, ΔH^o_f for $CO_2(g)$ is –394 kJ/mol, and ΔH^o_f for $H_2O(l)$ is –286 kJ/mol.

(continued on next page)

solution:

First we write balanced equations for the combustion of octane and ethanol:

$$C_8H_{18}(l) + \frac{25}{2}O_2(g) \rightarrow 8CO_2(g) + 9H_2O(l)$$

$$C_2H_5OH(l) + 3O_2(g) \rightarrow 2CO_2(g) + 3H_2O(l)$$

Calculate ΔH^o_{comb} (per mole and per gram of octane and ethanol):

$$\Delta H^o_{comb} = \sum m\Delta H^o_{f\ (products)} - \sum n\Delta H^o_{f\ (reactants)}$$

$$\Delta H^o_{comb}\ (octane) = [8 \times \Delta H^o_f\ CO_2(g) + 9 \times \Delta H^o_f\ H_2O(l)] - [\Delta H^o_f\ C_8H_{18}(l)]$$

$$= [(8 \times -394\ kJ/mol) + (9 \times -286\ kJ/mol)] - [1 \times (-269\ kJ/mol)]$$

$$= -5457\ kJ/mole\ of\ octane$$

Molar mass of octane = 114 g/mol so:

$$\frac{-5457\ kJ/mol}{114\ g/mol} = -47.9\ kJ\ per\ gram\ of\ octane$$

$$\Delta H^o_{comb}\ (ethanol) = [2 \times \Delta H^o_f\ CO_2(g) + 3 \times \Delta H^o_f\ H_2O(l)] - [\Delta H^o_f\ C_2H_5OH(l)]$$

$$= [(2 \times -394\ kJ/mol) + (3 \times -286\ kJ/mol)] - [(1 \times -279\ kJ/mol)]$$

$$= -1367\ kJ/mole\ of\ ethanol$$

Molar mass of ethanol = 46 g/mol so:

$$\frac{-1367\ kJ/mol}{46\ g/mol} = -29.7\ kJ\ per\ gram\ of\ ethanol$$

Octane provides ≈1.5 times more energy per gram than does ethanol.

SUMMARY

Energy

All mechanical energy can be classified as potential or kinetic:

$$\boldsymbol{E_{kinetic} + E_{potential} = constant}$$

$$\frac{1}{2}mv^2 + mgh = constant$$

Units of energy:

SI unit for energy is the Joule (kg m^2/s^2): 4.184 J = 1 cal

1 cal = amount of energy needed to raise the temperature of 1 g of water 1°C

Law of conservation of energy:

Energy cannot be created or destroyed.

$\boldsymbol{\Delta E_{sys} + \Delta E_{surr} = 0}$ where: surr = thermal surr + mechanical surr

$\Delta E = E_{final} - E_{initial}$

$\Delta E = q + w$ where: $q = -\Delta E_{\text{therm surr}}$ and $w = -\Delta E_{\text{mech surr}}$

$\Delta E = q_P - P\Delta V$ (constant pressure); $\Delta E = q_V$ (constant volume)

Energy is a state function; it is independent of path.

Enthalpy

$H = E + PV$

$\Delta H = q_P = \Delta E + P\Delta V$ (constant pressure, only $P\Delta V$ work possible)
($\Delta H = H_{\text{products}} - H_{\text{reactants}}$)

ΔH differs from ΔE by the work term ($P\Delta V$); generally, $\Delta H \approx \Delta E$

Exothermic: $\Delta H < 0$; heat given off to surroundings

Endothermic: $\Delta H > 0$; heat absorbed from surroundings

Enthalpy is a state function; it is independent of path.

Calorimetry (Measurement of heat flow)

Heat capacity (J/ K): the amount of heat required to raise the temperature of a substance 1 K.

Specific heat capacity (c, J/g K): the amount of heat required to raise the temperature of 1 gram of a substance by 1 K.

$$q = c \times m \times \Delta T$$

Molar heat capacity: (C, J/mol K) the amount of heat required to raise the temperature of 1 mole of a substance by 1 K.

Coffee cup calorimeter (constant pressure system, $q_P = \Delta H$): allows you to calculate ΔH by measuring the temperature change of water in a Styrofoam cup.

Bomb calorimeter (constant volume system, $q_V = \Delta E$): allows you to calculate ΔE by measuring the temperature change of water in a bomb calorimeter and by knowing the heat capacity of the calorimeter.

Hess's Law of Constant Heat Summation

ΔH can be calculated by adding together ΔH's for individual steps.

Bond Energies

Bond energies can be used to calculate ΔH_{rxn}.
Energy is added to break bonds; energy is released when bonds form.

Enthalpies of Reaction from Enthalpies of Formation

ΔH_f = Enthalpy of formation
1 mole of a compound is formed from its elements.

ΔH^o_{rxn} = Heat of reaction measured with all substances in their standard states.
The standard states are defined to be:
- for a gas: 1 atm.
- for an aqueous solution: 1 M concentration in mol/L.
- for an element or compound: the most stable form at 1 atm pressure and the temperature of interest, usually 25°C (298 K).

$$\Delta H^o_{\text{rxn}} = \sum m\Delta H^o_{f\,\text{(products)}} - \sum n\Delta H^o_{f\,\text{(reactants)}}$$

CHAPTER CHECK

Make sure you can...

- Write the relationship between kinetic and potential energy.
- Know the definition for a joule, and its relationship to a calorie.
- State the first law of thermodynamics; define a system and its surroundings.
- Explain how energy is transferred through heat and work.
- Define the term "state function."
- Define enthalpy and relate it to exothermic and endothermic reactions.
- Write the relationship between energy and enthalpy; know when $E = H$.
- Define heat capacity, specific heat capacity (c), and molar heat capacity (C).
- Calculate specific heat from data from a coffee cup calorimeter.
- Calculate heats of combustion from data from a bomb calorimeter.
- Use Hess's law to manipulate ΔH values.
- Use bond energy data to obtain ΔH_{rxn}.
- Determine ΔH_{rxn} from standard heats of formation.

Chapter 6 Exercises

6.1) a. If a system absorbs 35 kJ of heat from its surroundings, and does 25 kJ of work on its surroundings, what is the value of ΔE, the change in internal energy?

b. If the surroundings do 20 kJ of work on a system, and the system loses 8 kJ of heat to the surroundings, what is the change in internal energy of the system?

6.2) a. The relationship between the SI unit of energy, the joule, and the more traditional unit, the calorie, is: 1 calorie = 4.184 J. If one gram of water requires 1 calorie to raise its temperature by 1°C, how many joules are required to heat 1 kg water by 1°C?

b. Another, larger, unit of energy is the BTU, which is the energy required to heat one pound of water by 1°F. What is the relationship between one joule and one BTU? (1 pound = 453.6 g)

6.3) How much work is done when 18.0 grams of water vaporize at 100.°C under a constant external pressure of 1.00 atm?

6.4) Identify the following reactions or processes as either endothermic or exothermic:

a. steam condensing
b. natural gas burning
c. hydrogen and oxygen gases exploding
d. a gas fire radiating heat
e. water freezing
f. gunpowder exploding
g. molten lava cooling down
h. magnesium ribbon reacting with oxygen to produce a brilliant white flame
i. ether evaporating
j. water decomposing into hydrogen and oxygen
k. ice melting

l. decomposition of calcium carbonate
m. N_2O_4 decomposing to produce two NO_2 molecules
n. iodine crystals subliming

6.5) If an endothermic reaction takes place in an insulated container, at the end of the reaction, will the temperature of the contents of the container be higher than before the reaction began, lower, or will it remain the same?

6.6) How much heat is required to raise the temperature of 160. g of water from 18.8°C to 37.0°C?

6.7) 10.0 kJ of heat is added to a coil of aluminum wire at room temperature (25.0°C) with a mass of 100. g. What is the final temperature of the wire?

6.8) The value of ΔH for the reaction shown is –280. kJ:

$$C(s) + 1/2O_2(g) \rightarrow CO(g) \quad \Delta H = -280.\ kJ$$

a. What is the value of ΔH for the reaction: $2C(s) + O_2(g) \rightarrow 2CO(g)$
b. What is the value of ΔH for the reaction: $2CO(g) \rightarrow 2C(s) + O_2(g)$

6.9) When natural gas (methane) burns, the quantity of heat released is 890. kJ/mol. The standard molar heat of combustion of carbon (graphite) is –394 kJ/mol. The standard molar heat of combustion of hydrogen is –286 kJ/mol. Calculate the standard molar heat of formation of methane in kJ/mol.

6.10) Determine ΔH_{rxn} for the following reaction:

$$2CH_4(g) + 4Cl_2(g) + O_2(g) \rightarrow 2CO(g) + 8HCl(g)$$

given the following set of reaction data:

$H_2(g) + Cl_2(g) \rightarrow 2HCl(g)$ $\quad \Delta H^0 = -92.30\ kJ$
$2H_2(g) + O_2(g) \rightarrow 2H_2O(g)$ $\quad \Delta H^0 = -241.8\ kJ$
$2CO(g) + O_2(g) \rightarrow 2CO_2(g)$ $\quad \Delta H^0 = -566\ kJ$
$CH_4(g) + 2O_2(g) \rightarrow CO_2(g) + 2H_2O(g)$ $\quad \Delta H^0 = -802.3\ kJ$

6.11) When benzene is burned in air, considerable heat is released. Calculate the heat released (the enthalpy change) for the following reaction. How much heat is released per mole of benzene?

$$2C_6H_6(l) + 15O_2(g) \rightarrow 12CO_2(g) + 6H_2O(g)$$

ΔH_f^o for $H_2O(g) = -242$ kJ/mol; ΔH_f^o for $C_6H_6(l) = +49$ kJ/mol; ΔH_f^o for $CO_2(g) = -394$ kJ/mol

6.12) Determine ΔH_{rxn} for the following reaction using ΔH_f^o values in the text:

$$SiO_2(s) + 4HF(g) \rightarrow SiF_4(g) + 2H_2O(g)$$

6.13) a. A 30. gram block of lead at 90.°C is dropped into water at 20.°C in an insulated calorimeter. The final temperature of the water and the lead block is 22°C. How much heat was lost by the lead block? The specific heat of lead is 0.128 J/g K.

b. A 5.0 gram block of ice at 0°C is dropped into an insulated container of 100. grams of water at 70.°C. What is the temperature of the water in the container at the end of the experiment after all the ice has melted? Heat of fusion of ice = 334 J/g.

6.14) A coin appraiser determines the specific heat of some ancient coins to determine if they are pure gold. He heats the coins (113.4 g) to 100°C in a boiling water bath and then transfers them to a calorimeter containing 200. g of water at 20.0°C. The final temperature of the coins and water in the calorimeter is 25.9°C:

a. What is the specific heat of the coins? b. Are they pure gold?

6.15) When 20. grams of ammonium nitrate is dissolved in 50. grams of water in an insulated container, the temperature drops by 22.4 K. Calculate the heat of solution of ammonium nitrate. Assume that the specific heat capacity of the ammonium nitrate solution is 4.18 J/K g.

Chapter 6 Answers

6.1) $\Delta E = q + w$; q and w are both positive if heat is gained by the system and work is done on the system, i.e., if both processes increase the internal energy of the system.

a. q = +35 kJ, w = –25 kJ
ΔE = +10 kJ, the internal energy of the system increases by 10 kJ

b. q = –8 kJ, w = +20 kJ
ΔE = +12 kJ, the internal energy of the system increases by 12 kJ

6.2) a. 4184 J (or 4.184 kJ)

b. Energy required to heat 1 gram of water by 1°C = 1 cal = 4.184 J
Energy required to heat 453.6 grams of water by 1°C = 4.184 × 453.6 = 1898 J
Energy required to heat 453.6 grams of water by 1°F = 1898 × 5/9 = 1054 J
1 BTU is approximately equal to 1 kJ

6.3) 18 grams of water is one mole. At 100°C and 1.00 atm, the volume of one mole of an ideal gas is 30.62 L. This equals the change in volume when one mole of liquid is vaporized because the volume of the liquid (18 mL) is negligible compared to the volume of the vapor.

$w = -P\Delta V$ = –1.00 atm × 30.62 L = –30.62 L atm = –3.10 kJ
The negative sign indicates that the system does work on the surroundings.

(A useful conversion to remember is that 8.314 J = 0.08206 L atm. These are the values for R in the two different units.)

6.4) The first eight are exothermic and heat is liberated; the last six are endothermic, and heat is absorbed.

6.5) An endothermic reaction is a reaction that absorbs energy. The bonds that are broken in the reaction are stronger than the bonds that are formed. If the reaction occurs within an insulated container, where does this energy come from? It must come from the molecules (reactants and products) within the container itself. The kinetic energy of these molecules decreases as energy is withdrawn and the temperature drops.

6.6) $q = c \times m \times \Delta T$
q = 4.184 J/g°C × 160. g × 18.2°C = 1.22 × 104 J = 12.2 kJ
(Remember that since the size of a degree Celsius and a degree Kelvin are the same, either can be used for the units of c.)

6.7) $q = c \times m \times \Delta T$; $\Delta T = q / c \times m$
ΔT = 10.0 × 10^3 J / 0.900 J/g°C × 100. g = 111°C
$T_{final} = \Delta T + T_{initial}$ = 25.0°C + 111°C = 136°C

6.8) The enthalpy change, ΔH for a reaction is an extensive property; its value depends upon the reaction as it is written. If the reaction is reversed, then the sign of the enthalpy change, ΔH, changes; if the reaction is exothermic in one direction, then it must be endothermic in the reverse direction.

Likewise, if the coefficients of the equation are changed, then the value of ΔH changes in proportion:

a. The coefficients are all doubled; the value of ΔH doubles: $\Delta H = -560.$ kJ
b. The equation is reversed (and doubled); the sign of ΔH changes: $\Delta H = +560.$ kJ

6.9) It is often useful to write the equations for the reactions described:

1. $CH_4(g) + 2O_2(g) \rightarrow CO_2(g) + 2H_2O(l)$ $\Delta H^\circ = -890.$ kJ
2. $C(s) + O_2(g) \rightarrow CO_2(g)$ $\Delta H^\circ = -394$ kJ
3. $H_2(g) + 1/2O_2(g) \rightarrow H_2O(l)$ $\Delta H^\circ = -286$ kJ

The reaction for which the heat of reaction is required is the formation of methane from its constituent elements:

$C(s) + 2H_2(g) \rightarrow CH_4(g)$ $\Delta H^\circ = ??$ kJ

The ΔH for this reaction can be calculated by summing the ΔH's for a series of reactions which together add up to the same overall reaction (Hess's law of heat summation).

1. $CO_2(g) + 2H_2O(l) \rightarrow CH_4(g) + 2O_2(g)$ $\Delta H^\circ = +890.$ kJ (sign changed)
2. $C(s) + O_2(g) \rightarrow CO_2(g)$ $\Delta H^\circ = -394$ kJ
3. $2H_2(g) + O_2(g) \rightarrow 2H_2O(l)$ $\Delta H^\circ = -572$ kJ (value doubled)

These equations add up to the desired equation, so the ΔH values are added:

$\Delta H_f^\circ (CH_4) = +890. - 394 - 572$ kJ/mol $= -76$ kJ/mol

6.10) $\Delta H_{rxn} = -924$ kJ

6.11) The heat of reaction is the difference between the heat of formation of the products and the heat of formation of the reactants in the balanced equation:

$\Delta H^\circ = 12 \times (-384) + 6 \times (-242) - 2 \times (+49) = -6278$ kJ

There are two moles of benzene in the equation, so the heat released for one mole is 1/2 this amount, 3139 kJ/mol.

6.12) $\Delta H_{rxn} = -95.6$ kJ

6.13) a. Heat lost = 30. g × 0.128 J/K g × 68°C = 260 J

b. q(ice melting) + q(water warming) = $-q$(water cooling)
5.0 g × 334 J/g + 5.0 g × 4.184 J/g°C (T_f – 0°C) = –100 g × 4.184 J/g°C (T_f – 70°C)
1670 J + 20.92 T_f = –418.4 T_f + 29,288 J
439.32 T_f = 27,618
T_f = 62.9 = 63°C

6.14) a. q (water) = $-q$ (coins)
$c \times m \times \Delta T = -c \times m \times \Delta T$
4.184 J/g°C × 200. g × 5.9°C = $-c$ × 113.4 g × (–74.1°C)
Solving for c gives: 0.587 J/g°C

b. No, the specific heat capacity of pure gold is 0.129 J/g°C.

6.15) ΔH = 4.18 J/g K × 22.4 K × 70. g = 6550 J
For one mole of ammonium nitrate, ΔH = 6550 × 80.05/20 = 26 kJ/mol.

Quantum Theory and Atomic Structure

"Energy and Matter Merge"

7.1 The Nature of Light

7.2 Atomic Line Spectra

7.3 Wave Properties of Particles

7.4 The Quantum-Mechanical Model of the Atom

INTRODUCTION

When atoms and molecules react, they must bump into each other physically. In so doing, their outer electrons interact. Their reactivity depends on the arrangement of these outer (valence) electrons. Chemists like to have a model for atomic structure so they can understand and predict reactivity of elements and compounds. Some models are more successful than others. Experimental evidence can support models or cause them to be discarded. Some models are useful only for certain systems. Bohr's hydrogen atom model is useful for predicting energy levels in the hydrogen atom, but cannot be extended to more complicated systems.

This chapter discusses the theories and models for atomic structure (particularly the arrangement of electrons around an atom), which successfully account for the repetition of properties (periodicity) of elements.

Matter and Energy

In our macroscopic world, we usually think of matter and energy as being distinct from one another: matter has substance that we touch (has mass) and see (is localized); energy has no mass and is delocalized. However, in order to explain experimental observations made about the behavior of atoms, we need to consider that radiant energy and matter both have properties of waves and particles.

It may seem strange to think of matter as having a wavelength associated with it, and light as being composed of particles that have mass. Quantum mechanics blurs the distinction between matter and energy, and challenges us to expand our classical views on the matter.

(continued on next page)

In the macroscopic world, the wavelengths of baseballs, and dust particles are insignificant compared to the size of the objects, and these objects behave as particles. However, on an atomic level, where electrons weigh 9.11×10^{-31} kg, their wavelengths are about the size of an atom and are significant. The wavelike behavior of electrons can explain why atoms have only certain energy levels.

Light in our macroscopic world behaves as if it were a wave: it bends around the edges of objects (diffracts), and refracts when it passes into a new substance. However, as we will see, light also acts as if it were composed of particles of energy.

7.1 THE NATURE OF LIGHT

Light, the electromagnetic energy that we can see, is only a small part of the **electromagnetic spectrum.** Light has a wavelength of about 400 nm (violet light) to 750 nm (red light). Radio waves (lower energy than light) have wavelengths of several meters, and X rays (higher energy than light) have wavelengths of 10^{-10} m (1 Å).

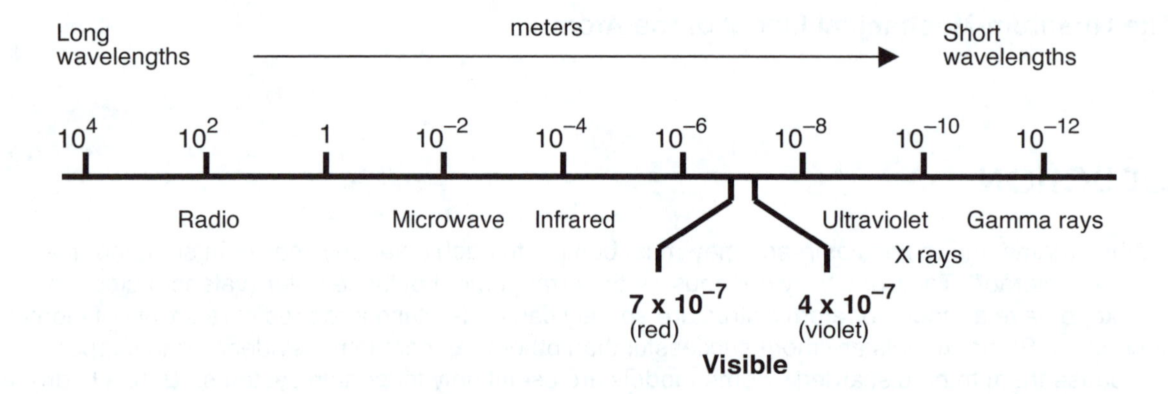

Light that we see is **polychromatic,** made up of many wavelengths. Lasers can generate **monochromatic** light, light at a narrow wavelength.

The wave nature of light

The diagram below depicts a wave of light:

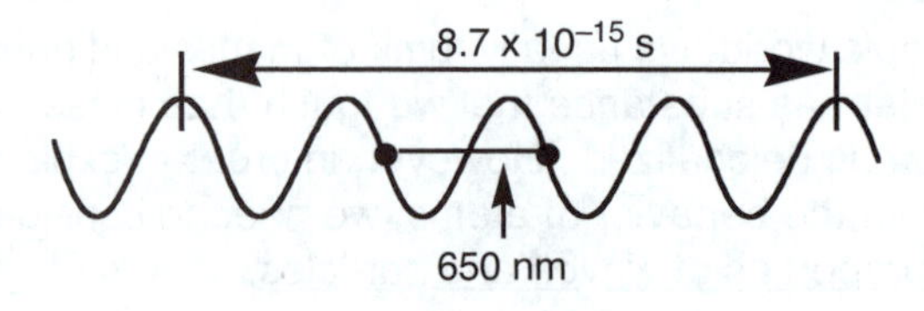

a) What is its wavelength?

b) What is the wave's frequency?

c) How fast is the wave traveling?

(continued on next page)

solutions:

a) The wavelength (λ) is the distance between corresponding points on the wave. It is the distance the wave travels during one cycle: (m/cycle) = 650 nm.

b) The frequency (υ) is the number of cycles per second (cycles/s):

$$= \frac{4 \text{ cycles}}{8.7 \times 10^{-15} \text{s}} = 4.6 \times 10^{14} \text{ cycles/s}$$

c) Picture a wave of light one second long. To obtain the distance traveled by the wave in one second (its speed), we need to count the waves (frequency) and multiply by the length of a wave. (The speed of the wave = wavelength × frequency.)

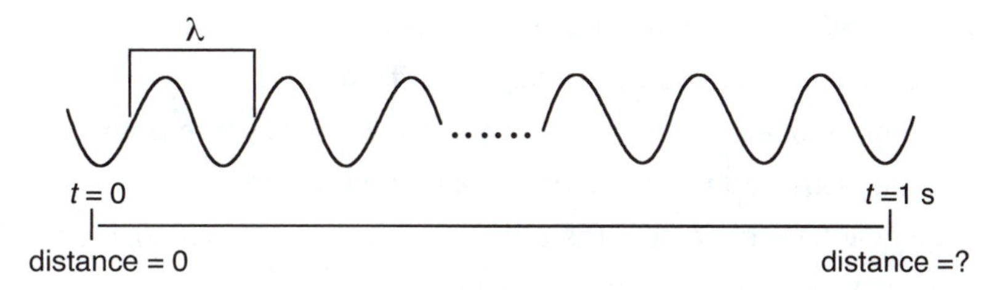

Therefore:

$$\frac{\text{m}}{\text{s}} = \frac{\text{m}}{\text{cycle}} \times \frac{\text{cycles}}{\text{s}}$$

$$= \frac{6.5 \times 10^{-7} \text{m}}{\text{cycle}} \times \frac{4.6 \times 10^{14} \text{ cycles}}{\text{s}}$$

$$= 3.0 \times 10^{8} \text{ m/s}$$

In a vacuum, all electromagnetic radiation travels at 3.00×10^8 m/s, which is the speed of light:

c = frequency × wavelength = 3.00 × 10^8 m/s

Exercise

Choose words from the list below to complete the following paragraph:

(a) amplitude
(b) color
(c) green
(d) frequency
(e) intensity
(f) speed
(g) wavelength

The wavelength of the light in the visible part of the electromagnetic spectrum determines its ______________(1) that we see. Light with a wavelength of 520 nm would have the color ______________(2). The ________________(3) of the wave, the height of the crest, corresponds to the color's ______________(4). All electromagnetic radiation travels at the same ______________(5) in a vacuum (the speed of light), but differs in ______________, and ______________(6 & 7).

answers: 1-b, 2-c, 3-a, 4-e, 5-f, 6-d, 7-g.

Classical thinking: wave vs. particle behavior

Classify the following as either wave or particle behavior:

1) Has mass
2) Has no mass
3) Is composed of oscillating electric and magnetic fields
4) velocity = frequency × wavelength
5) velocity = acceleration × time
6) Its amplitude is related to its intensity
7) Bends **(refracts)** when it hits a new medium
8) Bends around the edge of an object **(diffracts)**
9) Is either stopped by an object, or continues on its same path
10) Can be dispersed into component parts by a prism
11) Causes interference patterns after passing through two adjacent slits
12) Travels at 3.00×10^8 m/s in a vacuum

answers: wave behavior: 2, 3, 4, 6, 7, 8, 10, 11, 12; particle behavior: 1, 5, 9

Quantum Theory

We use quantum theory to explain the behavior of electrons and other small particles. Quantum theory challenges us to consider light as a wave or as a stream of particles, and to consider moving objects as particles or as waves. Three principles of the theory that are of chemical interest to us are listed below.

Postulates of the Quantum Theory

1) Energy is quantized. An atom or molecule has only certain states that are characterized by definite amounts of energy. Light exists as a stream of particles, or little packets of quantized energy called **photons,** which have mass. The energy of a photon is associated with its frequency or wavelength. An atom can change energy states by absorbing or emitting a photon:

$$E_{\text{photon}} = h\upsilon = h\frac{c}{\lambda} = \Delta E_{\text{atom}}$$

2) Moving objects have wavelengths associated with them which relate to their velocity and mass:

$$\lambda = \frac{h}{mv}$$

3) The allowed energy states of atoms and molecules are described by sets of numbers called quantum numbers. The quantum numbers are associated with individual electrons in the atom. The quantum numbers of all the electrons in an atom designate the energy level, or quantum state, of the atom.

Max Planck, studying blackbody radiation, was the first to suggest that the energy of matter is not continuous, but quantized into discrete units. Each small packet of energy is called a **quantum.** Energy can be gained or lost in integer multiples of $h\upsilon$:

$$\Delta E = nh\upsilon$$

where n = integer, h = Planck's constant, 6.626×10^{-34} J s, and υ = frequency.

A few years later in 1905, Albert Einstein postulated that light itself exists as a stream of particles, or photons which have mass.

Mass of a photon

1) Use Einstein's equation for the energy of a photon, and his famous equation relating mass and energy ($E = mc^2$), to derive an expression for the mass of a photon.

solution:

$$E_{photon} = \frac{hc}{\lambda} = mc^2$$

$$m_{photon} = \frac{hc}{\lambda c^2} = \frac{h}{\lambda c}$$

2) What is the mass of a photon of blue (λ = 400. nm) light?

solution:

$$m_{photon} = \frac{h}{\lambda c} = \frac{6.626 \times 10^{-34}\ \text{J s}}{400.\text{nm} \times (3.00 \times 10^{8}\ \text{m/s})}$$

$$= \frac{6.626 \times 10^{-34}\ \text{kgm}^2/\text{s}}{(4.00 \times 10^{-7}\ \text{m}) \times (3.00 \times 10^{8}\ \text{m/s})}$$

$$= 5.52 \times 10^{-36}\ \text{kg} = 5.52 \times 10^{-33}\ \text{g}$$

The mass of a photon is so small that it is difficult to think of it as a particle.

The particle behavior of light explains the **photoelectric effect.** Classical wave mechanics predicted that light of any frequency shining on a metal should cause conduction by freeing an electron once the metal absorbed enough energy. This prediction was not borne out by experiments. The photon theory predicts that a metal must absorb a photon of high enough energy in order to release an electron. Increasing the intensity of the light only increases the number of photons, but not their energy. Two Nobel Prizes resulted from work on the photoelectric effect: one for studying it (Millikan), and one for explaining it (Einstein).

On reflection, we can see evidence for the particle nature of light all around us. It explains why we tan with ultraviolet rays and not with infrared rays (no matter how long we might expose ourselves to them), and why film is not exposed with lower energy red light (a "safelight"), but is ruined with higher energy light.

7.2 ATOMIC LINE SPECTRA

Classical physics predicted that a negatively charged particle (an electron), would spiral around a positive particle (a nucleus), and emit a continuous spectrum of light as its energy decreased. However, excited atoms do not emit a continuous spectrum of light, but a line spectrum (when dispersed by a prism) that is characteristic of the element. Spectroscopists studied the line spectra of atomic hydrogen, and although they did not know the reason for the discrete spectra, they derived an equation (the Rydberg equation below) that would predict the pattern of the spectral lines:

$$\frac{1}{\lambda} = R\left(\frac{1}{n_1^2} - \frac{1}{n_2^2}\right) \qquad n_1, n_2 = \text{integers}$$

where λ = wavelength of a spectral line, n_1 and n_2 are positive integers with $n_2 > n_1$, and R = Rydberg constant = $1.096776 \times 10^7\ m^{-1}$. It is important to remember that the Rydberg equation applies only to the spectral lines of the hydrogen atom or other one electron systems such as He^+ and Li^{2+}.

Spectral lines in the hydrogen atom

1) What is the wavelength of the spectral line in the hydrogen atom when $n_1 = 2$ and $n_2 = 3$?

We plug in the values for n_1 and n_2 into the Rydberg equation and solve for wavelength, λ:

$$\frac{1}{\lambda} = 1.096776 \times 10^7\ m^{-1}\left(\frac{1}{(2)^2} - \frac{1}{(3)^2}\right) = 1.5245 \times 10^6\ m^{-1}$$

$$\lambda = 6.5595 \times 10^{-7}\ m = 656.5\ nm$$

2) What is the wavelength of the spectral line in the hydrogen atom when $n_1 = 2$ and $n_2 = 4$?

$$\frac{1}{\lambda} = 1.096776 \times 10^7\ m^{-1}\left(\frac{1}{(2)^2} - \frac{1}{(4)^2}\right) = 2.056 \times 10^6\ m^{-1}$$

$$\lambda = 4.864 \times 10^{-7}\ m = 486.3\ nm$$

3) What is the wavelength of the spectral line in the hydrogen atom when $n_1 = 2$ and $n_2 = 5$?

$$\frac{1}{\lambda} = 1.096776 \times 10^7\ m^{-1}\left(\frac{1}{(2)^2} - \frac{1}{(5)^2}\right) = 2.303 \times 10^6\ m^{-1}$$

$$\lambda = 4.342 \times 10^{-7}\ m = 434.2\ nm$$

4) What can we say about the solutions for the Rydberg equation when $n_1 = 2$?

solution:

The spectral lines for $n_1 = 2$ (i.e., 656.5, 486.3, and 434.2 nm, calculated above) fall in the visible region of the electromagnetic spectrum. It turns out that if you solve the Rydberg equation for $n_1 = 1$, it predicts where the spectral lines of a hydrogen atom fall in the ultraviolet region. When $n_1 = 3$, the equation predicts the lines fall in the infrared region.

Notice also that the lines become closer together as n_2 increases: the energy levels in an atom become closer together as their distance from the nucleus increases.

The Bohr Model of a Hydrogen Atom

In 1923, Niels Bohr presented a model of the hydrogen atom that predicted the observed line spectra. Bohr assumed a hydrogen atom consisted of a central proton, with an electron moving around it in a circular orbit. He used classical physics to relate the energy of the atom to the radius of the electron's orbit, but incorporated quantum theory into his model by proposing that the angular momentum was quantized. This quantum condition restricted the energy levels in a hydrogen atom to only certain allowable levels that were associated with the fixed orbits:

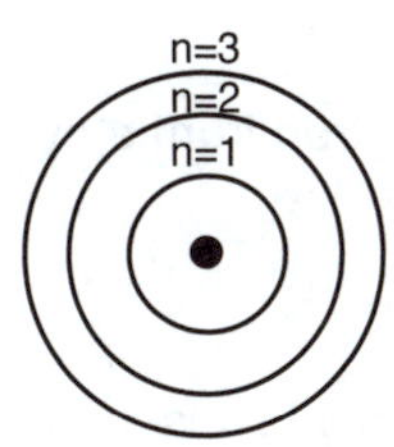

Remember that electrons do not actually circle protons in fixed orbits! This was Bohr's model, which was successful for its ability to predict the spectral lines in a one-electron system. The model broke down when it was extended to atoms with more than one electron.

Bohr's model for the hydrogen atom

Choose words from the list below to complete the following paragraph:

(a) energy	(c) one	(e) photon
(b) excited	(d) orbits	(f) radius

Bohr proposed that an atom can emit energy by emitting a ______________(1) whose energy equals the difference in energy between the ______________(2). The quantum number n = 1,2,3... is associated with the ______________(3) of the electron's orbit, which in turn, is directly related to the atom's ______________(4). Bohr calculated the energy levels for a hydrogen atom to be:

$$E = -2.18 \times 10^{-18}\ \text{J}\left(\frac{1}{n^2}\right)$$

The atom is in its ground state when n = ______________(5); all higher levels ($n > 1$) are ______________(6) states. When a sample of hydrogen is excited, the electrons in different atoms are excited to different levels. As they fall back to the ground state, they emit energy of specific wavelengths giving hydrogen its atomic line spectrum.

answers: 1-e, 2-d, 3-f, 4-a, 5-c, 6-b.

Energy levels of Bohr's orbits

What is the energy of the first excited state in a hydrogen atom?

solution:

$$E = -2.18 \times 10^{-18}\ \text{J}\left(\frac{1}{(2)^2}\right) = -5.45 \times 10^{-19}\ \text{J}$$

We define zero energy as the atom's energy when the electron is completely removed from the atom: $E = 0$ when $n = \infty$ (infinity). Energy is therefore negative for any n smaller than ∞. The energy becomes a more negative number as the electron moves from orbits farther from the nucleus to orbits closer to the nucleus. The energy of the most stable orbit (the ground state when $n = 1$) corresponds to the most negative number.

Ionization energy of a hydrogen atom

1) How much energy is required to remove an electron from a hydrogen atom?

$$H(g) \rightarrow H^+(g) + e^-$$

solution:

The energy difference between any two energy levels in the hydrogen atom is:

$$\Delta E = E_{final} - E_{initial} = 2.18 \times 10^{-18}\ J \left(\frac{1}{n_{final}^2} - \frac{1}{n_{initial}^2} \right)$$

An electron in the ground state of a hydrogen atom is in the $n = 1$ orbit. When it is completely removed from the atom, $n = \infty$:

$$\Delta E = E_{final} - E_{initial} = -2.18 \times 10^{-18}\ J \left(\frac{1}{\infty^2} - \frac{1}{1^2} \right)$$

$$= 2.18 \times 10^{-18}\ J$$

2) How much energy is required to remove an electron from a mole of hydrogen atoms?

solution:

$$\Delta E = \frac{2.18 \times 10^{-18}\ J}{1\ H\ atom} \times \frac{6.022 \times 10^{23}\ atom}{mol} \times \frac{1\ kJ}{10^3\ J} = 1.31 \times 10^3\ kJ/mol$$

This is the **ionization energy** of the hydrogen atom, the amount of energy required to form a mole of gaseous H^+ ions from a mole of gaseous H atoms.

Bohr's model of the atom failed to predict the line spectra of any atom other than hydrogen. His model does not take into account electron-electron repulsions, and nucleus-electron attractions of a multiple electron system. In addition, his assumption that electrons circle protons in fixed orbits was incorrect.

7.3 WAVE PROPERTIES OF PARTICLES

de Broglie

Bohr arbitrarily incorporated quantized angular momentum into his calculation for the energy levels of a hydrogen atom because it produced results that agreed with experimental observations. It wasn't until Louis de Broglie showed that particles could have wave properties, that there was a reasonable explanation for why atoms had only discrete energy levels. If electrons exhibit properties of waves, they would have only certain allowed energies since a wave of fixed radius can only have certain frequencies (energies).

The wavelength of particles

1) Use Einstein's equation relating mass and energy ($E = mc^2$), and Planck's equation for the energy of light ($E = \frac{hc}{\lambda}$) to solve for the wavelength of a particle of mass m moving at velocity, v.

solution:

$$E = \frac{hc}{\lambda} = mc^2 \text{ therefore, } \lambda = \frac{h}{mc}, \text{ or}$$

$$\lambda = \frac{h}{mv}$$ for a particle of any mass moving at velocity v.

2) Determine the wavelength of an electron (mass = 9.11×10^{-31} kg) moving at velocity of 8.00×10^6 m/s.

solution:

We'll use the equation $\lambda = \frac{h}{mv}$ (h = 6.626×10^{-34} J s = 6.626×10^{-34} kg m^2/s)

$$\lambda = \frac{6.626 \times 10^{-34} \text{ kg m}^2/\text{s}}{9.11 \times 10^{-31} \text{ kg} \times \left(8.00 \times 10^6 \text{ m/s}\right)} = 9.09 \times 10^{-11} \text{ m}$$

In Å, the wavelength of the electron is close to 1: $\lambda = 0.909$ Å

3) Determine the wavelength of a ball (mass = 120. g) moving at a velocity of 45.0 m/s (about 100 mi/hr).

solution:

$$\lambda = \frac{6.626 \times 10^{-34} \text{ kg m}^2/\text{s}}{0.120 \text{ kg} \times 45.0 \text{ m/s}} = 1.23 \times 10^{-34} \text{ m}$$

Notice that although both the wavelengths of the electron and the ball are incredibly short, the wavelength associated with the electron is many orders of magnitude larger than that of the ball, and it is on the same order as the spacing between the atoms in a typical crystal.

A qualitative rule is that if the wavelength of an object moving toward a target is small compared to the size of the target, the interaction will be particle-like. In the above example, the wavelength of the ball, 1.23×10^{-34} m, is much, much smaller than the size of the bat. If the wavelength of a moving object is about the same magnitude as the size of the target (or larger), the interaction is wavelike. The wavelength of the electron in the example above is approximately the same size as an atom.

Evidence for the Wave Behavior of Electrons

Waves that pass between adjacent slits cause interference patterns, particularly if the distance between slits is about the same distance as the wavelength. In 1927, Davisson and Germer at Bell Laboratories observed a diffraction pattern when they directed a beam of electrons at a nickel crystal. The atoms in a crystal are about the same distance apart as the wavelength of a moving electron, so the atoms act as "slits" to the beam of electrons which diffract around them and form interference patterns. Since diffraction patterns are usually explained in terms of waves, this result supported de Broglie's idea that moving particles have wavelengths.

Wavelength of particles

1) How fast would the ball need to be moving in the example on the previous page in order for it to cause a diffraction pattern between slits 1 meter apart?

solution:

Efficient interference patterns occur when the distance between slits is approximately the same as the wavelength. In order to have a wavelength of 1 meter, the speed of the ball would need to be:

$$\lambda = 1 \text{ m} = \frac{6.626 \times 10^{-34} \text{ kg m}^2/\text{s}}{0.120 \text{ kg} \times v \text{ m/s}}$$

$v = 5.52 \times 10^{-33}$ m/s, not a plausible speed!

2) How close together would the slits need to be to produce effective interference patterns for a particle with a wavelength of 1.23×10^{-34} m (the wavelength of the ball in the above example)?

solution:

The slits would need to be approximately the same distance apart as the wavelength of the particle (1.23×10^{-34} m, or 1.23×10^{-23} Å) to produce efficient diffraction interference patterns. This distance is many orders of magnitude smaller than the distance between atoms in crystals.

We can see from these examples that it is not possible to determine experimentally if large pieces of matter have wavelengths. However, scientists believe that all matter obeys de Broglie's equation.

The Heisenberg Uncertainty Principle

If electrons have wavelike properties, and waves are spread out in space, what can we determine about the position of an electron in the atom? Heisenberg's uncertainty principle states that we cannot know simultaneously both the exact position and velocity of a particle. If you try to locate an electron, you must use high-energy radiation to "see" it. The high-energy (short wavelength) radiation acts like a particle to the electron and knocks it away from where it was.

The uncertainty in the momentum of the particle $\Delta(mv)$, multiplied by the uncertainty in its position, Δx, must be equal to, or greater than $h/4\pi$ (h = Planck's constant) :

$$(\Delta x) \times \Delta(mv) \geq \frac{h}{4\pi}$$

The uncertainty principle

1) What is the uncertainty in the position of an electron ($m = 9.11 \times 10^{-31}$ kg) moving at $8 \times 10^6 \pm 1\%$ m/s?

solution:

We calculate the uncertainty in the velocity of the electron, and substitute this value into Heisenberg's equation and solve for the uncertainty in its position (Δx):

$\Delta v = 8 \times 10^6 \times 0.01 = 8 \times 10^4$ m/s

(continued on next page)

$$\Delta x \geq \frac{h}{4\pi m \Delta v} \geq \frac{6.626 \times 10^{-34}\ \text{kg m}^2/\text{s}}{4\pi(9.11 \times 10^{-31}\ \text{kg})(8 \times 10^4\ \text{m/s})} \geq 7 \times 10^{-10}\ \text{m}$$

2) What is the uncertainty in the position of a ball (m = 120 g) moving at 45 ± 1% m/s?

solution:

$$\Delta v = 45\ \text{m/s} \times 0.01 = 0.45\ \text{m/s}.$$

$$\Delta x \geq \frac{h}{4\pi m \Delta v} \geq \frac{6.626 \times 10^{-34}\ \text{kg m}^2/\text{s}}{4\pi(0.120\ \text{kg})(0.45\ \text{m/s})} \geq 9.8 \times 10^{-34}\ \text{m}$$

The uncertainty in the position of the ball is incredibly tiny compared to the size of a playing field, so you can feel confident that its position can be calculated. The uncertainty in the position of the electron (7×10^{-10} m, or 7 Å) is about 7 times greater than the diameter of an atom, so we do not know exactly where it is located.

7.4 THE QUANTUM-MECHANICAL MODEL OF THE ATOM

Schrödinger: The Wave Mechanical Model of the Atom

Schrödinger applied classical equations for waves to the problem of an electron in an atom and developed an equation that allowed him to determine the energy levels and wave properties of the hydrogen atom, and a few other simple systems. The wave equation appears to apply exactly to atoms other than hydrogen, although the solutions are too complex to compute. We want to find functions that satisfy the differential equation:

$$H\psi = E\psi$$

H is a set of mathematical instructions, an operator. E is the energy of the atom, and ψ (psi) is a wave function, a mathematical description of the motion of the electrons as a function of time and position. The wave function, ψ, has no physical meaning, but ψ^2 tells us the probability of finding an electron at a given point.

Main Ideas from the Schrödinger Wave Equation

- It does not give information on the detailed path of an electron.
- For one-electron systems, each solution to the equation is associated with a particular wave function, which is also called an atomic orbital. The square of the wave function gives the probability of finding the electron at any point.
- The electron could be anywhere; the probability of finding it at any point never reaches zero. To visualize where the electron spends most of its time, we draw diagrams to show where the electron spends 90% of its time.

Despite the similarity between the terms "orbit" used for Bohr's model, and "orbital" used from Schrödinger's model, they have distinct meanings. Electrons in orbits circle the nucleus at fixed distances; orbitals are regions of space where the electron might be found.

Representations of a 1*s* Orbital

1) Electron density diagram: The density of the dots in a certain region is proportional to the probability of finding an electron there.

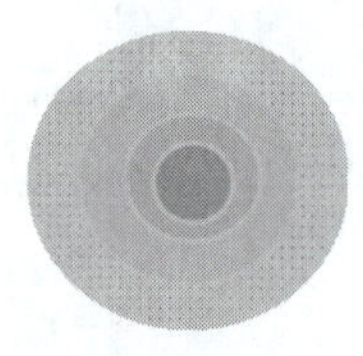

2) Radial probability plot: A graph of probability plotted as a function of distance from the nucleus. The peak for the ground-state H atom appears at the same distance from the nucleus (0.529 Å) as the closest Bohr orbit. The Schrodinger model predicts that the electron spends most of its time at the same distance that the Bohr model predicted it spent all of its time.

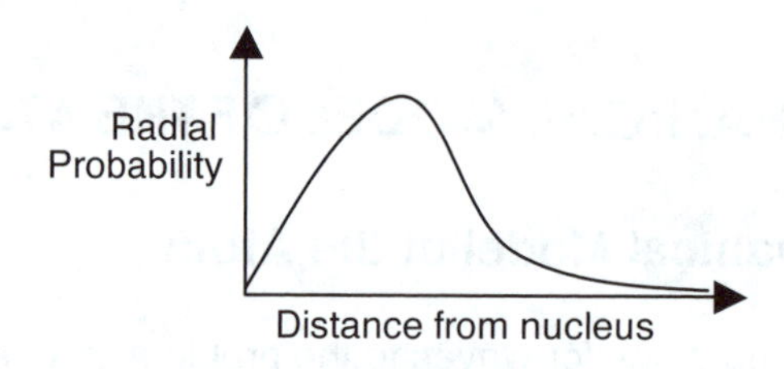

3) 90% probability contour: Contains 90% of the total probability of finding the electron.

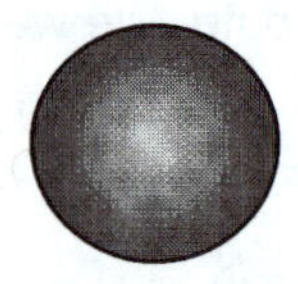

Orbital questions

1) A hydrogen atom in its ground state has a 90% probability contour diameter of 0.74 Å. What would be the size of the diameter for a 100% probability contour?

   answer:

   The diameter would be infinite, since there is a chance that the electron could exist anywhere.

2) How can you reconcile the apparent paradox that the electron density diagram for a 1*s* orbital shows the highest electron density at the nucleus, while the radial probability plot shows the electron density decreasing to 0 at the nucleus?

   answer:

   The probability of finding a 1*s* electron decreases as we move away from the nucleus (ψ^2 decreases), but the square of the radius increases:

(continued on next page)

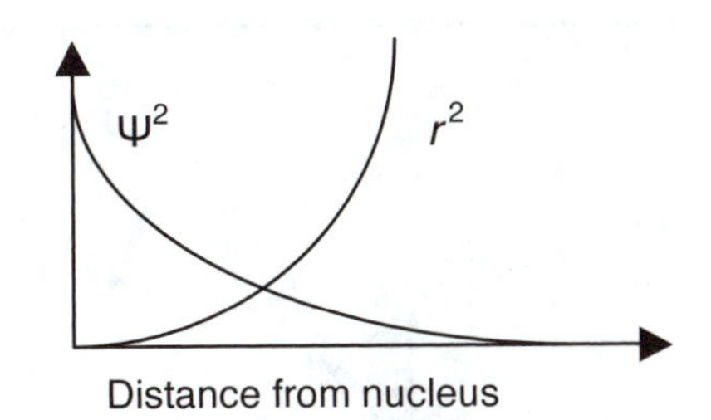

The radial probability curve is a plot of the product of $\left(\psi^2 \times r^2\right)$ vs. distance from the nucleus:

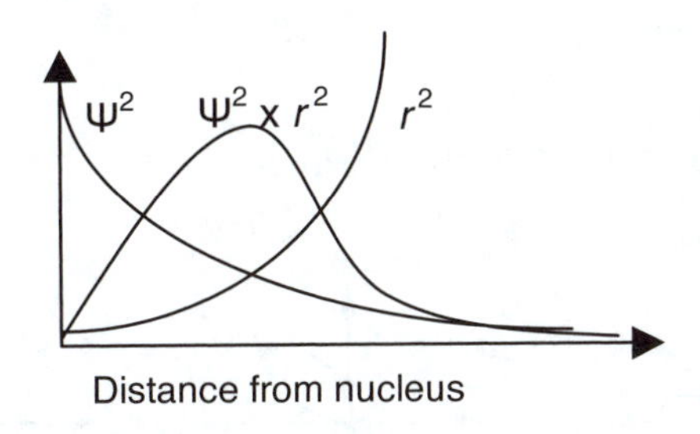

We use 3 quantum numbers to characterize an electron's wave function (atomic orbital). The quantum numbers are summarized below.

Quantum Numbers

Name	Letter	Integer	Relates to orbital's:
Principal (energy level*)	n	positive	size, energy
Azimuthal (sublevels)	ℓ	0 to $n - 1$	shape
Magnetic (orbitals)	m_ℓ	$-\ell,...0,...+\ell$	orientation

*The energy state of an atom with more than one electron depends on both the n and ℓ values of the occupied orbital.

Azimuthal orbitals:
(sublevels)

number	letter	#orientations ($-\ell,...0,...+\ell$)	shape
$\ell = 0$	s	1	1s = sphere
$\ell = 1$	p	3	dumbbell
$\ell = 2$	d	5	clover leaf, donut dumbbell
$\ell = 3$	f	7	multilobed

Orientation of Orbitals

s sublevel, $m_\ell = 0$; one orientation

s

p sublevel, $m_\ell = -1, 0, 1$; three orientations (p_x, p_y, p_z)

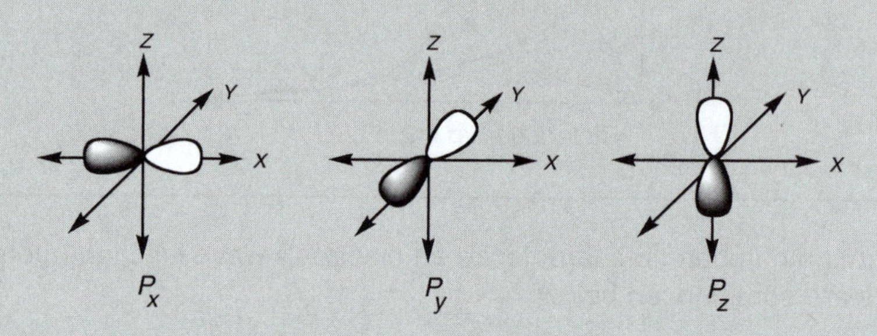

P_x P_y P_z

d sublevel, $m_\ell = -2, -1, 0, 1, 2$; five orientations (d_{xy}, d_{xz}, d_{yz}, $d_{x^2-y^2}$, d_{z^2})

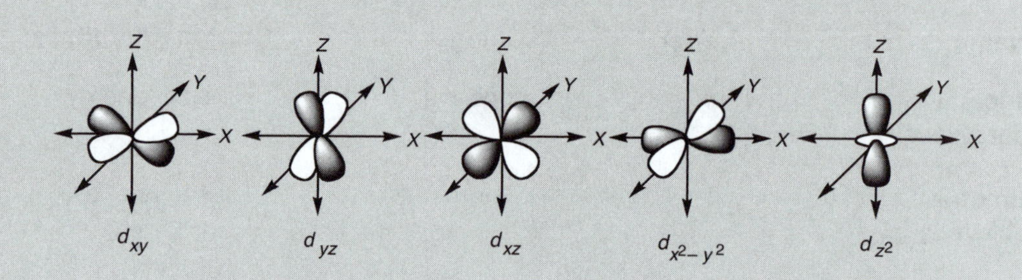

d_{xy} d_{yz} d_{xz} $d_{x^2-y^2}$ d_{z^2}

Quantum numbers

1) When $n = 1$, how many values are there for ℓ? How many orbitals are in the first energy level?

   <u>solution</u>:

   $\ell = 0$ to $n - 1$, so when $n = 1$, ℓ can only have one value (0).
   $m_\ell = -\ell, \ldots, 0, \ldots, +\ell$, so when $\ell = 0$, m_ℓ can only have one value (0).

   Each m_ℓ value corresponds to one atomic orbital, so there is one orbital in the first energy level. The $\ell = 0$ sublevel is designated by the letter "*s*," which has only one orientation in space, designated by $m_\ell = 0$.

2) When $n = 2$, how many values are there for ℓ? How many orbitals are in the second energy level?

(continued on next page)

solution:

When $n = 2$, ℓ can have the values 0, and $n - 1 = 1$. The second energy level therefore has 2 sublevels, $\ell = 0$ (the "s" sublevel) and $\ell = 1$ (the "p" sublevel).

For $\ell = 0$ (s orbital), m_ℓ has only one value (0); for $\ell = 1$ (p orbital), m_ℓ can have 3 values ($-\ell$, 0, $+\ell$ = −1, 0, +1).

The number of orbitals in the $n = 2$ energy level is 4 (an "s" orbital with one orientation, and "p" orbitals with three orientations in space often called p_x, p_y, and p_z).

3) When $n = 3$, how many values are there for ℓ? How many orbitals are in the third energy level?

solution:

When $n = 3$, ℓ can have the values 0, 1, and 2. The third energy level therefore has 3 sublevels, $\ell = 0$ (the "s" sublevel), $\ell = 1$ (the "p" sublevel), and $\ell = 2$ (the "d" sublevel).

For $\ell = 0$ (s orbital), m_ℓ has only one value (0);
for $\ell = 1$ (p orbital), m_ℓ can have 3 values (−1, 0, +1);
for $\ell = 2$, m_ℓ has 5 values (−2, −1, 0, +1, +2).

The "s" orbital has only one orientation ($m_\ell = 0$), the "p" orbitals have three orientations ($m_\ell = -1, 0, +1$), and the "d" orbitals have five orientations ($m_\ell = -2, -1, 0, +1, +2$). The total number of orbitals in the $n = 3$ energy level is therefore: $1 + 3 + 5 = 9$.

4) What is the possible number of ℓ sublevels for a given n?

solution:

When $n = 1$, there is one sublevel (s).
When $n = 2$, there are two sublevels (s and p).
When $n = 3$, there are three sublevels (s, p, and d).
When $n = n$, there are "n" sublevels.

5) What is the total number of orbitals for a given n?

solution:

When $n = 1$, there is one orbital (1s).
When $n = 2$, there are 4 orbitals (2s, $2p_x$, $2p_y$, $2p_z$).
When $n = 3$, there are 9 orbitals
($3s$, $3p_x$, $3p_y$, $3p_z$, $3d_{xy}$, $3d_{xz}$, $3d_{yz}$, $3d_{x^2-y^2}$, $3d_{z^2}$).

When $n = n$, there are n^2 orbitals.

Summary Exercise

On the following page, match the appropriate scientists, equations, or concepts from the three lists at the top of the exercise to the statements that follow.

(continued on next page)

Scientists

a) Niels Bohr

b) Louis de Broglie

c) Albert Einstein

d) Werner Heisenberg

e) Max Planck

f) Erwin Schrödinger

Equations

g) $\Delta E = nh\upsilon$

h) $E_{photon} = h\upsilon = h\frac{c}{\lambda} = \Delta E_{atom}$

i) $E = mc^2$

j) $\frac{1}{\lambda} = R\left(\frac{1}{n_1^2} - \frac{1}{n_2^2}\right)$ n = integers

k) $E = -2.18 \times 10^{-18}\,\text{J}\left(\frac{1}{n^2}\right)$

l) $\lambda = \frac{h}{mv}$

m) $H\psi = E\psi$

n) $(\Delta x) \times \Delta(mv) \geq \frac{h}{4\pi}$

Concepts

o) atomic line spectra

p) blackbody radiation

q) Bohr model

r) diffraction patterns

s) photoelectric effect

t) quantum numbers

u) uncertainty principle

1) Experimental results, observed by the end of the 1900's, which were unexplainable with classical physics. (concepts)

2) Could mathematically predict radiation emitted from blackbodies if he assumed that energy can be gained or lost in integer multiples of $h\upsilon$. (name and equation)

3) His equation was the first to suggest that energy of matter is not continuous but quantized into discrete packets. (name)

4) Postulated that light itself exists as a stream of particles or photons. (name, equation)

5) His idea that mass and energy are alternate forms of the same entity led to a famous equation. (name, equation)

6) Predicted that photons have mass. (name)

7) An equation (named the Rydberg equation), which predicted the wavelengths of emission lines from the hydrogen atom.

8) Developed a model for the hydrogen atom that predicted the observed line spectra for a hydrogen atom; it broke down for more complicated systems. (name, equation)

9) His model for the hydrogen atom restricted the allowed energy levels in a hydrogen atom to only certain allowable levels which were associated with a fixed circular orbit of the electron around the nucleus. (name)

10) Gave the energy levels of a hydrogen atom as: $E = -2.17 \times 10^{-18}\,\text{J}\,(1/n^2)$. (concept)

11) Postulated that electrons exhibit wave properties that could explain why electrons have quantized energy levels in an atom. (name, equation)

12) Experimental result that supported the theory that electrons have wave behavior. (concept)

13) Developed the wave mechanical model of the atom, which allowed the determination of energy levels and wave properties of the hydrogen atom, and a few other simple systems. (name, eqn.)

14) Stated that it is impossible to know both the exact position and velocity of a particle. (name, eqn. con.)

15) Numbers used to specify an electron's energy and wave function (atomic orbital). (concept)

answers: 1-o,p,s; 2-e,g; 3-e; 4-c,h; 5-c,i; 6-c; 7-j; 8-a,k; 9-a; 10-q; 11-b,l; 12-r; 13-f,m; 14-d,n,u; 15-t.

Evolution of our Mental Image of an Atom

Bowling ball (Democritus, Dalton) →

Plum pudding.(Thomson) →

Planetary model (Rutherford) →

Planetary model with quantized shells (orbits) (Bohr) →

Nucleus surrounded by an electron cloud (orbitals)
(contributions from Planck, Einstein, de Broglie, Schrodinger, Heisenberg)

The atom through about the turn of the 20th century (see Chapter 2):

Democritus (about 400 BC): The Greek philosopher proposed that small, indivisible particles make up all material things. He called these irreducible spheres *atoms*. Other Greek philosophers, including Aristotle, rejected Democritus' theory, instead believing that all nature consisted of four elements: air, earth, fire, and water.

John Dalton (1803): More than 2000 years later, Dalton, based on experiments showing that mass is always conserved in chemical transformations, postulated that matter is composed of indestructible particles which merely rearrange during chemical reactions. He imagined these particles as tiny bowling balls, and retained the Greek name *atom*.

J.J. Thomson (1898): Thomson, after discovering the negatively charged *electron*, proposed that atoms were positive spheres with negative electrons embedded in them (the "plum pudding" model).

Ernest Rutherford (1909): Rutherford realized from experiments beaming alpha particles at gold foil that an atom is mainly empty space with a small positively charged nucleus containing most of the mass. Low mass negatively charged electrons orbit this heavy nucleus, much like the planets orbit the sun. Because this model did not always account for the mass of the nucleus, he postulated the existence of neutral particles (which he called *neutrons*) with mass nearly the same as the proton. His colleague, James Chadwick later detected the neutron experimentally.

The 20th century vision of the atom evolves (Chapter 7):

Niels Bohr (1913): Rutherford's planetary model could not explain why negatively charged electrons orbiting a positively charged nucleus did not radiate a continuous spectrum of energy and spiral down into the nucleus. Bohr came up with an explanation that fit with experimental results: he suggested that electrons move in orbits of fixed size and energy and that radiation can occur only when the electron jumps from one orbit to another. When the electrons are in the lowest orbit available, the atom is in the "ground state" and cannot radiate energy.

de Broglie, Schrödinger, Heisenberg (1924-1926): Bohr's model broke down when it was extended to atoms with more than one electron. Louis de Broglie built on Planck and Einstein's work showing that light can exist as both particles and waves to suggest that particles could have wave properties. Now there was a reasonable explanation for why atoms had only discreet energy levels, since a wave of fixed radius can have only certain frequencies. Schrödinger applied classical equations for waves to the model of an electron in an atom and developed an equation whose solutions give us a way to determine the probability of finding an electron at any point. Bohr's idea of fixed orbits for electrons had now evolved to picturing regions of space where the electron might be found (orbitals). Heisenberg theorized that no experiment could measure the position and momentum of a quantum particle simultaneously (the "Heisenberg uncertainty principle").

So, from bowling balls, to plum puddings, to planetary orbits, we arrive at today's visual concept of an atom: an electron cloud surrounding a positively charged nucleus containing protons and neutrons. Probability distribution maps determine the likely location of the electrons, and quantum numbers characterize an electron's atomic orbital.

CHAPTER CHECK

Make sure you can...

- List the approximate wavelengths of radio, microwave, infrared, visible, UV, and X ray radiation.
- Relate the wavelength, frequency, and speed of light (know c, the speed of light).
- Describe wave and particle behavior in terms of classical thinking.
- List the postulates of quantum theory.
- Interconvert E_{photon} with its frequency and/or wavelength.
- Use the Rydberg equation to find wavelengths for the spectral lines in the hydrogen atom.
- Describe Bohr's model for a hydrogen atom and his calculated energy levels for a H atom.
- Find the ionization energy of a hydrogen atom.
- Use de Broglie's equation to find the wavelength of a particle.
- Explain the physical meaning of ψ^2.
- Apply the uncertainty principle to determine the uncertainty in the position of an atomic-size particle.
- Relate quantum numbers to the energy level, shape, and orientation of orbitals.
- Complete the summary exercise on page 149.

Chapter 7 Exercises

7.1) a. If it takes 4.0×10^2 ms for one wavelength of a wave to pass a stationary point, what is its frequency?

b. What is the wavelength of an infrared wave with velocity of 3.0×10^{10} cm/s and frequency of 8.0×10^{11} s^{-1}?

c. What is the frequency of a gamma ray wave with velocity of 3.0×10^{10} cm/s and wavelength of 1.2×10^{-12} cm?

7.2) How many joules of energy are contained in one mole of violet light quanta of frequency 7.0×10^{14} s^{-1}?

7.3) The energy (in joules) of an electron energy level in the Bohr atom is given by the expression: $E_n = -2.179 \times 10^{-18}$ J/n^2, where n is the quantum number that specifies the energy level. What is the frequency in Hz of the radiation emitted when an electron falls from level $n = 3$ to level $n = 2$?

7.4) What is the wavelength of a beam of electrons traveling at 2.00×10^4 cm/s? Assume the mass of an electron at this velocity is 9.1094×10^{-31} kg.

7.5) a. What are the allowed values of the magnetic quantum number m_ℓ associated with the $\ell = 3$ sublevel?

b. If the quantum number n equals 4, what are the allowed values of ℓ associated with this value of n?

7.6) a. Which orbital quantum number specifies the shape of an atomic orbital?

b. If an orbital has a single planar node through the nucleus, what kind of orbital is it?

c. If, in addition to the planar node through the nucleus, the orbital has a single spherical node, to what principal quantum level does the orbital belong?

d. As the value of the principal quantum number increases, what happens to the energy of the orbital?

7.7) Describe, in as much detail as possible, a 2*p* orbital.

7.8) a. What is the total number of electrons that can occupy the principal quantum level $n = 4$ in one atom?

b. Derive a general expression to determine the total number of electrons that will completely fill all the orbitals in the principal quantum level n of an atom.

7.9) Which of the following are not acceptable sets of quantum numbers?

a. $n = 4,\ \ell = 0,\ m_\ell = 0$ c. $n = 0,\ \ell = 1,\ m_\ell = 0$
b. $n = 2,\ \ell = 1,\ m_\ell = -2$ d. $n = 1,\ \ell = 0,\ m_\ell = 0$

7.10) Which of the following are not acceptable sets of quantum numbers?

a. $n = 3,\ \ell = 1,\ m_\ell = -1$ c. $n = 2,\ \ell = 2,\ m_\ell = 0$
b. $n = 7,\ \ell = 3,\ m_\ell = +3$ d. $n = 5,\ \ell = 1,\ m_\ell = 0$

Chapter 7 Answers

7.1) a. The time taken for one wavelength of a traveling wave to pass a stationary point is called the period; it is the reciprocal of the frequency. Frequency = $1/4.0 \times 10^2$ ms = $2.5\ s^{-1}$

b. The velocity of a traveling wave equals the frequency × the wavelength.
Wavelength = $3.0 \times 10^{10}\ cm\ s^{-1}/8.0 \times 10^{11}\ s^{-1} = 3.8 \times 10^{-2}$ cm

c. Frequency = velocity/wavelength = $3.0 \times 10^{10}\ cm\ s^{-1}/1.2 \times 10^{-12}\ cm = 2.5 \times 10^{22}\ s^{-1}$

7.2) $E = h\upsilon = 6.626 \times 10^{-34}\ J\ s \times 7.0 \times 10^{14}\ s^{-1} = 4.64 \times 10^{-19}$ J
The energy in one mole of photons = 280 kJ.

7.3) $$\Delta E = E_{final} - E_{initial} = -2.18 \times 10^{-18}\ J\left(\frac{1}{n_{final}^2} - \frac{1}{n_{initial}^2}\right) = -2.18 \times 10^{-18}\ J\left(\frac{1}{(2)^2} - \frac{1}{(3)^2}\right)$$

$$= -3.026 \times 10^{-19}\ J$$

To find frequency of energy emitted, we use the relationship, $E = h\upsilon$

$$\upsilon = \frac{E}{h} = \frac{3.026 \times 10^{-19}\ J}{6.626 \times 10^{-34}\ J\ s} = 4.568 \times 10^{14}\ s^{-1}$$

7.4) $\lambda = \dfrac{h}{mv} = 6.626 \times 10^{-34}$ J s/9.1094 $\times 10^{-31}$ kg $\times$ 2.00 $\times 10^{4}$ m/s

(Note the change in units from cm/s to m/s); wavelength, λ, = 3.64 $\times 10^{-8}$ m

7.5) a. The total number of values of m_ℓ is equal to $2\ell + 1$ and is called the orbital degeneracy. If, as in this case, $\ell = 3$, then $2\ell + 1 = 7$, so there are seven different values of m_ℓ for this value of ℓ. $m_\ell = -3, -2, -1, 0, +1, +2, +3$

b. The number of possible values of ℓ is equal to the value of n; in this case there are four possible values of ℓ: 0, 1, 2, and 3.

7.6) a. The secondary or azimuthal quantum number ℓ specifies the shape of an orbital.

b. *p* orbital

c. The total number of nodes is related to the value of the principal quantum number, and therefore to the energy of the orbital. There are two nodes in this orbital—one planar and one spherical. The value of the principal quantum number is therefore 3 (the number of nodes +1). Since there is a single planar node, $\ell = 1$, so the orbital is a 3*p* orbital.

d. As the value of *n* increases, the energy of the orbital increases. For the hydrogen atom, the energy of each principal quantum level is given by Bohr's equation:

$E_n = -2.179 \times 10^{-18}/n^2$ J

7.7) The value of the principal quantum number is 2.

The value of the azimuthal quantum number is 1.

The orbital is one of a set of three orbitals, all of equal energy, and all at right angles to one another.

The orbital has a planar node running through the nucleus.

The orbital has no spherical node.

For atoms other than hydrogen, the orbital is the highest in energy within its principal quantum level; the only other type of orbital in this principal quantum level is an *s* orbital.

The orbital has a capacity for two electrons, each of opposite spin.

7.8) a. In the $n = 4$ principal quantum level there are *s, p, d,* and *f* orbitals, in sets of 1, 3, 5, and 7, respectively. The total number of orbitals in the $n = 4$ equals 1 + 3 + 5 + 7 = 16. The total number of electrons that can be accommodated is 32 (2 in each orbital).

b. A general expression for the total number of electrons that can be accommodated by any principal quantum level *n* of an atom = $2n^2$.

7.9) b: m_ℓ cannot equal −2; it must be −1, 0, or +1; c: *n* cannot equal 0, it must be greater than 0

7.10) c: ℓ cannot equal 2; it must be 0 or 1

Electron Configuration and Chemical Periodicity

INTRODUCTION

In the second half of the 19th century, chemists searched for a way to organize the more than 60 elements known at the time. The first to come up with the idea of a periodic table was John Newlands. He noticed that if the elements were arranged in order of increasing atomic weight in rows of seven, elements with similar properties tended to end up underneath one another. His "Law of Octaves" was not accepted because, although his arrangement worked well for elements through calcium, beyond calcium his scheme fell apart.

Dmitri Mendeleev, who in 1867 was organizing material on the elements for a textbook in chemistry, solved the problem Newlands faced by leaving gaps in the table after Ca for elements that had yet to be discovered. He predicted in detail the properties of three of these undiscovered elements, and 20 years later when the elements had been isolated, the amazing agreement with Mendeleev's predictions was convincing support for his periodic table. The German chemist, Julius Lothar Meyer, arrived at virtually the same organization by showing that physical properties such as molar volume were a periodic function of atomic weight. In 1882, Mendeleev and Meyer were jointly awarded the Davy Medal, the highest honor of the Royal Society.

8.1 MANY-ELECTRON ATOMS

This chapter will discuss the rules for filling orbitals with electrons and how the resulting electron configurations can explain periodic patterns of chemical reactivity.

Approximate solutions to the Schrödinger equation show that the atomic orbitals of many-electron atoms are hydrogen like, so we use the same quantum numbers to describe all atoms. We saw in Chapter 7 that three quantum numbers describe an atom's orbitals:

n (size)

ℓ (shape)

m_ℓ (orientation)

A fourth quantum number, m_s, describes a property of the electron itself. The **spin quantum number,** as it is called, cannot be described classically, but it is associated with the spin of the electron about its own axis.

The spin quantum number (m_s)

Choose words from the list below to complete the paragraph that follows the list:

(a) direction
(b) filled
(c) four
(d) opposite
(e) paired
(f) spins
(g) two

Experiments show that an electron acts as though it _____________(1) on its axis. The spin quantum number (m_s) indicates the _____________(2) of the electron spin and has the value $+\frac{1}{2}$ or $-\frac{1}{2}$.

No two electrons in the same atom can have the same set of _____________(3) quantum numbers **(Pauli exclusion principle).** A consequence of the exclusion principle is that an atomic orbital can hold a maximum of _____________(4) electrons and they must have _____________(5) spins. When there are two electrons in an orbital, the electrons are said to be _____________(6) and the orbital is _____________(7).

answers: 1-f, 2-a, 3-c, 4-g, 5-d, 6-e, 7-b.

We saw in Chapter 7 that orbitals closer to the nucleus are more stable (lower energy) than those farther away from the nucleus. The only electrostatic interactions in a hydrogen atom are those between the proton and the electron. For many-electron systems, we have to consider both proton-electron attractions and electron-electron repulsions to determine the energy level of an orbital.

Electrostatic interactions

Determine for each set of charges below, which configuration has the lower energy with respect to infinite separation:

1)

(+) (-)　　(+2) (-)

(a)　　(b)

2)

(+) (-)　　(+)　　　(-)

(a)　　(b)

3)

(+) (-)　　(-) (-)

(a)　　(b)

(continued on next page)

4)

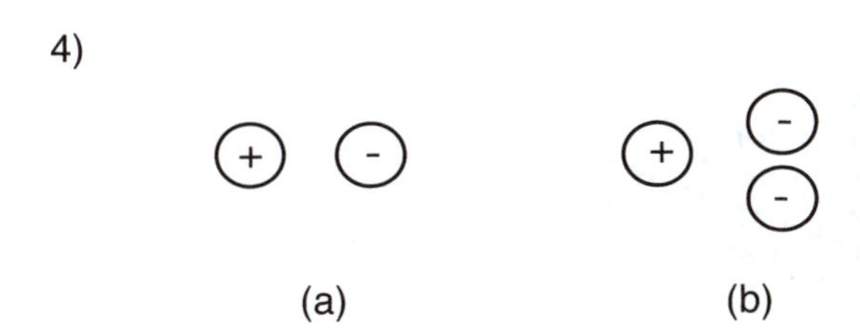

5) Given that the lower the ℓ value of an orbital, the more time the electron spends near the nucleus, order the sublevels from lowest energy to highest (for a given "*n*").

answers:

1-b: A charge of +2 attracts a negative charge more strongly than a charge of +1; stronger attractions make a system lower in energy.

2-a: The energy of a system is lower when opposite charges are closer together.

3-a: Attractions make a system lower in energy than repulsions. Like charges repel each other. Energy must be added in (b) to bring the like charges together from infinite separation.

4-a: The two negative charges repel each other in system (b) raising the energy of the system above that of (a). Configuration (b) may be thought of as configuration (a) with an added negative charge, which being closer to the other negative charge than to the positive charge, raises the energy of the system.

5: $s < p < d < f$
Note: This order does not apply to the hydrogen atom where the energy of sublevels is equal. For many-electron atoms, energy of the 2*s* sublevel is lower than the 2*p* sublevel due to penetration/screening effects.

Summary of Electrostatic Effects on Orbital Energies

- Electron-electron repulsions raise energy.
- Greater nuclear charge lowers orbital energy.
- Electrons in outer orbitals are higher in energy, owing to their greater distance from nuclear charges.
- For a given *n* value, the energy of the sublevels in many-electron atoms is: $s < p < d < f$.

8.2 CONSTRUCTING A PERIODIC TABLE

We can use our knowledge of energy levels of orbitals to write ground-state electron configurations by adding one electron per element to the lowest energy orbital available (the **aufbau principle**). The recurring pattern in electron configurations is the basis for recurring "periodic" patterns in chemical reactivity. A useful diagram for remembering the order of orbital energies is shown on the following page.

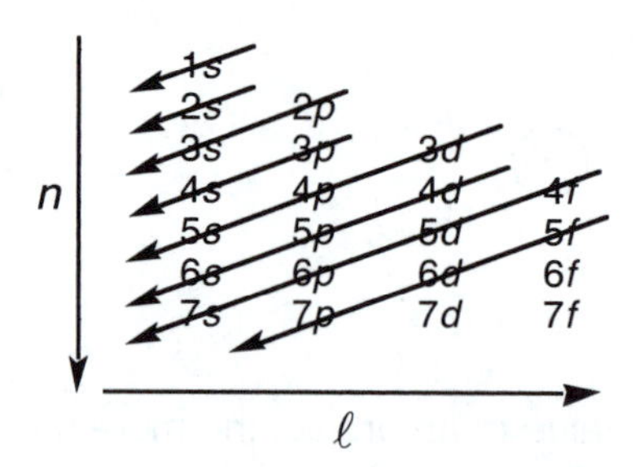

The order of filling is: 1*s* 2*s* 2*p* 3*s* 3*p* 4*s* 3*d* 4*p* 5*s* 4*d* 5*p* 6*s* 4*f* 5*d* 6*p* 7*s*, etc.

Order of filling orbitals

Consider the line connecting 3*d*, 4*p*, and 5*s*. What is the value for $n + \ell$ in each case?

solution:

Since *s*, *p*, *d* stand for $\ell = 0, 1, 2$, respectively:

3*d*: $n = 3$, $\ell = 2$, $n + \ell = 5$

4*p*: $n = 4$, $\ell = 1$, $n + \ell = 5$

5*s*: $n = 5$, $\ell = 0$, $n + \ell = 5$

We can see that the lines in the diagram above connect orbitals that have $n + \ell$ = a constant. For a given $n + \ell$ line, the smallest *n* value fills first.

Rules for Filling Atomic Orbitals

- **Hund's Rule:** When orbitals of equal energy are available, the electron configuration of lowest energy has the maximum number of unpaired electrons with parallel spins.
- **Pauli Exclusion Principle:** No two electrons in an atom can have the same set of four quantum numbers. Another, in some ways stronger statement is: two electrons with the same spin cannot be at the same place at the same time.
- Filled and half-filled sublevels are unusually stable.

Electron configurations show the principal energy level (*n* value), the sublevel letter (ℓ value), and the number of electrons in the sublevel. The electron configuration for a helium atom is $1s^2$.

Orbital diagrams give additional information about paired vs. unpaired electrons. An orbital diagram depicts an orbital with a line (or box) grouping the orbitals by sublevel. Arrows indicate an electron and its direction of spin.

Electron configurations and orbital diagrams

1) Show the electron configurations and orbital diagrams for He, Ne, Ar, and Kr.

solution:

(continued on next page)

He: $1s^2$

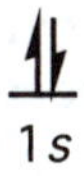

↑↓
1s

Ne: $1s^2 2s^2 2p^6$

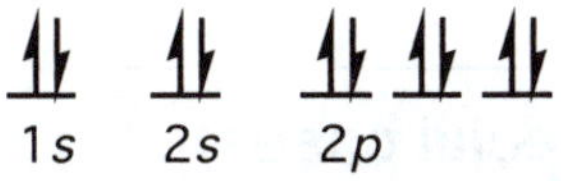

↑↓ ↑↓ ↑↓ ↑↓ ↑↓
1s 2s 2p

Ar: $1s^2 2s^2 2p^6 3s^2 3p^6$

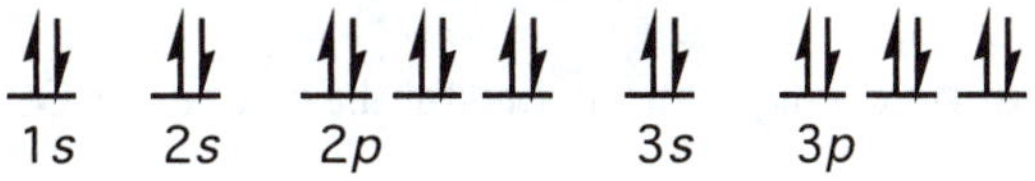

↑↓ ↑↓ ↑↓ ↑↓ ↑↓ ↑↓ ↑↓ ↑↓ ↑↓
1s 2s 2p 3s 3p

Kr: $1s^2 2s^2 2p^6 3s^2 3p^6 4s^2 3d^{10} 4p^6$

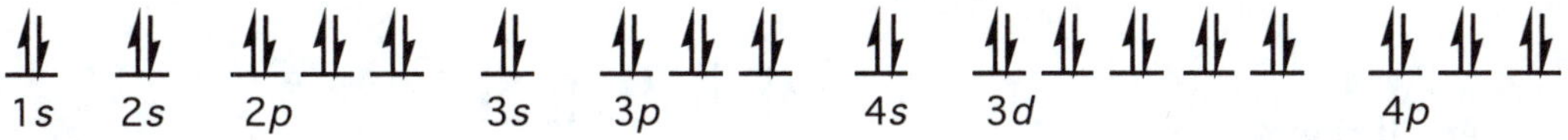

↑↓ ↑↓ ↑↓ ↑↓ ↑↓ ↑↓ ↑↓ ↑↓ ↑↓ ↑↓ ↑↓ ↑↓ ↑↓ ↑↓ ↑↓ ↑↓ ↑↓ ↑↓
1s 2s 2p 3s 3p 4s 3d 4p

2) What do the electron configurations for Ne, Ar, and Kr have in common, and what is different about He?

solution:

Ne, Ar, and Kr have filled "*p*" subshells as their last occupied subshell. He has a filled "*s*" subshell.

3) Why are noble gases so unreactive (stable)?

solution:

Their outer ("valence") subshells are filled (no vacancies), and the next available subshell has a higher principal quantum number.

similar electron configurations = similar chemical properties

4) What are the electron configurations for C and Si in group 4A?

solution:

C: $1s^2 2s^2 2p^2$

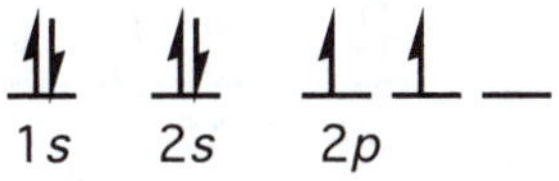

↑↓ ↑↓ ↑ ↑ _
1s 2s 2p

Si: $1s^2 2s^2 2p^6 3s^2 3p^2$

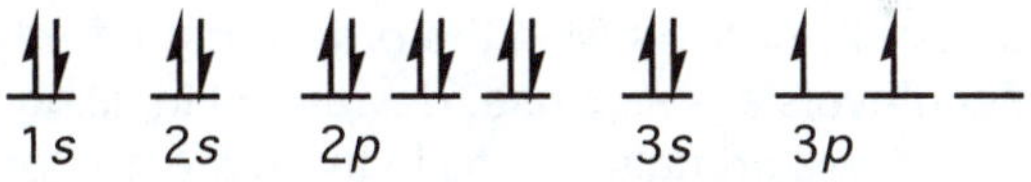

↑↓ ↑↓ ↑↓ ↑↓ ↑↓ ↑↓ ↑ ↑ _
1s 2s 2p 3s 3p

5) Which are the inner shell electrons and which are the valence electrons for Si?

solution:

The inner shell electrons are all those in the 1st and 2nd shells. The valence electrons are those in the 3*s* and 3*p* orbitals.

6) What do the electron configurations for C and Si have in common?

solution:

They each have 4 valence electrons.

7) Use a short-hand method to write the electron configurations for Si that eliminates the writing of all the inner electrons.

solution:

[Ne]$3s^2 3p^2$

There are exceptions (see below) to the order of filling orbitals. These deviations arise because half-filled and filled shells are unexpectedly stable, and because the energies of the *ns* and $(n-1)d$ sublevels are so close.

Electron configurations (special cases)

1) Show the electron configurations orbital diagrams for Cu, Ag, and Au:

Cu: $[Ar]4s^13d^{10}$ — ↑ (4*s*) ↑↓ ↑↓ ↑↓ ↑↓ ↑↓ (3*d*)

Ag: $[Kr]5s^14d^{10}$ — ↑ (5*s*) ↑↓ ↑↓ ↑↓ ↑↓ ↑↓ (4*d*)

Au: $[Xe]6s^14f^{14}5d^{10}$ — ↑ (6*s*) ↑↓ ↑↓ ↑↓ ↑↓ ↑↓ ↑↓ ↑↓ (4*f*) ↑↓ ↑↓ ↑↓ ↑↓ ↑↓ (5*d*)

Note: Half-filled and filled shells are unexpectedly stable.

2) Show the orbital diagrams for Cr, Mo, and W:

Cr: $[Ar]4s^13d^5$ — ↑ (4*s*) ↑ ↑ ↑ ↑ ↑ (3*d*)

Mo: $[Kr]5s^14d^5$ — ↑ (5*s*) ↑ ↑ ↑ ↑ ↑ (4*d*)

W: $[Xe]6s^24f^{14}5d^4$ — ↑↓ (6*s*) ↑↓ ↑↓ ↑↓ ↑↓ ↑↓ ↑↓ ↑↓ (4*f*) ↑ ↑ ↑ ↑ __ (5*d*)

Note: Generally orbitals fill in the order: *ns*, $(n-1)d$, *np*. However, the energies of the *ns* and $(n-1)d$ sublevels are so close, deviations in filling patterns occur. Notice also that W, unlike Cr and Mo, contains orbitals that fill in the expected order.

We may use our plot of *n* vs. ℓ (below, left) to construct a periodic table, in two steps. First, flip the diagram, horizontally, to obtain its mirror image (below, right):

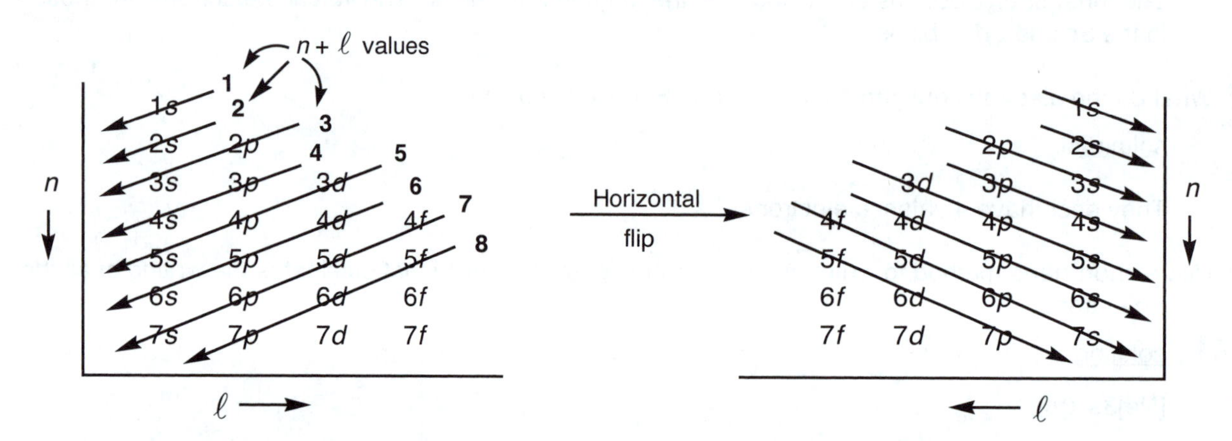

Then, second, tilt the diagram to make the lines of constant $n + \ell$ values horizontal:

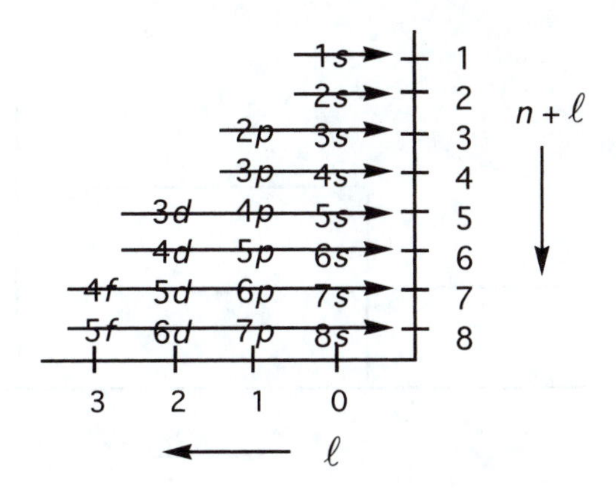

Obtained is a compressed periodic table. Columns in the compressed periodic table correspond to "blocks" of families of elements (called the *s*-, *p*-, *d*-, or *f*-blocks). The numbers of families in the blocks—the blocks' horizontal lengths—are given by the expression $2(2\ell + 1)$, as explained below:

m_ℓ values: $-\ell, \ldots\ldots, -1,\ 0,\ +1, \ldots\ldots, +\ell$

($-\ell$ … -1: ℓ values; 0: 1 value; $+1$ … $+\ell$: ℓ values; altogether $2\ell + 1$ values)

Each ℓ value corresponds to an orbital. 2 electrons may occupy each orbital.

Summary of ℓ values and block lengths

ℓ symbol	value	number of orbitals $2\ell + 1$	max. no. of e^- in subshell $2(2\ell + 1)$
s	0	1	2
p	1	3	6
d	2	5	10
f	3	7	14

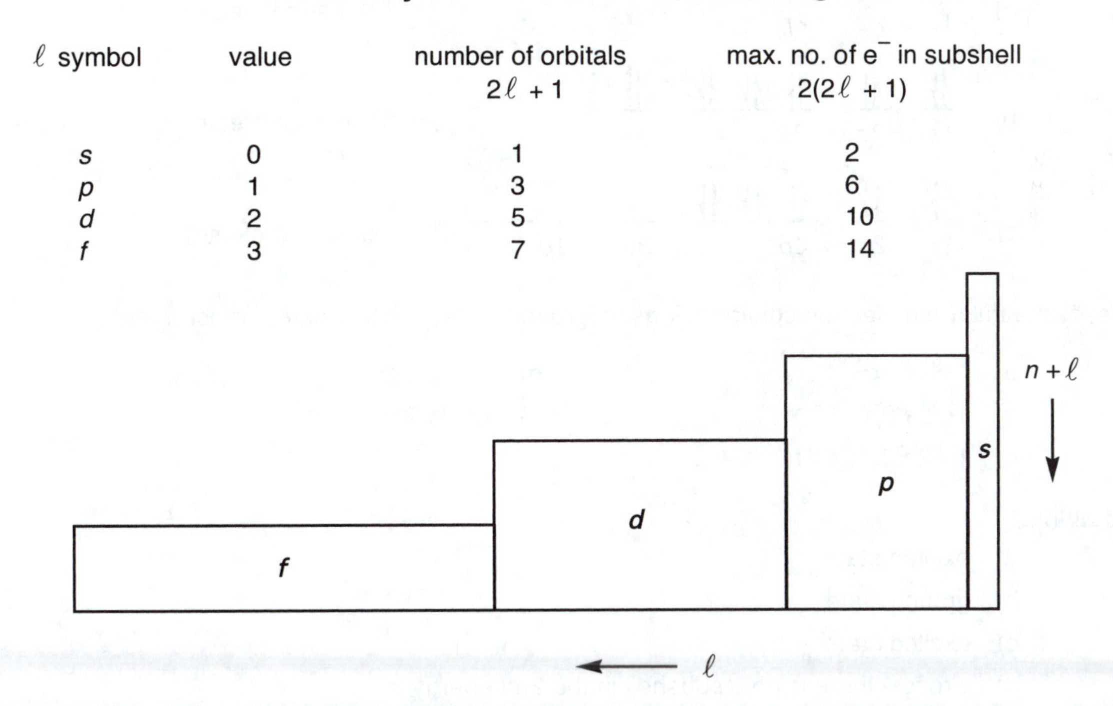

Usually the *s*-block is moved to the left to place all the metals together on the left side of the table. The *s*-block is shifted down one element, and, He, although an ns^2 family member, is generally placed in the np^6 family above Ne.

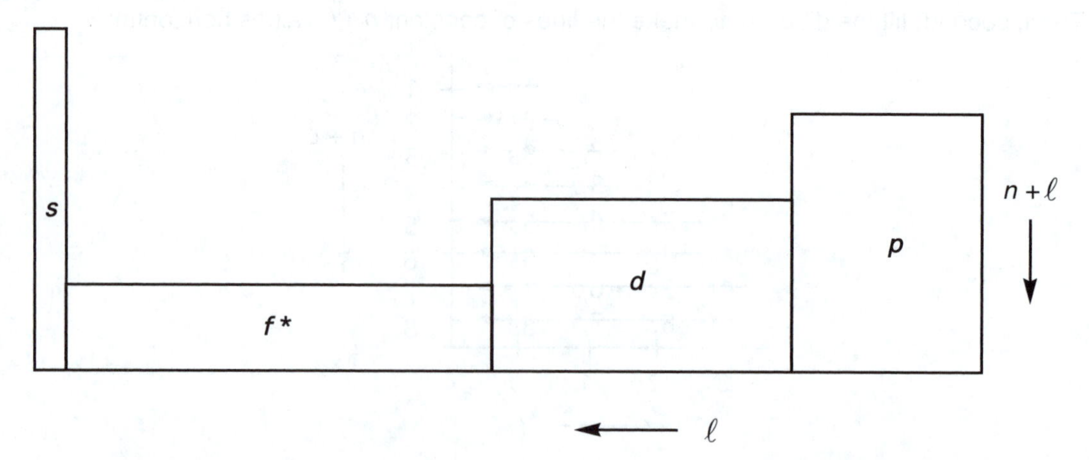

* This block is usually removed and placed below the table, to save space, horizontally.

Electron configuration practice problems

1) Draw orbital diagrams for atoms with the following electron configurations. How many unpaired electrons does each atom have?

   a) $1s^22s^22p$

   b) $1s^22s^22p^63s^2$

   c) $1s^22s^22p^63s^23p^4$

   solutions:

   a) ↑↓ (1s) ↑↓ (2s) ↑ _ _ (2p) — 1 unpaired electron

   b) ↑↓ (1s) ↑↓ (2s) ↑↓ ↑↓ ↑↓ (2p) ↑↓ (3s) — 0 unpaired electrons

   c) ↑↓ (1s) ↑↓ (2s) ↑↓ ↑↓ ↑↓ (2p) ↑↓ (3s) ↑ ↑ _ (3p) — 2 unpaired electrons

2) Classify the following electron configurations as ground state, excited state, or incorrect:

   a) $1s^22s^12p^2$

   b) $1s^22s^22p^6$

   c) $1s^22s^22p^63s^14f$

   d) $1s^22s^12d^1$

   e) $1s^32s^22p^6$

   solutions:

   a) excited state

   b) ground state

   c) excited state

   d) wrong—there is no *d* subshell in the 2nd energy level

   e) wrong—there cannot be 3 electrons in an "*s*" orbital (or any orbital).

(continued on next page)

3) What is the maximum number of electrons in an atom that can have the following quantum numbers:

a) $n = 3$

b) $n = 3$, $m_s = +\frac{1}{2}$

c) $n = 3$, $\ell = 2$

d) $n = 3$, $\ell = 2$, $m_\ell = -2$

e) $n = 3$, $\ell = 2$, $m_\ell = -2$, $m_s = +\frac{1}{2}$

f) $n = 4$, $m_\ell = 1$

solutions:

a) 18 electrons:

The 3rd shell contains the following orbitals:

$$3s,\ 3p_x,\ 3p_y,\ 3p_z,\ 3d_{xy},\ 3d_{xz},\ 3d_{yz},\ 3d_{x^2-y^2},\ 3d_{z^2}$$

each of which can hold 2 electrons.

b) 9 electrons: Half of the electrons in the 3rd shell orbitals have a spin of $+\frac{1}{2}$, and half have a spin of $m_s = -\frac{1}{2}$.

c) 10 electrons: The five orbitals with $\ell = 2$ (the 5 "*d*" sublevels), can hold 10 electrons.

d) 2 electrons: The m_ℓ value (–2) defines one orbital (a specific orientation of a "*d*" orbital), and 2 electrons may exist in one orbital.

e) 1 electron: The four quantum numbers define a unique electron state in an atom.

f) 6 electrons: When $n = 4$, ℓ may equal 0, 1, 2, or 3. For the three sublevels $\ell = 1$, 2, or 3, m_ℓ may have the value 1, so there are 3 orbitals (which can hold 6 electrons) in the 4th shell that contain an m_ℓ value of 1.

4) What is wrong with the following sets of quantum numbers?

a) $n = 1$, $\ell = 1$, $m_\ell = 0$, $m_s = +\frac{1}{2}$

b) $n = 1$, $\ell = 0$, $m_\ell = 1$, $m_s = +\frac{1}{2}$

c) $n = 5$, $\ell = 2$, $m_\ell = -3$, $m_s = -\frac{1}{2}$

d) $n = 3$, $\ell = 0$, $m_\ell = -1$, $m_s = -\frac{1}{2}$

solutions:

a) For the first shell, ℓ has only one value (0 = "*s*" orbital).

b) $m_\ell = -\ell, ...0,...+\ell$, so when $\ell = 0$, m_ℓ can only equal 0. (This one value for m_ℓ indicates the *s* orbital's one orientation in space.)

c) For $\ell = 2$, m_ℓ can equal –2, –1, 0, 1, 2 (not –3).

d) For $\ell = 0$, m_ℓ can equal only 0.

Experimental support for electron configurations

1) What is the electron configuration for the sodium atom, Na?

   answer: $1s^2 2s^2 2p^6 3s^1$.

2) Which electron(s) would you predict from theory (its electron configuration) would be most easily removed from a sodium atom?

   answer:

   One would expect that electrons farthest away from the positively charged nucleus would be the easiest to remove. The one electron in the outermost 3rd energy level of a sodium atom should be substantially easier to remove than the electrons closer to the nucleus in the 1st and 2nd energy levels.

   In fact, experiments show that the ionization energies (the energy required to remove electrons from a gaseous atom) for a sodium atom fall into three groups with one very low value (the 3*s* electron), eight which gradually increase in magnitude (the 8 electrons for which $n = 2$), and two very large value (the 2 electrons in the 1*s* shell which are bound very tightly to the nucleus).

8.3 TRENDS IN THREE ATOMIC PROPERTIES

Atomic size and electron configuration influence ionization energy (the energy required to remove an electron from a gaseous atom), and electron affinity (the energy involved when an electron adds to an atom). The size of an atom is the main factor that determines the lengths of covalent bonds; bond lengths influence bond strength, and bond strength determines chemical reactivity (the breaking and forming of bonds).

How do we determine an atom's size? We assume that atoms closest to each other in elementary substances are touching.

Atomic size

What is the atomic radius of a cesium atom and a fluorine atom based on the following data?

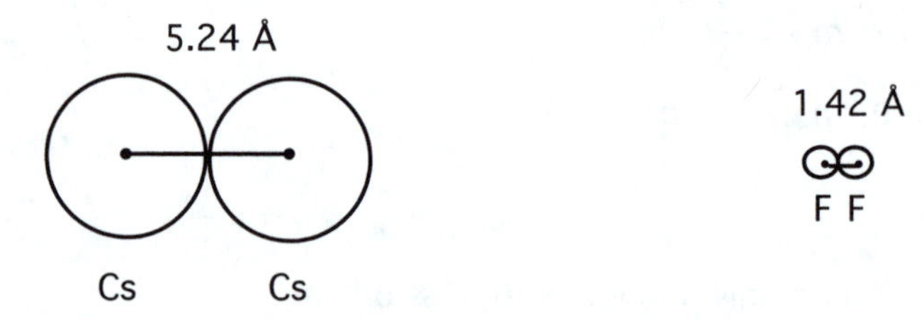

solution:

radius of Cs = 1/2 diameter = 5.24/2 = 2.62 Å

radius of F = 1/2 diameter = 1.42/2 = 0.71 Å

Keep in mind that because an electron cloud is not rigid, the size of an atom will depend to some extent on the nature of the neighboring atom(s).

Periodic Trends in Atomic Size

To determine the size of atoms, we must consider both the principal quantum number of its valence electrons, and the effective nuclear charge on those electrons.

Atomic size trends

1) Down a group
Consider the first three elements of group 1A (the alkali metals), H, Li, and Na:

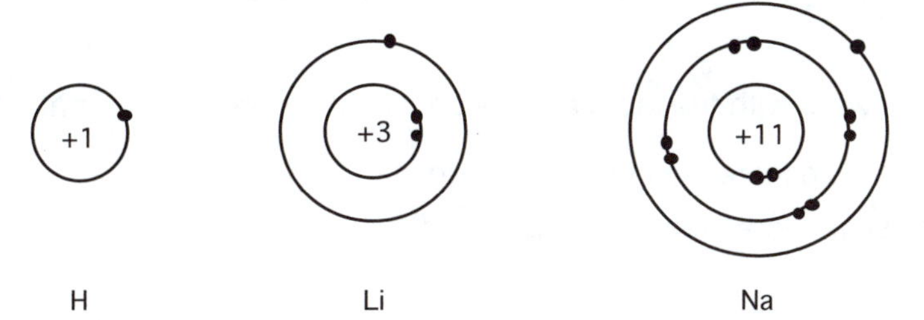

What is the effective nuclear charge (Z_{eff}) on the valence electron for each atom, and how does atomic radius change as we move down the group?

solution:

Electrons in inner shells "shield" outer shell electrons from the positive charge of the nucleus. To a very rough first approximation, each inner shell electron cancels the charge of one proton in the nucleus, so the outer "*s*" electron for all elements in group 1A is attracted by the same nuclear charge (≈ +1).

The average distance of an electron from a nucleus of constant charge increases with the principal quantum number, so **atomic radius increases as we move down a group in the periodic table.**

2) Across a period
Consider the first three elements of the 3rd period (Na, Mg, and Al):

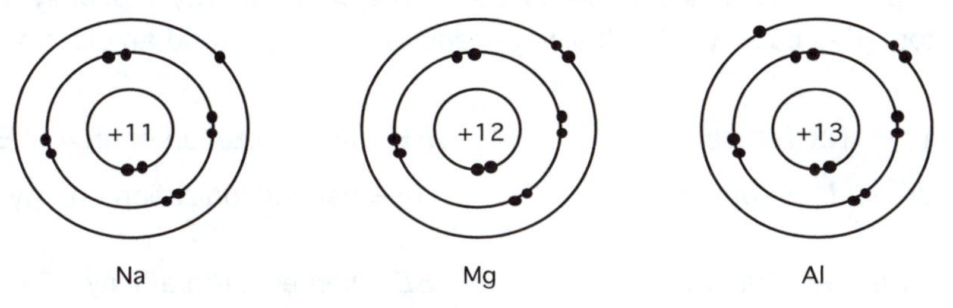

What is the approximate effective nuclear charge (Z_{eff}) on the outer electrons for each atom, and how does atomic radius change as we move across the period?

solution:

Electrons that are approximately the same distance from the nucleus are not effective at shielding one another from the nuclear charge. Therefore, as we move across a period, as each additional proton is added to the nucleus, the effective charge on the outer electrons increases by approximately one. The increasing effective nuclear charge pulls outer electrons in more tightly so **atomic radius decreases as we move across a period.**

Note: The size trends hold most regularly for the main-group elements; minor variations occur in the transition elements.

Atomic size exercises

1) Arrange the following groups of atoms in order of increasing size:

a) Mg, Ba, Ca
b) Ca, Br, Kr
c) C, O, F
d) B, Al, Ga

solutions:
(Compare your answers with the atomic sizes shown in Figure 8.13 on page 337 in your textbook.)

a) Mg < Ca < Ba (size increases down a group)
b) Kr < Br < Ca (size decreases across a period)
c) F < O < C
c) B < Ga < Al

(Size normally increases down a group, so you might expect gallium to be larger than aluminum. However, the effective nuclear charge is large for gallium, see the next problem.)

2) Write the electron configuration for Ga. Use the configuration to explain the high Z_{eff} felt by gallium's 4*p* electron.

solution: $1s^22s^22p^63s^23p^64s^23d^{10}4p^1$

The 3*d* electrons are not effective at shielding the nuclear charge from the 4*p* electrons, so the 4*p* electron in gallium feels an increased Z_{eff} from all the protons added during the transition series.

Ionization Energy and Electron Affinity

It requires energy to *remove* electrons from atoms ($IE_1 > 0$); energy is usually released when an electron *adds* to an atom (EA_1 usually < 0). It always requires energy to add an electron to an anion ($EA_2 > 0$).

$Na(g) \rightarrow Na^+(g) + e^-$ — ΔE = first ionization energy = $IE_1 > 0$

$Na^+(g) \rightarrow Na^{2+}(g) + e^-$ — ΔE = second ionization energy = $IE_2 > IE_1$

$Na(g) + e^- \rightarrow Na^-(g)$ — ΔE = first electron affinity = EA_1 usually < 0

$Na^-(g) + e^- \rightarrow Na^{2-}(g)$ — ΔE = second electron affinity = $EA_2 > 0$

Trends in ionization energy

In general, the closer an electron is to the nucleus the more difficult it is to remove, so **ionization energy decreases down a group and increases across a period.**

1) Arrange the following groups of atoms in order of increasing first ionization energy:

a) Mg, Ba, Ca
b) Ca, Br, Kr
c) C, O, F
d) B, Al, Ga

(continued on next page)

solutions: (compare these to the solutions for atomic radii in the previous exercise)

a) Ba < Ca < Mg
b) Ca < Br < Kr
c) C < O < F
d) Al < Ga < B

2) What element in period three would have the following series of ionization energies (kJ/mol)?

580	1815	2740	11,600
IE_1	IE_2	IE_3	IE_4

solution:

The large jump in ionization energy occurs when all the valence electrons have been removed from a subshell. For this atom, the jump occurs after three electrons have been removed, so the element is aluminum with 3 valence electrons: $1s^22s^22p^6\mathbf{3s^23p^1}$. Compare aluminum's ionization energies with boron (electron configuration $1s^2\mathbf{2s^22p^1}$) from Table 8.5 on page 342 in your textbook.

Trends in Electron Affinity

1) Elements in groups 6A and 7A (the halogens) attract electrons strongly and have highly negative (exothermic) electron affinities. In ionic compounds, they form negative ions.

2) Elements in groups 1A and 2A lose electrons readily and do not attract them, and have positive (endothermic), or slightly negative, electron affinities. In ionic compounds, they form positive ions.

3) The noble gases (group 8A) in general neither lose nor gain electrons and have positive (endothermic) electron affinities.

8.4 ATOMIC PROPERTIES AND CHEMICAL REACTIVITY

Large atoms with low ionization energies exhibit more metallic behavior than small atoms with high ionization energies. Therefore, metallic behavior increases as you go down a group (as the atoms get larger), and decreases across a period (as the atoms get smaller).

Atomic size, ionization energy, and metallic behavior

1) Indicate whether each of the trends listed increases or decreases as you move down, and from right to left across the periodic table:

a) metallic behavior
b) atomic size
c) ionization energy

(continued on next page)

answers:

a) metallic behavior increases, b) atomic size increases, c) ionization energy decreases

2) Given that metallic character increases down a group and decreases across a period, draw a line along which metallic character is approximately constant.

solution:

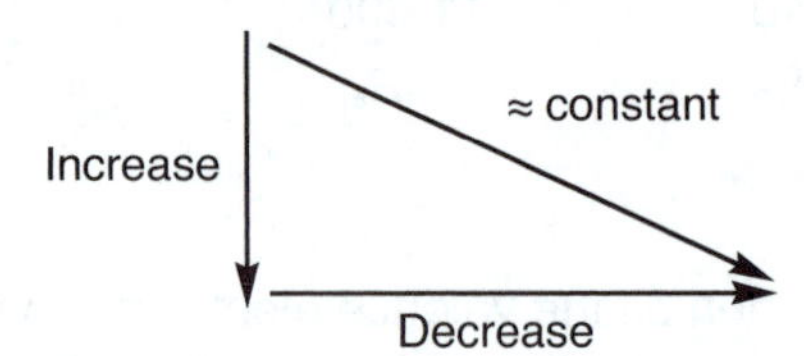

Notice the similar diagonal line on the periodic table that separates metals from nonmetals.

Element oxides

Consider the reactions of a metal oxide, CaO, and a nonmetal oxide, NO_2 with water:

$$CaO(s) + H_2O \rightarrow Ca(OH)_2(aq)$$

$$3NO_2(g) + H_2O \rightarrow 2HNO_3(aq) + NO(g)$$

Which of the oxides formed an acid and which formed a base?

answer:

CaO acts as a base by removing protons from water to produce OH^- ions. NO_2 reacts with water to form nitric acid (HNO_3). As metallic behavior increases down a group, element oxides become more basic. As metallic behavior decreases across a period, element oxides become more acidic.

Monatomic Ions

Main-group atoms gain or lose electrons to attain noble gas configurations (filled outer *p* subshells). It often requires too much energy to remove all the electrons necessary for transition metals to attain noble gas configurations, so they often attain a **pseudo-noble gas configuration,** in which valence "*s*" electrons are lost and a filled *d* shell remains.

Although the 4*s* orbitals are lower in energy than the 3*d* orbitals when they are filled by the aufbau process, once the orbitals are filled with electrons, the 3*d* orbital becomes lower in energy than the 4*s*. The 4*s* electrons are lost first in the formation of period 4 transition metal ions. Except for the lanthanides, electrons with the highest *n* value are always lost first.

Electron configuration of ions

1) Predict the ions that Na, Mg, and Al will form and give their electron configurations.

answers: Na^{1+}: $1s^22s^22p^6$, Mg^{2+}: $1s^22s^22p^6$, Al^{3+}: $1s^22s^22p^6$.

2) Which noble gas is **isoelectronic** (the same electron configuration) with the above ions?

answer: Ne.

(continued on next page)

3) Arrange the above ions and Ne in order of increasing atomic size.

answer: $Al^{3+} < Mg^{2+} < Na^{1+} < Ne$

Since the electron configurations are the same for all four species, it is the nuclear charge that determines the atomic size. The highest nuclear charge of aluminum (+13) pulls the electrons in the most tightly followed by magnesium (+12), sodium (+11), and neon (+10).

4) Write the electron configuration for the transition metal, Zn. What would you predict would be the ion it forms in compounds?

answer: $[Ar]4s^2 3d^{10}$

Zinc loses its two 4*s* electrons to form Zn^{2+} with an electron configuration of $[Ar]3d^{10}$. This configuration is called a pseudo-noble gas configuration.

5) Write the electron configurations thallium (Tl), lead (Pb), and bismuth (Bi) and predict the ions they would likely form in compounds.

answers:

Tl: $[Xe]6s^2 4f^{14} 5d^{10} 6p$

Pb: $[Xe]6s^2 4f^{14} 5d^{10} 6p^2$

Bi: $[Xe]6s^2 4f^{14} 5d^{10} 6p^3$

To obtain pseudo-noble gas configurations, Tl, Pb, and Bi would need to lose their 6*s* and 6*p* electrons forming: Tl^{3+}, Pb^{4+}, and Bi^{5+}. These ions do in fact form, but the most common ions formed by these elements are the ones which retain the two "*s*" electrons (sometimes called an **inert pair**) and lose the "*p*" electrons: Tl^{1+}, Pb^{2+}, Bi^{3+}.

Magnetic Properties

The magnetic properties of transition metals and their compounds can support or refute a given electron configuration. Elements or compounds with all paired electrons are **diamagnetic** and are not attracted by a magnetic field. Elements or compounds with unpaired electrons are **paramagnetic,** and are attracted by an external magnetic field.

Predicting magnetic properties

Write the electron configuration for each of the following transition metals and predict if it would be paramagnetic or diamagnetic:

a) Fe^{3+} ($Z = 26$) b) Fe^{2+} c) Cu^{2+} ($Z = 29$) d) Cu^{1+}

answers:

a) Fe: $[Ar]4s^2 3d^6$; Fe^{3+}: $[Ar]3d^5$

There is one electron in each of the five *d* orbitals, so we would predict that Fe^{3+} is paramagnetic with five unpaired electrons.

b) Fe: $[Ar]4s^2 3d^6$; Fe^{2+}: $[Ar]3d^6$

Four of the *d* orbitals contain one unpaired electron; the remaining *d* orbital contains two paired electrons. Fe^{2+} is paramagnetic with four unpaired electrons.

(continued on next page)

c) Cu: [Ar]$4s^1 3d^{10}$; Cu^{2+}: [Ar]$3d^9$

Four of the *d* orbitals contain paired electrons; the remaining *d* orbital contains an unpaired electron. Cu^{2+} is paramagnetic with 1 unpaired electron.

d) Cu^{1+}: [Ar]$3d^{10}$

All five *d* orbitals contain paired electrons; Cu^{1+} is diamagnetic.

A word of caution: As you will see in Chapter 23, the magnetic properties of transition metal complexes depend both on the electron configuration of the metal, and the attached ligands. For example, Fe^{2+} with attached water molecules is paramagnetic with four unpaired electrons, but Fe^{2+} with attached CN^- molecules is diamagnetic with all electrons paired.

Ionic size

Choose words from the list below to complete the following paragraph:

(a) anion
(b) decrease
(c) increase(s)
(d) larger
(e) more
(f) proton
(g) smaller

Cations, since they have fewer electrons than their parent atoms, are _______________(1) than their parent atoms; anions which have more electrons than their parent atoms are _______________(2) than their parent atoms. Consider the hydrogen atom. If it loses its electron to form the cation, H^+, all its orbitals are empty and it is only as large as a _______________(3). If it gains an electron into its 1*s* orbital to form the anion, H^-, the electron-electron repulsions cause the electrons to occupy _______________(4) space. Ionic size _______________(5) down a group just as does atomic size. Across a period, we move from cations of metals to anions of nonmetals. There is a consistent _______________(6) in ionic size among the cations, followed by a large _______________(7) in size when we reach the anions, and then another _______________(8) in size among the anions.

<u>answers</u>: 1-g, 2-d, 3-f, 4-e, 5-c, 6-b, 7-c, 8-b.

CHAPTER CHECK

Make sure you can...

- Explain why the periodic table is "periodic."
- Summarize electrostatic effects on orbital energies.
- Apply Hund's rule and the Pauli exclusion principle to orbital filling.
- Write electron configurations and orbital diagrams for elements (full and condensed).
- Relate electron configurations to atomic size, ionic size, ionization energies, and magnetic properties.
- Explain effective nuclear charge.
- Identify an element from its successive ionization energies.
- Explain how trends in metallic character relate to ion formation and oxide acidity.

Chapter 8 Exercises

8.1) a. Electrons in which type of atomic orbital are best at shielding other electrons from the nuclear charge?

b. Electrons in which type of atomic orbital (*s*, *p*, *d*, or *f*) are the most shielded by electrons in other orbitals from the nuclear charge?

c. Electrons in which type of orbital penetrate nearest to the nucleus?

d. In a many-electron atom, electrons in which type of orbital within a single principal quantum level experience the lowest effective nuclear charge?

8.2) The energy required to excite the 2*s* electron of a lithium atom to a 2*p* orbital is far greater than the energy required to move an electron in the 2*s* orbital of the Li^{2+} ion to a 2*p* orbital. Explain why.

8.3) What are the electron configurations of:

a. germanium b. krypton c. zinc d. iridium

8.4) State, in your own words, the meaning of the aufbau principle.

8.5) Write out the electron configuration and give the set of four quantum numbers for a valence electron for each of the following elements or ions.

a. Sb b. Br^- c. Al d. Fe^{2+} e. Sm

8.6) According to Hund's rule of maximum multiplicity, how many singly-occupied orbitals are there in the valence shell of the ground state of:

a. a carbon atom b. a nitrogen atom c. an iron atom d. a sulfur atom

8.7) Pair each element with its classification:

a. chromium
b. strontium
c. bromine
d. neon
e. sulfur
f. lithium
g. lead
h. gadolinium

1. noble gas
2. representative metal
3. transition metal
4. alkali metal
5. halogen
6. nonmetal
7. lanthanide
8. alkaline earth metal

8.8) How many valence electrons do the following elements have in their valence shells?

a. sulfur
b. argon
c. magnesium
d. silicon
e. chlorine
f. aluminum

8.9) Which of the sixteen representative elements in periods 2 and 3 of the periodic table have:

a. the smallest atomic radius
b. the highest ionization energy
c. the highest electron affinity
d. the most metallic character

8.10) Which of the elements in the third period of the periodic table has a relatively low first ionization energy, a higher second ionization energy, but a very high third ionization energy?

8.11) The first ionization energy for the second period elements are given below:

Li	Be	B	C	N	O	F	Ne
0.52	0.90	0.80	1.09	1.40	1.31	1.68	2.08

a. Explain why the first ionization energy generally increases from left to right.
b. Explain the variations of the trend at the positions of B and O.

8.12) Which is the most probable ion produced by the removal of valence electron(s) from, or the addition of electrons to:

a. an aluminum atom
b. a phosphorus atom
c. a chlorine atom
d. a calcium atom
e. a sulfur atom

8.13) Which is the largest ion?

sodium Na^+ rubidium Rb^+ chloride Cl^- iodide I^-

Chapter 8 Answers

8.1) a. *s* orbital; b. *f* orbital; c. *s* orbital; d. *f* orbital

8.2) In a lithium atom the 2*s* and 2*p* orbitals are separated in energy because of the shielding effect. The 2*s* orbital is pulled down in energy because it experiences a greater effective nuclear charge. Therefore it requires considerable energy to excite an electron from the 2*s* to the 2*p* level.

In the Li^{2+} ion, there is only one electron and there cannot be any shielding. As a result the 2*s* and 2*p* orbitals are approximately equal in energy and it is easy to move an electron between them.

8.3) a. Ge: $1s^22s^22p^63s^23p^64s^23d^{10}4p^2$
b. Kr: $1s^22s^22p^63s^23p^64s^23d^{10}4p^6$
c. Zn: $1s^22s^22p^63s^23p^64s^23d^{10}$
d. Ir: $1s^22s^22p^63s^23p^64s^23d^{10}4p^65s^24d^{10}5p^66s^24f^{14}5d^7$

8.4) The aufbau principle states that the atomic orbitals of an atom fill in order of increasing energy. The lower energy orbitals fill first. The aufbau principle allows the easy determination of the most probable ground-state electron configuration of an atom. There are a few instances where Hund's rule of maximum multiplicity takes precedence, for example, in copper, where the ground-state configuration is $[Ar]4s^13d^{10}$ not $[Ar]4s^23d^9$.

One way to determine the order of filling orbitals is to use the periodic table, horizontally row by row. Another method is to use the "$n + \ell$ rule" which states that the atomic orbitals of an atom fill in order of increasing "$n + \ell$" value, and if "$n + \ell$" is the same for two orbitals, the orbital with the lowest *n* value fills first. For example, 3*d* is filled before 4*p*.

8.5) a. $1s^22s^22p^63s^23p^64s^23d^{10}4p^65s^24d^{10}5p^3$ $n = 5$ $\ell = 1$ $m_\ell = +1$ $m_s = +\frac{1}{2}$
b. $1s^22s^22p^63s^23p^64s^23d^{10}4p^6$ $n = 4$ $\ell = 1$ $m_\ell = +1$ $m_s = -\frac{1}{2}$
c. $1s^22s^22p^63s^23p^1$ $n = 3$ $\ell = 1$ $m_\ell = -1$ $m_s = +\frac{1}{2}$
d. $1s^22s^22p^63s^23p^64s^23d^4$ $n = 3$ $\ell = 2$ $m_\ell = +2$ $m_s = +\frac{1}{2}$
e. $1s^22s^22p^63s^23p^64s^23d^{10}4p^65s^24d^{10}5p^66s^24f^6$ $n = 4$ $\ell = 3$ $m_\ell = +2$ $m_s = +\frac{1}{2}$

8.6) a. two; valence shell configuration is $2s^22p^2$
b. three; valence shell configuration is $2s^22p^3$

c. four; valence shell configuration is $4s^23d^6$
d. two; valence shell configuration is $3s^23p^4$

8.7) a-3; b-8; c-5; d-1; e-6; f-4; g-2; h-7
Some elements fit into more than one category.

8.8)
a. 6 valence electrons
b. 8 valence electrons
c. 2 valence electrons
d. 4 valence electrons
e. 7 valence electrons
f. 3 valence electrons

8.9)
a. Atomic size decreases left to right, due to the increase in the nuclear charge, and increases top to bottom, due to the accumulation of more and more principal quantum levels. Neon is the smallest of the sixteen elements.

b. Neon also has the highest ionization energy; it is difficult to remove a tightly bound electron from a small atom.

c. Chorine has the highest electron affinity. Fluorine would have been the logical choice, but the fluorine atom is very small and addition of another electron is inhibited by a large electron repulsion in the crowded space around the nucleus.

d. The most metallic element is the one furthest to the left and lowest in the periodic table: sodium in this example.

8.10) The very high third ionization indicates that this ionization breaks up a stable "noble gas configuration." This situation would occur in an atom that originally has two valence electrons. The element is magnesium.

8.11)
a. The increase in nuclear charge moving from left to right across a period results in smaller and smaller atoms and therefore an increase in the energy needed to remove an electron from the valence shell.

b. Electrons gain additional energy stabilization when an energy shell or subshell is filled or half-filled. Be has a higher ionization energy because an electron must be removed from a filled *s* subshell. N has a higher ionization energy because an electron must be removed from a half-filled *p* subshell.

8.12) Sufficient electrons should be added or removed to achieve a valence shell configuration equivalent to that of a noble gas.

a. removal of three electrons: Al^{3+}
b. addition of three electrons: P^{3-}
c. addition of one electron: Cl^-
d. removal of two electrons: Ca^{2+}
e. addition of two electrons: S^{2-}

8.13) Although for atoms there is a general decrease in size from left to right across the representative elements in the periodic table, there is no such trend for ions because the charges invariably differ. Monatomic ions on the left side typically have positive charges whereas monatomic ions on the right have negative charges.

For example, which is larger, rubidium Rb^+ or iodine I^-? It is sometimes convenient to consider isoelectronic series around the end of one period to the beginning of the next. In this case, consider: Se^{2-} Br^- Kr Rb^+ Sr^{2+}. The nuclear charge increases, but the number of electrons remains the same, so the ions progressively get smaller and smaller through the series. Rb^+ is smaller than Br^-. Since Br^- is smaller than I^-, then Rb^+ must be smaller than I^-. I^- is the largest ion, Na^+ is the smallest.

Models of Chemical Bonding

9.1 CHEMICAL BONDS

Atoms bond together to obtain stable (filled shell) electron configurations. Electrical conductivity is the best guide to determine the type of bonding. We usually classify chemical bonds in one of three ways:

"Ionic"	Conducts in the molten state by motion of ions
"Covalent"	Non-conductor
"Metallic"	Conducts in the solid state by electron flow

We can picture bonds as valence electrons attracted by atomic "cores." A "core" or "kernel" is an atom without its valence electrons.

1) Ionic (electrons held tightly by small core):

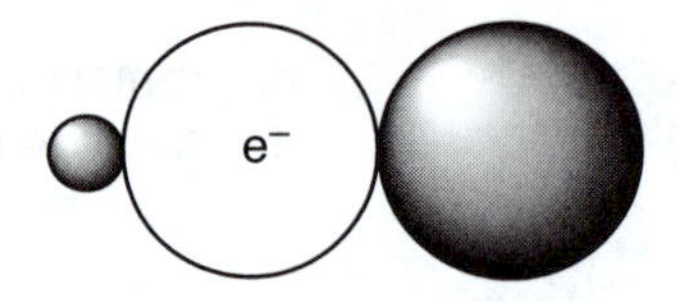

examples: Cl^{7+} and Na^{+}

2) Covalent (tightly held electrons by small cores):

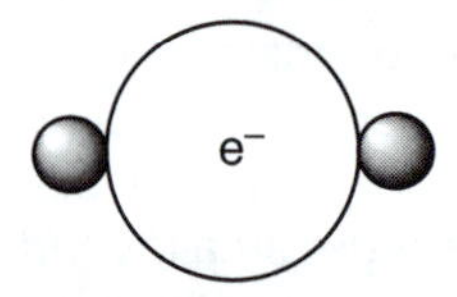

examples: F^{7+}, C^{4+}

3) Metallic (loosely held electrons by large cores):

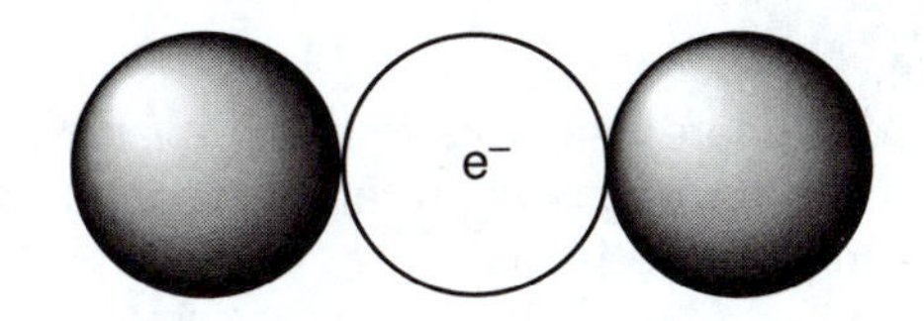

examples: Ca^{2+}, Na^+

Ionic, covalent, and metallic bonding

Classify the following as typical of ionic bonding (i), covalent bonding (c), or metallic bonding (m):

1) Nonmetals share a localized pair of electrons between atoms.
2) Valence electrons from many atoms are pooled and delocalized to form a sea of electrons.
3) Reactive metals in group 1A or 2A transfer electrons to reactive nonmetals in group 7A (or the top of group 6A).
4) The electrostatic attraction between positive and negatively charged particles causes a rigid three-dimensional array of atoms to form.
5) Exist as separate molecules with relatively weak intermolecular forces.
6) The chemical formula represents the cation to anion ratio in the compound (empirical formula).
7) The chemical formula reflects the number of atoms in the molecule (molecular formula).
8) The bonding in H_2O.
9) The bonding in NaCl.
10) The bonding in diamond.
11) The bonding in solid silver.

answers: 1-c, 2-m, 3-i, 4-i, 5-c, 6-i, 7-c, 8-c, 9-i, 10-c, 11-m.

Warm-up to the octet rule

1) What is the electron configuration for sodium?

   answer: $[Ne]3s^1$.

2) What is the electron configuration for chlorine?

   answer: $[Ne]3s^2sp^5$.

3) Describe the easiest way for sodium and chlorine to obtain noble gas configurations.

   answer:

   Sodium loses one electron to obtain the electron configuration of the noble gas neon. Chlorine gains one electron to obtain the electron configuration of the noble gas argon. Na^+ and Cl^- both have noble gas electron configurations.

The Octet Rule

Atoms lose, gain, or share electrons to attain a filled outer level (a noble gas configuration). An outer level contains eight (an octet of) electrons: two from the "*s*" orbital, and six from the three "*p*" orbitals.

Lewis Electron-Dot Symbols

Lewis symbols (after chemist G. N. Lewis) help us keep track of valence electrons. It is these outer, or valence, electrons that interact during bonding, and Lewis symbols allow us to see at a glance how many unpaired electrons are available for covalent bonding (in the case of nonmetals), or how many electrons will be lost from metals to form a cation during ionic bonding.

Electron-dot practice

1) a) Write the Lewis symbol for nitrogen.
Nitrogen is in group 5A, so it has five valence electrons. Place one dot at a time on the four sides of the symbol. There is one dot left over, so we pair it up on any side of the nitrogen:

·N̈·

b) How many bonds is nitrogen likely to form in its compounds?

Nitrogen has three electrons available for sharing, so it forms three bonds in its compounds. The bonds may be single, double, or triple:

H–N̈(–H)–H   ⁻:Ö–N̈=O: :N≡N:

2) a) Write the Lewis symbol for magnesium.
Magnesium is in group 2A, so it has two valence electrons:

·Mg·

b) What is the charge of the cation magnesium forms in its compounds?
Magnesium loses its two valence electrons to form a +2 cation.

3) Write the Lewis symbol for phosphorus.

Since phosphorus is in the same group as nitrogen, it has the same number of valence electrons and its Lewis symbol is the same:

·P̈·

Phosphorus, like nitrogen, often forms three bonds in its compounds. However, phosphorus is bigger than nitrogen and can "expand" its octet to accommodate 5 bonding groups.

9.2 IONIC BONDING MODEL

We can show ionic bonding as electrons held tightly by a small core:

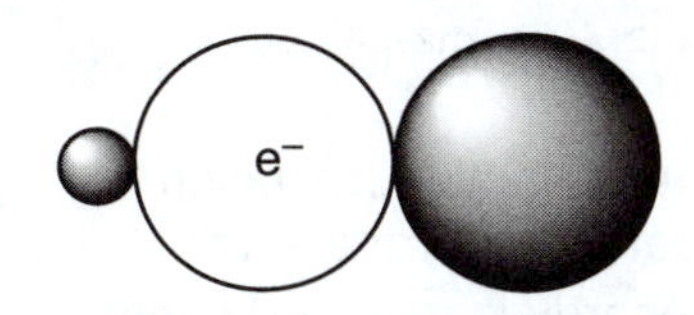

example: Cl^{7+} and Na^{+}

Ionic compounds form when metals transfer electrons to nonmetals to form ions with noble gas electron configurations; the electrostatic attraction between the cations and anions draws them into a three-dimensional array of ions.

Formation of magnesium oxide

1) What is the electron configuration for magnesium?

   answer: [Ne]$3s^2$

2) How does magnesium obtain a noble gas configuration in an ionic compound?

   answer:

   Mg loses its two "*s*" electrons to become Mg^{2+}, and attains the same electron configuration as neon. ([Ne]$3s^2 - 2e^- =$ [Ne])

3) What is the electron configuration for oxygen?

   answer: [He]$2s^2 2p^4$

4) How does oxygen obtain a noble gas configuration in its compounds?

   answer:

   Oxygen gains two electrons to become O^{2-} and attains the same electron configuration as Ne. ([He]$2s^2 2p^4 + 2e^- =$ [He]$2s^2 2p^6 =$ [Ne]).

   Magnesium transfers two of its electrons to oxygen to form Mg^{2+} and O^{2-}, both with the electron configuration of neon. These ions form an ionic solid with an empirical formula of MgO.

5) Is an electron transfer process endothermic or exothermic?

   answer:

   It requires energy to remove two electrons from a gaseous magnesium atom (2180 kJ/mol). It also requires energy to add two electrons to a gaseous oxygen atom. Addition of the first electron is exothermic (–141 kJ/mol), but addition of the second electron is endothermic (878 kJ/mol). Therefore, the electron transfer process requires 2917 kJ/mol.

   Why does MgO form at all if it requires so much energy to transfer the electrons? As you will see in the next problem, the large amount of energy released when separated gaseous ions are packed together to form an ionic solid, the **lattice energy,** makes the entire process energetically favorable.

Lattice Energy and the Born-Haber Cycle

The change in energy for the following reaction is known as its lattice formation energy:

$$Mg^{2+}(g) + O^{2-}(g) \rightarrow MgO(s)$$

Notice that the ions are in their gaseous state. Lattice formation energy is the amount of energy released when ions in their gaseous state come together to form a solid. $\Delta H_{\text{lattice formation energy}}$ is usually a large negative number since a large amount of energy is released during the process. (The reverse of this process, lattice dissociation energy, is therefore usually a large positive number.) We can calculate the lattice formation energy by breaking the reaction down into steps, the sum of which gives the overall reaction. (Review Hess's law, Chapter 6: $\Delta H_{\text{total}} = \Delta H_1 + \Delta H_2 + \Delta H_3 ...$).

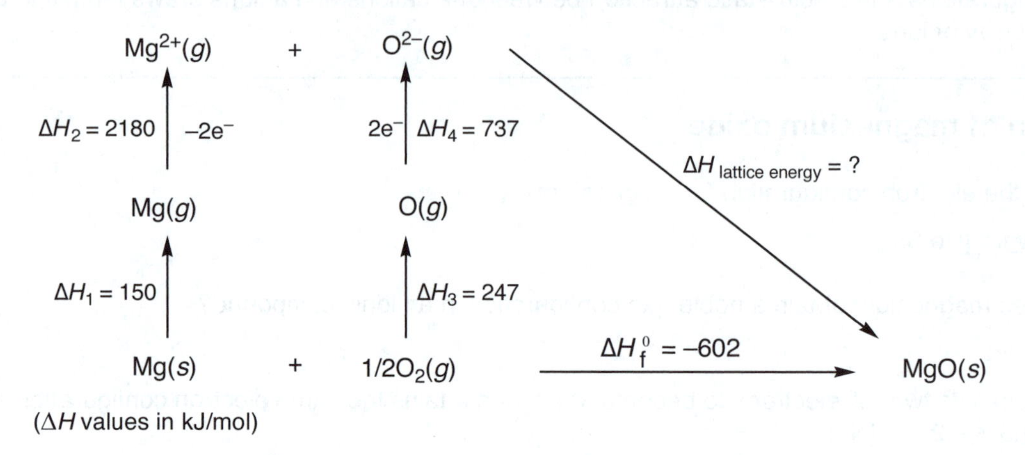

Lattice energy of MgO

Use Hess's law to determine the lattice formation energy of MgO.

solution:

Using Hess's law: $\Delta H_f^o = \Delta H_1 + \Delta H_2 + \Delta H_3 + \Delta H_4 + \Delta H_{\text{lattice formation energy}}$

The heat of formation of magnesium oxide is an exothermic process: heat is released ($\Delta H_f^o = -602$ kJ/mol). Notice that all the steps taking elemental magnesium and oxygen to gaseous ions are endothermic (positive ΔH values). The lattice formation energy releases a large amount of energy to make the overall process energetically favorable.

$$\Delta H_{\text{lattice formation energy}} = \Delta H_f^o - \Delta H_1 - \Delta H_2 - \Delta H_3 - \Delta H_4$$

$$= -602 - 150 - 2180 - 247 - 737 = -3920 \text{ kJ/mol}$$

Periodic Trends in Lattice Energy

The lattice energy is greater as the charges on the ions increase, and the radii of the cation and anion become smaller:

$$\Delta H^o_{\text{lattice}} \propto \frac{\text{cation charge} \times \text{anion charge}}{\text{cation radius} + \text{anion radius}}$$

Ionic radii increase as we move down a group of metals or nonmetals. Figure 9.8 on page 366 in your textbook shows the decreases in lattice energy as ionic radii increase. Find the ionic compound with the highest lattice energy (LiF). Notice that LiF has the smallest ions. Follow the curve from LiF down to LiI. The lattice energy decreases as the anions increase in size from F^- to Cl^- to Br^- to I^-. Now start back at LiF and run your finger through NaF, KF, and RbF. Lattice energy decreases as the cation increases in size.

Ionic charge has a large effect on lattice energy. Electrostatic attractions increase as charge increases, so higher charges on the ions increase lattice energy.

Lattice energy prediction

Would you expect the lattice energy to be greater for MgO or for NaCl?

solution:

MgO with +2 and –2 charges on its ions makes the numerator in the equation on page 180 larger than NaCl with +1 and –1 cation and anion charges.

The smaller ionic radii of Mg^{2+} and O^{2-} make the denominator smaller than for Na^+ and Cl^-, also making $\Delta H^{o}_{\text{lattice}}$ larger, so the lattice energy is greater for MgO.

Properties of Ionic Compounds

Most ionic solids have the following properties:

- hard, rigid, and brittle
- high melting and boiling points
- do not conduct electricity in the solid state
- conduct electricity when melted or when dissolved in water

How do these properties fit in with our molecular model of ionic solids?

Picture sodium chloride as a three-dimensional array of alternating sodium and chloride ions. The powerful attractive forces between the sodium and chloride ions hold the ions in specific positions in the crystal lattice and a lot of energy is required to break up the lattice. Since the ions are not easily moved out of position (they don't slide over one another), rock salt cannot be dent or bent. However, if a strong force is applied, the sample cracks as ions of like charge are forced next to one another. The ions in the crystal lattice are not free to move and carry electric current until they are melted or dissolved in water, and high temperatures are required to break up the crystal lattice (melting). Even higher temperatures are required to separate the ions enough for them to vaporize (boiling).

9.3 COVALENT BONDING MODEL

We can picture covalent bonds as tightly held electrons by small cores:

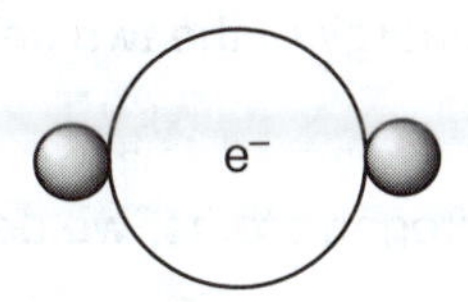

examples: F^{7+}, C^{4+}

In the previous section we learned to picture ionic solids as an array of alternating cations and anions. Now we consider atoms that share electrons to form covalently bonded molecules. Covalent compounds usually exist as discreet molecules, and the physical properties of these compounds depend on how strongly the molecules interact with one another. When we boil water, we do not put in nearly enough energy to break hydrogen and oxygen covalent bonds. The energy increases the motion of the water molecules until they overcome their attraction for each other and escape into the gaseous state. In some cases, covalent bonds may form a three-dimensional array of atoms **(network covalent solids)** similar to an ionic solid. Picture a three-dimensional array of covalently bonded carbon atoms. If each carbon has a tetrahedral environment, we have diamond, the hardest substance known. The melting point of diamond is around 3550°C.

The H_2 Molecule

Why is a hydrogen molecule more stable than two hydrogen atoms? Consider hydrogen atoms 1 and 2 separated by a large distance (no interactions, 0 energy state):

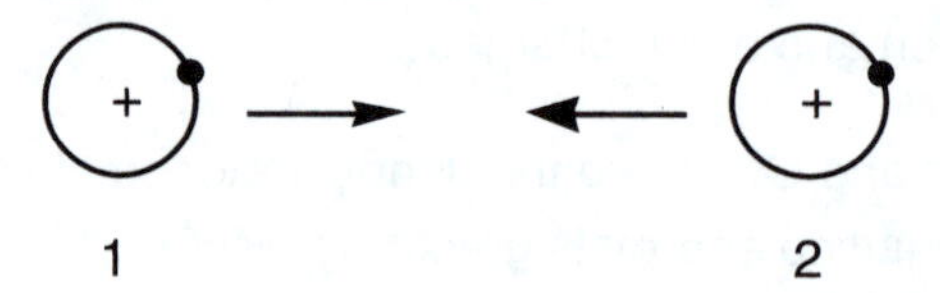

As the two atoms approach one another, what are the forces of attraction between the atoms? (Answer: Attractive forces occur between particles of opposite charge: proton 1 attracts electron 2; proton 2 attracts electron 1.) What are the forces of repulsion between the atoms? (Answer: Repulsive forces occur between particles of like charge: electron 1 repulses electron 2; proton 1 repulses proton 2.) The hydrogen molecule exists when the attractive forces balance the repulsive forces.

A graph of the energy of an H_2 molecule as a function of the distance between the two nuclei looks like this:

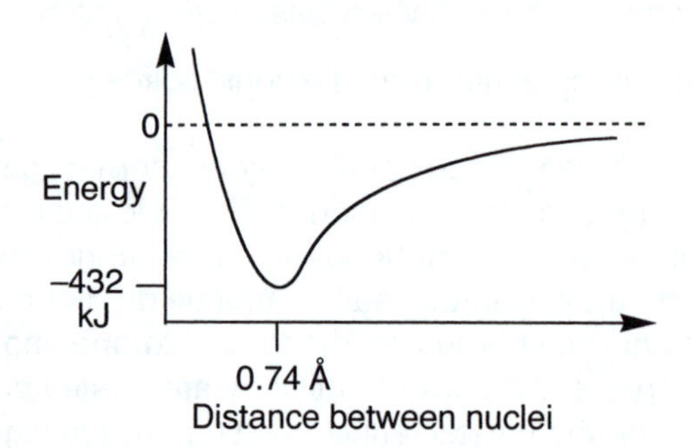

(The calculation is based on an accurate, if approximate, solution of the Schrödinger equation.) This graph is shown in more detail in Figure 9.12 on page 367 in your textbook. Study the graph and then answer the questions below.

The H_2 molecule

1) Based on the previous drawing, what is the energy of the two separated hydrogen atoms?

answer:

When there is no interaction between hydrogen atoms, we define the energy to be 0.

2) How long is a bond in an H_2 molecule?

(continued on next page)

answer:

0.74 Å: the distance between bonded hydrogen atoms represents an energy minimum.

3) What can you say about the attractive and repulsive forces when the hydrogen atoms are 0.74 Å apart?

answer:

At a distance of 0.74 Å (the bond length of a hydrogen molecule), attractive forces balance repulsive forces. If we attempt to bring the atoms any closer together, the repulsive forces exceed the attractive forces, and the system becomes less stable.

Bond Order, Bond Length, and Bond Energy

Covalent bonds between atoms may be single (one shared electron pair) or multiple (two or more shared electron pairs). In general, single bonds are longer and weaker than multiple bonds.

Bond type	Bond order	Bond energy (kJ/mol)	Bond length (Å)
single C−C	1	347	1.54
double C=C	2	614	1.34
triple C≡C	3	839	1.20

Bond energy (BE) is the energy required to break the bonds in 1 mole of gaseous molecules. This process always takes energy, so bond energy is always positive:

$$\text{A–B}(g) \longrightarrow \text{A}(g) + \text{B}(g) \qquad \Delta H^{\circ}_{\text{bond breaking}} = \text{BE}_{\text{A–B}} > 0$$

The opposite of this process is bond formation. Forming a bond releases energy, so bond formation energy is always negative:

$$\text{A}(g) + \text{B}(g) \longrightarrow \text{A–B}(g) \qquad \Delta H^{\circ}_{\text{bond forming}} = -\text{BE}_{\text{A–B}} < 0$$

Table 9.2 on page 371 in your textbook lists bond energies and bond lengths for some common covalent bonds.

Find the weakest bond listed in Table 9.2 and observe its bond length. Compare the bond length you found to others in the table. What can you say about bond length and bond strength? (Answer: Long bonds are generally weak bonds.) The big fluffy electron clouds around iodine form long, weak bonds between atoms. The bond length is 266 pm (or 2.66 Å) for I_2, the longest in the table; the bond strength is 151 kJ/mol, the weakest in the table.

Find the strongest bond listed in Table 9.2. Now find the strongest single bond listed. Explain your finds. (Answer: High bond order results in high bond energy and short bonds.) It is not surprising that the triple bond between carbon and oxygen in carbon monoxide is the strongest in the table (1070 kJ/mol). The tiny hydrogen atom forms the strongest single bond with a small, electronegative fluorine atom. The bond energy for a HF molecule is 565 kJ/mol, and its bond length is short (92 pm, less than 1Å).

Properties of Covalent Compounds

Most covalent compounds have the following properties:

- low melting and boiling points (many exist as liquids and gases)
- poor electrical conductors, even when melted or dissolved in water

Physical properties of covalent compounds usually do not reflect the strength of the covalent bond because the intramolecular covalent bonds do not break during melting and boiling. It is the weak, intermolecular forces between molecules which determine the physical properties of covalent compounds. The exception is network covalent solids such as quartz and diamond. All the atoms in these substances are connected through covalent bonds, and therefore the covalent bonds must break before these substances will melt or boil.

9.4 BOND ENERGY AND CHEMICAL CHANGE

We saw in Chapter 6 that Hess's law of heat summation could be used to calculate heats of reaction using bond energy data when ΔH_f° data is unavailable. We can think of reactions as two-step processes in which heat is required (absorbed) to break reactant bonds and is released when the atoms come together to form product bonds. The sum of these enthalpy changes is the heat of reaction, ΔH_{rxn}°:

$$\Delta H_{rxn}^\circ = \sum \Delta H_{f\ \text{reactant bonds broken}}^\circ + \sum \Delta H_{f\ \text{product bonds formed}}^\circ$$

Remember that bond energies are average values obtained from many different compounds, so the energy of a bond in a particular substance is usually close, but not equal to this average.

Bond energy example

Use the bond energies from Table 9.2 on page 371 to calculate ΔH_{rxn}° for the reaction of methane with chlorine and fluorine gas to give CF_2Cl_2 (Freon-12):

$$CH_4(g) + 2Cl_2(g) + 2F_2(g) \rightarrow CF_2Cl_2(g) + 2HF(g) + 2HCl(g)$$

Why might your calculated value differ from the experimentally determined ΔH value for this reaction?

solution:

We calculate the energy required to break the reactant bonds, and the energy released when the product bonds form (in kJ/mol):

4 C–H bonds broken: $4 \times 413 = 1652$ kJ/mol

2 Cl–Cl bonds broken: $2 \times 243 = 486$

2 F–F bonds broken: $2 \times 159 = 318$

Total energy required: 2456 kJ/mol

2 C–Cl bonds formed: $2 \times -339 = -678$ kJ/mol

2 C–F bonds formed: $2 \times -453 = -906$

2 H–F bonds formed: $2 \times -565 = -1130$

2 H–Cl bonds formed : $2 \times -427 = -854$

Total energy released: –3568 kJ/mol

(continued on next page)

$$\Delta H^o_{rxn} = \Delta H^o_{reactant\ bonds\ broken} + \Delta H^o_{products\ bonds\ formed}$$

$$\Delta H_{rxn} = 2456\ kJ/mol - 3568\ kJ/mol = -1112\ kJ/mol$$

A driving force for this reaction is the large amount of energy released on formation of the H–F and C–F bonds. The bond energies used for this calculation represent the average bond energy values obtained from many different compounds. The bond energies in any particular substance will usually be close, but not equal to the average.

Recall from Chapter 3 that hydrocarbons (molecules containing C–C and C–H bonds) react with oxygen to produce carbon dioxide and water ($CO_2 + H_2O$) in combustion reactions. Our cars and our bodies are both fueled by large organic molecules that release energy as they react with oxygen. The fewer bonds to oxygen in a fuel, the more energy it releases when burned since it requires more energy to break apart the C–O or O–H bonds. Fats contain more energy per gram than carbohydrates because fats contain fewer of the stronger C–O and O–H bonds than do carbohydrates.

9.5 ELECTRONEGATIVITY AND BOND POLARITY

Most bonds do not fit neatly into either ionic or covalent bonding models; they have a little of each character. We saw that two identical hydrogen atoms share electrons between them to form a hydrogen molecule, and that magnesium transfers two electrons to oxygen to form magnesium oxide. Between these extremes, we have atoms that do not transfer electrons, but are different enough so that they do not share them equally. In these cases a **polar covalent bond** forms. In a polar covalent bond, the shared electrons spend more time around one atom than the other, giving the atoms partial positive and negative charges.

Polar covalent bonds

1) Predict which of the following molecules contain polar covalent bonds, and show the partial charges:

a) HCl b) O_2 c) HF d) H_2O

The symbol "δ" means "partial"; δ+ means that atom has a partial positive charge, δ− means that atom has a partial negative charge.

answers:

a) polar: $\overset{\delta+}{H}-\overset{\delta-}{Cl}$ b) nonpolar c) polar: $\overset{\delta+}{H}-\overset{\delta-}{F}$ d) polar: $\underset{\delta+}{H}-\overset{\delta-}{O}-\underset{\delta+}{H}$

2) Can you predict the bond energy for a **H–Cl** bond given that the bond energy for **H–H** is 432 kJ/mol, and the bond energy for **Cl–Cl** is 239 kJ/mol?

answer:

You might expect that the bond energy for **H–Cl** would be the average of the **H–H** and **Cl–Cl** bond energies:

$$\text{Expected H-Cl bond energy} = \frac{432\ kJ/mol + 239\ kJ/mol}{2} = 336\ kJ/mol$$

However, the measured bond energy for **H–Cl** is 427 kJ/mol. We attribute this discrepancy to the ionic character in the polar **H–Cl** bond. The electrostatic attraction between the partially charged atoms leads to greater bond strength, and the actual bond energy is higher than expected.
Linus Pauling developed a scale of **electronegativity** of atoms based on the difference between the expected and actual bond energies:

$$\Delta = (H\text{–}Cl)_{actual} - (H\text{–}Cl)_{expected}$$

Electronegativity

Choose words from the list below to complete the following paragraph:

(a) attract (b) covalent (c) decreases (d) fluorine (e) increases (f) ionic (g) polar covalent (h) same

Electronegativity is the relative ability of bonded atoms to _______________(1) the shared electrons. Atoms with very different electronegativities form _______________(2) bonds, those with intermediate electronegativity differences form _______________(3) bonds, and those with no electronegativity difference form _______________(4) bonds. A purely covalent bond exists only between two of the _______________(5) atoms. Electronegativity, in general, increases as an atom's size _______________(6), so electronegativity _______________(7) as you go up a group in the periodic table and _______________(8) from left to right across the periodic table. The most electronegative element is _______________(9), to which Pauling assigned an electronegativity of 4.0.

answers: 1-a, 2-f, 3-g, 4-b, 5-h, 6-c, 7-e, 8-e, 9-d.

Electronegativity and Oxidation Number

Electronegativity is used when assigning oxidation numbers to atoms in a molecule. The more electronegative element is assigned all the shared electrons. The oxidation number is the difference between the number of valence electrons and the number assigned to the atom from shared and unshared electrons:

O.N. = number of valence electrons – (shared + unshared electrons)

Assigning oxidation numbers

1) What are the oxidation numbers for hydrogen and oxygen in water?

answer:

Oxygen is more electronegative than hydrogen, so the shared electrons are all assigned to oxygen.

Oxygen has four unshared electrons, therefore:
(shared + unshared electrons) = 4 + 4 = 8.

Oxygen has 6 valence electrons; its oxidation number in water is:
O.N. = number of valence electrons – (shared + unshared electrons) = 6 – 8 = –2

Hydrogen has one valence electron and no shared or unshared electrons.
O.N. = number of valence electrons – (shared + unshared electrons) = 1 – 0 = +1

2) What are the oxidation numbers for carbon and oxygen in CO_2?

answer:

Oxygen is more electronegative than carbon, so shared electrons are again assigned to oxygen.

Each oxygen has four unshared electrons, therefore:

(continued on next page)

(shared + unshared electrons) = 4 + 4 = 8.

Oxygen has 6 valence electrons; its oxidation number in water is:
O.N. = number of valence electrons – (shared + unshared electrons) = 6 – 8 = –2

Carbon has four valence electrons and no shared or unshared electrons.
O.N. = number of valence electrons – (shared + unshared electrons) = 4 – 0 = +4

Polarity and boiling points

Match the following compounds with their values for ΔEN (the difference in electronegativity between their two atoms), and their expected boiling points:

	ΔEN	bp, °C
1) PCl_3	a) 0	i) 1413
2) Cl_2	b) 0.5	ii) 1412
3) NaCl	c) 0.9	iii) 183
4) $SiCl_4$	d) 1.2	iv) 76
5) $MgCl_2$	e) 1.5	v) 59
6) $AlCl_3$	f) 1.8	vi) 56
7) S_2Cl_2	g) 2.1	vii) –34.6

answers:

Expected:	Cl_2	S_2Cl_2	PCl_3	$SiCl_4$	$AlCl_3$	$MgCl_2$	NaCl
	a	b	c	d	e	f	g
	vii	vi	v	iv	iii	ii	i
Observed:	vii	v	iv	vi	iii	ii	i

All of the elements in the above compounds are in the third period of the periodic table. Sodium and chlorine have the greatest difference in electronegativity and they combine to form NaCl, an ionic solid with high melting and boiling points. As we move from sodium across the table towards chlorine, the electronegativity of the elements increases bringing them closer in value to the electronegativity of chlorine, and the resulting compounds show less ionic character as ΔEN decreases.

Figure 9.27 on page 385 in your text shows a photo of a sample of each of the third period chlorides (NaCl, $MgCl_2$, $AlCl_3$, $SiCl_4$, PCl_3, S_2Cl_2, and Cl_2). Notice that as ΔEN decreases, the bond becomes more covalent and the properties of the compounds change from an ionic solid (NaCl) to a gas consisting of individual molecules (Cl_2).

As bonds become less polar, boiling points tend to decrease. You probably could not predict exactly the order of the boiling points, because the boiling point for $SiCl_4$ is unusually low (56°C) due to weak forces between molecules, and it falls out of order.

Figure 9.25 on page 383 in your textbook shows a plot of the percentage of ionic character in a bond as a function of electronegativity difference for some simple gaseous diatomic molecules. Notice that diatomic molecules containing two of the same atom have 0% ionic character, but no diatomic molecule has 100% ionic character. Electron sharing occurs to some extent in every bond. Notice also that HF, the molecule we found in Table 9.2 with the highest covalent bond energy, has almost 50% ionic character.

9.6 AN INTRODUCTION TO METALLIC BONDING

We can picture metallic bonding as loosely held electrons by large cores:

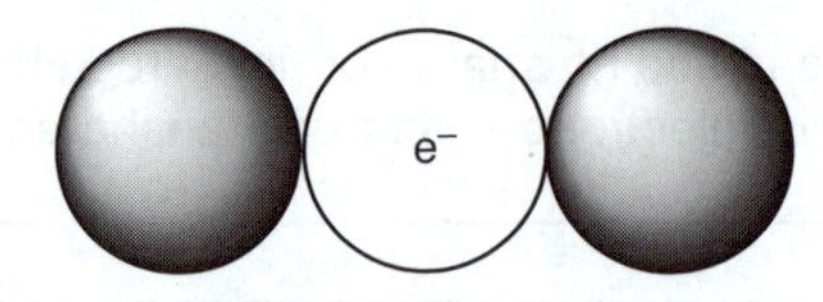

examples: Ca^{2+}, Na^{+}

The electrons are delocalized throughout the substance and the nuclei with their core electrons are submerged in the **electron-sea** in an orderly array. The loosely held electrons "flow" and metals conduct in the solid state by this electron flow. The metal ions are not held in a rigid array; rather they can slide past one another through the electron-sea to new lattice positions. Typically metals exist as alloys, solid mixtures with variable composition.

Properties of Metallic Compounds

Metals tend to have the following properties:

- bend or dent rather than crack or shatter
- moderate to high melting points and much higher boiling points (most exist as solids)
- conduct heat and electricity well in both the solid and liquid states
- malleable and ductile

Can you use the electron-sea model of metallic bonding to explain these observed properties of metals? It may help to use our model from Chapter 1 of liquids (atoms touching and disordered) and gases (atoms not touching) to explain the moderately high melting points and much higher boiling points of metals.

Summary of the three bonding models

The properties of an ionic solid, a metal, and a covalent compound

Consider the familiar ionic solid, sodium chloride (NaCl), and the metal and covalent nonmetal from which it is composed, sodium (Na) and chlorine (Cl_2). Identify each of the following as a property of Na, NaCl, or Cl_2:

a) BP = 883°C, MP = 97.8°C
b) BP = –35°, MP = –101°C
c) BP = 1413°C, MP = 801°C
d) shatters under stress
e) deforms under stress
f) conducts electricity when dissolved in water
g) conducts electricity when melted
h) conducts electricity as a solid
i) does not conduct electricity
j) a pale green gas at room temperature
k) a soft, silvery-white metal at room temperature
l) a white solid at room temperature
m) violently decomposes water producing a basic solution and hydrogen gas
n) burns with a yellow flame

(*continued on next page*)

answers:

sodium, Na (metallic): a, e, g, h, k, m, n
Metals do not usually crack, but deform with stress because metal ions can slide past each other through the electrons to move to new lattice positions. Metals are good conductors of heat and electricity as both solids and liquids due to their mobile electrons.

chlorine, Cl_2 (covalent): b, i, j
Chlorine, like most covalent substances, consists of individual molecules which have relatively weak attractions for each other (intermolecular attractions). These compounds have low melting and boiling points because the individual molecules stay intact (no covalent bonds are broken); it is only the relatively weak intermolecular forces between molecules which must be overcome.

> Note: Remember that covalent compounds with low melting and boiling points are not breaking covalent bonds. Covalent bonds are extremely strong (240 kJ/mol for the bond in chlorine, 347 kJ/mol for a typical single bond between carbon atoms), and covalent substances which form a network of covalent bonds throughout the sample have extremely high melting points and boiling points. Diamond, which consists of a three-dimensional array of carbon atoms covalently bonded together, melts at 3550°C and is used to test the hardness of other substances because it is so hard.

Covalent compounds are poor conductors of electricity because the electrons are localized (not free to move), and no ions are present.

sodium chloride, NaCl (ionic): c, d, f, g, l
Ionic compounds usually have high melting points and boiling points because it requires large amounts of energy to free ions from lattice positions in ionic compounds.

Ionic compounds are hard and brittle because strong electrostatic forces hold the ions in place throughout the crystal. If enough pressure is applied to overcome the attractions, ions of like charge are brought close together and their repulsions crack the sample.

The ions of an ionic compound are generally immobilized in a solid, especially at ordinary temperatures, and do not conduct electricity. When an ionic solid melts or dissolves, the ions can move and carry an electric current.

CHAPTER CHECK

Make sure you can...

- Visualize the three types of bonding: ionic, covalent, and metallic.
- Draw Lewis electron-dot symbols for main-group atoms.
- Calculate lattice energy by showing the Born-Haber cycle and using Hess's law.
- Explain periodic trends in lattice energy (Coulomb's law).
- Predict relative sizes of lattice energy in ionic compounds.
- Explain why ionic compounds are typically brittle, high melting, and conduct electricity when molten or dissolved in water.
- Describe how energy changes as a function of distance between nuclei for a hydrogen molecule.
- Relate bond energies, bond orders, and bond lengths.

- Explain why covalent compounds have low melting and boiling points, and are poor electrical conductors, even when melted or dissolved in water.
- Find heats of reaction from bond energy data and Hess's law.
- Predict which molecules contain polar covalent bonds.
- Relate bond polarity to ΔEN and the boiling points of compounds.
- Use electronegativities to assign oxidation numbers.
- Explain why metals tend to bend or dent, have moderate to high melting points, and much higher boiling points, conduct heat and electricity well as both solids and liquids, and are malleable and ductile.

Chapter 9 Exercises

9.1) Explain, in terms of the relative electronegativities of the participants, the difference between covalent, ionic, and metallic bonding.

9.2) a. Although molecular compounds are always covalent, some covalent substances are not described as molecular. Explain with examples.

b. Is it possible for a compound to contain both covalent and ionic bonds?

c. What is the difference between a metallic bond and an ionic bond?

9.3) In the Lewis electron-dot symbols for atoms of the following elements, how many electron dots are placed around the symbol for the element?

a. germanium (Ge)
b. silicon (Si)
c. phosphorus (P)

9.4) Calculate the lattice formation energy of sodium chloride if:

the sublimation energy of sodium metal is 108 kJ/mol
the dissociation energy for chlorine is 242 kJ/mol of molecules
the first ionization energy for sodium is 495 kJ/mol
the electron affinity of a chlorine atom is –348 kJ/mol
the standard enthalpy of formation of sodium chloride is –411 kJ/mol

9.5) Which salt has the highest lattice energy?

RbCl RbBr KI $MgCl_2$ $MgBr_2$ CaI_2

9.6) a. In which of the following diatomic molecules is the bond the strongest? In which molecule is the bond the longest?

H_2 F_2 Cl_2 Br_2 I_2

b. Which carbon-carbon bond is the longest?

C—C C=C C≡C

c. Which triple bond is the strongest?

N≡N C≡C C≡N C≡O

9.7) Calculate the enthalpy of formation of ammonia ΔH_f^o (NH_3) if:
the bond energy of an N—H bond = 391 kJ/ mol
the bond energy of an H—H bond = 436 kJ/ mol
the bond energy of an N≡N bond = 945 kJ/ mol

9.8) Classify the following compounds as covalent, polar covalent, ionic but with some covalent character, or ionic:

HF KCl MgS Br_2 CCl_4 $CaCl_2$ O_2 SO_2

9.9) Which of the following elements, arranged in the order in which they appear in the periodic table, is the most electronegative? Which is the least electronegative?

B	N	F
Al	P	Cl
Ga	As	Br

Chapter 9 Answers

9.1) Electronegativity is a measure of the tendency of an element in a compound to attract electrons to itself. It is therefore a predictor of how an element will behave when involved in bonding with other elements.

If two elements involved in a bond are both electronegative, they both tend to attract the bonding electrons and the compromise is to share them—a covalent bond.

If one element is electronegative and the other electropositive, the one that is electropositive tends to give electrons to the element that is electronegative. The result is the formation of ions and an ionic bond.

If both elements are electropositive, then neither wants the electrons. The result is a delocalization of the electrons from their respective atoms and the formation of a metallic bond.

9.2) a. Substances that exist with their atoms bonded covalently in a three-dimensional lattice are not molecules in the accepted sense and are often referred to as covalent networks. Examples are: diamond C_n, silicon carbide $(SiC)_n$, quartz $(SiO_2)_n$, boron nitride $(BN)_n$.

b. Yes—polyatomic ions contain covalent bonds, but are involved in ionic bonding with other ions. An example is ammonium nitrate. This is an ionic compound containing ammonium ions and nitrate ions. However, both the ammonium ion NH_4^+ and the nitrate ion NO_3^- contain polar covalent bonds.

c. Metallic bonding occurs in a metallic lattice and involves the delocalization of valence electrons throughout the lattice. This global sharing of electrons holds the lattice together. Ionic bonding involves the transfer of electron(s) from one species to another and the formation of oppositely charged ions. Attraction between the opposite charges holds the ions together in the solid ionic lattice.

9.3) A Lewis symbol for an element indicates the number of electrons an atom of the element has in its valence shell: a. 4; b. 4; c. 5

9.4) The standard enthalpy of formation, ΔH_f^0, of NaCl is the change in enthalpy when the elements sodium and chlorine, in their standard states, combine to form NaCl. Lattice formation energy is the amount of energy released when gaseous sodium ions and chloride ions come together to form a solid. To form a gaseous ion from an elemental sodium atom, first it must be sublimated, and then it must lose an electron to form $Na^+(g)$. To form a gaseous ion from an elemental chlorine molecule, first it must dissociate into atoms, and then it must gain an electron to form $Cl^-(g)$. We are given the enthalpy changes associated with these individual steps, so we can use Hess's law to calculate the lattice formation energy (we expect a fairly large, negative value since energy is usually released during the process of gaseous ions forming a solid lattice). Notice that the dissociation energy for chlorine is given for one mole of chlorine; we have written the reaction with half of a mole, so we use 1/2 of 242 kJ/mol for ΔH_{diss}: (ΔH values are in kJ/mol).

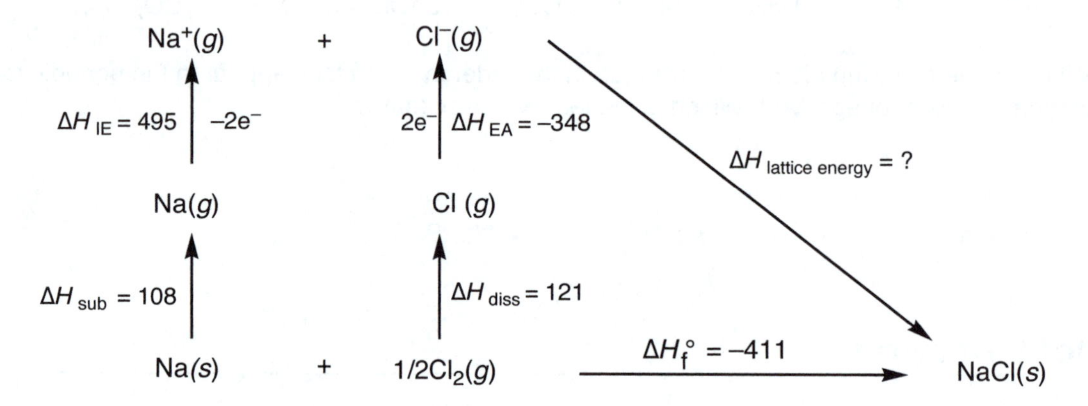

Using Hess's law we can write:

$$\Delta H_f^o = \Delta H_{sub} + \Delta H_{IE} + \Delta H_{diss} + \Delta H_{EA} + \Delta H_{lattice\ energy}$$

$$\Delta H_{lattice\ form.\ energy} = \Delta H_f^o - \Delta H_{sub} - \Delta H_{IE} - \Delta H_{diss} - \Delta H_{EA}$$

$$\Delta H_{lattice\ form.\ energy} = (-411\ \text{kJ/mol}) - (108\ \text{kJ/mol}) - (495\ \text{kJ/mol}) - (121\ \text{kJ/mol}) - (-348\ \text{kJ/mol})$$

$$\Delta H_{lattice\ form.\ energy} = -787\ \text{kJ/mol}$$

9.5) The lattice energy of a salt depends upon the charges on the ions, the sizes of the ions, and the crystal symmetry. The charge on the ion is usually the predominant factor. Since magnesium and calcium form ions with a 2+ charge, these salts will have a higher lattice energy than the salts of rubidium and potassium. Magnesium is smaller than calcium, and chlorine is smaller than bromine or iodine, so, of the salts listed, magnesium chloride will have the highest lattice energy.

9.6) a. H_2 has the strongest bond; I_2 has the longest bond.

b. The C–C single bond is the longest bond. As more pairs of electrons are shared, the closer the two nuclei are drawn together.

c. The difference in electronegativity contributes to the bond strength. A triple bond between C and O is the strongest.

9.7) ΔH_f^o (NH_3) = bonds broken – bonds made =

$$945 + 3 \times 436 - 6 \times 391\ \text{kJ} = -93\ \text{kJ for 2 mol ammonia} = -46\ \text{kJ/mol}$$

9.8) There is a continuum in the character of bonding from pure covalent (equal sharing of electrons) to ionic (transfer of electrons). The description of a particular bond is based upon the difference in the electronegativities of the participants. Note that there is no such thing as a pure ionic bond, some sharing always takes place. However, in cases where the electronegativity difference is large (e.g., NaCl), the bond is usually said to be ionic:

HF	polar covalent
KCl	ionic
MgS	ionic but with some covalent character
Br_2	covalent
CCl_4	polar covalent (although the molecule is nonpolar)
$CaCl_2$	ionic
O_2	covalent
SO_2	polar covalent

9.9) Fluorine is the most electronegative element in the periodic table. Of the elements listed, you might have expected gallium to be the least electronegative. In fact, compared to gallium, aluminum has a slightly lower electronegativity, a slightly lower first ionization energy, and a larger atomic radius—all contrary to the expected trends. This anomaly is due to the position of gallium immediately following the first transition series. Filling the first set of *d* orbitals leads to an increase in nuclear charge of 10+ without a compensating increase in the shielding of the nuclear charge. The 4*s* and 4*p* electrons are therefore held more tightly than might otherwise have been expected.

10

The Shapes of Molecules

10.1 LEWIS STRUCTURES

The **Lewis structure** (after G.N. Lewis) of a molecule shows atoms with their valence electrons. The Lewis structure does not necessarily show the correct geometry of the atoms, but their sequence (or connectivity) must be correct. A Lewis structure for water drawn as:

$$\mathbf{H{-}\ddot{\underset{\cdot\cdot}{O}}{-}H}$$

is a correct Lewis structure for water even though water is actually a bent molecule. In Lewis structures, lines indicate electron pairs in covalent (shared electron-pair) bonds.

The Lewis structure of water

1) Show an oxygen atom with its valence electrons.

   Oxygen, in group 6A, has 6 valence electrons. We place these around four sides of oxygen one at a time, and then pair up electrons as needed:

   The number of unpaired electrons shows the number of bonds the nonmetal usually forms. Oxygen usually forms two bonds, which gives it the two electrons it needs to attain a noble gas configuration.

2) Show two hydrogen atoms with their valence electrons.

   Each hydrogen atom needs one electron to fill its 1*s* shell:

   H·　　　　　　**·H**

3) Show, using dots for electrons, how oxygen and hydrogen combine to form water, by sharing electrons.

$$\mathbf{H\cdot\cdot\ddot{\underset{\cdot\cdot}{O}}\cdot\cdot H}$$

Method for Drawing Lewis Structures

1) Decide on the sequence of atoms: usually place the atom with the lowest electronegativity in the center, except for hydrogen, which never goes in the center:

H O H

2) Add up the total number of valence electrons.
(for main-group atoms, # valence e^- = group number)

valence electrons in H_2O = valence electrons from O + 2 × valence electrons from H

$$= 6 + 2(1) = 8$$

3) Draw single bonds between atoms and subtract two valence electrons for each bond:

H–O–H

It requires two electrons for each single bond, so electrons left to place are: 8 – 4 = 4.

4) Distribute remaining electrons to give octets to all atoms except hydrogen (central atom last):

H–Ö–H

5) If there are not enough electrons for the central atom, make multiple bonds. Oxygen has an octet, so multiple bonds are not needed.

Lewis octet structure practice problems

1) Draw the Lewis electron-dot symbol for the atoms C, N, O, and F.

solutions:

·C· ·N· ·O· ·F:

2) How many hydrogen atoms would be needed for each atom to give it an octet of electrons? Draw each structure.

solutions:

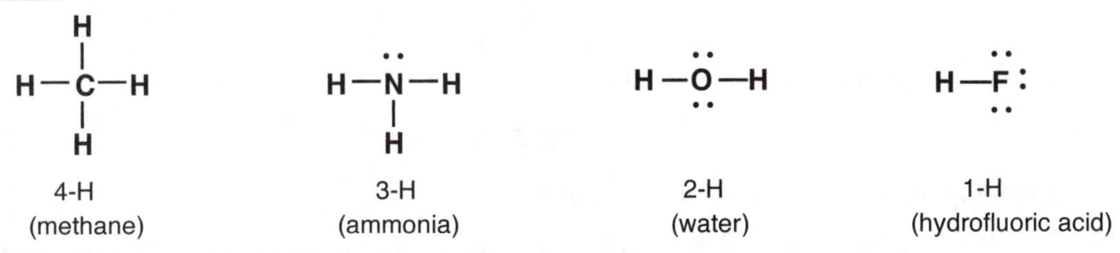

3) Draw Lewis structures for the following compounds or ions:

a) CO_2 c) CN^- e) NOF g) N_2

b) CO d) NO^+ f) H_2CO h) C_2N_2 (atom order is NCCN)

(continued on next page)

solutions:

a) CO_2 valence electrons = 4 + (2 × 6) = 16

Carbon is placed in the middle because it has the lowest electronegativity. If we use four electrons to make single bonds between carbon and oxygen, we would need 16 more electrons to complete octets around the carbon and oxygen atoms:

:Ö—C̤̈—Ö:

CO_2 has only 16 valence electrons, so we must make multiple bonds. Double bonds between carbon and oxygen use 8 electrons and leave 8 to place around the two oxygen atoms as lone pairs to complete their octets:

:Ö=C=Ö:

b) CO valence electrons = 4 + 6 = 10

The only way to give carbon and oxygen octets with 10 electrons are to use 6 electrons to make a triple bond, and use the other 4 electrons to give each atom a lone pair:

:C≡O:

c) CN^- valence electrons = 4 + 5 + 1 = 10

$[:C\equiv N:]^-$

Notice that CN^- is **isoelectronic** (same number of electrons) with CO and has essentially the same structure.

d) NO^+ valence electrons = 5 + 6 – 1 = 10

$[:N\equiv O:]^+$

NO^+ is isoelectronic with both CO and CN^-. If we start with CO, and replace carbon with nitrogen, nitrogen brings with it an extra valence electron. The +1 charge of NO^+ "cancels" the extra electron, and we can predict, without counting all valence electrons, that CO and NO^+ are isoelectronic.

e) NOF valence electrons = 5 + 6 + 7 = 18

:Ö=N̈—F̤̈:

f) H_2CO valence electrons = (2 × 1) + 4 + 6 = 12

H₂C=Ö: (O double bonded to C; C bonded to two H)

g) N_2 valence electrons = (2 × 5) = 10

:N≡N:

A nitrogen molecule contains a triple bond.

h) C_2N_2 valence electrons = (2 × 4) + (2 × 5) = 18

The only way to obtain octets on all the atoms is to use 12 electrons to make triple bonds between the carbon and nitrogen atoms, and use the last four electrons to give each nitrogen a lone pair:

:N≡C—C≡N:

Exceptions to the Octet Rule

Be and B are often electron deficient; they have fewer than eight electrons around the Be or B atom. Atoms with low-lying (valence-shell) *d* orbitals available can expand their valence shells and accept more than eight electrons. Phosphorus and sulfur can accommodate 12 electrons; iodine can accommodate 14. Molecules with an odd number of electrons form a free radical with an unpaired electron and do not obey the octet rule.

Lewis structures

Show Lewis structures for the following compounds:

a) SO_2 b) SF_6 c) PCl_5

solutions:

a) SO_2 valence electrons = 6 + (2 × 6) = 18

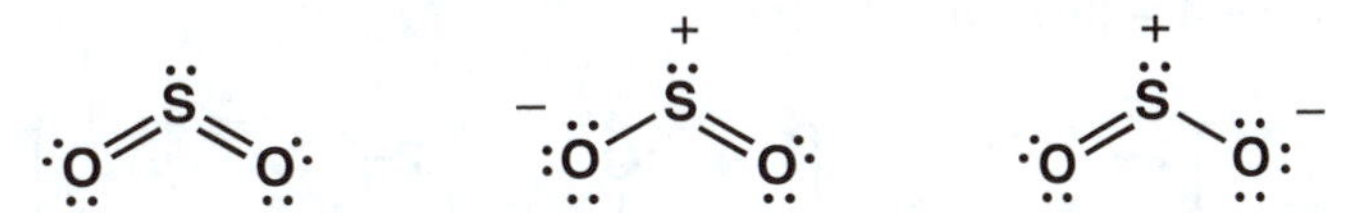

Sulfur expands its octet to hold 10 electrons. Measurements show that the sulfur oxygen bonds in SO_2 are all approximately the S=O bond length, so the first structure above is the best representation of the bonding in SO_2. Notice that the first structure also has 0 formal charges on all of its atoms.

b) SF_6 valence electrons = 6 + (6 × 7) = 48

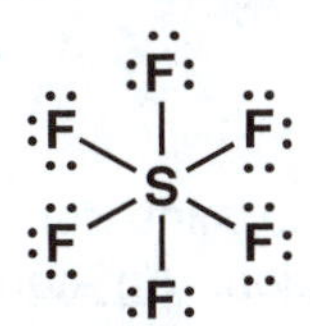

Sulfur expands its octet to hold 12 electrons.

c) PCl_5 valence electrons = 5 + (5 × 7) = 40

Phosphorus expands its octet to hold 10 electrons.

Resonance (Delocalized Bonding)

What is the Lewis structure for the nitrite anion, NO_2^-? NO_2^- has 18 valence electrons. In order to achieve octets around each atom, a double bond is needed between the nitrogen and one of the oxygen atoms:

Experiments show that NO_2^- has only one bond type, not one shorter (and stronger) double bond and one longer single bond. In order to represent this idea, we use **resonance** structures, with a two headed resonance arrow between them:

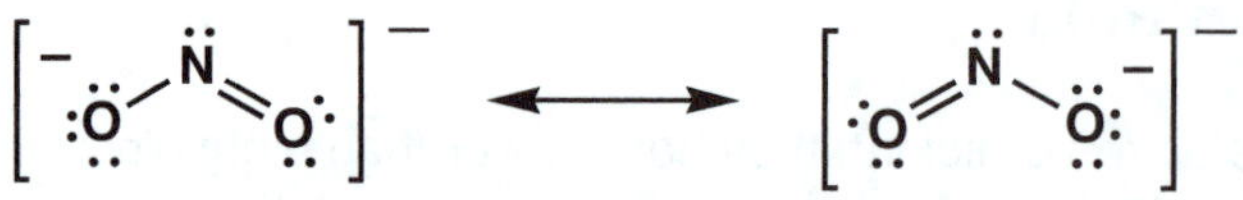

Both the drawings and the term resonance can be misleading because they seem to imply that the molecule switches back and forth, or resonates between structures. The actual structure is a hybrid of the two resonance structures. The bonds in the nitrite anion can be thought of as containing (on average) 1.5 bonds (3 electrons).

Resonance structure

Show resonance structures for CO_3^{2-}.
How many electrons, on average, are in each bond of CO_3^{2-}?

solution:

Valence electrons = 4 + (3 × 6) + 2 = 24

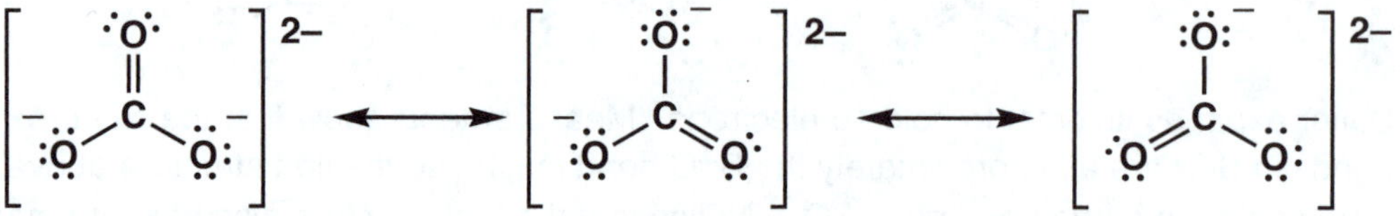

Each bond can be thought of as being 1 1/3 bond containing 2 2/3 electrons.

Formal Charge

We use "formal charges" as a bookkeeping device to determine, for example, which resonance structures contribute the most to a molecule's electron configuration. An atom has a formal charge if the number of electrons it "controls" does not equal its number of valence electrons. We consider an atom to control its lone pair electrons (or any single, not shared, electrons), and to control one of the electrons in any shared electron-pair bond to other atoms. For example, consider the Lewis structure for ammonia:

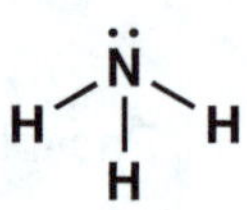

The nitrogen atom "controls" its lone pair electrons plus one electron from each of the three bonds to hydrogen for a total of five electrons. Nitrogen, in group 5A, has five valence electrons and so has no formal charge in an ammonia molecule.

Now consider the Lewis structure for one resonance form of CO_3^{2-} :

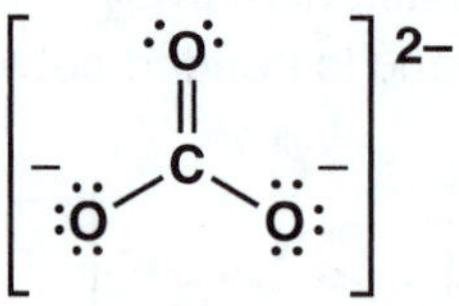

The carbon atom "controls" four electrons (one electron from each of the 4 bonds to oxygen). Carbon, in group 4A, has four valence electrons and so has no formal charge in CO_3^{2-}.

The doubly bonded oxygen atom controls two lone pairs, and one electron from each double bond for a total of six electrons. Since oxygen has six valence electrons, it has no formal charge.

The singly bonded oxygen atoms each control three lone pairs, and one electron from the single bond for a total of 7 electrons. Since oxygen has six valence electrons, each singly bonded oxygen has a formal charge of –1. Carbon, with four single bonds, has a formal charge of 0. The formal charges in an ion must add up to the charge on the ion (in this case, –2).

Formal charge practice problem

1) Draw resonance forms for the molecule SCN^- and assign formal charges to each atom.

solution:

$[\text{:S=C=N:}]^-$ (a) $[\text{:S—C≡N:}]^-$ (b) $[\text{:S≡C—N:}]^-$ (c)

Resonance structure "a" formal charges: S = 0; C = 0; N = –1
Resonance structure "b" formal charges: S = –1; C = 0; N = 0
Resonance structure "c" formal charges: S = +1; C = 0; N = –2

2) Given that:

- smaller formal charges are preferable to larger ones,
- a more negative formal charge should reside on a more electronegative atom,

which resonance form above is the most important contributor to the resonance hybrid of the thiocyanate ion?

solution:

We eliminate resonance structure "c" because of the –2 formal charge on nitrogen, and the positive formal charge on sulfur. We pick "a" over "b" because nitrogen is more electronegative than sulfur, and therefore should be given the negative charge.

Odd-Electron Molecules

Some molecules, called **free radicals,** contain a central atom with an odd number of valence electrons. Obviously, these molecules cannot follow the octet rule. The lone, unpaired electron makes the species paramagnetic (they show a weak attraction toward a magnetic field) and extremely reactive. Formal charges may be used to decide where the lone electron resides. Both nitric oxide, NO, and nitrogen dioxide, NO_2, contain an odd number of valence electrons and are therefore free radicals. The NO molecule is considered to be a resonance hybrid with two contributing structures:

·N=O: ⟷ :N=O·

We would guess that the structure on the left would be the more important resonance structure since both the nitrogen and oxygen in this structure have a 0 formal charge. The structure on the right has a –1 formal charge on the nitrogen, and a +1 formal charge on the oxygen. Sometimes formal charges do not correctly predict the primary resonance structure, as discussed in the case of nitrogen dioxide in your textbook on page 310.

Elementary oxygen (O_2) is also paramagnetic, suggesting which of the following Lewis structures?

:O=O: :O—O:

Although the structure on the right, with two unpaired electrons, would explain the observed paramagnetism, that structure does not satisfy the octet rule, nor does it agree with experimental data for bond distances. The distance between oxygen atoms in O_2 (1.21 Å) is much smaller than would be expected for an O–O single bond (1.48Å). The valence bond theory is useful for predicting molecular geometry, but it has its limitations. As we will see in the next chapter on bonding theories, the molecular orbital approach leads to a more satisfactory picture of the electron distribution in the oxygen molecule.

10.2 VALENCE-SHELL ELECTRON-PAIR REPULSION (VSEPR) THEORY AND MOLECULAR SHAPE

Lewis structures give us a method for drawing molecules with their bonds and lone pairs, but they do not predict the geometry of a molecule. We can predict the shapes of molecules by drawing them in a way that minimizes repulsions between groups of valence electrons while allowing them to approach the nucleus as close as possible. (We define a "group" of electrons as a single, double, or triple bond, a lone pair, or a single electron.) As you will see, it is essential to know the geometry of electron groups in order to predict the geometry of molecules.

VSEPR

What geometry minimizes valence-shell electron-pair repulsions for the following compounds?

a) CH_4 b) NH_3 c) H_2O d) HF

solution:

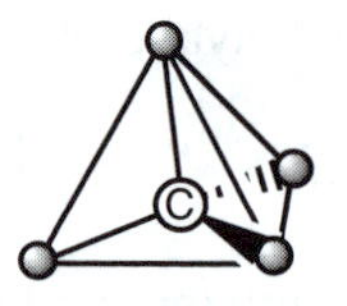

a) tetrahedral (109.5°)

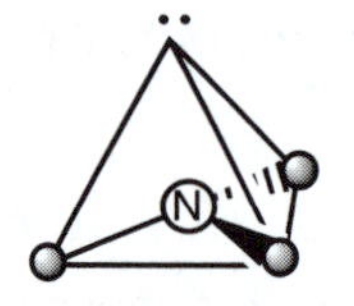

b) trigonal pyramid (107.3°)

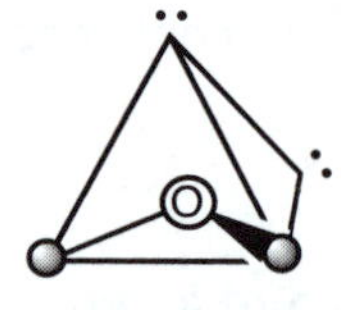

c) bent (104.5°)

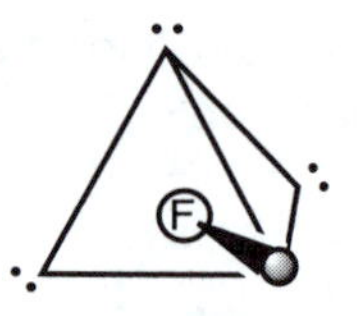

d) linear

(The terms describing the geometry refer to the nuclei, not the electron clouds.)

In each case, the bond angles are close to tetrahedral.

In the drawings, the shaded balls represent hydrogen atoms, and the lines connecting the balls to the central atom represent two-electron covalent bonds. The lines connecting hydrogen atoms are only there to emphasize the tetrahedral geometry of the electron clouds (they do not represent electrons). We can draw the molecules without these lines, and use "rabbit ears" for the lone pairs, to come up with the following structures:

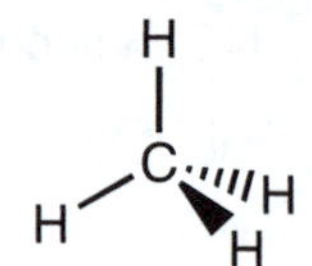

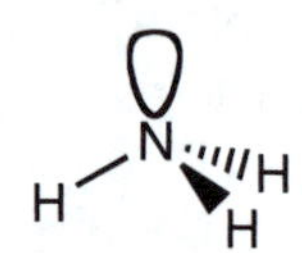

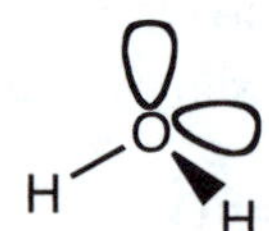

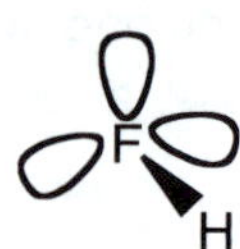

The above example demonstrates several things:

1) The decrease in bond angles from methane to ammonia to water shows that lone pairs take up more room than bonded electron pairs. It may be useful to draw a lone pair as a large loop to remind yourself of the extra space needed for a lone pair.

2) A tetrahedral arrangement allows four electron pairs to be as close to the nucleus as possible.

3) We can talk about the geometry of the electrons (tetrahedral for all cases above), or the structure (the geometry) of the molecules. In order to predict the structure of a molecule, we need to know the arrangement of the electrons.

Trigonal Planar vs. Trigonal Pyramid

The atoms in an ammonia molecule form a trigonal pyramid because the lone pair on nitrogen takes up one of four tetrahedral sites. Imagine (hypothetically) removing the lone pair from nitrogen: the three hydrogen atoms would spread out into a trigonal planar arrangement to minimize electron repulsion. Which element has two fewer electrons than nitrogen? (Answer: boron.)

3 groups around a central atom

1) What is the geometry of a boron trichloride molecule (BCl_3)?

solution:

The number of valence electrons in BCl_3 = 3 + (3 × 7) = 24 (= 12 pairs):

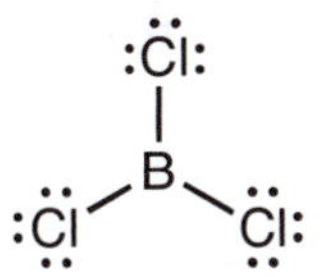

Boron is electron deficient with six electrons. The geometry of boron trichloride is trigonal planar. The bond angles are 120°.

2) What is the geometry of a carbonate molecule (CO_3^{2-})?

solution:

The number of valence electrons in CO_3^{2-} = 4 + (3 × 6) + 2 = 24 (= 12 pairs):

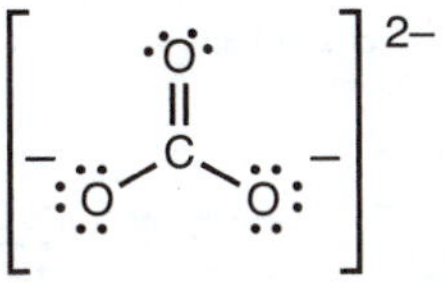

A double bond counts as one "domain." Three domains around carbon make its arrangement of atoms (and electron pairs) trigonal planar. The bond angles are 120°.

3) What is the geometry of a molecule of formaldehyde (CH_2O)?

solution:

The number of valence electrons in CH_2O = 4 + (2 × 1) + 6 = 12 (= 6 pairs):

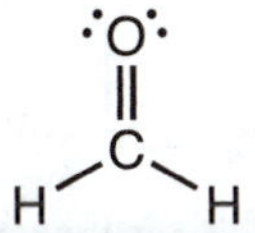

The arrangement of atoms is trigonal planar, but the bond angles are distorted from 120° because of the different atoms around carbon.

Bent vs. Linear

Water is a bent molecule because its arrangement of electron pairs is tetrahedral: the two lone pairs on oxygen take up tetrahedral sites and the two bonds to hydrogen take up the other two tetrahedral sites. Imagine (hypothetically) removing the two lone pairs from the oxygen. The two hydrogen atoms would spread out into a linear arrangement to minimize electron repulsion. What element has four fewer electrons than oxygen? (Answer: Beryllium.)

2 groups around a central atom

1) What is the geometry of gaseous beryllium chloride ($BeCl_2$)?

solution:

The number of valence electrons in $BeCl_2$ is $2 + (2 \times 7) = 16$ (= 8 pairs):

:Cl—Be—Cl:

Beryllium is electron deficient with four electrons around it. The shape of $BeCl_2$ is linear.

2) What is the shape of carbon dioxide (CO_2)?

solution:

The number of valence electrons in CO_2 is $4 + (2 \times 6) = 16$ (= 8 pairs):

O=C=O

Carbon dioxide is linear.

3) What is the shape of ozone (O_3)?

solution:

O_3 has $6 \times 3 = 18$ valence electrons (= 9 pairs):

$O{=}O^{+}{-}O^{-} \longleftrightarrow {}^{-}O{-}O^{+}{=}O$

Ozone is a bent molecule.

Five groups around a central atom produce an interesting situation where not all the five sites are geometrically equivalent. Phosphorus pentachloride forms a **trigonal bipyramid** with three chlorine atoms in axial positions, and two chlorine atoms in equatorial positions:

equatorial positions → Cl—P(Cl)(Cl)(Cl)(Cl) ← axial positions

The exercise on the next page demonstrates four molecular structures that may arise from trigonal bipyramidal arrangements of electron pairs.

5 groups around a central atom

Given the bond angles in a trigonal bipyramid:

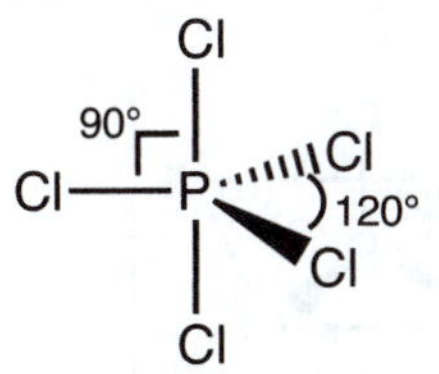

1) Where would a lone pair exist to minimize electron-electron repulsions?

solution:

A lone pair takes an equatorial site, which has only two neighbors 90° from it; an axial site has three neighbors 90° away.

2) What is the shape of sulfur tetrachloride (SCl_4)?

solution:

The number of valence electrons in SCl_4 is $6 + (4 \times 7) = 34$ (= 17 pairs):

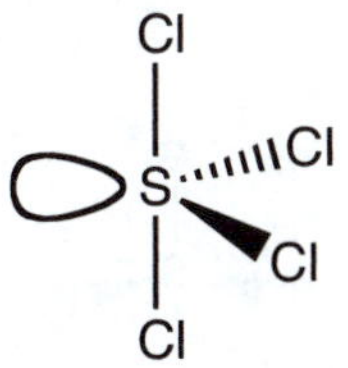

The lone pair on sulfur takes an equatorial site of the trigonal bipyramid; the geometry of the atoms is something like a seesaw. Lone pairs distort bond angles from the ideal 90° and 120°.

3) What is the shape of bromine trifluoride?

solution:

The number of valence electrons in BrF_3 is $7 + (3 \times 7) = 28$ (= 14 pairs):

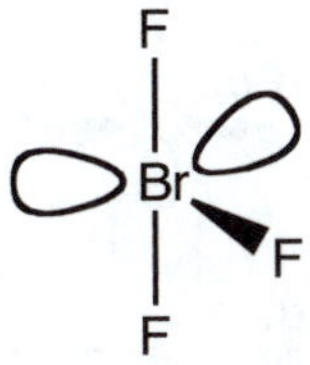

The two lone pairs on bromine take two equatorial sites. The atoms form a distorted "T" shape.

4) What is the shape of the triiodide ion, I_3^-?

solution:

The number of valence electrons in I_3^- is $(3 \times 7) + 1 = 22$ (= 11 pairs):

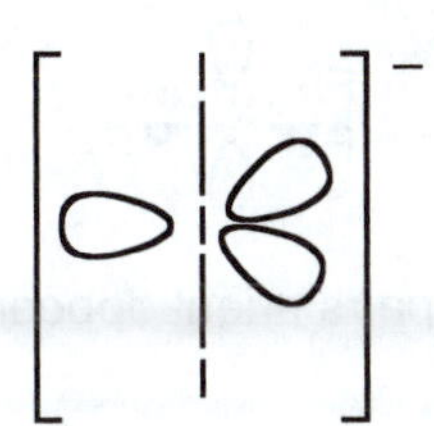

The three lone pairs all take equatorial positions. The ion is linear.

Six groups around a central atom form an octahedral arrangement. Don't get confused by the term octahedral for six groups: if we filled in the sides, there would be eight to form an octahedron. Notice that all the positions are equivalent. Once an electron pair takes a site, however, the second electron pair minimizes its interactions with the first electron pair by taking the site opposite (180°) from the first lone pair:

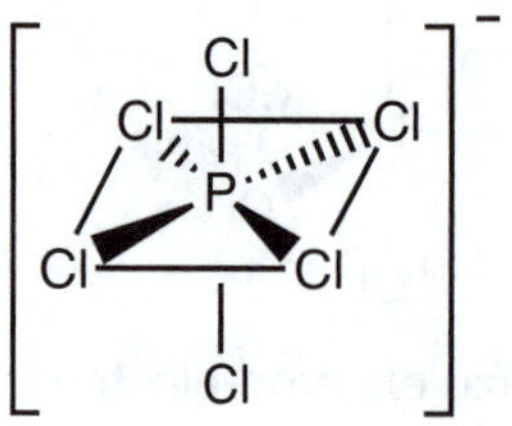

Cl
Cl Cl
P
Cl Cl
Cl

6 groups around a central atom

1) What is the shape of sulfur hexafluoride (SF_6)?

solution:

The number of valence electrons in $SF_6 = 6 + (6 \times 7) = 48$ (= 24 pairs, of which $3 \times 6 = 18$ are lone pairs on F, not shown below):

F
F S F
F F
F

2) What is the shape of iodine pentafluoride (IF_5)?

solution:

The number of valence electrons in $IF_5 = 7 + (5 \times 7) = 42$ (= 21 pairs):

F
F I F
F F

Iodine pentafluoride is a **square pyramid.**

3) What is the shape of xenon tetrafluoride (XeF_4)?

solution:

The number of valence electrons in $XeF_4 = 8 + (4 \times 7) = 36$ (= 18 pairs):

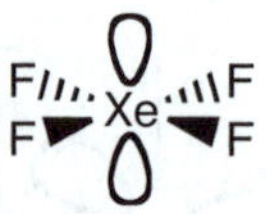

F Xe F
F F

XeF_4 is **square planar.** The electron pairs orient opposite each other to minimize lone-pair/lone-pair repulsions.

Additional exercises

1) Describe the environment of the carbon atoms in acetone (CH_3COCH_3).

solution:

The number of valence electrons in acetone is:

carbons: $(3 \times 4) = 12$; hydrogens: $(6 \times 1) = 6$; oxygens: 6; Total: 24 (= 12 pairs)

The Lewis structure of acetone is:

The structure of the carbon atoms bonded to 3 hydrogen atoms and the central carbon atom is tetrahedral.

The structure of the central carbon atom with three groups (a double bond and two single bonds) is trigonal planar.

2) What would you predict the geometry of phosphine (PH_3) to be?

solution:

The number of valence electrons in $PH_3 = 5 + (3 \times 1) = 8$.
The Lewis structure of phosphine is the same as ammonia:

We would predict that the electron group arrangement of phosphine with four groups around it (three single bonds and a lone pair), like ammonia, would be tetrahedral and therefore would have bond angles close to 109°. (Remember the bond angles in ammonia (107°) are slightly less than 109° because the lone pair takes up more room than a bonded electron pair, and pushes the bonded atoms together.) However, unexpectedly, the bond angles in phosphine are 94°, almost right angles! In general, as atoms get larger, they contain a more "spread out" lone pair:

3) What is the Lewis structure for XeO_3? What are the formal charges on Xe and the O's?

solution:

The number of valence electrons in XeO_3 is: $8 + (3 \times 6) = 26$
The best Lewis structure is: (Xe expands its octet to give a structure with no formal charges.)

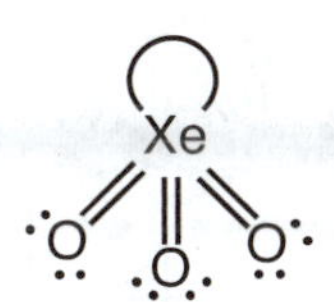

Summary of Molecular Geometry

Number of electron groups	Arrangement of electron groups	Bond angle(s)
2	**linear**	**180°**
Examples: O=C=O	Cl—Be—Cl	H—C≡N
3	**trigonal planar**	**120°**
Examples: BCl_3	$H_2C{=}O$	SO_3
4	**tetrahedral**	**109°**
Example: CH_4		
5	**trigonal pyramid**	**90° & 120°**
Example: PCl_5		
6	**octahedral**	**90°**
Example: SF_6		

The Common Cases of 4 Electron Pair Arrangements

2 electron groups: CO_2, HCN

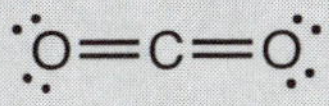

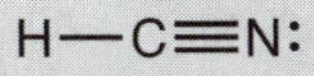

:Ö=C=Ö: H—C≡N:

3 electron groups: CO_3^{2-} (trigonal planar), O_3 (bent)

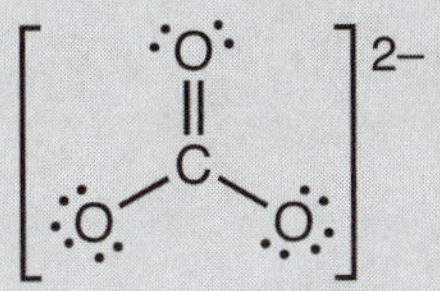

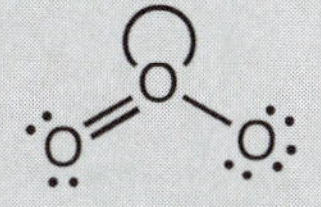

(where ⌒ = lone pair)

4 electron groups: CH_4, NH_3, H_2O, HF

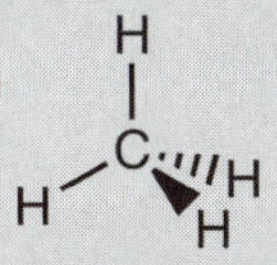

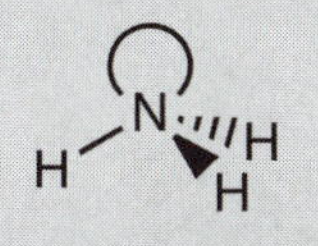

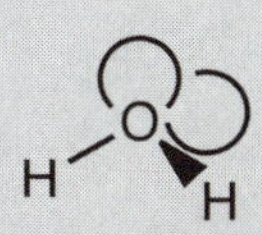

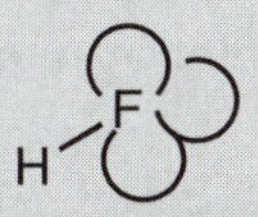

The Common Cases of 5 Electron Pair Arrangements

3 electron groups: FSN, SO_2

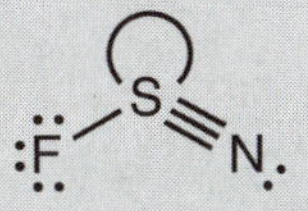

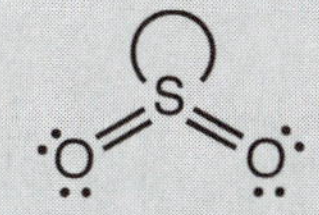

4 electron groups: SOF_2

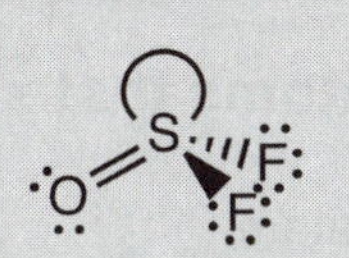

5 electron groups: PF_5, SF_4, ClF_3, XeF_2

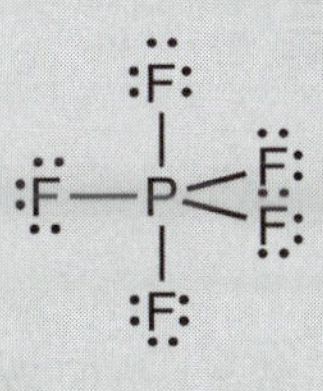

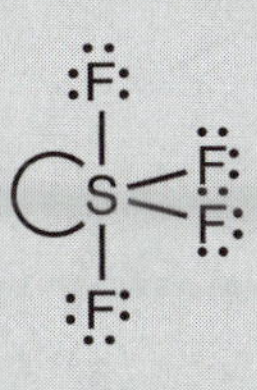

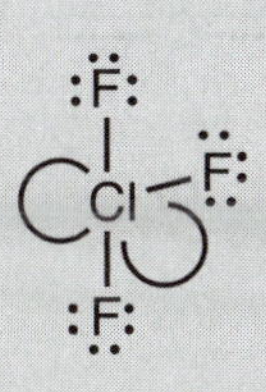

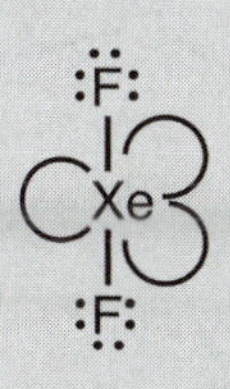

(continued on next page)

The Common Cases of 6 Electron Pair Arrangements

3 electron groups: SO_3

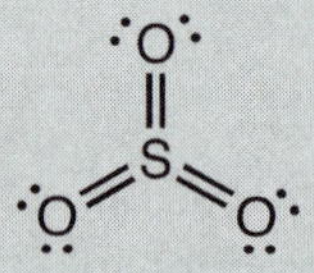

4 electron groups: SO_2F, NSF_3

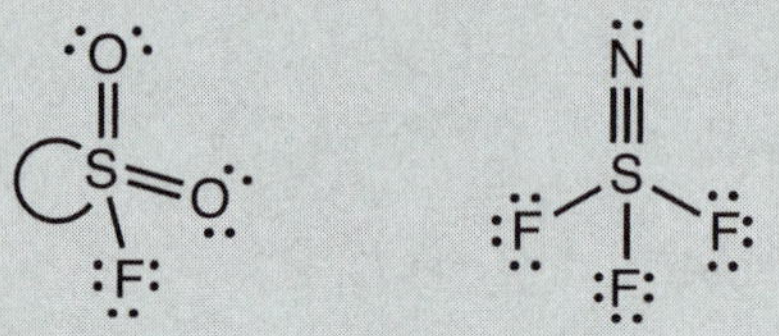

5 electron groups: SOF_4

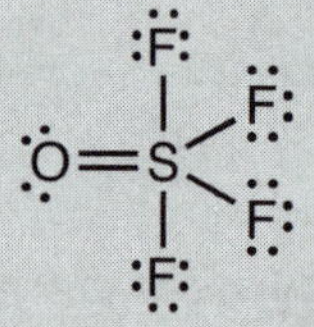

6 electron groups: SF_6, ClF_5, XeF_4

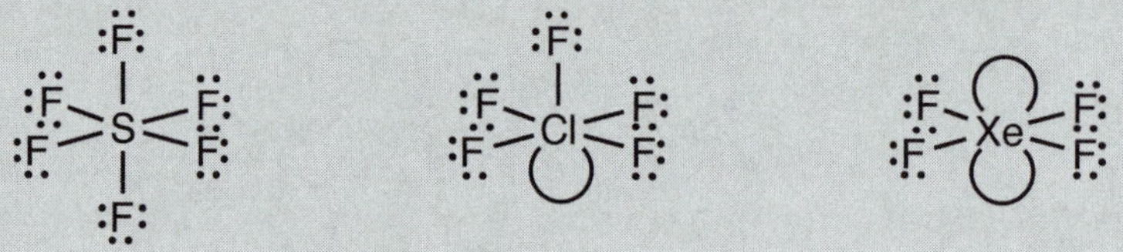

0 formal charge environments for atoms that satisfy the octet rule

Determine the formal charge on each atom.

1) Carbon

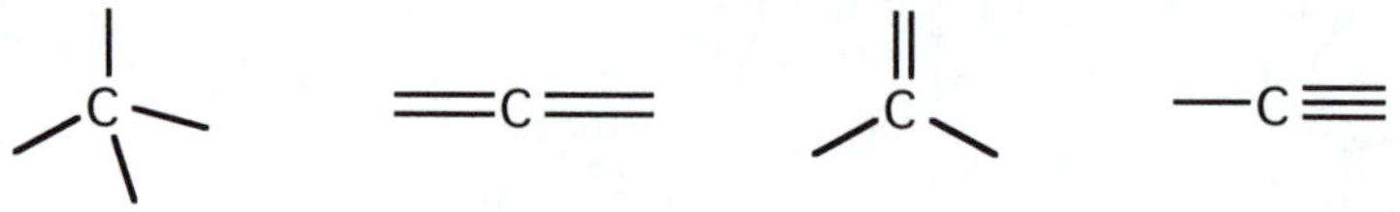

a) b) c) d)

(continued on next page)

2) Nitrogen

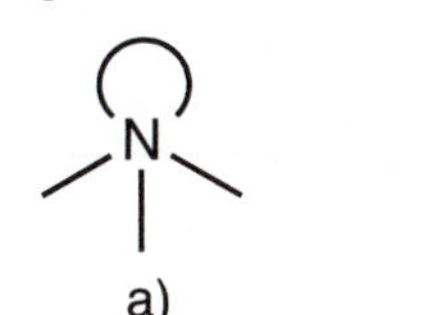
a)

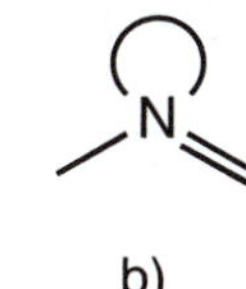
b)

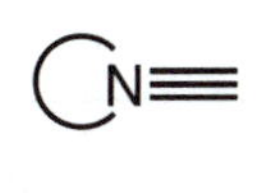
c)

3) Oxygen

a)

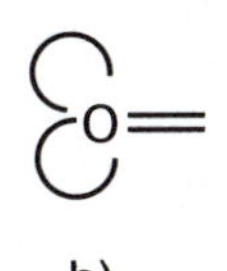
b)

4) Fluorine

solutions:

All atoms are in 0 formal charge environments. Can you think of specific molecules with electron arrangements as depicted in 1-4 above?

Some answers:
1. a) CH_4, b) CO_2, c) H_2CO, d) HCN
2. a) NH_3, b) HNO, c) N_2
3. a) H_2O, b) O_2
4. HF

10.3 MOLECULAR SHAPE AND MOLECULAR POLARITY

The geometry of some molecules causes the molecules to be polar. We saw in Chapter 9 that atoms of different electronegativity share electrons unequally. The polar covalent bond in H–F causes that molecule to have a **dipole moment:**

H—F:

The polar bonds in a linear molecule such as BeF_2 cancel each other out so the molecule has no net dipole:

:F—Be—F:

If water were a linear molecule, it would have no net dipole. However, the bent geometry of water causes it to have polarity, regions of positive and negative charge:

H—O—H

Your textbook shows an electron density model of a water molecule on page 417. You can see from this model that the negative part of the molecule is around the oxygen atom, and the more positive region is between the hydrogen atoms.

Polar molecules

Which of the following molecules would you expect to be polar?

a) CO_2
b) $SiCl_4$
c) SCl_4
d) XeF_4

solutions:

a) The structure of CO_2 is:

O=C=O

Since it is a linear molecule, its polar bonds cancel each other out, and there is no net dipole. Page 417 in your textbook shows an electron density model of carbon dioxide. Regions of high negative charge around the oxygens are distributed equally on either side of the central region of positive charge around the carbon.

b) The structure of $SiCl_4$ is tetrahedral:

:Cl: Si Cl: Cl: :Cl

There is no net dipole.

c) The structure of SCl_4 is a seesaw:

:Cl: S Cl: Cl: :Cl:

This molecule has a dipole moment.

d) The structure of XeF_4 is square planar:

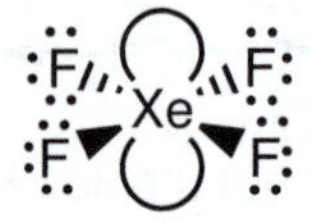

The symmetry of this molecule causes it to have no net dipole.

The Effect of Molecular Polarity on Physical Properties

Polar molecules have stronger intermolecular forces than nonpolar molecules and therefore have higher boiling points. As your textbook describes on page 419, the *cis* isomer of dichloroethylene ($C_2H_2Cl_2$), which is a polar molecule, has a higher boiling point by over 10°C than the *trans* isomer, a nonpolar molecule. Polar molecules will best dissolve in polar solvents, while nonpolar molecules are best solvated by nonpolar solvent molecules.

CHAPTER CHECK

Make sure you can...

- Draw Lewis structures using the octet rule and your knowledge of valence electrons.
- Identify molecules whose Lewis structures are most accurately shown using resonance structures.
- Determine formal charges of atoms in molecules.
- Use VSEPR theory to predict molecular geometry.
- Show the common cases of 4, 5, and 6 electron pair arrangements.
- Use molecular geometry to predict which molecules have dipole moments.

Chapter 10 Exercises

10.1) a. What is the formal charge on the nitrogen atom in the Lewis electron-dot structure for the nitrite ion NO_2^-?

b. What is the formal charge on the carbon atom in the Lewis electron-dot structure for the carbonate ion CO_3^{2-}?

c. What is the formal charge on the sulfur atom in the electron-dot structure for the molecule of sulfur dioxide in which one of the bonds is written as a single bond and the other is written as a double bond?

10.2) For which of the following molecules or polyatomic ions is the drawing of resonance structures desirable in order to achieve a more realistic picture of the bonding?

N_2 SO_4^{2-} C_2H_2 CCl_4 CO_3^{2-} H_2O CH_4 NaCl NO_2^- BF_4^-

10.3) Which of the following twelve molecules do not obey the octet rule?

HF	PF_3	OF_2	XeF_2	$CHCl_3$	PCl_5
CH_4	BCl_3	H_2S	NH_3	CO_2	SF_6

10.4) How many nonbonding pairs of electrons do the following compounds have in valence orbitals on the central atom?

NH_3	H_2O	BF_3	$BeCl_2$	HCN
H_2S	CH_4	OF_2	PCl_3	$CHCl_3$

10.5) Which of the following species have a trigonal bipyramidal arrangement of electron pairs around the central atom?

PCl_5	SF_4	BrF_3	SOF_4	I_3^-
TeF_5^-	ICl_4^-	IOF_5	BrF_5	NH_3

10.6) What is the arrangement of electron pairs, bonding and nonbonding, around the central atoms, and what are the shapes of these molecules or ions?

ICl_4^- SOF_2 NO_3^-

10.7) With which one of the five VSEPR electron pair arrangements are the "square planar" and "square pyramidal" molecular geometries associated?

10.8) Write the formula for a molecule that you would expect to have the same Lewis structure as:

NH_4^+ ClO^- NO_3^-

10.9) Which two of the following molecules have dipole moments?

CH_4	XeF_4	SO_3	SF_6
H_2S	PCl_3	BF_3	XeF_2

Chapter 10 Answers

10.1) a. 0; b. 0; c. +1

10.2) Resonance structures are drawn when different arrangements of single and double bonds are possible. The actual structure, and the actual distribution of electrons in the molecule, is somewhere between the extremes represented by the resonance structures.

The sulfate ion, the carbonate ion, and the nitrite ion are all molecules for which different arrangements of bonds can be drawn. For example, the sulfate ion is usually drawn with two double S=O bonds and two S–O single bonds. These four bonds can be arranged in six different ways making six different resonance structures. In reality, the electrons are distributed so that each bond has a bond order of 1.5, and each bond is the same.

10.3) The octet rule is often not obeyed. Although terminal atoms in a molecule almost always have a complete set of 8 electrons in their valence shells (except H), elements other than C,N,O, and F often disobey the rule when they are the central atom in a molecule or polyatomic ion:

XeF_2 has 10 electrons in the valence shell of Xe

PCl_5 has 10 electrons in the valence shell of P

BCl_3 has only 6 electrons in the valence shell of B

SF_6 has 12 electrons in the valence shell of S

10.4) Use the VSEPR approach to determine the number of bonding and nonbonding pairs of electrons around the central atom in each molecule:

NH_3	one lone pair on the N	H_2S	two lone pairs on the S
H_2O	two lone pairs on the O	CH_4	no lone pairs on the C
BF_3	no lone pairs on the B	OF_2	two lone pairs on the O
$BeCl_2$	no lone pairs on the Be	PCl_3	one lone pair on the P
HCN	no lone pairs on the C	$CHCl_3$	no lone pairs on the C

10.5) Do not be misled by the number of terminal atoms in the molecule or ion. Use the VSEPR approach to determine the total number of bonding *and* nonbonding pairs of electrons around the central atom. It is the *total* number of electron pairs that ultimately determines the geometry of the molecule or polyatomic ion:

PCl_5	trigonal bipyramid	TeF_5^-	octahedral
SF_4	trigonal bipyramid	ICl_4^-	octahedral
BrF_3	trigonal bipyramid	IOF_5	octahedral
SOF_4	trigonal bipyramid	BrF_5	octahedral
I_3^-	trigonal bipyramid	NH_3	tetrahedral

10.6)

ICl_4^-	octahedral arrangement	two lone pairs	square planar molecule
SOF_2	tetrahedral arrangement	one lone pair	trigonal pyramidal molecule
NO_3^-	trigonal planar arrangement	no lone pairs	trigonal planar molecule

10.7) Both the square planar and the square pyramidal geometries arise from an octahedral arrangement of electron pairs around the central atom. In the square planar geometry there are two lone pairs opposite one another on two sides of the plane. In the square pyramidal geometry there is one lone pair.

10.8) Nitrogen with an electron removed has the same number of electrons as carbon: CH_4
Oxygen with an electron added has the same number of electrons as chlorine: Cl_2
Nitrogen with an electron added has the same number of electrons as oxygen. There is no such species as O_4, but sulfur has the same number of valence electrons as oxygen: SO_3

10.9)

CH_4	tetrahedral	nonpolar
SO_3	trigonal planar	nonpolar
BF_3	trigonal planar	nonpolar
H_2S	V-shaped	*polar* (dipole moment)
XeF_4	square planar	nonpolar
SF_6	octahedral	nonpolar
PCl_3	trigonal pyramid	*polar* (dipole moment)
XeF_2	linear	nonpolar

11

Theories of Covalent Bonding

11.1 VALENCE BOND (VB) THEORY AND ORBITAL HYBRIDIZATION

We have used the valence-shell electron-pair repulsion theory to help us predict the shapes of molecules. Now we will look at another model that describes the bonds in molecules as the overlap of a half-filled atomic orbital of one atom with a half-filled atomic orbital from another atom. The space formed by the overlapping orbitals can hold two electrons with opposite spins.

Valence bonds in simple molecules

Consider a hydrogen molecule, H_2. Describe the orbitals that overlap to form its bond.

solution:

First we will write the electron configuration for the two hydrogen atoms:

H: $\frac{\uparrow}{1s}$ H: $\frac{\uparrow}{1s}$

The two atomic orbitals overlap to form a bond:

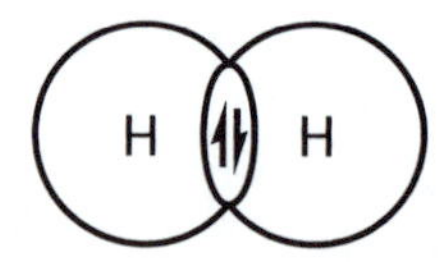

This bond, which is cylindrically symmetric, is called a sigma (σ) bond. Its cross section resembles an *s* orbital.

Valence Bonds in More Complex Molecules (Hybridization)

We saw that VSEPR theory predicts that methane, CH_4, has a tetrahedral geometry. Let's consider the orbitals that overlap to form the bonds in methane (CH_4).

The electron configurations for the carbon and hydrogen atoms are:

C: ↑↓ (1*s*) ↑↓ (2*s*) ↑ ↑ __ (2*p*)

H's: ↑ (1*s*)

It is not obvious how to make four bonds to carbon since there are only <u>two</u> half-filled orbitals. If we excite one electron into the empty 2*p* orbital from the 2*s* orbital, we would have four half-filled orbitals:

↑↓ (1*s*) ↑↓ (2*s*) ↑ ↑ __ (2*p*) ⟶ ↑↓ (1*s*) ↑ (2*s*) ↑ ↑ ↑ (2*p*)

What would you expect the bond angles to be of bonds formed between hydrogen's 1*s* orbital, and the 2*p* orbitals from carbon? (<u>Answer</u>: Since the carbon 2*p* orbitals are perpendicular to one another, the C–H bonds formed with these orbitals would be oriented at 90° angles from one another.)

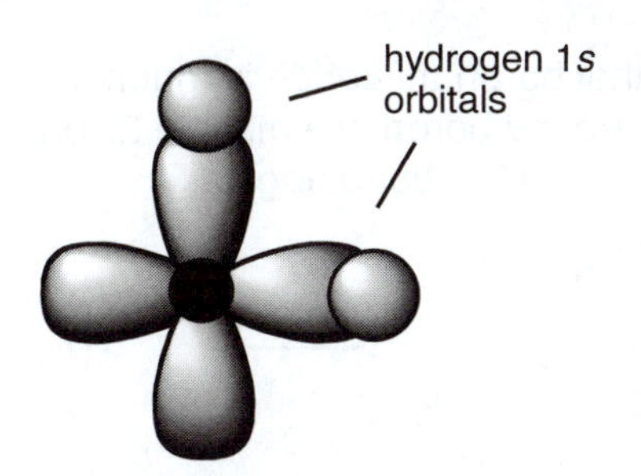

In the drawing, the black ball represents a carbon core with 2 of its 2*p* orbitals. The hydrogen 1*s* orbitals overlap to form bonds oriented at 90° angles.

However, we know from experiment that a methane molecule is tetrahedral with bond angles of 109.5°.

Linus Pauling works us around this problem with his proposal that <u>orbitals in a molecule are different from the orbitals in the isolated atoms</u>. We can combine the 2*s* orbital and the three 2*p* orbitals mathematically to give new **hybrid** atomic orbitals, which have 109.5° bond angles between them. We call these hybrid orbitals ***sp*³**:

↑↓ (1*s*) ↑↓ (2*s*) ↑ ↑ __ (2*p*) ⟶ ↑↓ (1*s*) [↑ ↑ ↑ ↑ (*sp*³)]

four equivalent hybrid *sp*³ orbitals

Now four hydrogen atoms can combine with the four <u>equivalent</u> half-filled *sp*³ atomic orbitals on carbon to form a tetrahedral methane molecule:

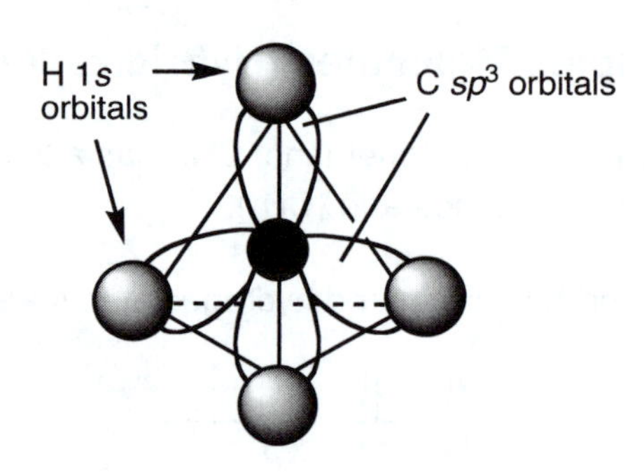

Valence bond theory and hybridization

1) How would the valence bond theory describe the atomic orbitals of nitrogen in ammonia (NH_3) and the atomic orbitals of oxygen in water (H_2O)?

   solution:

   Since the bond angles in ammonia and water are close to tetrahedral, we use sp^3 hybrids to describe the bonding in these molecules. The lone pair on nitrogen in ammonia, and the two lone pairs on oxygen in water reside in nonbonding sp^3 orbitals.

2) Describe the orbitals that overlap to form the bonds in boron trifluoride (BF_3).

   solution:

   We know from VSEPR theory that boron trifluoride is trigonal planar with 120° bond angles. To make three equivalent orbitals around boron, we mix the 2*s* orbital and two of the 2*p* orbitals to obtain three hybrid ***sp***^{2} orbitals with 120° bond angles:

   B: ↑↓ (1*s*) ↑↓ (2*s*) ↑ _ _ (2*p*) ⟶ ↑↓ (1*s*) [↑ ↑ ↑ (sp^2)] _ (2*p*)

   F's: ↑↓ (1*s*) ↑↓ (2*s*) ↑↓ ↑↓ ↑ (2*p*)

   The half-filled $3p^2$ orbitals on boron overlap the half-filled 2*p* orbitals of the fluorine atoms to form three sigma bonds:

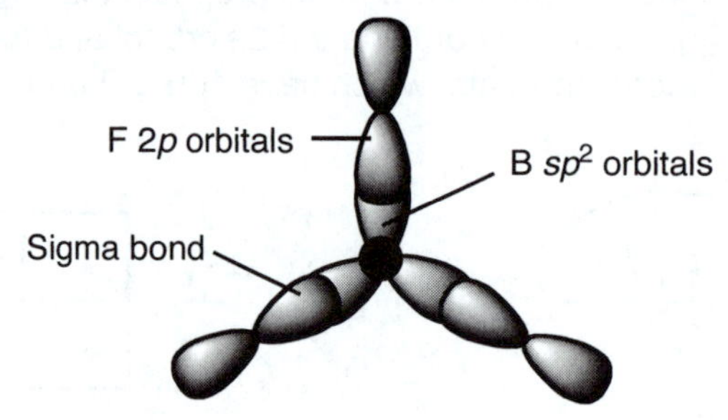

   Notice that two of boron's *p* orbitals are hybridized, but the third *p* orbital remains unhybridized and lies perpendicular to the plane made by the hybridized bonds. Interaction between this empty *p* orbital and a lone pair of electrons on fluorine causes the B–F bond distance to be shorter than expected for a purely single B–F bond.

(continued on next page)

3) Describe the bonding in carbon dioxide (CO_2).

<u>solution</u>:

Carbon dioxide is a linear molecule containing double bonds between carbon and the oxygen atoms:

O=C=O

Carbon needs a pair of atomic orbitals oriented in opposite directions. This requires a new type of hybridization which mixes one *s* orbital and one *p* orbital to form 2 ***sp*** hybrid orbitals oriented 180° from each other:

C: ↑↓ (1*s*) ↑↓ (2*s*) ↑ ↑ _ (2*p*) ⟶ ↑↓ (1*s*) [↑ ↑ (*sp*)] ↑ ↑ (2*p*)

Oxygen has three groups around it (a double bond and two lone pairs) and so requires sp^2 hybridization to obtain a trigonal planar structure:

O: ↑↓ (1*s*) ↑↓ (2*s*) ↑↓ ↑↓ _ (2*p*) ⟶ ↑↓ [↑↓ ↑↓ ↑ (sp^2)] ↑ (2*p*)

The sigma bonds in carbon dioxide form from overlap of the carbon half-filled *sp* hybrid orbitals with the half-filled oxygen sp^2 orbital on each oxygen (see drawing below). The pi bonds form from the side-to-side overlap of the electrons in the unhybridized *p* orbitals from carbon and oxygen:

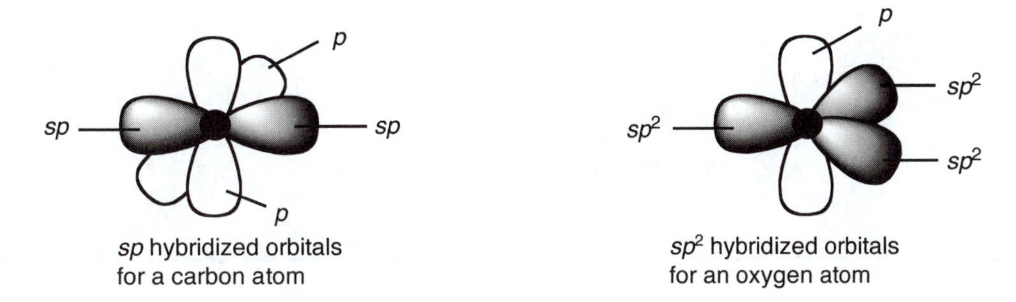

sp hybridized orbitals for a carbon atom

sp^2 hybridized orbitals for an oxygen atom

4) The shapes of molecules with expanded octets have atoms with atomic *d* orbitals available to mix with *s* and *p* orbitals. A trigonal bipyramidal shape requires hybrid orbitals mixed from one "*s*," three "*p*," and one "*d*" orbital. An octahedral shape requires hybrid orbitals mixed from one "*s*," three "*p*," and two "*d*" orbitals. What would be the name for these two hybrids?

<u>solution</u>:

Trigonal bipyramid: five sp^3d hybrid orbitals

Octahedral: six sp^3d^2 hybrid orbitals

On the following page, we summarize the shapes and bond angles of hybrid orbitals.

Summary of Hybrid Orbitals

Hybrid orbital	Number of groups	Shape	Bond angles	Example
sp	2	linear	180°	O=C=O
sp^2	3	trigonal planar	120°	SO_3
sp^3	4	tetrahedral	109.5°	CH_4
sp^3*d*	5	trigonal bipyramid	120° and 90°	PCl_5
sp^3*d*^{2}	6	octahedral	90°	SF_6

11.2 TYPES OF COVALENT BONDS

P orbitals can form two types of bonds. Two *p* orbitals can overlap end-to-end to form a sigma (σ) bond, which has its highest electron density along the bond axis:

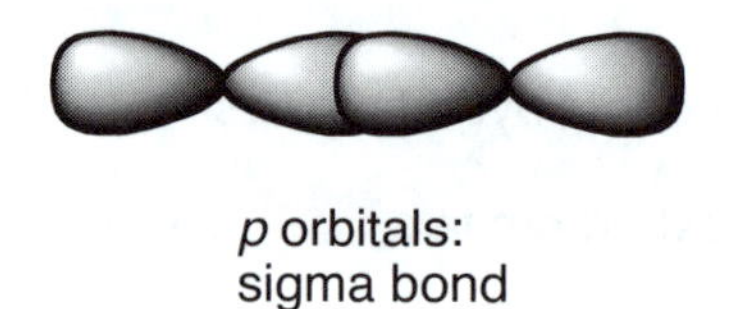

p orbitals:
sigma bond

P orbitals can also overlap side-to-side to form a "pi" (π) bond. In cross section the pi orbital resembles a *p* orbital:

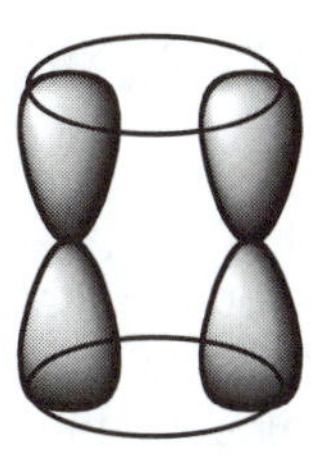

p orbitals:
"pi" bond

A double bond, in this picture of bonding, contains one σ bond and one π bond. For example, a carbon-carbon double bond (represented below) consists of a σ bond, in which the shared electrons exist between the atoms, and a π bond, in which the shared electrons spend time above and below the σ bond. The π bond forms from *p* orbitals on the two carbon atoms. The σ bond forms from hybridized "sp^2" orbitals on the carbon atoms (discussed later):

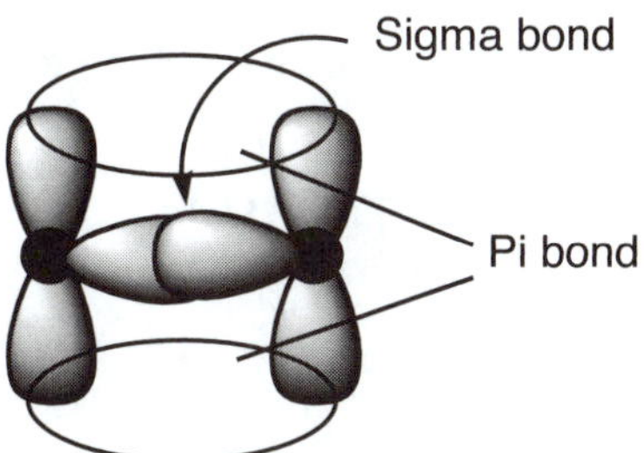

A carbon-carbon bond consisting of a sigma bond and a pi bond.

A triple bond contains one σ bond and two π bonds.

Summary of Covalent Bond Types

Sigma (σ) bonds form when orbitals overlap end-to-end. The orbitals that overlap may be hybridized such as sp^3, or they may be pure *s* or *p* orbitals. All single bonds are sigma bonds. The highest electron density exists along the bond axis. A sigma bond allows free rotation of the groups around the bond because the extent of orbital overlap is not affected as the groups rotate. Figure 11.9 on page 437 in your textbook shows the sigma bonds in ethane, C_2H_6. What are the two types of sigma bonds in this molecule? (Answer: The bond between the carbon atoms is a sigma bond formed by overlap of two sp^3 hybridized orbitals. Each of the six carbon-hydrogen bonds is a sigma bond formed by overlap between a sp^3 hybridized orbital from carbon and a 1*s* orbital from a hydrogen atom. All bonds in ethane rotate freely at room temperature.)

Pi (π) bonds form when parallel *p* orbitals overlap side-to-side. A pi bond has two regions of electron density, one above and one below the sigma bond axis. Pi bonds restrict rotation because rotation would require breaking the pi bond. Double bonds consist of one sigma and one pi bond. Triple bonds consist of one sigma and two pi bonds. Figure 11.10 in your textbook shows the sigma and pi bonds in the double bond of ethylene, C_2H_4, and Figure 11.11 shows the one sigma and two pi bonds in the triple bond of acetylene (C_2H_2). Be able to explain why separate *cis* and *trans* structures can exist in molecules containing double bonds.

Double bonds are approximately twice as strong as single bonds, and triple bonds are approximately three times as strong.

11.3 MOLECULAR ORBITAL (MO) THEORY AND ELECTRON DELOCALIZATION

In valence bond theory, an atom forms hybrid orbitals from combinations of its atomic orbitals. These orbitals overlap with orbitals from other atoms to form bonds containing localized electrons. In MO theory, as two atoms come together to form a molecule, the molecule forms new orbitals that are delocalized over the entire molecule. Each molecular orbital can be occupied by two electrons and has a particular energy (and shape). We can write electron configurations for the molecule just as we write electron configurations for atoms. MO theory is useful for explaining the magnetic and spectral properties of molecules, but it is more difficult to picture molecular orbitals than the hybrid orbitals of valence bond theory, which are sometimes called Localized Molecular Orbitals (LMO).

The MO Description of the Bond in a Hydrogen Molecule

When two hydrogen atoms bond, their 1*s* orbitals combine to form two molecular orbitals:

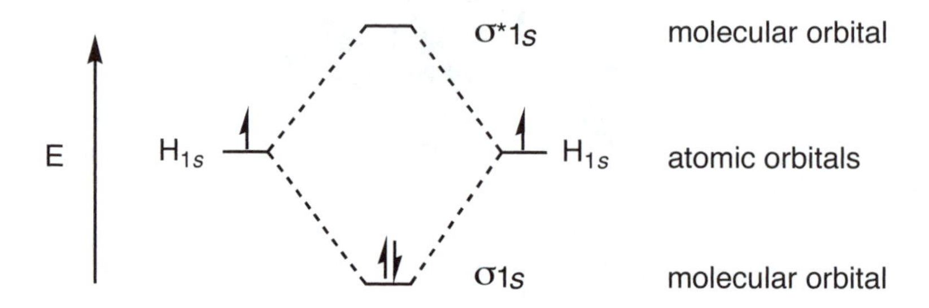

The $\sigma 1s$ orbital (the bonding orbital) is lower in energy than the atomic orbitals from which it is composed; the $\sigma^* 1s$ orbital (the antibonding orbital) is higher in energy than the original atomic orbitals. The bonding MO is spread mostly between the nuclei of the hydrogen atoms; the antibonding MO has a node between the nuclei and when it is occupied, the molecule is less stable than the separate atoms. Atomic orbitals must have similar energy and orientation to overlap effectively.

We fill molecular orbitals with electrons the same way we fill atomic orbitals:

- Orbitals fill in order of increasing energy.
- An orbital can hold a maximum of two electrons with opposite spins.
- Electrons do not pair up until all orbitals of equal energy contain one electron.

Molecular orbitals

Choose words from the list below to complete the following paragraph:

(a) antibonding	(c) between	(e) energy	(g) lower
(b) atomic	(d) beyond	(f) higher	(h) valence

When a molecule forms, atomic orbitals of similar ______________(1), and hence of similar size, and the right symmetry overlap to form molecular orbitals. Inner atomic orbitals are too low in energy (and therefore too small) to overlap sufficiently to form molecular orbitals; it is the ______________(2) atomic orbitals which combine to form molecular orbitals. Some combinations of atomic orbitals form bonding

(continued on next page)

molecular orbitals, orbitals that are ______________(3) in energy than the separate atomic orbitals. Some combinations of atomic orbitals form antibonding molecular orbitals, orbitals that are ______________(4) in energy than the separate atomic orbitals. Every bonding molecular orbital has a corresponding ______________(5) orbital. Electrons in bonding orbitals have a high probability of being found ______________(6) the nuclei; electrons in antibonding orbitals have a high probability of being found ______________(7) the nuclei. The number of molecular orbitals formed always equals the number of ______________(8) orbitals combined to form them.

answers: 1-e, 2-h, 3-g, 4-f, 5-a, 6-c, 7-d, 8-b.

Hydrogen's MO's

Revisit the drawing at the beginning of this section for the molecular orbitals of a H_2 molecule.

1) What is the electron configuration for the H_2 molecule?

solution: $(\sigma 1s)^2$

The σ symbol indicates the orbital's symmetry. "1*s*" indicates the atomic orbitals used to form the molecular orbital. (An asterisk (*) indicates an antibonding orbital.)

2) What is the bond order for this molecule?

solution:

bond order = 1/2[(# of e^- in bonding MO's) – (# of e^- in antibonding MO's)]
= 1/2 (2) = 1

Note: A bond order > 0 implies the molecule is more stable than the separate atoms. In general, the higher the bond order, the stronger the bond.

3) Would you predict that H_2^+ would exist? H_2^-? H_2^{2-}? What are their electron configurations?

solution:

If we remove an electron from the MO diagram of H_2 to form H_2^+, the bond order is:
1/2 (1) = 1/2; electron configuration = $(\sigma 1s)^1$

If we add an electron to the MO diagram of H_2 to form H_2^-, the bond order is:
1/2 (2 – 1) = 1/2; electron configuration = $(\sigma 1s)^2(\sigma^* 1s)^1$

If we add two electrons to the MO diagram of H_2 to form H_2^{2-}, the bond order is:
1/2 (2 – 2) = 0; electron configuration = $(\sigma 1s)^2(\sigma^* 1s)^2$

H_2^+ and H_2^- have bond orders > 0, so they should exist (in fact, they have both been detected spectroscopically). The bond order of H_2^{2-} = 0, so this molecule is predicted not to exist.

The MO Description of the Bonding in an Oxygen Molecule

An oxygen atom contains valence electrons in its 2*s* and 2*p* orbitals. When two oxygen atoms bond, their 2*s* orbitals overlap to form σ bonding and antibonding orbitals. One 2*p* orbital from each oxygen atom overlaps end-to-end to form σ bonding and antibonding orbitals; the other two 2*p* orbitals from each oxygen atom overlap side-to-side to form two sets of π bonding and antibonding orbitals. We fill the orbitals with the valence electrons from the oxygen atoms using the filling rules. There are 12 valence electrons (6 from each oxygen atom):

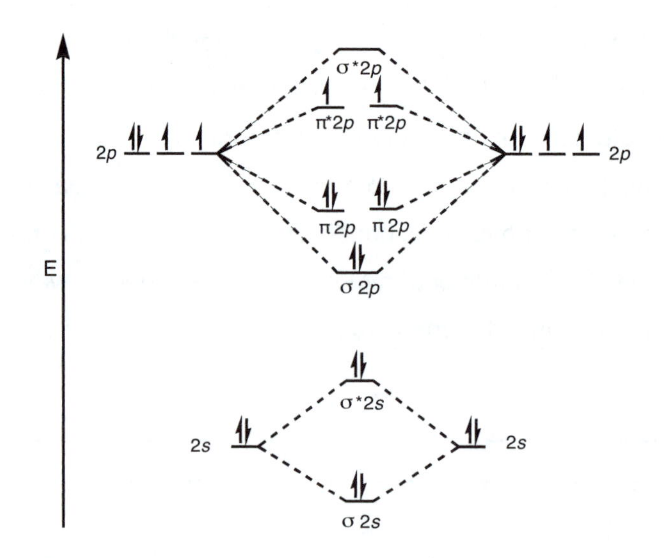

Oxygen's MO's

Consider the above drawing of the molecular orbitals of an oxygen molecule.

1) What is the valence electron configuration for the O_2 molecule?

solution:

$(\sigma 2s)^2(\sigma^* 2s)^2(\sigma 2p)^2(\pi 2p)^4(\pi^* 2p)^2$

2) What is the bond order for this molecule?

solution:

bond order = 1/2 [(# of e⁻ in bonding MO's) – (# of e⁻ in antibonding MO's)]
= 1/2 (8 – 4) = 2

Note: A bond order of 2 implies a bond that is approximately twice as strong as a bond order of 1 (the solution implies a double bond).

3) Is oxygen paramagnetic or diamagnetic?

solution:

Oxygen contains two unpaired electrons, so it is paramagnetic. Remember from Chapter 10, that we noted that a weakness of the valence bond theory is its failure to predict oxygen to be paramagnetic. We see here that one of the strengths of MO theory is its ability to explain the paramagnetism in molecules such as oxygen.

4) Would O_2^+ have a stronger or weaker bond than O_2? What about O_2^-?

solution:

If we remove an electron from the MO diagram, we remove it from an antibonding orbital, so the bond order becomes larger:

bond order = 1/2 (8 – 3) = 2.5.

If we add an electron to the MO diagram for O_2^-, we add it to an antibonding orbital, and the bond order becomes smaller:

bond order = 1/2 (8 – 5) = 1.5.

We would predict O_2^+ would have a stronger bond than O_2, and O_2^- would have a weaker bond than O_2.

The Bonding in Heteronuclear Diatomic Molecules

In polar covalent molecules, bonding MO's are closer in energy to the atomic orbitals of the more electronegative atoms. When two atoms of a diatomic molecule are very different, we cannot use the energy level diagram for homonuclear molecules. We devise a new diagram for each molecule. The main strengths of the molecular orbital model are that it predicts relative bond strengths and magnetism of simple diatomic molecules. It is difficult to apply qualitatively to polyatomic molecules.

In practice, we often combine the localized electron and molecular orbital models. Let's consider the molecule benzene (C_6H_6), which requires resonance structures since all C–C bonds are known to be equivalent:

We consider the sigma bonds between carbon atoms to be localized electrons in sp^2 hybrid orbitals giving the benzene ring its planar structure (valence bond theory). The remaining *p* orbital on each carbon atom is perpendicular to the plane of the benzene ring and contains the remaining electrons. These 6 *p* orbitals can combine to form π molecular orbitals (MO theory). The electrons in these π molecular orbitals are delocalized above and below the ring. You will often see the structure of benzene written with a circle inside the carbon six-membered ring to indicate its delocalized π bonding.

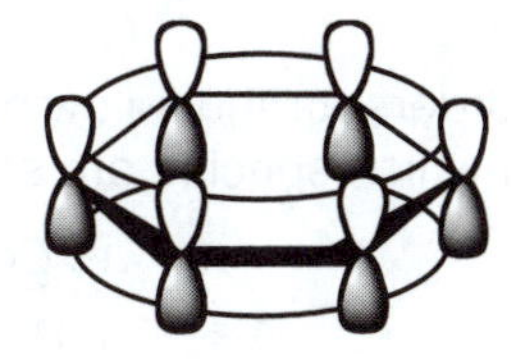

CHAPTER CHECK

Make sure you can...

- Describe how, in valence bond theory, atomic orbitals mix to form atomic hybrid orbitals with new spatial orientations.
- Name the hybrid orbitals used in bonding to form linear, trigonal planar, tetrahedral, trigonal bipyramidal, and octahedral geometries.
- Distinguish between sigma and pi bonds.
- Explain why pi bonds restrict the rotation in molecules with double bonds.
- Describe how, in molecular orbital theory, atomic orbitals mix to form delocalized molecular orbitals; describe the difference between bonding and antibonding molecular orbitals.
- Draw MO diagrams for simple diatomic molecules and use them to calculate bond orders and to predict whether a species is diamagnetic or paramagnetic.

Chapter 11 Exercises

11.1) a. If one 2*s* orbital and two 2*p* orbitals combine to form hybrid orbitals, which type of hybrid orbitals are formed?

b. If hybrid orbitals form from one *d* orbital, one *s* orbital, and two *p* orbitals from the same principal quantum level, how many orbitals would be created and what would be their designation?

11.2) The number of hybrid orbitals on a central atom in a molecule always equals:
a. the number of electrons that need to be accommodated around the atom
b. the number of sigma bonds that the atom will form
c. the number of atomic orbitals used to make the set
d. the number of π bonds that need to be catered for
e. the total number of sigma and pi bonds that will be formed by the atom

11.3) What orbital hybridization is necessary to describe the molecular geometry of:

a. XeF_2 b. BrF_5 c. PCl_5

11.4) When the valence orbitals of carbon are sp^2 hybridized, what would you predict would be the geometry of the molecule, and what type(s) of bonds would carbon have around it?

11.5) If the *z*-axis is the internuclear axis between two bonded atoms, which *p* orbitals on the atoms could be used to form a π bond between them?

11.6) Using the bond energies from Table 9.2 on page 371 in your textbook, estimate the barrier to rotation around a carbon-carbon double bond.

11.7) What are the bond orders in the diatomic ions OF^+, F_2^+, and N_2^{2-}?

11.8) Determine the electron configurations and bond orders for the following diatomic molecules according to molecular orbital theory, and predict which species are stable:

a. He_2 c. He_2^{2+} e. Be_2
b. He_2^+ d. Li_2 f. B_2

11.9) a. How many orbitals are involved in the delocalized π system of the carbonate ion?

b. What is the advantage of using MO theory to describe the π bonds in molecules and ions such as the carbonate ion, as electrons occupying delocalized π molecular orbitals?

11.10) Describe the bonding in nitric oxide, NO, using the localized electron model and the molecular orbital model. Which model gives a better description for the bonding in this molecule?

11.11) Describe the bonding in NO_3^- without using resonance by combining the localized electron and the molecular orbital models.

Chapter 11 Answers

11.1) The number of hybrid orbitals formed equals the number of atomic orbitals used to create them. The designation incorporates letters corresponding to the orbitals used, with superscripts denoting the number of each orbital used:

a. Three sp^2 hybrid orbitals
b. Four sp^2d hybrid orbitals

11.2) The correct answer is c: The number of hybrid orbitals formed equals the number of atomic orbitals used to generate them. Hybridization of orbitals in valence bond theory is a procedure usually reserved for the orbitals involved in sigma bonding and the orbitals containing lone pairs of electrons around the central atom (b is incorrect). In other words, hybrid orbitals contain sigma bonding or nonbonding electrons. There is no relation between the hybridization and the number of electrons that need to be accommodated—some electrons might be π bonding (a is incorrect). It is the number of sigma bonds and lone pairs that determines the number of hybrid orbitals required (d and e are incorrect).

11.3) a. XeF_2 has a trigonal bipyramidal arrangement of electron pairs around the xenon; it is a linear molecule. The hybridization is that required for a trigonal bipyramidal arrangement, sp^3d.

b. BrF_5 has an octahedral arrangement of electron pairs around the bromine; it is a square pyramidal molecule. The hybridization is that required for an octahedral arrangement, sp^3d^2.

c. PCl_5 has a trigonal bipyramidal arrangement of electron pairs around the phosphorus; it is a trigonal bipyramidal molecule; there are no lone pairs. The hybridization is that required for a trigonal bipyramidal arrangement, sp^3d.

11.4) Carbon, in group 4, has four valence electrons, each available to form a bond. When carbon is sp^2 hybridized, the sp^2 hybrid orbitals participate in three sigma bonds, and the remaining *p* orbital forms a π bond. We would expect the molecule to be trigonal planar with one double, and 2 single bonds.

11.5) Pi bonds form when parallel *p* orbitals overlap side-to-side. The π bond will have a *p* orbital shape when viewed down the internuclear axis (the *z*-axis in this case). An *s* orbital can never participate in π bond formation because there is no orientation of an *s* orbital that gives it a double-lobed *p* orbital shape.

Possible combinations are:

p_x and p_x, p_y and p_y

(not p_x and p_y, and not p_z and p_z, which would lead to sigma bonds)

11.6) In order for rotation to occur around a double bond, the π bond must break while the sigma bond remains intact. The bond energy for a single C–C bond is 347 kJ/mol; for a double C=C bond, the bond energy is 614 kJ/mol. The difference in bond energy between the C=C double bond and the C–C single bond is the bond energy due to the π bond: 614 – 347 = 267 kJ/mol.

11.7) Use the molecular orbital energy level diagram for a diatomic molecule. Fill the energy levels according to the aufbau principle, Hund's rule of maximum spin multiplicity, and the Pauli exclusion principle. The bond order is the number of bonding electrons minus the number of antibonding electrons divided by two. It is often useful to consider isoelectronic species. For example, OF^+ is isoelectronic with O_2 and has a bond order of 2.0. F_2^+ has one more antibonding electron and therefore must have a bond order of 1.5 (0.5 less), and N_2^{2-} is isoelectronic with O_2 and must also have a bond order of 2.0.

11.8)
a. He_2: $(\sigma 1s)^2(\sigma^*1s)^2$; bond order = 0; not stable
b. He_2^+: $(\sigma 1s)^2(\sigma^*1s)^1$; bond order = 1/2; stable
c. He_2^{2+}: $(\sigma 1s)^2$; bond order = 1; stable
d. Li_2: $(\sigma 1s)^2(\sigma^*1s)^2(\sigma 2s)^2$; bond order = 1; stable
e. Be_2: $(\sigma 1s)^2(\sigma^*1s)^2(\sigma 2s)^2(\sigma^*2s)^2$; bond order = 0; not stable
f. B_2: $(\sigma 1s)^2(\sigma^*1s)^2(\sigma 2s)^2(\sigma^*2s)^2(\pi 2p)^2$; bond order = 1; stable

11.9) a. The delocalized π system of the carbonate ion, CO_3^{2-}, involves p_π orbitals on each of the four atoms, and each perpendicular to the plane of the molecule. The four atomic orbitals result in four molecular orbitals. The pair of π bonding electrons is in the lowest molecular orbital, an orbital that extends over all four atoms. The electrons in this orbital are therefore delocalized over the entire molecule.

b. The MO theory, with its use of delocalized π molecular orbitals, eliminates the need to draw resonance structures required in valence bond theory to show when electrons are delocalized.

11.10) NO contains 11 valence electrons for bonding. We can draw a Lewis structure with a double bond between the oxygen and nitrogen, or one with a triple bond:

$$\cdot N \equiv \ddot{O}: \qquad \dot{\ddot{N}} = \ddot{O}:$$

The first structure contains formal charges on the nitrogen (+1) and the oxygen (–1).

The electron configuration in the molecular orbital description of the bonding in NO is:

$$(\sigma 1s)^2(\sigma^* 1s)^2(\sigma 2s)^2(\sigma^* 2s)^2(\pi 2p)^4(\sigma 2p)^2(\pi^* 2p)^1$$

The bond order is: 1/2 (6 – 1) = 2.5, which gives a better description of the bonding in NO.

11.11) We assume each atom is sp^2 hybridized to produce a trigonal planar molecule. The remaining *p* orbital on each atom is perpendicular to the plane of the molecule. These *p* orbitals combine to form a π molecular orbital system in a three-pronged shape above and below the plane of the NO_3^- ion.

12

Intermolecular Forces: Liquids, Solids, and Phase Changes

INTRODUCTION

In the last three chapters we saw how chemical bonds form from electrostatic attractions between particles.

Chemical bonds

Determine if the following statements describe 1) covalent, 2) ionic, or 3) metallic bonds.

a) electrostatic attraction between anions and cations
b) electrostatic attraction between cations and delocalized valence electrons
c) electrostatic attraction between positive nuclei and electron pairs
d) non-conductor
e) conducts in the molten state by motion of ions
f) conducts in the solid state by electron flow
g) no large cores
h) both small and large cores
i) no small cores

answers: 1 (covalent)-c, d, g; 2 (ionic)-a, e, h; 3 (metallic)-b, f, i.

The strength of bonds between atoms in a molecule determines chemical reactivity, since bonds break and form during chemical reactions. In this chapter, we discuss the electrostatic attractions between molecules (intermolecular attractions), which determine physical properties of substances.

Intermolecular attractions

Which electrostatic attraction must be overcome (a or b in the drawing below) in order for a water molecule to pass from a liquid state into the vapor state?

solution:

When water evaporates, no hydrogen and oxygen covalent bonds (a) break; the water molecules must overcome intermolecular attractions (b) to one another. These intermolecular attractions are much smaller (usually 5% or less) than the attractions which form a typical covalent bond because there is more distance between the charged particles, and the charges are generally smaller.

12.1 PHYSICAL STATES AND PHASE CHANGES

We saw in Chapter 1 (and again in Chapter 5), that, broadly speaking, there are three ways which we classify matter:

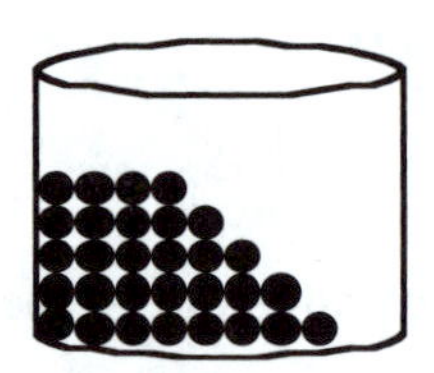

Touching & ordered
(solid)

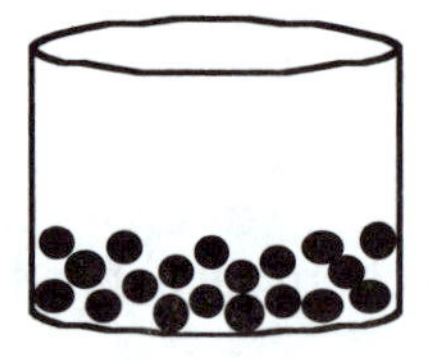

Touching & disordered
(liquid)

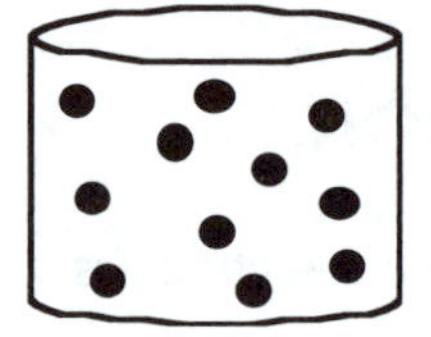

Not touching
(gas)

States of matter

Use the above drawings as a guide to classify the following as characteristics of solids, liquids, or gases:

a) molecules are far apart
b) molecules touch and are fixed in position
c) molecules touch but have no long range order
d) essentially no interparticle interactions
e) interparticle forces are strong
f) condensed states of matter
g) easily compressed

answers:

a-gas, b-solid, c-liquid (also amorphous solids), d-gas, e-solid & liquid, f-solid & liquid, g-gas.

Each physical state is called a **phase,** and **phase changes** depend on temperature and pressure. Increases in temperature increase the average kinetic energy of particles so they move faster and are more likely to overcome attractions between them (intermolecular forces). An increase in pressure generally makes it more difficult for particles to separate and disperse.

Particles in a substance change their distance from each other and their freedom of motion when the substance changes its state of matter. It always takes more energy to completely separate particles (vaporize a liquid) than to overcome the forces holding the molecules in their fixed positions (melt a solid). Figure 12.2 on page 457 in your textbook compares the heat of vaporization and the heat of fusion for some common substances. Notice the large difference between these values, especially in the case of water.

Phase changes

Match the words with the phase change it describes and indicate whether it is an exothermic or endothermic process:

1) condensation	a) solid → liquid
2) deposition	b) solid → gas
3) freezing	c) liquid → gas
4) melting	d) liquid → solid
5) sublimation	e) gas → liquid
6) vaporization	f) gas → solid

answers: 1-e, 2-f, 3-d, 4-a, 5-b, 6-c. 1,2,3 - endothermic; 4,5,6 – exothermic.

(These phase changes and their enthalpy changes are summarized in Figure 12.3 on page 458 in your textbook.)

Heat of deposition

What is the heat of deposition for one mole of water given:

$H_2O(g) \rightarrow H_2O(l)$ $\quad \Delta H = -\Delta H^o_{vap} = -40.7$ kJ/mol (100°C)

$H_2O(l) \rightarrow H_2O(s)$ $\quad \Delta H = -\Delta H^o_{fus} = -6.02$ kJ/mol (0°C)

solution:

If we add the above two equations we obtain:

$H_2O(g) \rightarrow H_2O(s)$ $\quad \Delta H$ = heat of deposition

$$\Delta H^o_{dep} = -\Delta H^o_{vap} - \Delta H^o_{fus}$$

$$\Delta H^o_{dep} = -40.7 \text{ kJ/mol} - 6.02 \text{ kJ/mol} = -46.7 \text{ kJ/mol}$$

Ice crystals form on a cold window from the deposition of water vapor.

12.2 QUANTITATIVE ASPECTS OF PHASE CHANGES

If we add energy at a constant rate to a substance, such as ice, and plot the temperature vs. time, we obtain a heating curve. The heating curve for water is shown below. Study the heating curve and then work out the solutions for the questions that follow. Notice that it takes more heat to separate molecules from each other as water changes from the liquid to the vapor phase than to disorder the molecules as water melts from a solid to a liquid (check your answers to #2 and #4 below).

Notice also that it takes less time for steam to increase in temperature than it takes for water to increase in temperature. The heat capacity of water vapor is smaller than the heat capacity of liquid water, so it takes less energy to raise 1 mole of water vapor 10°C than to raise 1 mole of liquid water 10°C.

Heating curve for water

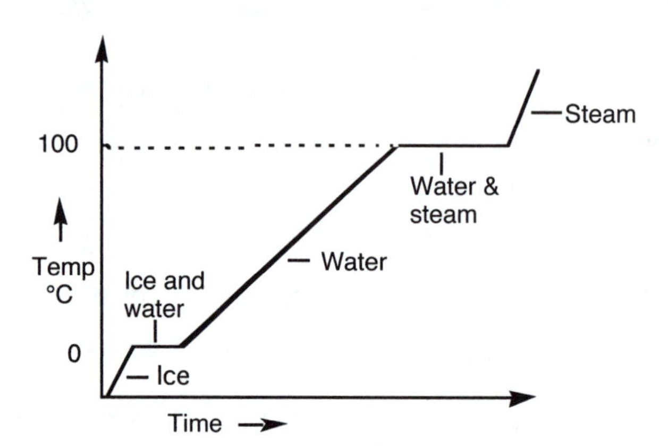

(For the corresponding cooling curve for conversion of gaseous water to ice, see Figure 12.4 on page 459 in your textbook.)

1) How much energy is required to take 18 grams of water (1 mole) from –10°C to its melting point?

solution:

$$q = n \times C_{\text{ice}} \times \Delta T$$

$$= (1 \text{ mol}) \times (37.6 \text{ J/mol°C}) \times (0°\text{C} - (-10°\text{C})) = 376 \text{ J} \ \mathbf{(0.376 \ kJ)}$$

2) How much energy is required to melt 18 grams of water at 0°C?

solution:

$$q = n \times \Delta H^{\text{o}}_{\text{fus}}$$

$$= (1 \text{ mol}) \times (6.02 \text{ kJ/mol}) = \mathbf{6.02 \ kJ}$$

3) How much energy is required to take 18 grams of water from its melting point to its boiling point?

solution:

$$q = n \times C_{\text{water}} \times \Delta T$$

$$= (1 \text{ mol}) \times (75.4 \text{ J/mol°C}) \times (100°\text{C} - 0°\text{C}) = 7540 \text{ J} \ \mathbf{(7.54 \ kJ)}$$

4) How much energy is required to vaporize 18 grams of water at 100°C?

solution:

(continued on next page)

$$q = n \times \Delta H^{o}_{vap}$$

$$= (1 \text{ mol}) \times (40.7 \text{ kJ/mol}) = \mathbf{40.7\ kJ}$$

5) How much energy is required to raise the temperature of 18 grams of water vapor from 100°C to 110°C?

   solution:

$$q = n \times C_{\text{water vapor}} \times \Delta T$$

$$= (1 \text{ mol}) \times (33.1 \text{ J/mol°C}) \times (110°\text{C} - 100°\text{C}) = 331 \text{ J } (\mathbf{0.331\ kJ})$$

6) What is the total heat required to take 18 grams of solid water from –10° to water vapor at 110°C?

   solution:

   The total heat required equals the sum of the heat required for each step of the process (Hess's law):

   Total heat = 0.376 kJ + 6.02 kJ + 7.54 kJ + 40.7 kJ + 0.331 kJ = **55.0 kJ**

   Notice that of all the steps, it takes by far the most heat to separate molecules from each other as they change from the liquid to the vapor phase (step #4).

Within a given state of matter (solid, liquid, or gas), added energy increases the temperature of the system and so increases the average kinetic energy of the molecules. During a change in state, added energy increases the potential energy of the particles, and in the case of water, breaks hydrogen bonds (solid to liquid), or breaks all hydrogen bonds so that the molecules can move apart (liquid to gas).

Vapor Pressure

Liquid-gas equilibria

Choose words from the list below to complete the following paragraph:

(a) condense	(d) higher	(g) kinetic energy
(b) constant	(e) increases	(h) vapor pressure
(c) equilibrium	(f) pressure	

Water evaporates as surface water molecules with enough _____________(1) overcome attractions to other water molecules and escape from the liquid. If water is in a closed container, molecules that enter the vapor phase collide with the walls and create _____________(2) inside the container. At the same time, some of the gaseous water molecules that collide with the liquid surface are attracted strongly enough to remain there (they _____________(3)). When the rate of evaporation = the rate of condensation, the pressure of the vapor is _____________(4). The pressure exerted by the vapor in the flask at equilibrium is called the _____________(5). If we disturb the system, it adjusts itself to regain an _____________(6) state. The vapor pressure of water inside a large container is the same as the vapor pressure of water inside a small container. If we increase the temperature, the vapor pressure _____________(7). The vapor pressure of a liquid is affected by intermolecular forces. Molecules with weak intermolecular forces vaporize more easily than molecules with strong intermolecular forces and therefore have _____________(8) vapor pressures.

answers: 1-g, 2-f, 3-a, 4-b, 5-h, 6-c, 7-e, 8-d.

Vapor pressure increases as temperature increases, but the relationship is not linear. We can plot ln *P* vs. $1/T$ to obtain a straight line.

The Clausius-Clapeyron equation

$$\ln P = \frac{-\Delta H_{vap}}{R} \times \frac{1}{T} + C$$

If you plot the natural log of the vapor pressure (ln *P*) vs. $1/T$ (Kelvin), what is the slope of the line? What is the *y* intercept?

solution:

The **Clausius-Clapeyron** equation is the equation of a straight line ($y = mx + b$).
$y = \ln P$, and $x = 1/T$:

$$\text{The slope } (m) = \frac{-\Delta H_{vap}}{R}$$

The *y*-intercept (b) = C, a constant characteristic of the liquid in question.

Vapor pressure of water at different temperatures

The vapor pressure of water at room temperature (25°C) is 23.8 torr and the heat of vaporization of water is 41.2 kJ/mol. What is the vapor pressure of water at 80.°C?

solution:

The constant, C, does not depend on temperature, so we can solve the Clausius-Clapeyron equation for C at both temperatures and set them equal to one another:

$$\ln P_{T_1} + \frac{\Delta H_{vap}}{RT_1} = C = \ln P_{T_2} + \frac{\Delta H_{vap}}{RT_2}$$

$$\ln P_{T_1} - \ln P_{T_2} = \frac{\Delta H_{vap}}{R}\left(\frac{1}{T_2} - \frac{1}{T_1}\right)$$

$$\ln \frac{P_{T_1}}{P_{T_2}} = \frac{\Delta H_{vap}}{R}\left(\frac{1}{T_2} - \frac{1}{T_1}\right)$$

Now we can substitute in our values for water:

P_{T_1} = 23.8 torr

$T_1 = 25 + 273 = 298$ K; $T_2 = 80. + 273 = 353$ K

$\Delta H_{vap} = 41.2$ kJ/mol = 41,200 J/mol; $R = 8.3148$ J/K mol

$$\ln \frac{23.8 \text{ torr}}{P_{T_2}} = \frac{41{,}200 \text{ J/mol}}{8.3148 \text{ J/K mol}}\left(\frac{1}{353 \text{ K}} - \frac{1}{298 \text{ K}}\right)$$

$$P_{T_2} = 317 \text{ torr}$$

Check: We would expect a higher vapor pressure for water at 80°C than at room temperature (23.8 torr). Our answer (317 torr at 80°C) appears to be reasonable.

Boiling

Choose words from the list below to complete the following paragraph:

(a) atmosphere
(b) lower
(c) 1 atm
(d) 100°C
(e) 94°C

A liquid boils when the vapor pressure equals the pressure of the _______________(1). No bubbles can form in the interior of the liquid if the vapor pressure inside the bubbles is _______________(2) than the atmospheric pressure. The **normal boiling point** is the temperature at which the vapor pressure of the liquid is _______________(3). The normal boiling point of water is _______________(4); water boils at _______________(5) in New York City. In Boulder, Colorado (5430 ft above sea level), water boils at _______________(6). It takes longer to boil an egg in Boulder than in New York because atmospheric pressure is less at 5430 ft than at sea level, and the water boils at a _______________(7) temperature.

answers: 1-a, 2-b, 3-c, 4-d, 5-d, 6-e, 7-b.

Melting

Choose words from the list below to complete the following paragraph:

(a) compress
(b) denser
(c) higher
(d) larger
(e) less
(f) pressure
(g) volume

Atoms or molecules in solids and liquids are close together, and it is therefore difficult to _______________(1) them. Since there is not a large volume change between liquid and solid states, the melting point of solids is only slightly affected by _______________(2). For most substances, it takes slightly _______________(3) temperatures to reach the melting point when pressure is applied. Water is an exception: large intermolecular spaces in ice make it _______________(4) dense than liquid water. Ice takes up a _______________(5) volume than liquid water (anyone who has had their pipes burst during a cold snap can attest to this). When we apply pressure, ice can reduce its _______________(6) by changing to liquid water. We can glide over ice on skates because the _______________(7) from our blades melts the ice and provides lubrication. We are so used to seeing ice float in water, that it doesn't seem strange, but in almost every other substance, the solid state is _______________(8) than the liquid.

answers: 1-a, 2-f, 3-c, 4-e, 5-d, 6-g, 7-f, 8-b.

Sublimation

Up to their melting points, solids always have a lower vapor pressure than the corresponding liquids. Substances which have high enough vapor pressures to sublime under normal conditions are not too common: dry ice, moth balls, and solid room deodorizers are familiar examples.

Phase Diagrams

Phase diagrams show the phase changes of a substance at different combinations of temperature and pressure in a closed container. The phase diagram for carbon dioxide, CO_2, is shown below. Study the diagram and then answer the questions in the exercise below:

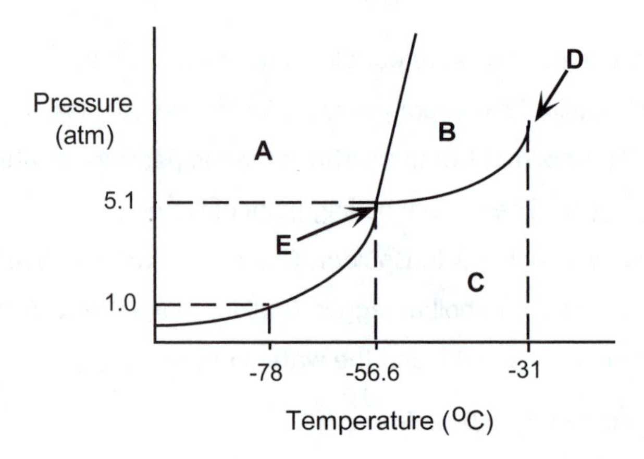

Phase diagram exercise

Answer the following questions using the phase diagram above.

1) Identify regions A, B, and C as solid, liquid, or gas.

   answers: A = solid, B = liquid, C = gas.

2) Identify the critical point (the point where vapor can no longer be condensed to a liquid regardless of the applied pressure).

   answer: D.

3) Identify the triple point (the point at which solid, liquid, and gas are all at equilibrium).

   answer: E.

4) What would the atmospheric pressure need to be in order for CO_2 to exist as a liquid?

   answer: at least 5.1 atm.; CO_2 exists as a liquid in fire extinguishers.

5) What is the temperature of dry ice?

   answer: –78°C.

Compare the phase diagram of carbon dioxide above with the phase diagram of water shown in Figure 12.11 on page 466 in your textbook and notice the negative slope of the solid-liquid line in the water diagram. This seemingly subtle difference in the diagram's appearance reflects a unique property of water: its solid phase is *less dense* than its liquid phase. We are so used to seeing ice float on water, or knowing that it expands when it freezes (bursting frozen water bottles or pipes), or that the pressure from ice skates melts the ice allowing us to glide, that we don't realize how unusual this situation is. Almost every other substance is denser as a solid than as a liquid, freezing from the bottom up. It is the geometry of strong hydrogen bonds, which form only at lower temps that gives water this unique property.

12.3 TYPES OF INTERMOLECULAR FORCES

Forces between molecules are relatively weak compared to bonding forces because they typically involve smaller charges that are farther apart. The closest distance that two molecules can approach each other (the point at which attractions balance electron-cloud repulsions) is called the **van der Waals** distance (after Dutch physicist, Johannes van der Waals). The van der Waals radius is one-half the closest distance between nuclei of identical nonbonded atoms. Several types of van der Waals forces are discussed below.

Dispersion (London) Forces

Bromine is a liquid at room temperature and ordinary pressure. What causes the nonpolar bromine molecules to attract one another? The boiling point of bromine is fairly low (58.8°), so the attractive forces must not be all that strong. **Dispersion forces** are named after Fritz London, a physicist who provided a quantum mechanical explanation for why two noble gas atoms attract each other at a large distance, but repel each other at short distances. These dispersion forces are weak, intermolecular attractions between nonpolar molecules caused by momentary oscillations of electron density that create transient dipoles. These dipoles induce dipoles in neighboring molecules, and the induced dipoles attract each other:

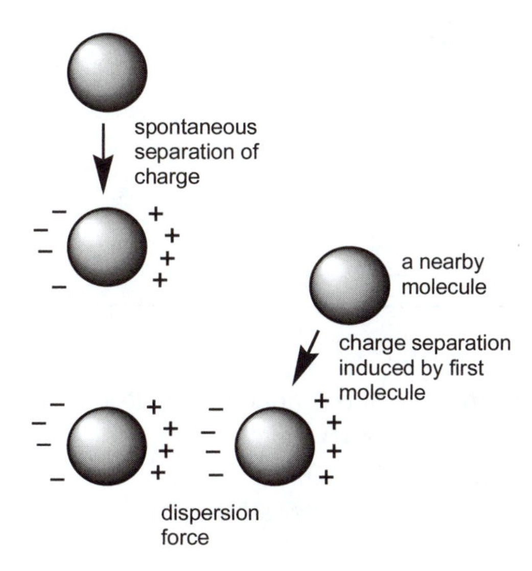

Dispersion forces exist in all molecules, but are overpowered by stronger dipole-dipole interactions in polar molecules. Dispersion forces generally increase with molar mass because electrons farther from the nucleus are easier to distort (larger atoms are more **polarizable**).

Ion-Dipole Forces

An **ion-dipole** attraction results when an ion and a nearby polar molecule (such as water) attract each other. A familiar example occurs when an ionic compound dissolves in water: the attraction between the ions and the oppositely charged poles of water molecules overcomes the attraction between the ions themselves.

Charge-Induced Dipole Forces

A charge from an ion, the electrodes of a battery, or the partial charge of a polar molecule, can induce or enhance a charge in a nearby molecule. The resulting attraction is called a charge-induced dipole force. Atoms that hold their electrons more tightly (smaller atoms) have electron clouds, which are more difficult to distort and are less **polarizable** than larger atoms, which have larger, more easily distorted electron clouds. Ion-induced dipole forces are important in solutions, discussed in the following chapter.

Dipole-Dipole Attractions

Dipole-dipole attractions exist when polar molecules (molecules with dipole moments) line up so their partial positive and negative charges attract each other. In a collection of polar molecules, the dipoles arrange themselves to maximize the attractions between molecules.

The **hydrogen bond** is a special type of particularly strong dipole-dipole attraction, and it stands out as the most important intermolecular force. Hydrogen bonds occur between molecules that have a hydrogen atom bound to a small, highly electronegative atom with lone pairs (such as oxygen, nitrogen, or fluorine). The hydrogen bond is particularly strong due to the great polarity of the bond, and the small size of hydrogen that allows close approach of the dipoles. Water has a much higher boiling point than expected from trends for hydrides due to hydrogen bonding. Figure 12.5 on page 470 in your textbook shows that without hydrogen bonding, we would expect water to boil at about −100°C!

Mental note

What is the most important intermolecular force, depicted below?

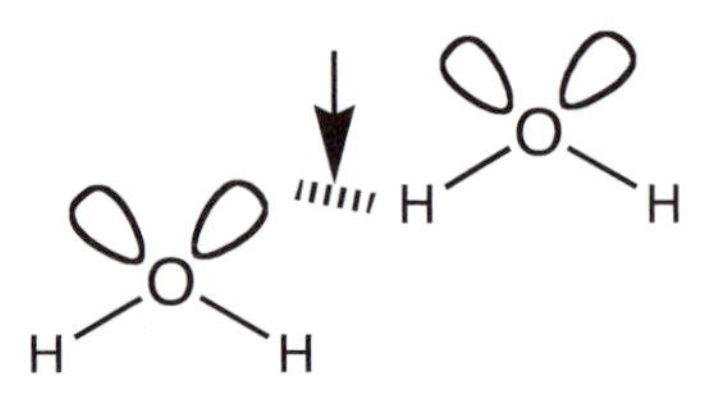

answer: the hydrogen bond.

Intermolecular forces

Rationalize the difference in boiling points for each of the following pairs of substances:

a) Cl_2 (–34.6°C) and Br_2 (58.8°C)
b) H_2O (100.°C) and H_2S (–60.7°C)
c) pentane (36.1°C) and cyclopentane (49.2°C)
d) HF (20.°C) and HCl (–85°C)
e) CH_3OCH_3 (35°C, **M** = 46 g/mol) and CH_3CH_2OH (79°C, **M** = 46 g/mol)

solutions:

a) Both chlorine and bromine are nonpolar molecules with weak dispersion intermolecular forces. At room temperature, chlorine is a gas and bromine is a dark red liquid due to its higher molar mass and greater number of electrons.

b) Hydrogen sulfide has a higher molar mass than does water, but hydrogen bonding in water molecules makes it have a much higher boiling point. Sulfur is not small enough (or electronegative enough) to allow hydrogen bonding.

c) The molar mass of pentane is 72 g/mol and the molar mass of cyclopentane is 70. g/mol. They are both nonpolar molecules. Cyclopentane has the higher boiling point because its disc shape makes more intermolecular contact possible causing it to favor the liquid state.

d) Both HCl and HF are polar molecules with dipole-dipole interactions. The higher boiling point of HF is due to hydrogen bonding.

e) The higher boiling point of ethanol over diethyl ether is due to hydrogen bonding between the hydrogen from an O–H group on one molecule and the oxygen on another molecule.

Hydrogen bonding

Hydrogen bonding is the most important intermolecular force.

1) Give explanations for the boiling points (°C) of the two sets of compounds:

A		B	
CH_4	–164	CH_3CH_3	–88.6
NH_3	–33.35	CH_3NH_2	–6.3
H_2O	100	CH_3OH	65
HF	19.54	CH_3F	–78.4

answer:

Hydrogen bonding in ammonia, water, and hydrogen fluoride cause these compounds to have higher boiling points than methane which has no hydrogen bonding.

Methyl amine (CH_3NH_2) and methanol (CH_3OH) have higher boiling points than ethane (CH_3CH_3) and fluoromethane (CH_3F) because of hydrogen bonding. (Fluoromethane has no hydrogen bonding because there are no H–F bonds.)

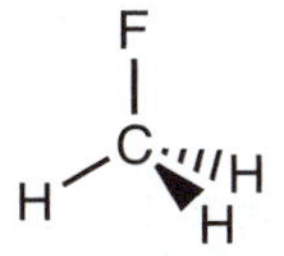

F
C
H H H

2) Acetic acid (CH_3CO_2H, structure on next page) dimerizes in nonpolar solvents:

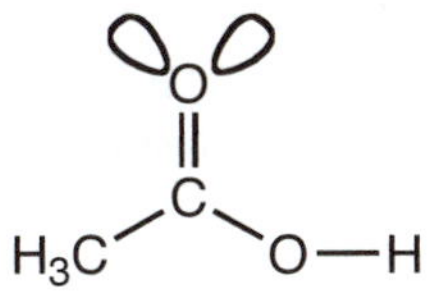

O
C
H_3C O—H

What would you predict the structure of its dimer to be?

solution:

Hydrogen bonding causes the oxygen atom on one molecule to hydrogen bond with the hydrogen atom of the OH bond on another molecule:

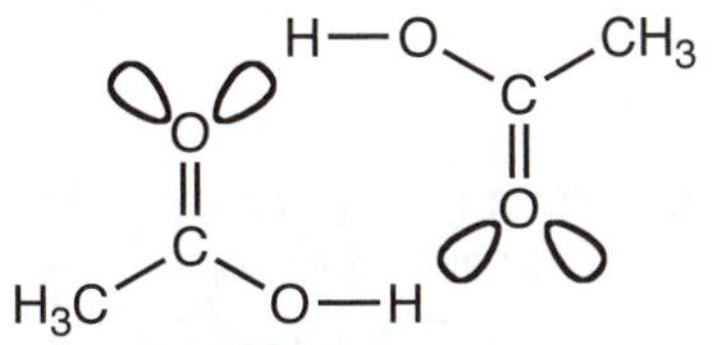

H—O CH_3
O C
C O
H_3C O—H

3) What organic compounds are soluble in water?

answer:

In general, organic compounds that contain a group capable of hydrogen bonding with water are soluble in water.

4) What intermolecular force is responsible for the pairing of bases in DNA?

answer:

the hydrogen bond.

12.4 THE LIQUID STATE

There are three properties of liquids that we can explain by their intermolecular forces: surface tension, capillarity, and viscosity.

Surface Tension

Surface tension is the energy required to increase the surface area by a given amount. The stronger the intermolecular forces, the stronger the surface tension. Molecules in the interior of the liquid are attracted by molecules in all directions; however, molecules at the surface are pulled into the interior of the liquid by attractions from molecules below and to their sides:

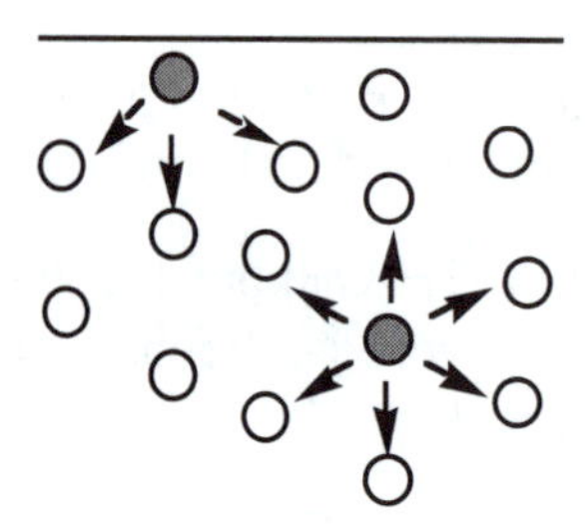

Capillarity

The ability of a liquid to rise in a tube is due to the competition between internal forces within a liquid (cohesive forces) and those between the liquid and the wall of a container (adhesive forces).

Exercise

Which example shows a liquid with stronger cohesive forces than adhesive forces?

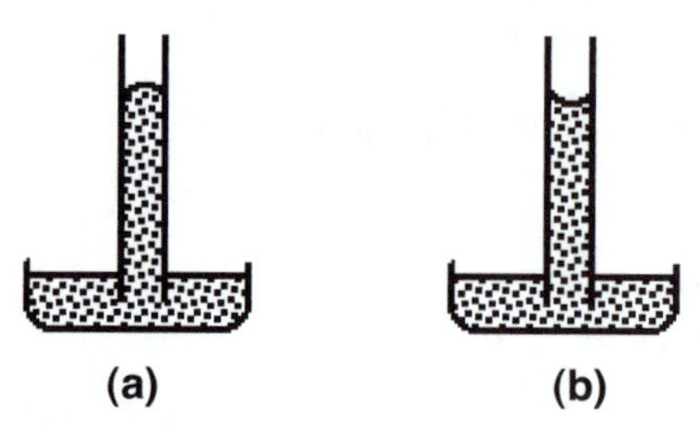

solution:

Liquid "a" has stronger attractions to itself (cohesive forces) than to the glass walls (adhesive forces) causing the liquid to form a convex meniscus. Liquid "b" has stronger attractions to the glass walls (adhesive forces) than to itself (cohesive forces) causing the liquid to move up the tube and form a concave meniscus. Glass is a polar substance. Liquid "a" is a nonpolar liquid such as mercury. Liquid "b" is a polar substance such as water.

Viscosity

Both gases and liquids exhibit viscosity, the resistance to flow, which depends both on the strength of intermolecular interactions, and the shape of the molecule. Viscosity decreases with an increase in temperature. We warm up chocolate fudge sauce to make it become less viscous and flow more easily. Large molecules, which can become entangled, make liquids more viscous than smaller, less tangled molecules.

12.5 A UNIQUE SUBSTANCE

Ubiquitous and unique

What common substance has the following properties?

It is the chief constituent of our bodies.
It reacts violently with sodium metal to produce aqueous sodium hydroxide and hydrogen gas.
It boils at about 90°C at an elevation of 10,000 ft.
It expands when it freezes.
It has a high specific heat capacity, higher than almost all liquids.
It has a high heat of vaporization.
It has high surface tension and high capillarity.
Its solid state is less dense than its liquid state.
It is a good solvent for table salt and for table sugar.
A cupful of its liquid occupies a volume of approximately 100 gallons at 100.°C and atmospheric pressure, as a gas.
It does not mix with oil.
It is one of the products when natural gas burns.
It is the only product of combustion of hydrogen.
It has a density of 1.0 g/mL at 25°C and 1 atm.
It can be electrolytically decomposed into the elements hydrogen and oxygen.

answer: water.

Read section 12.5 in your textbook on the uniqueness of water, and then make sure you can answer the following questions:

1) Why is water able to dissolve ionic compounds, polar nonionic substances, and nonpolar gases (to a limited extent)? Why is the latter property of water important for aquatic life?

2) Why is water's specific heat capacity higher than almost all liquids? How does this characteristic of water help to support life on earth?

3) Why does water have a high heat of vaporization, high surface tension, and high capillarity? Explain how the high heat of vaporization of water powers the winds and storms on earth. Why is the high surface tension of water vital for surface aquatic life? Why is water's high capillarity crucial to survival of land plants?

4) Why does the solid state have a lower density than the liquid state? Explain why this property of water is crucial for the survival of aquatic life. Explain how this property of water helped to produce earth's sand and soil.

5) At what temperature is water most dense? Explain how the change in the density of water with temperature distributes nutrients and dissolved oxygen throughout a lake.

12.6 THE SOLID STATE

In **crystalline** solids, atoms pack together in an orderly fashion. The smallest repeating unit of the lattice is its **unit cell.** Three common unit cells are shown on the next page:

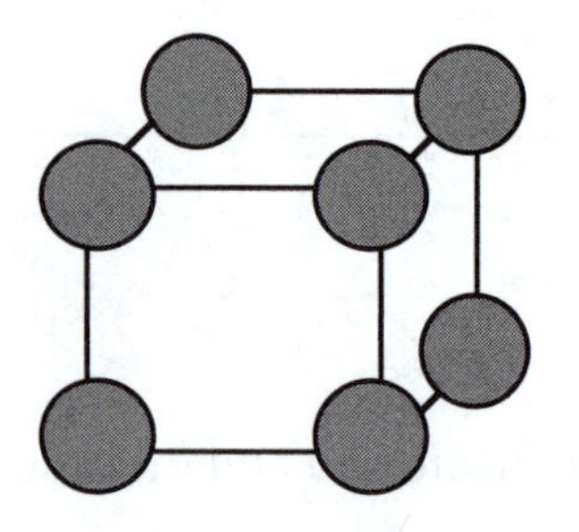

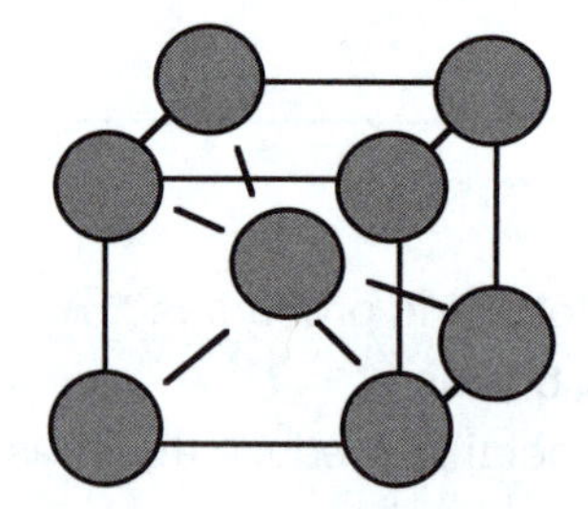

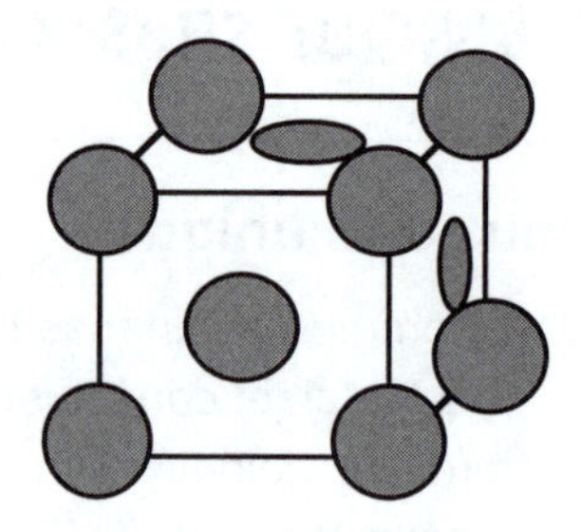

Simple cubic　　Body-centered cubic　　Face-centered cubic

If the unit cell is repeated in three directions, it forms a crystal lattice. Crystal structures can be determined by **X-ray diffraction,** a technique that allows us to reconstruct the structure of a crystalline solid by analyzing the diffraction patterns made by X rays scattering off the atoms in the crystal.

The balls in the above drawings can represent ions, atoms, or molecules.

- When balls = ions, we have an ionic solid such as sodium chloride, NaCl.
- When balls = molecules we have a molecular solid, such as ice.
- When balls = atoms, there are three types of solids possible:

1) a network covalent solid, a network of covalently bonded atoms. (example: diamond)

2) a metallic solid, an array of metal atoms bonded by metallic bonding forces. (example: gold)

3) an atomic solid, a network of atoms held together by dispersion forces. (example: a solidified noble gas)

Properties of crystalline solids

For each type of solid listed, determine whether it must overcome intermolecular forces or intramolecular forces (break chemical bonds) to pass into the gas phase. What would you predict would be the boiling points (high or low) for each type of solid?

1) Ionic solid
2) Molecular solid
3) Network covalent solid
4) Metallic solid
5) Atomic solid

solution:

1 (ionic solids), 3 (network covalent solids), and 4 (metallic solids) must break chemical bonds in order to pass into the vapor phase. They have high boiling points reflecting the strength of the bonds that must be broken.

2 (molecular solids), and 5 (atomic solids) must overcome weak intermolecular (or interatomic) forces to pass from a liquid to a vapor.

Molecular solids have lattice points occupied by individual molecules. The molecules stay intact, but separate as they pass into the gas phase. These solids have a wide range of boiling points reflecting the many types of intermolecular forces that exist between molecules. Since intermolecular forces are relatively weak, the boiling points for molecular solids are generally fairly low.

Atomic solids have lattice points occupied by atoms held together by dispersion forces. They have low boiling points due to the weakness of London dispersion forces.

Determining the number of atoms per unit cell

Determine the number of atoms per unit cell for the three common types of unit cells drawn at the beginning of this section. What is the coordination number for each atom in the solids formed by these unit cells?

solution:

1) **simple cubic:** There is one atom at each corner of a cube making 8 atoms. Each of the atoms is shared by eight other unit cells, so only 1/8th of it belongs to the unit cell in question.
   8 atoms $\times$ 1/8/unit cell = 1 atom/unit cell.

   Each atom in the solid has 6 closest neighbors, so the coordination number = 6.

2) **body-centered:** Just as with the simple cubic unit cell, the atoms on the corners of the cube each contribute 1/8th of an atom making 1 atom/unit cell. The atom in the center of the cube belongs only to that unit cell, so there are 1 + 1 = 2 atoms/unit cell.

   Each atom in the solid has 8 nearest neighbors so the coordination number = 8.

3) **face-centered:** There is one atom/unit cell from the 8 atoms on the corners. The 6 atoms on the faces of the cube are each shared by two unit cells making 6 $\times$ 1/2 = 3 more atoms/unit cell. The total atom per unit cell is therefore 3 + 1 = 4 .

   Each atom in the solid has 12 nearest neighbors so the coordination number = 12.
   The face-centered cube with 4 atoms/unit cell is the most efficiently packed.

There are two packing arrangements that pack spheres as efficiently as possible. To see why this is so, imagine making three layers of marbles. We make the first layer so all the marbles are as close together as possible. We place the second layer of marbles in the triangular spaces made by the first layer. We can then stack the third layer so the marbles are directly over marbles in the first layer, or we can stack the third layer so the marbles lie over spaces in the first layer. The first arrangement in which the layers are stacked in an ABABAB pattern is called **hexagonal closest packing.** It is based on the hexagonal unit cell. The second arrangement has the layers stacked in an ABCABCABC pattern and is called **cubic closest packing.** It is based on the face-centered cubic cell.

Atomic radius from unit cell measurements

Metallic gold crystallizes in a face-centered cubic lattice with its atoms touching along the diagonal of the face of the unit cell. If the unit cell measures 4.07×10^{-10} m on a side, what is the apparent radius of the gold atom?

solution:

We can use the Pythagorean theorem to calculate the diagonal across the unit cell from the edge length of the unit cell:

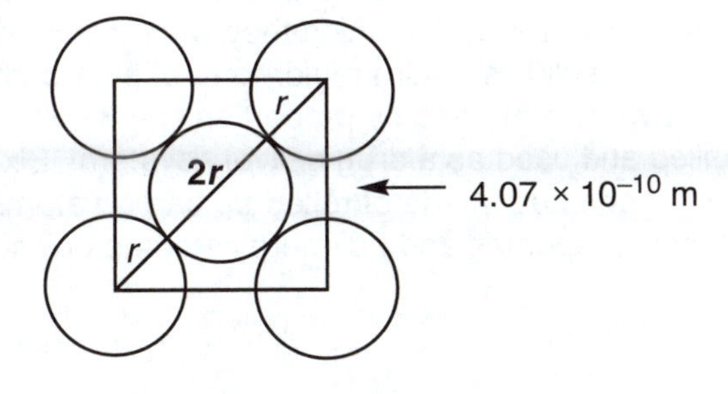

(continued on next page)

$$(\text{edge})^2 + (\text{edge})^2 = (\text{diagonal})^2$$

$$2 \times (4.07 \times 10^{-10}\text{m})^2 = (\text{diagonal})^2$$

$$5.75 \times 10^{-10}\text{ m} = \text{diagonal}$$

We can see from the drawing that the diagonal equals the radius of a gold atom at each corner plus the diameter of the gold atom in the center of the face, or $4r$:

$$\text{radius} = \frac{\text{diagonal}}{4} = \frac{5.75 \times 10^{-10}\text{ m}}{4} = 1.44 \times 10^{-10}\text{ m}$$

Check: If we change our answer to angstroms (1.44 Å), we can see that our answer is the right order of magnitude for an atomic radius. The smallest atomic radius is hydrogen (0.37 Å); the largest is Cs (2.62 Å). The atomic radius of gold falls just about in the middle of these two extremes.

Atomic Solids

Extreme conditions are required to solidify the noble gases. Under low temperature and/or high-pressures, the unreactive atoms form a network of individual atoms held together only by dispersion forces. These weak forces between atoms mean very low melting and boiling points for these atomic solids, which rise with increasing molar mass.

Network Covalent Solids

Strong covalent bonds link atoms together in a three-dimensional network to form network covalent solids. The silicates are an important class of network covalent solids with an extended array of covalently bonded silicon and oxygen atoms. Silicates form the structure of clays, rocks, and many minerals. Diamond is a network covalent solid of tetrahedrally linked carbon atoms, forming the hardest substance known. It is an insulator because the bonding electrons are localized. Graphite, by contrast, conducts electricity well and is soft. This marked difference from diamond is due to graphite's arrangement of carbon atoms in stacked flat sheets of six-membered carbon rings, with the π electrons delocalized over the entire sheet, and the sheets' ability to slide past each another. In Chapter 16 we will consider the conditions under which diamond or graphite is the more stable form of carbon.

Molecular Solids

Molecular solids have strong covalent bonds within the molecules, but weak forces between molecules. Ice is the most common example of a molecular solid. It requires 6 kJ of energy to melt 18 grams (1 mole) of ice, however, it takes ≈47 kJ of energy to break a mole of covalent O–H bonds. Some molecular solids have only London dispersion forces holding nonpolar molecules together, and so we would expect it would take even less than 6 kJ of energy to melt a mole of these solids.

Until the mid 1980's, pure carbon was known only in two forms: graphite and diamond. In 1985, however, British and Texas chemists discovered cages of carbon atoms, including the famous 60-carbon "buckyballs" (that look like miniature soccer balls), for which they were jointly awarded the 1996 Nobel Lauriate for chemistry. C60, a molecular solid, is a third major form of pure carbon. It is the roundest and most symmetrical large molecule known to man, and its properties continue to amaze. Because it is a hollow sphere, it has the potential to be filled and used as a drug delivery system. Six years later in 1991 a Japanese scientist discovered tiny "nanotubes" made of rolled-up carbon atoms. These carbon nanotubes have been explored for their electronic properties and as energy storage devices.

Perhaps the most exciting form of carbon for use as an electronic material is the superflat material called graphene, discovered in 2004. Graphene is nothing more than a sheet of honeycombed carbon atoms, which you can find in flakes from pencil lead. The scientists who won the Nobel Prize in physics for graphene used the high-tech equipment of Scotch tape to pull apart flakes of graphite. By repeatedly folding and opening up a piece of tape with graphite stuck on, the scientists managed to peel off single graphene layers. It was stunning to discover that, unlike other known materials, graphene sometimes behaves according to the rules of quantum mechanics. In most materials, the speed of electrons changes with their energy. In graphene, however, electrons behave as if they have no mass; they move at a constant speed no matter their energy, and they cannot be stopped. Because graphene is cheap, can be molded over surfaces, and can be stacked in different ways to produce different electronic properties, its potential use in new kinds of gadgets is exciting and could lead to a new age of graphene electronics.

Ionic Solids

Ionic solids are held together by the electrostatic attraction between anions and cations. Usually one ion lies in the spaces formed by the packing of the other ion. From a ball-and-stick model of a sodium chloride crystal lattice, it may look as though the unit cell is simple cubic unless you look at <u>one type of ion</u> only. The sodium ions form a face-centered unit cell. The chloride ions form an interlaced face-centered cubic cell. The drawing below shows the front face of the face-centered cube. For clarity, not all atoms are shown on the back face of the crystal lattice:

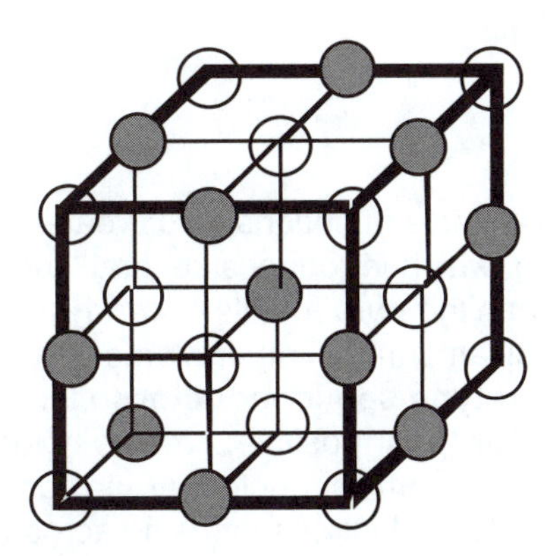

Metallic Solids and Molecular Orbital Band Theory

We can picture metals as a regular array of metal cations in a "sea" of valence electrons. The mobile electrons conduct heat and electricity, and the cations move easily as the metal is shaped into a sheet or a wire. The more quantitative **band model** describes the sea of electrons as a continuous band of molecular orbitals formed from the valence atomic orbitals of the many metal atoms in a metal crystal. The drawing on the next page shows electrons in filled molecular orbitals **(valence band)** that can be excited into empty orbitals **(conduction band)** where they are mobile and are free to conduct electricity or transmit thermal energy:

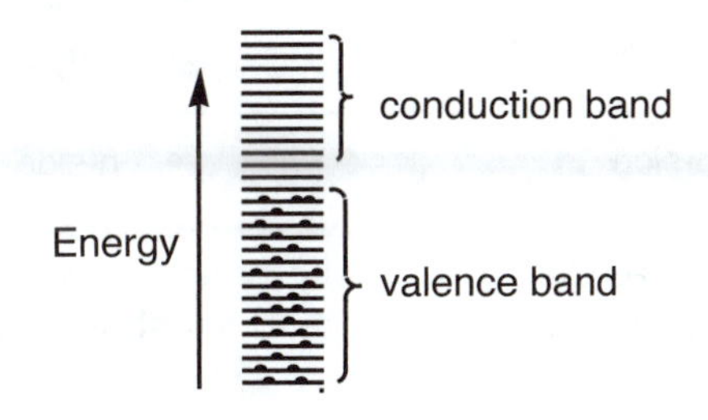

Semiconductors have a small energy gap between the valence and conduction bands. Thermally excited electrons can pass the gap and allow a small current to flow.

Insulators have a large gap between the bands that is too large for electrons to jump even when heated.

Superconductors can transmit electricity with little loss of energy. When metals conduct at ordinary temperatures, moving electrons collide with vibrating atoms and produce heat with a resulting loss of energy. Very low temperatures minimize atomic vibrations and allow certain materials to become superconductors, metals able to transmit electricity with little energy loss. Certain metal-oxides have been found to be good superconductors. Unfortunately, they tend to be brittle, making them hard to machine, and they may permanently lose their superconductivity when warmed.

Amorphous Solids

Amorphous solids are noncrystalline. Their structure can be compared to a liquid (which has no long-range molecular order), whose atoms are frozen in place. They often have small, ordered regions connected by large disordered regions. Rubber, charcoal, and glass are familiar amorphous solids. Quartz glass, formed when crystalline quartz is melted and the disordered liquid structure is cooled rapidly to "freeze" the disordered atoms in place, is sometimes referred to as a *supercooled liquid.*

12.7 ADVANCED MATERIALS

Electronic Materials

Defects may be introduced into crystals intentionally to create materials with increased strength, hardness, or conductivity. Metal alloying, which introduces several kinds of defects into the crystal lattice, often forms a substance that is harder than the pure metals from which the alloy was formed. Semiconductors may be doped with small amounts of other elements to change the number of valence electrons and increase conductivity. An **n-type semiconductor** (**n** for **n**egative) adds electrons to the conduction band; a **p-type semiconductor** (**p** for **p**ositive) creates holes (or vacancies) in the valence band. A p-n junction forms when p-type and n-type semiconductors are in contact with each other. The unequal concentration of electrons and holes results in electrons diffusing across the junction from the n-side to the p-side, and holes diffusing from the p-side to the n-side until equilibrium is established:

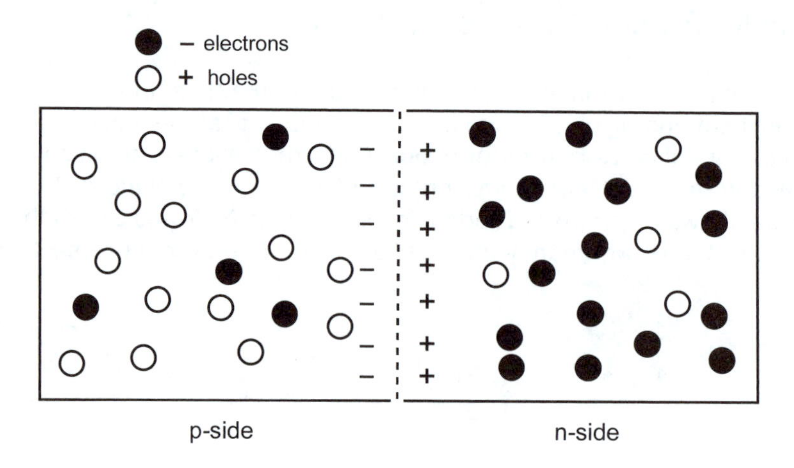

P-n junctions may act as rectifiers, devices that convert alternating current (AC) to direct current (DC), or as transistors, which amplify a signal. Figure 12.40 in your textbook shows how a p-n junction results in unidirectional current flow.

Liquid Crystals

Liquid crystals form when molecules pack with a high degree of order but like liquids, flow. Molecules that form liquid crystal phases usually have long, rod-like shapes that align with each other through intermolecular interactions along their long axes. Three common types of order that these molecules may exhibit are *nematic* (molecules lie in the same direction but ends are not aligned), *cholesteric* (a twisted nematic phase), and *smectic* (the most ordered phase, molecules lie parallel to each other with their ends aligned). Figure 12.42 on page 496 in your textbook illustrates these three common types of liquid crystal phases. Liquid crystal phases may form as the result of temperature changes (*thermotropic* phases) or as a result of concentration changes (*lyotropic* phases). Imagine a thermotropic liquid crystal in the liquid state. If we cool the liquid, the disordered molecules begin to align themselves and form a liquid crystal. The liquid crystal may transition from the somewhat disordered nematic phase to the more ordered smectic phase as it cools. Finally, if cooled enough, the molecules become an ordered crystal.

Liquid crystal displays (LCDs) in watches, calculators, and computers depend on changes in molecular orientation in an electric field. A current from a battery controls the orientation of the molecules. When the current is on, molecules align, block the light and form a dark region in the liquid crystal display. When the current is off, the molecules are not aligned, and light passes through to form a bright region on the liquid crystal display. Cholesteric liquid crystals may involve color changes with temperature. The helical twist unwinds with heating and the extent of unwinding determines the color.

Ceramics

Think back to the structure of diamond, the hardest substance known, discussed in Chapter 9. Diamond is a three-dimensional array of tetrahedrally bound carbon atoms (a network covalent solid). If you were to replace every other carbon atom (4 valence electrons) with a silicon atom (also 4 valence electrons), you would have silicon carbide, SiC. Silicon carbide is a *ceramic* material of great strength that can be made into thin fibers, called whiskers, to reinforce other ceramic materials. Silicon carbide forms as a silicone polymer is heated to 800°C to drive off methane and hydrogen gas:

$$[(CH_3)_2Si]_n \longrightarrow nCH_4(g) + nH_2(g) + nSiC(s)$$

Imagine taking diamond and removing a proton and electron from every other carbon atom and adding them to the remaining carbon atoms. You would have a new diamond-like structure formed from the elements boron (3 valence electrons) and nitrogen (5 valence electrons). This material, boron nitride (BN) is an extremely hard and durable ceramic. Boron nitride may also exist in a graphite-like form, in which case it has extraordinary properties as an electrical insulator.

Much of the interest in ceramic materials is directed towards the high-temperature superconducting ceramic oxides (such as $YBa_2Cu_3O_7$ discussed in your textbook). Research has begun to overcome the inherent brittleness of these materials and allow them to be machined into films and ribbons.

Polymeric Materials

Polymers are large molecules made up of covalently linked chains of smaller molecules called **monomers.** Polymers may have anywhere from hundreds to hundreds of thousands of repeat units. Polymers are ubiquitous in our lives both as synthetic polymers made in laboratories, and as biopolymers produced in organisms.

We can find the molecular mass of a polymer chain by multiplying the number of repeat units (*n*) by the molecular mass of the repeat unit:

$$\mathbf{M}_{\text{polymer}} = \mathbf{M}_{\text{repeat}} \times n$$

Synthetic polymers all have a distribution of chain lengths, requiring polymer chemists to use average molecular masses.

The long axis of a polymer chain is called its **backbone.** The length of an extended backbone is the number of repeat units times the length of each repeat unit. Most polymers adopt a **random coil** shape, which is determined by the nature of intermolecular forces between chain portions, and between the chain and solvent (if the polymer is in solution). The radius of gyration, R_g, (the average distance from the center of mass to the outside edge of the chain) describes the coiled size of a polymer chain. For many polymers, the radius of gyration can be determined experimentally with light-scattering techniques. Typical polymers often have lengths more than 100 times the size of their diameters.

When portions of polymer chains align regularly with neighboring chains (or with other parts of the same chain) the polymer becomes **semicrystalline.** Intermolecular interactions such as dipole-dipole, hydrogen bonding, or dispersion forces between different functional groups in polymer chains cause these orderly alignments to occur.

Many polymers exist as viscous liquids. As temperature decreases, the polymer may form a rigid solid called a **polymer glass.** This transition of a polymer from a liquid to a glass occurs over a narrow (10-20°) temperature range and is called its glass transition temperature, T_g. Like glass itself, many polymer glasses are transparent. Polymers are often used to increase the viscosity of many common materials, such as motor oil, paint, and salad dressing. Dissolved polymers increase viscosity by their interaction with the solvent. Many polymers can be deformed when they are warm and they retain their shape when cool. We count on this **plastic** behavior to produce everything from milk jugs to car parts.

Polymer chemists tailor polymer properties such as viscosity, strength, flexibility, and toughness through changes in the polymer's architecture. Highly branched polymers (**dendrimers** are the extreme example) prevent polymer chains from packing together as well, and are more flexible. **Crosslinks,** branches that link one chain to another, affect polymer flow. Lightly crosslinked chains produce **thermoplastic** polymers (still flow at high temperatures); highly crosslinked chains form **thermoset** polymers (cannot flow because they are single networks). Thermoset polymers are extremely rigid and strong below their glass transition temperatures. Above their glass transition temperatures, many thermosets become **elastomers,** polymers that spring back to their initial shapes after being stretched.

Different monomer sequences also influence polymer properties. A **homopolymer** consists of only one type of monomer; a **copolymer** consists of two or more types. Two simple types of **block copolymers** are the **AB** block copolymer and the **ABA** block copolymer:

...A–A–A–A–A–B–B–B–B–B...

...A–A–A–B–B–...–B–B–A–A–A...

AB block copolymers may be used as adhesives to join two polymer surfaces covalently. Some ABA block copolymers are examples of thermoplastic elastomers, materials shaped at high temperatures that become elastomers at room temperature (think footware).

Nanotechnology

Nanotechnology is the manufacture of machines from atomic and molecular parts. Nanoscale materials might be made one atom at a time through self-assembly (using building blocks of small molecules with complementary shapes that would "stick" together from intermolecular forces such as dipole-dipole and H-bonding) and controlled orientation (positioning two molecules near each other long enough for the intermolecular forces to take effect).

CHAPTER CHECK

Make sure you can...

- List phase changes that are exothermic and phase changes that are endothermic.
- Calculate the overall enthalpy change when a substance changes temperature and/or its state of matter.
- Define vapor pressure.
- Determine how vapor pressure changes with temperature (the Clausius-Clapeyron eqn.).
- Describe the relationship between vapor pressure and boiling point.
- Use a phase diagram to predict the physical state (or phase change) of a substance.
- Describe the types and relative strengths of intermolecular forces.
- Predict when H–bonds will form and their effects on the physical properties of a substance.
- Define surface tension, capillarity, and viscosity.
- Explain why water is a unique substance.
- Draw the three common unit cells and find the number of atoms per unit cell.
- Describe the properties of atomic, metallic, ionic, molecular, and network covalent solids.
- Calculate atomic radii from unit cell measurements; know the size range for atomic radii.
- Describe the electron-sea and band theory bonding models and use them to explain metallic properties such as electrical conductivity, luster, malleability, and thermal conductivity.
- Use band theory to describe the differences between conductors, semiconductors, insulators, and superconductors.

Chapter 12 Exercises

12.1) a. The heat of vaporization of water at 100°C plus the heat of fusion of ice at 0°C does not equal the heat of sublimation of ice. Why?

b. Does $\Delta H^o_{fus} + \Delta H^o_{vap} = \Delta H^o_{sub}$?

12.2) a. What are the three principal types of intermolecular forces?

b. What is the difference between an intermolecular force and an intramolecular force? Is a hydrogen bond in water an intramolecular or an intermolecular force?

12.3) What is the enthalpy change when 20.0 grams of steam initially at 110.0°C and 760. torr is cooled to ice at –10.0°C?

Heat capacity of ice = 2.00 J/g°C
Heat of fusion of ice = 334 J/g
Heat capacity of water = 4.18 J/g°C
Heat of vaporization of water = 2.26 kJ/g
Heat capacity of steam = 1.84 J/g°C

Is heat liberated or absorbed when this change takes place?

12.4) What amount of heat is released when 75.0 g of steam at 115°C cools to 37.0°C, body temp.? Why does a burn due to steam tend to be more severe than one caused by boiling water?

12.5) How much heat is needed to melt 160. g of snow at –12.0°C and warm the resulting water to 37.0°C? Why might it be dangerous to eat large quantities of snow in the winter?

12.6) A system of ice and water at 0.0°C has a mass of 5.0 grams. The system is converted completely to water at 0.0°C by supplying 1.34 kJ of heat. How much water was initially present? The heat of fusion of ice = 334 J/g.

12.7) If 40.0 g of ice at –8°C is added to 100. g of water at 25°C, what is the final temperature of the mixture?

12.8) Ammonia boils under atmospheric pressure at –33°C. If the molar heat of vaporization of ammonia is 24 kJ/mol, at what temperature would ammonia boil under a pressure of 6.0 atm?

12.9) What do the following represent on a temperature-pressure phase diagram for a single substance?

a. the areas between lines on the diagram
b. a line on the diagram
c. a triple point where lines meet on the diagram
d. the critical point on the diagram

12.10) Calculate how much of the space within a body-centered unit cell is actually occupied by the atoms, i.e., what is the occupancy of available space within the cube.

The edge length of a body-centered cubic unit cell is $4r/\sqrt{3}$, where r = radius of the atom involved. The volume of a sphere = $4\pi r^3/3$.

12.11) An unknown element crystallizes in a face-centered cubic lattice with a unit cell edge length of 352 pm. The density of the metal is 8.92 g/cm^3. What is the element?

12.12) X-ray crystallography reveals that strontium has a face-centered cubic lattice with a unit cell length of 608 pm. What is the atomic radius of Sr?

12.13) Tungsten, W, is used in the filaments of light bulbs. It forms a body-centered metallic lattice and has an atomic radius of 141 pm. What is the length of the W unit cell?

12.14) Thorium, Th, has the largest molar mass of any element with a face-centered cubic lattice. X-ray crystallography studies indicate the atomic radius of the element is 180. pm. The molar mass of thorium is 232 g/mol. What is the volume of the Th unit cell? What is the density?

12.15) Europium, Eu, is the only rare-earth element that has a body-centered cubic lattice. The density of Eu is 5.26 g/cm^3. The molar mass of europium is 152 g/mol. What is the volume of the Eu unit cell? What is the atomic radius of Eu?

Chapter 12 Answers

12.1) Hess's law of constant heat summation states that the enthalpy change for a reaction is independent of the route taken between the reactants and products. It is a specialized statement of the first law of thermodynamics. To apply Hess's law, the starting point and ending point for each route must be the same and the routes must be complete (no missed steps).

a. The heat of vaporization of water at 100°C plus the heat of fusion of ice at 0°C does not equal the heat of sublimation of ice because the heat required to heat the water from 0°C to 100°C has not been included. The temperature of the sublimation is also not specified.

b. Under conditions of 1 bar pressure and 25°C for all the enthalpy changes:
$\Delta H^{o}_{fus} + \Delta H^{o}_{vap} = \Delta H^{o}_{sub}$

12.2) a. The strongest intermolecular force is hydrogen bonding. The strength of a hydrogen bond lies between 10 and 40 kJ/mol.

The second strongest intermolecular force is the attraction between molecules possessing a dipole moment (dipole-dipole attraction). The strength of this type of intermolecular force lies between 5 and 25 kJ/mol.

The third type of intermolecular force is London dispersion. The strength of this force varies from very weak (about 0.05 kJ/mol) to quite strong (about 40 kJ/mol) depending upon the size of the molecule.

b. An intermolecular force is an attraction between molecules; an intramolecular force is a bond within a molecule. A hydrogen bond is an intermolecular force between water molecules. The oxygen atom of one water molecule hydrogen bonds to the hydrogen atom of another water molecule.

12.3) Heat liberated when the steam is cooled to 100.°C = 20.0 g × 1.84 J/g °C × 10.0°C = 368 J

Heat liberated when the steam is condensed to water = 20.0 g × 2260 J/g = 45,200 J

Heat liberated when the water is cooled to 0°C = 20.0 g × 4.18 J/g °C × 100.°C = 8360 J

Heat liberated when the water is frozen = 20.0 g × 334 J/g = 6680 J

Heat liberated when the ice is cooled to –10.0°C = 20.0 g × 2.00 J/g °C × 10.0°C = 400. J

Adding all the steps involved, the total enthalpy change = 61.0 kJ.

Heat is liberated in each step of the process; energy is released when a substance cools and energy is also released when bonds are formed.

12.4) 1.91×10^5 J; The heat released by the condensation of the steam represents the major fraction of the heat transferred.

12.5) 8.20×10^4 J; The heat required to melt and warm the ice comes from the body core. This heat loss could cause hypothermia.

12.6) If the heat of fusion of ice = 334 J/g, then 1.34 kJ is sufficient heat to melt 4.0 grams of ice. If the total mass of the system is 5.0 grams, there must have been 1.0 gram of water originally present.

12.7) 0°C; 14,000 J of heat are needed to melt the 40.0 g of ice, but only 10,400 J of heat is available in the 100 g of H_2O as it cools from 25°C to 0°C.

12.8) The relationship between the vapor pressure and temperature is provided by the Clausius-Clapeyron equation. Recall that a liquid will boil when its vapor pressure equals the external (atmospheric) pressure. We expect ammonia to boil at a higher temperature when it is under 6 atm. of pressure than when it is at atmospheric pressure.

Remember to convert the temperatures into degrees Kelvin:

$$\ln (p_2/p_1) = -\Delta H_{vap}/R(1/T_2 - 1/T_1)$$

$$\ln (6.0/1.0) = -24{,}000/8.314(1/T_2 - 1/240.)$$

The new boiling point T_2, under an increased external pressure of 6.0 atm, is 9.0°C.

12.9) a. the conditions under which only one phase is the thermodynamically stable phase

b. the conditions under which two phases can coexist in equilibrium together

c. the conditions under which three phases can coexist in equilibrium with one another

d. the temperature above which it is impossible to liquefy a gas no matter how high the pressure

12.10) The number of atoms within a body-centered unit cell is 2; one at the center and one on the corners (eight corners each contributing one-eighth). The atoms touch along the diagonal of the cube so that the length of the body diagonal is equal to $4r$, where r is the radius of the atom involved. Thus, as stated in the question, the edge length of the unit cell is $\left(4r/\sqrt{3}\right)$.

The volume of the unit cell is $\left(4r/\sqrt{3}\right)^3$.

The volume of the two atoms occupying the unit cell is $2 \times 4\pi r^3/3$.

The fraction of the cell occupied by the atoms is $8\pi r^3/3 \div \left(4r/\sqrt{3}\right)^3 = 0.68$ or 68%.

12.11) There are four atoms in a face-centered cubic unit cell. The volume of this unit cell is $(352)^3$ pm^3. The density of the metal is 8.91 g/cm^3. Since the density equals the mass/volume, the mass of four atoms must be equal to the density $\times$ volume:

The mass of four atoms = 8.92 $g/cm^3 \times (352)^3 \times (10^{-10})^3$ $cm^3 = 3.89 \times 10^{-22}$.

The mass of one atom = $3.89 \times 10^{-22}/4 = 9.73 \times 10^{-23}$ g.

The mass of one mole of atoms = 9.73×10^{-23} g/atom $\times$ 6.022×10^{23} atom/mol = 58.6 g/mol. The element is probably nickel.

12.12) The diagonal across the face of a face-centered unit cell = 4 $\times$ the radius. Each side of the unit cell has a length of 608 pm. Therefore:

$$(608)^2 + (608)^2 = (4r)^2$$

$$r = 215 \text{ pm}$$

12.13) 326 pm

12.14) 4 $\times$ the atomic radius = 720. pm, the length of the diagonal of a unit cell face
The edge length (l) of the unit cell is:

$$l^2 + l^2 = (720.)^2$$

$$l = 509 \text{ pm}$$

Volume of the unit cell = $(\text{length})^3 = 1.32 \times 10^8$ pm = 1.32×10^{-22} cm^3 = volume of 4 Th atoms (there are 4 atoms per unit cell in a face-centered cube)

Find the volume for a mole of Th atoms:

$$\frac{1.32 \times 10^{-22} \text{ cm}^3}{\text{unit cell}} \times \frac{\text{unit cell}}{4 \text{ atoms}} \times \frac{6.02 \times 10^{23} \text{ atoms}}{\text{mol}} = 19.87 \text{ cm}^3\text{/mol}$$

Now we can find density:

$$d = \frac{m}{V} = \frac{232 \text{ g/mol}}{19.87 \text{ cm}^3\text{/mol}} = 11.7 \text{ g/cm}^3$$

12.15) First, find the volume for a mole of Eu atoms:

$$V = \frac{m}{d} = \frac{152 \text{ g/mol}}{5.26 \text{ g/cm}^3} = 28.897 \text{ cm}^3\text{/mol}$$

Now find the volume of a unit cell:

$$\frac{28.897 \text{ cm}^3}{\text{mol}} \times \frac{\text{mol}}{6.02 \times 10^{23} \text{ atoms}} \times \frac{2 \text{ atoms}}{\text{unit cell}} = 9.60 \times 10^{-23} \text{ cm}^3\text{/unit cell}$$

atomic radius = 198 pm

The Properties of Mixtures: Solutions and Colloids

13.1 TYPES OF SOLUTIONS

In Chapter 2 we saw that elements and compounds are pure substances, and that mixtures are impure substances. This chapter will concentrate on homogeneous mixtures (solutions).

Review exercise

Choose words from the list below to complete the following paragraph:

(a) aqueous	(c) gaseous	(e) homogeneous	(g) solid
(b) colloid	(d) heterogeneous	(f) mixture	(h) solution

A combination of pure substances is called a ______________(1). There are two kinds of mixtures: a mixture that has visible boundaries is a ______________(2) mixture; a mixture with no visible boundaries is said to be a ______________(3) mixture. A homogeneous mixture is also called a ______________(4). A substance dissolved in water is an ______________(5) solution. Air is a ______________(6) solution consisting of mainly oxygen and nitrogen molecules. Brass is a ______________(7) solution consisting of mainly copper and zinc. A heterogeneous mixture consisting of particles in suspension that do not settle out is called a ______________(8).

answers: 1-f, 2-d, 3-e, 4-h, 5-a, 6-c, 7-g, 8-b.

When sugar dissolves in a cup of water, we say that sugar is the **solute** and water is the **solvent.** The **solubility (*S*)** of the sugar is the maximum amount that dissolves in the cup of water at a specified temperature. The solubility of sugar in water increases as the temperature of the water increases. From a molecular point of view, when sugar molecules "dissolve" in water, water molecules surround them. It is therefore the intermolecular forces between solute and solvent that determine the extent to which solutes will dissolve in solvents.

Figure 13.1 on page 518 in your textbook uses space-filling models to show examples of the major types of intermolecular forces in solutions. These intermolecular forces are summarized in the box below.

Intermolecular Forces

Ion-dipole forces: the strongest intermolecular force. When table salt (sodium chloride) dissolves in water, sodium and chloride ions separate as water molecules cluster around the ions forming **hydration shells.**

Dipole-dipole forces: attractions between positive and negative regions of polar molecules. The most important dipole-dipole force is the hydrogen bond and is the reason water can dissolve so many compounds. When sugar dissolves in water, –OH groups on sugar molecules hydrogen bond with the water molecules.

Charge-induced dipole forces: ions or polar molecules distort the electron clouds of nearby nonpolar molecules and cause or induce dipoles in the nonpolar molecules.

Dispersion forces: attractions between positive and negative regions of nonpolar molecules.

Liquid-Liquid Solutions

Liquids with similar structures and therefore similar intermolecular forces of about the same type and magnitude are **miscible** (soluble in all proportions).

Like dissolves like

1) Which of the following pairs of liquids would you predict would be miscible with each other?

1)

a) carbon tetrachloride (CCl_4)

b) chloroform ($CHCl_3$)

2)

a) ethyl alcohol (CH_3CH_2OH)

b) methyl alcohol (CH_3OH)

(continued on next page)

3)

```
    H  H  H  H  H  H              H  H  H  H  H
    |  |  |  |  |  |              |  |  |  |  |
 H—C—C—C—C—C—C—H               H—C—C—C—C—C—H
    |  |  |  |  |  |              |  |  |  |  |
    H  H  H  H  H  H              H  H  H  H  H
```

a) hexane (C_6H_{14}) b) pentane (C_5H_{12})

4)

```
    H  H  H  H  H  H  H  H              H  H  H  H
    |  |  |  |  |  |  |  |              |  |  |  |
 H—C—C—C—C—C—C—C—C—O—H               H—C—C—C—C—O—H
    |  |  |  |  |  |  |  |              |  |  |  |
    H  H  H  H  H  H  H  H              H  H  H  H
```

a) octyl alcohol ($C_8H_{17}OH$) b) *n*-butyl alcohol (C_4H_9OH)

answers:

Each of the above pairs contains liquids that are miscible with each other.

1) Chloroform and carbon tetrachloride have similar intermolecular forces despite the fact that carbon tetrachloride is nonpolar, and chloroform is polar. Dipole forces make only a minor contribution to the intermolecular forces in chloroform.

2) Ethyl alcohol and methyl alcohol each contain an –OH group which causes the molecules to hydrogen bond and gives them similar intermolecular forces.

3) Molecules of pentane and hexane are held together by nonpolar dispersion forces of about the same magnitude.

4) Octyl alcohol and *n*-butyl alcohol each contain a nonpolar hydrocarbon region and a polar –OH group. They have similar intermolecular forces and are miscible.

2) Which of the above pairs would you predict to be miscible with water?

answers

Only pair "2", ethyl alcohol and methyl alcohol, are miscible with water. The –OH group in these alcohols allows them to hydrogen bond, and therefore the intermolecular forces between the alcohols and water in a solution are approximately the same as those of the pure liquids.

Note that octyl alcohol and *n*-butyl alcohol also each contain an –OH group, but the longer carbon-hydrogen chains in these molecules require many hydrogen bonds in water to break in order for them to insert themselves into the water structure. In general, solubility in water decreases with longer hydrocarbon chain length. Table 13.2 on page 520 in your textbook lists the solubilities of alcohols containing one to six carbon-hydrogen chains. The alcohols containing 4, 5 and 6 carbons (butanol, pentanol, and hexanol) are not infinitely soluble in water.

Solid-Liquid Solutions

Solids always have limited solubilities in liquids. The intermolecular forces in solids are much stronger than those of liquids at the same temperature. Compare the solubility of iodine (a bluish black solid with a violet vapor) to bromine (Br_2, a dark red liquid) in carbon tetrachloride at 25°C. Bromine, a liquid, is infinitely soluble in carbon tetrachloride; iodine, a solid, forms a saturated solution when the mole fraction of

I_2 is only 0.011. The dispersion forces in $I_2(s)$ are an order of magnitude greater than those in $CCl_4(l)$. Low-melting solids are more soluble than high-melting solids of similar structure.

Solid-liquid solution

Choose words from the list below to complete the following paragraphs:

(a) saturated
(b) solubility
(c) solute
(d) solvent
(e) temperature

When we dissolve table salt (sodium chloride) in water, we say that sodium chloride is the ______________(1), and water is the ______________(2). The **solute** is the part of the solution that we consider to be dissolved in the other, the **solvent.** The ______________(3) is usually present in larger amounts than the ______________(4).

In 100 mL of hot water (100°C), we could add 39 g of sodium chloride before salt would no longer go into solution. The ______________(5) of NaCl at 100°C is therefore 39 g/100 mL water, and we say we have a ______________(6) solution, one that contains (at a given temperature) the maximum amount of dissolved solute in the presence of excess ______________(6). At 0°C, only 36 grams of sodium chloride dissolves in 100 mL of water. Solubility depends on the system's ______________(8).

answers: 1-c, 2-d, 3-d, 4-c, 5-b, 6-a, 7-b, 8-e.

Solubilities

The structures of water, isopropyl alcohol (rubbing alcohol), *n*-hexane (a component of gasoline), and carbon tetrachloride are:

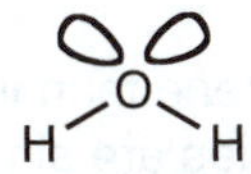

water 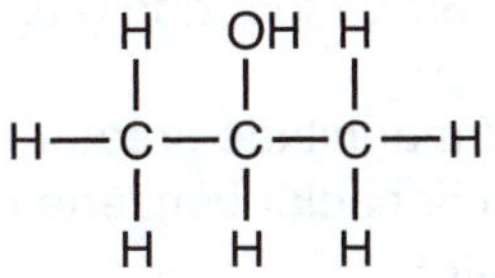 | isopropyl alcohol 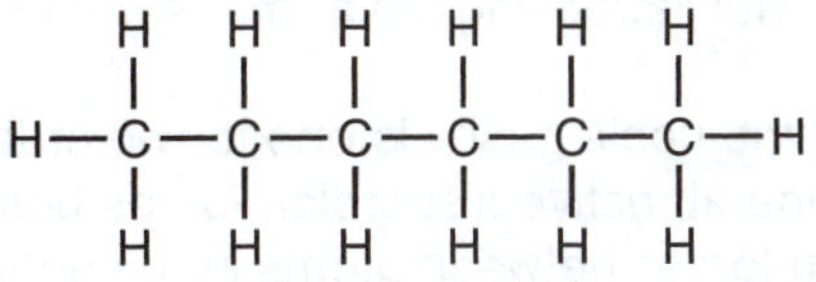 | *n*-hexane 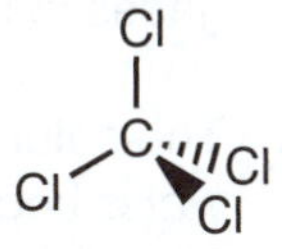 | carbon tetrachloride

Using the general rule that **like dissolves like,** which solvent would you choose to dissolve the following?

1) bike grease
2) oil-based paint
3) latex paint
4) soap
5) candle wax
6) DDT

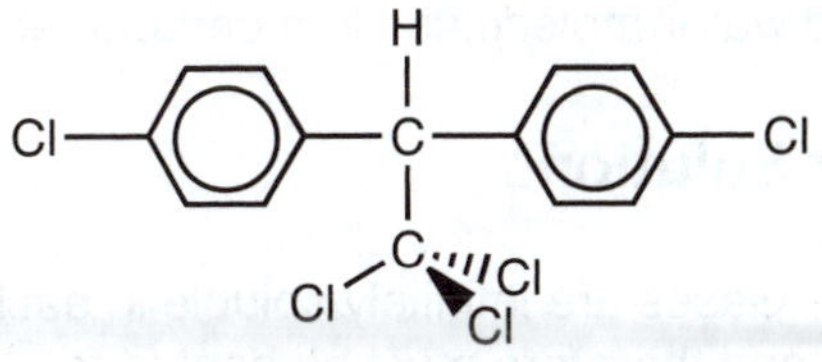

answers:

1) Grease is a mixture of nonpolar hydrocarbon molecules. Both gasoline (or *n*-hexane) and carbon tetrachloride would dissolve it.

(continued on next page)

2) An oil-based paint would require a nonpolar solvent such as *n*-hexane or carbon tetrachloride.

3) A latex (water-based) paint dissolves in water, or an alcohol such as isopropyl alcohol.

4) Soap is a molecule with a long nonpolar hydrocarbon chain, and a polar, ionic head. The hydrocarbon tails surround grease molecules, and the polar ionic heads hydrogen bond and dissolve in water.

5) Waxes dissolve in nonpolar solvents such as hexane and carbon tetrachloride.

6) DDT has a structure similar to carbon tetrachloride, and therefore is soluble in CCl_4. It tends to concentrate in the fatty tissue of animals. Its insolubility in water prevents it from being washed out of soil by rainfall.

Gas-Liquid Solutions

In general, the higher the boiling point of a gas, the closer its intermolecular forces will be to the magnitude of intermolecular forces of a liquid, and the more soluble it will be in a liquid. The best solvent for a gas will be the one whose intermolecular forces are the most similar to those of the gas.

Solubility of noble gases

Consider the noble gases He, Ne, Ar, Kr, Xe, and Rn:

1) Which noble gas is the most soluble in water?

2) Are the noble gases more soluble in water or in benzene?

answers:

1) Rn, with the highest boiling point, is the most soluble noble gas. The boiling points of the noble gases increase with increasing molar mass from He (–269°C) to radon (–62°C).

2) The solubility of the noble gases is one to two orders of magnitude greater in benzene than in water because the attractive dispersion forces between nonpolar benzene molecules are similar to the dispersion forces between atoms in the noble gases.

Oxygen dissolves in water and allows fish and other aquatic life to live. Tap water is slightly acidic (it has a pH of about 6) because carbon dioxide dissolves in the water and forms carbonic acid:

$$CO_2(g) + 2H_2O(l) \rightarrow H_2CO_3(aq) \rightleftharpoons H_3O^+(aq) + HCO_3^-(aq)$$

Gases are typically not very soluble in water because gas molecules interact weakly with highly hydrogen bonded water molecules. In a carbonated beverage, bottling the drink under pressure dissolves more CO_2.

Other Solutions

Gases are infinitely soluble in each other to form gas-gas solutions. Solids diffuse to only a small extent, and therefore must be heated to high temperatures and cooled to form solutions. Brass, sterling silver, and steel are examples of metal mixtures called **alloys.** Gases can dissolve into solids by filling spaces between the closely packed particles. Oxygen reacts with iron to form iron oxide, or rust.

13.2 BIOLOGICAL MACROMOLECULES: THE IMPORTANCE OF THE H–BOND

The structures of biopolymers – **proteins, nucleic acids,** and **polysaccharides** – depend on intermolecular forces. Because these molecules exist in an aqueous environment in biological systems, their structures are determined partially by their polar and ionic groups interacting with the surrounding water.

Chains of **amino acids,** organic compounds with amine ($-NH_2$) and carboxyl ($-COOH$) groups in the same molecule, form **proteins.** Anywhere from 50 to several thousand amino acids may bond together to form proteins, and it is the sequence of these amino acids that determines the protein's shape. You can recognize an amino acid by looking for a carbon atom (called the alpha carbon) that has the following four groups bonded to it: an H atom, a charged carbonyl group ($-COO^-$), an amine ($-NH_3^+$), and an R group side chain. The side chain may be a hydrogen atom, or a large chain:

$$H_3\overset{+}{N}-\underset{H}{\overset{R}{\overset{|}{\underset{|}{C}}}}-C\begin{smallmatrix}\nearrow O \\ \searrow O^-\end{smallmatrix}$$

Page 524 in your textbook shows examples of nonpolar (leucine), polar (serine and cysteine), and ionic (lysine and glutamic acid) side chains. Proteins fold in such a way that these polar and ionic side chains interact with water through ion-dipole forces and hydrogen bonds, and nonpolar interactions occur in the proteins' interior. The shape of a protein determines its function. You can see in Figure 13.7 on page 524 that *H–bonds*, salt links, disulfide (–S–S–) bonds, and dispersion forces determine the shape of this protein molecule.

Intermolecular forces are also responsible for the assembly of **cell membranes.** A cell membrane consists of a **phospholipid bilayer** with embedded proteins. The phospholipid bilayer contains a double sheet of molecules with nonplar tails, and polar ionic heads lined up as shown in Figure 13.10 in your textbook. The nonpolar tails of the two layers face each other, leaving the polar heads to face the aqueous cell exteriors, and the aqueous cell interiors. The embedded proteins tend to expose their nonpolar groups to the nonpolar hydrocarbon tails of the membrane lipids. These embedded proteins can act as pumps, moving ions and specific molecules in or out, or they may act as receptors for small biomolecules.

Most people have heard of **DNA** referred to as a **double helix.** What most people may not know is that it is millions of *H–bonds* that link base pairs together and hold the two chains together forming this double helix. The backbone of each DNA chain consists of an alternating pattern of a sugar molecule and a phosphate group. Attached to each sugar molecule is a nitrogen-containing base, a flat ring structure that dangles off the sugar-phosphate chain. In Chapter 15, we'll see how these base pairs are essential to protein synthesis and DNA replication as their *hydrogen bonds* zip and unzip.

13.3 WHY SUBSTANCES DISSOLVE

The tendency for all systems is for their disorder to increase. **Entropy (*S*)** is the measure of a system's disorder; entropy increases as systems become more disordered:

$$S_{\text{disorder}} > S_{\text{order}}$$

The entropy of the universe is always increasing. A spontaneous process may occur in which the <u>system</u> becomes more ordered, but the change in the entropy of the <u>system</u> + the <u>surroundings</u> (the universe) must be greater than zero:

$$\boldsymbol{\Delta S_{\text{universe}} = \Delta S_{\text{sys}} + \Delta S_{\text{surr}} > 0}$$

Entropy favors solution formation, since entropy increases as pure substances mix together and become more disordered. However, the **heat of solution (ΔH_{soln})** may prevent two substances from forming a solution because the entropy of the surroundings decreases if it requires heat to form a solution, and this decrease in the entropy of the surroundings may outweigh the increase in the entropy of the system.

When table salt, NaCl dissolves in water, polar water molecules attract the sodium and chloride ions breaking the crystal structure apart. The water molecules make **hydration shells** around the ions. For a sodium ion, Na^+, six water molecules octahedrally surround it. These water molecules hydrogen bond to water molecules farther away, forming a second, less structured hydration shell, and these in turn hydrogen bond to other molecules in the bulk solvent.

We can consider a solute and solvent forming a solution in steps:

1) The solid breaks apart into ions:

$$M^+X^-(s) \rightarrow M^+(g) + X^-(g) \qquad \Delta H > 0$$

2) Solvent molecules surround the ions (**solvation,** or **hydration** when the solvent is water):

$$M^+(g) \xrightarrow{H_2O} M^+(aq) \qquad \Delta H < 0$$

$$X^-(g) \xrightarrow{H_2O} X^-(aq) \qquad \Delta H < 0$$

The first step is the lattice dissociation energy of an ionic solid and always requires energy ($\Delta H > 0$). (Remember from Chapter 9 that the lattice formation energy is the reverse of this process and usually releases a large amount of energy.) The second step is the heat of hydration of the cation and anion, and this step releases energy ($\Delta H < 0$). The heat of solution for ionic compounds in water is the sum of these steps:

$$\Delta H_{soln} = \Delta H_{lattice\ diss.} + \Delta H_{hydration\ of\ the\ ions}$$

Remember that the solution process involves both the change in enthalpy and the change in entropy. The increase in entropy that occurs when a substance dissolves in another always favors solution formation, but the heat of solution may hinder solution formation if it is positive and if it outweighs the entropy factor.

13.4 SOLUBILITY AS AN EQUILIBRIUM PROCESS

The solubility of one substance in another may depend on the external conditions of temperature and pressure. We can determine the effects of these two variables on solubility if we think of the solution process as an equilibrium:

$$NaCl(s) \xrightleftharpoons{H_2O} Na^+(aq) + Cl^-(aq)$$

Temperature

If a solute does not dissolve, one of the first things we might try to get it to go into solution is to heat it. Most solids require heat for them to dissolve because heat must be absorbed to break down the crystal lattice. Consequently, the solubility of a solid in a liquid usually increases with temperature. Several sulfate salts are exceptions, one of which is shown on the graph in Figure 13.19 on page 534 in your textbook. You can see from the graph that the solubility of cerium sulfate, unlike most solids, decreases at higher temperatures.

Dissolving a gas in a liquid usually gives off heat, and as a result, gas solubility ordinarily decreases with an increase in temperature. Gases are only weakly attracted to solvents, and can overcome these weak forces as their kinetic energy increases with temperature. You observe this phenomenon each time you heat water and see bubbles form. Dissolved oxygen is less soluble at the higher temperature, and it escapes from the liquid when we heat a beaker of water.

The effect of temperature on solubility

Which of the following equilibrium equations near saturation represents sodium hydroxide (NaOH) dissolving in water, and which represents ammonium nitrate (NH_4NO_3) dissolving in water?
Do the solubilities of NaOH and NH_4NO_3 increase or decrease with temperature?

$$\text{solute} + \text{solvent} \rightleftharpoons \text{solution} + \textbf{heat}$$

$$\text{solute} + \text{solvent} + \textbf{heat} \rightleftharpoons \text{solution}$$

solution:

In the first reaction, heat is a product, and so it is an exothermic reaction ($\Delta H < 0$). If you dissolve sodium hydroxide in water, the flask feels hot as heat is released. In the second reaction, heat is a reactant, and so it is an endothermic reaction ($\Delta H > 0$). If you dissolve ammonium nitrate in water, the flask feels cold as heat is absorbed from the environment. You might expect the solubility of sodium hydroxide to decrease with temperature, and the solubility of ammonium nitrate to increase, but both compounds are more soluble at higher temperatures. Solubility is a complex behavior that cannot be predicted solely from ΔH_{soln} values.

Pressure

Pressure has little effect on the solubility of liquids and solids since they are almost incompressible, but it has a major effect on the solubility of gases. At moderate pressures, the solubility of a gas is proportional to the partial pressure of the gas above the solution **(Henry's law):**

$$S_{gas} = k_H \times P_{gas}$$

P_{gas} is the partial pressure of the gas over the solution, S_{gas} is its concentration in solution, and k_H is a constant characteristic of the particular gas-solvent combination.

Solubility of carbon dioxide

The solubility of carbon dioxide in water at 25°C is 1.2×10^{-5} mol/L. If an unopened soft drink contains a CO_2 pressure of 4.0 atm over the liquid, calculate the concentration of CO_2 in the soda before the bottle is opened. The partial pressure of CO_2 in the atmosphere is 4.0×10^{-4} atm.

solution:

First we use the solubility data to calculate Henry's law constant k_H:

$$S_{CO_2} = k_H \times P_{CO_2}$$

$$k_H = \frac{S_{CO_2}}{P_{CO_2}} = \frac{1.2 \times 10^{-5}\ \text{mol/L}}{4.0 \times 10^{-4}\ \text{atm}} = 0.030\ \text{mol/L atm}$$

Now we can use the value for k_H to solve for the concentration of CO_2 at 4 atm pressure:

$$S_{CO_2} = k_H \times P_{CO_2}$$

$$S_{CO_2} = 0.030\ \text{mol/L atm} \times 4.0\ \text{atm}$$

$$= 0.12\ \text{mol/L}$$

Notice the large difference in the concentration of CO_2 in the open container (1.2×10^{-5} mol/L) and the pressurized container (0.12 mol/L). If a soft drink is left uncapped, the carbon dioxide slowly diffuses out of solution, and the beverage becomes "flat."

13.5 WAYS TO EXPRESS CONCENTRATION

In Chapter 3 we saw that concentration (c) is the amount per unit volume (n/V). For the specific case when n = moles, and V = liters, concentration is given the special name **molarity** (M):

$$M = \frac{\text{moles solute}}{\text{liter solution}}$$

Molarity is not always a convenient unit to use since volume (liters of solution) depends on temperature, and volumes are not necessarily additive. **Molality** (m) expresses the concentration of a solution by moles of solute per kg of solvent:

$$m = \frac{\text{moles solute}}{\text{kg solvent}}$$

Mass percent, %(w/w), is a common way to record impurities in a solid. **Volume percent**, %(v/v), is often used for liquids and gases. **Mole percent** gives us a picture of the actual number of solute particles present in the solution:

$$\text{Mass \%} = \frac{\text{mass solute}}{\text{mass solution}} \times 100\%$$

$$\text{Volume \%} = \frac{\text{volume solute}}{\text{volume solution}} \times 100\%$$

$$\text{Mole \%} = \frac{\text{moles solute}}{\text{moles solution}} \times 100\%$$

$$\text{Mole Fraction } (X) = \frac{\text{mol solute}}{\text{mol solute} + \text{mol solvent}}$$

Lead poisoning

Lead and lead compounds inhibit development of cognitive processes in children. Recent guidelines have lowered the acceptable blood lead level from 24 micrograms of lead per deciliter of whole blood (24 μg/dL) to 9 μg/dL. The Merck Index defines lead poisoning as a blood Pb content of > 0.05 mg %. Compare this value to the 9 μg/dL guideline issued by pediatricians. (mg % = mg/100mL)

solution:

We need to convert 0.05 mg/100mL (mg %) to μg/dL:

$$\frac{0.05 \text{ mg}}{100 \text{ mL}} \times \frac{100 \text{ mL}}{\text{dL}} \times \frac{1000\ \mu\text{g}}{1 \text{ mg}} = 50\ \mu\text{g/dL}$$

Recent guidelines consider lower levels of lead to be more dangerous than was once thought. The level in The Merck Index refers to lead levels that would give noticeable symptoms such as weight loss, weakness, and anemia. The pediatrician guidelines are thought to indicate the level where slowed development may occur.

Radon limits

Radon is a colorless, odorless, inert gas that enters houses from the underlying soil. The United States Environmental Protection Agency recommends keeping indoor air levels below 4.0 pCi/L (picocuries per liter). The Merck Index lists the maximum permissible concentration of radon in air as 10^{-8} μCi/cc. How does this value compare to the value recommended by the EPA?

solution:

We will convert the units used by the EPA to the units listed in The Merck Index:

$$\frac{4.0 \text{ pCi}}{\text{L}} \times \frac{1\ \mu\text{Ci}}{10^6 \text{ pCi}} \times \frac{1 \text{ L}}{1000 \text{ mL}} \times \frac{1 \text{ mL}}{1 \text{ cc}} = 4 \times 10^{-9}\ \mu\text{Ci/cc}$$

The acceptable levels, according to the EPA, are less than half the limit listed in The Merck Index. Since no one can know for sure what constitutes "safe" levels for people, the EPA tries to set levels that they consider to be acceptable exposures over a long period of time.

Methylene chloride exposure

Many industrial processes use the solvent dichloromethane (or methylene chloride, CH_2Cl_2). It is a solvent for cellulose acetate (used for photographic films, waterproofing fabrics, phonograph records, insulating electric wires), degreasing and cleaning fluids, and as a solvent in food processing.

Many chlorinated organics are carcinogens, or suspected carcinogens. The EPA sets acceptable exposure to methylene chloride as 50 ppm (8 hour exposure). If the methylene chloride of your workplace is measured to be 150 mg/m^3, is this an acceptable level according to EPA guidelines? (The density of air is 1.18 g/L.)

solution:

We will start with 150 mg/m^3, and convert to ppm and compare the value to the EPA limit of 50 ppm:

$$\frac{150 \text{ mg } CH_2Cl_2}{\text{m}^3 \text{ air}} \left(\frac{\text{m}}{10 \text{ dm}}\right)^3 \times \frac{1 \text{ dm}^3}{\text{L air}} \times \frac{1 \text{ L air}}{1.18 \text{ g air}} \times \frac{1 \text{ g } CH_2Cl_2}{1000 \text{ mg } CH_2Cl_2}$$

$$= 1.27 \times 10^{-4} \text{ g } CH_2Cl_2/\text{g air}$$

To convert to parts per million, we multiply the answer above by $10^6/10^6$:

$$\frac{1.27 \times 10^{-4} \text{ g } CH_2Cl_2}{1 \text{ g air}} \times \frac{10^6}{10^6} = \frac{127 \text{ g } CH_2Cl_2}{10^6 \text{ g air}} = 127 \text{ ppm or } \frac{\text{mg}}{\text{kg}}$$

The measured level for the workplace (150 mg/m^3) is an unacceptable level for 8 hours of exposure according to EPA limits.

Hydrochloric acid

Concentrated hydrochloric acid, HCl(*aq*), contains approximately 36.0% HCl by mass.

1) What is the molarity of concentrated hydrochloric acid? The density of concentrated hydrochloric acid is 1.18 g/mL.

solution:

In 100. g of concentrated hydrochloric acid, there is 36 g of HCl:

$$\frac{36.0 \text{ g HCl}}{100. \text{ g soln}} \times \frac{\text{mol HCl}}{36.5 \text{ g HCl}} \times \frac{1.18 \text{ g HCl}}{\text{mL soln}} \times \frac{1000 \text{ mL}}{\text{L}} = 11.6 \text{ mol/L}$$

2) What is the molality of concentrated hydrochloric acid?

solution:

We need to find the moles of solute per kg of solvent. In 100. g of solution, there is 36.0 g of HCl, and 64.0 g (.0640 kg) of water:

$$\frac{36.0 \text{ g HCl}}{0.0640 \text{ kg water}} \times \frac{\text{mol HCl}}{36.5 \text{ g HCl}} = 15.4 \text{ mol/kg}$$

3) What is the mole fraction of HCl?

solution:

In 100. g of concentrated hydrochloric acid, there is 36.0 g of HCl and 64.0 g of water:

$$\text{moles HCl} = 36.0 \text{ g} \times \frac{\text{mol}}{36.5 \text{ g}} = 0.986 \text{ mol}$$

$$\text{moles } H_2O = 64.0 \text{ g} \times \frac{\text{mol}}{18.0 \text{ g}} = 3.55 \text{ mol}$$

$$\text{mole fraction HCl} = \frac{\text{mol HCl}}{\text{total moles of solution}}$$

$$= \frac{0.986 \text{ mol HCl}}{3.55 \text{ mol } H_2O + 0.986 \text{ mol HCl}} = 0.217$$

13.6 COLLIGATIVE PROPERTIES OF SOLUTIONS

Colligative properties depend only on the number of solute particles in an ideal solution, not on their identity.

Vapor Pressure Lowering

Vapor pressures

Arrange the following equilibrium solutions (25°C) in order of decreasing vapor pressure:

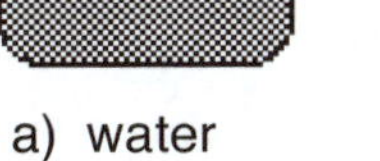

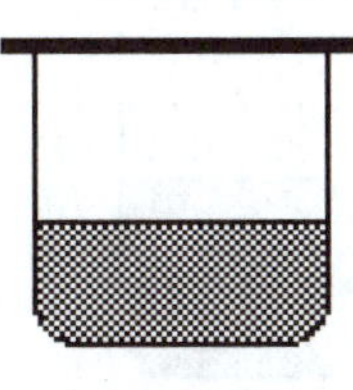
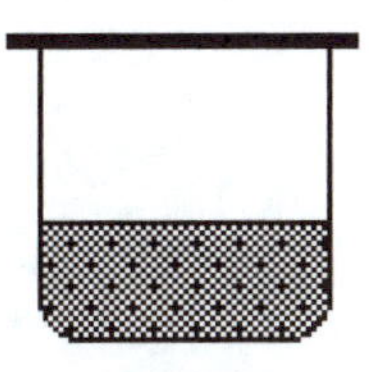

a) water b) water c) acetone d) sugar water

answer:

$c > a = b > d$

Acetone (c) is a volatile organic compound that has the highest vapor pressure. Water (a & b) is less volatile than acetone and has a lower vapor pressure. Water with a solute (sugar water, d) as you will see, has a lower vapor pressure than pure water.

Evaporation experiment

What would happen if you left a cup of water and a cup of water with dissolved sugar in a covered container for several days?

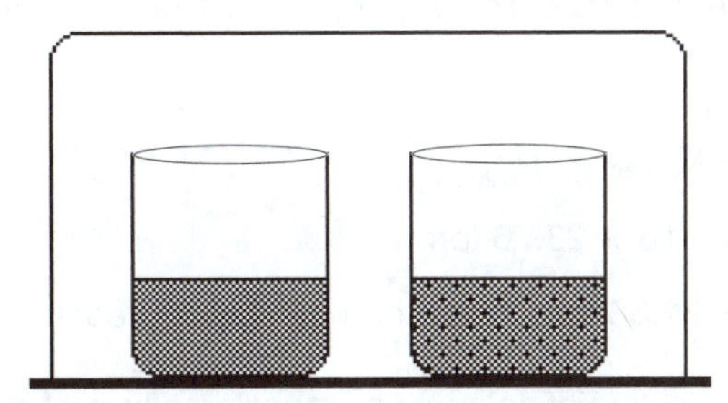

answer:

The vapor pressure of the pure water is higher than that of the sugar water. As the pure water vaporizes to attempt to reach equilibrium, the sugar solution absorbs the vapor to try to lower the vapor pressure to reach its equilibrium value. Water transfers from the pure water to the sugar water (via the vapor phase) until all the water transfers to the sugar solution, and the system reaches equilibrium:

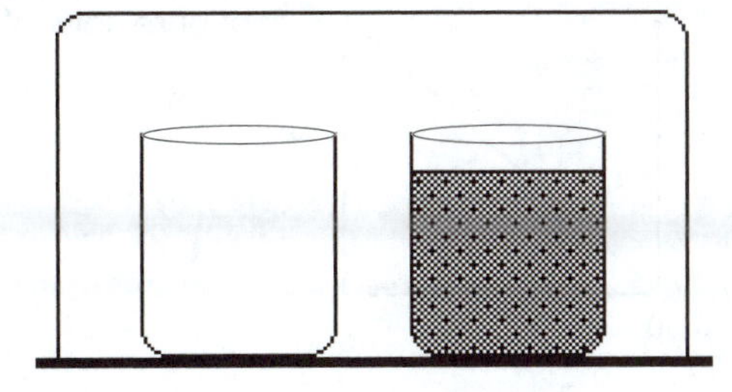

Picture a covered glass of water at equilibrium **(A).** Water molecules escape from the surface at the same rate water molecules in the vapor state return to the liquid. Now picture a covered glass of water with a dissolved solid (such as sugar) at equilibrium **(B).** Sugar is a nonvolatile solvent, and does not escape into the vapor state. Sugar molecules at or near the surface, block water molecules from escaping from the liquid, and so lower the vapor pressure of the solvent:

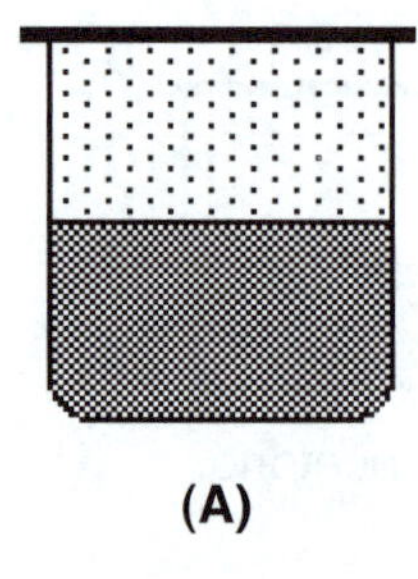

(A)

(B)

If half of the molecules are solute molecules, the vapor pressure is approximately half of that of the pure solvent. If one quarter of the molecules are solute molecules, they lower the pure solvent vapor pressure by approximately one fourth. **Raoult's law** expresses this relationship mathematically:

$$P_{\text{solution}} = X_{\text{solvent}} \times P^{\circ}_{\text{solvent}}$$

P_{solution} is the vapor pressure of the solution, X_{solvent} is the mole fraction of solvent, and $P^{\circ}_{\text{solvent}}$ is the vapor pressure of the pure solvent.

Vapor pressure of a solution

1) If one out of every ten molecules in a sugar water solution is a sugar (solute) molecule, what is the vapor pressure of the solution at 25°C? The vapor pressure of water at 25°C is 23.76 torr.

answer:

$$P_{\text{solution}} = X_{\text{solvent}} \times P^{\circ}_{\text{solvent}}$$

$$P_{\text{solution}} = 9/10 \times 23.76 \text{ torr}$$

$= 21.38$ torr, lower than the vapor pressure of pure water at 25°C

2) What does a plot of the mole fraction of solvent vs. vapor pressure of solution look like?

answer:

When mole fraction of the solvent = 0 (only solute present), there is no vapor pressure. When mole fraction of the solvent = 1 (only solvent present), the vapor pressure is that of the pure solvent. When there is half solute and half solvent, the vapor pressure is half that of the pure solvent:

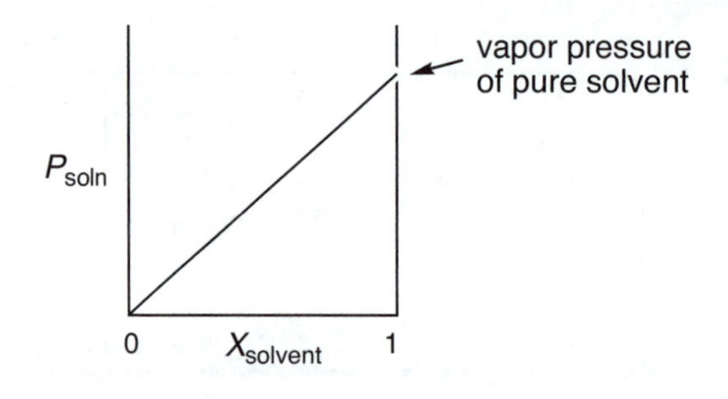

Raoult's law applies only to solutions that do not have strong interactions between the solvent and the solute ("ideal" solutions). We can imagine a case where a solute hydrogen bonds to the water molecules and prevents them from escaping as easily into the vapor state. In that case, the vapor pressure would be <u>lower</u> than Raoult's law predicts.

Now consider a mixture of two volatile components such as water and alcohol. Each liquid lowers the vapor pressure of the other liquid. The total vapor pressure is the sum of the two partial vapor pressures. In this case, Raoult's law becomes:

$$P_{\text{solution}} = P_{\text{water}} + P_{\text{alcohol}}$$

$$= (X_{\text{water}} \times P^{o}_{\text{water}}) + (X_{\text{alcohol}} \times P^{o}_{\text{alcohol}})$$

For an ideal solution of two volatile components, we can show the total vapor pressure as the sum of the partial pressures of each component:

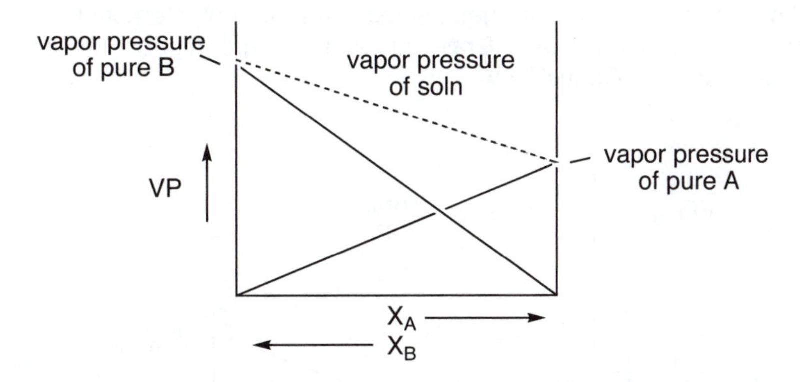

Vapor pressure lowering

The diagram above shows the vapor pressure of an "ideal" solution where there is no interaction between components. What would you predict a similar diagram of a water-ethanol solution would look like? What might the diagram of an ethanol/hexane solution look like? The structures of ethanol and hexane are shown below:

```
  H  H              H  H  H  H  H  H
  |  |              |  |  |  |  |  |
H-C--C-O-H        H-C--C--C--C--C--C-H
  |  |              |  |  |  |  |  |
  H  H              H  H  H  H  H  H
```

ethanol   hexane

<u>answer</u>:

Ethanol and water molecules form hydrogen bonds with their –OH groups. The attraction between these solvents lowers the vapor pressure of each component from its ideal value, and therefore lowers the vapor pressure of the solution from that predicted by Raoult's law. The graph on the following page contains dashed lines for the vapor pressures for ethanol, water, and the total vapor pressure <u>if there were no interaction between the solvents</u>. Draw curves on the graph below the dashed lines to represent the <u>lowered</u> vapor pressure that results from interactions between ethanol and water. The curves originate at the same points along the axes as the dashed lines since these points represent pure ethanol (left vertical axis) and pure water (right vertical axis):

(continued on next page)

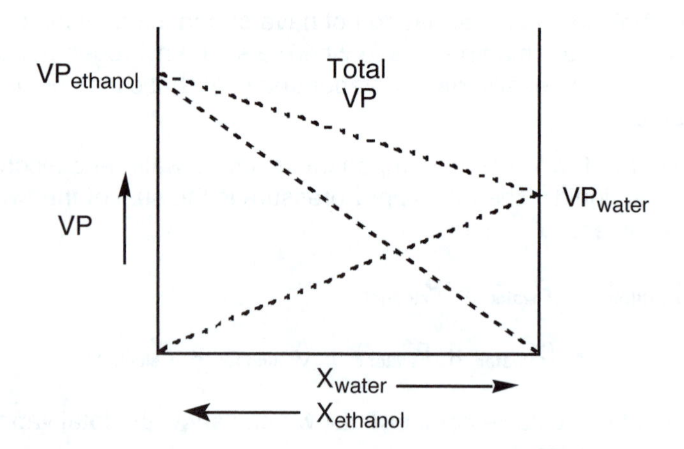

Polar ethanol molecules and nonpolar hexane molecules do not interact effectively. The solution shows a positive deviation from Raoult's law. Draw curves on the graph above the dashed lines to represent a positive deviation from Raoult's law:

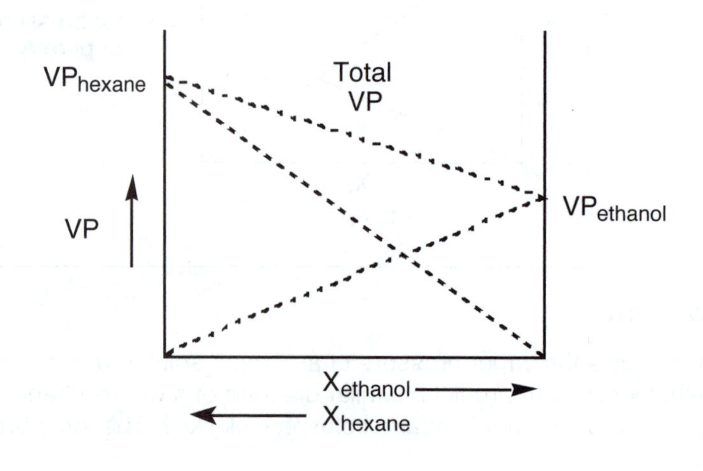

Vapor pressure of a benzene toluene mixture

What is the vapor pressure of a 50:50 mixture of benzene and toluene at 25°C?

$P^{o}_{benzene}$ = 95.1 torr, and $P^{o}_{toluene}$ = 28.4 torr

solution:

$$P_{solution} = (X_{benzene} \times P^{o}_{benzene}) + (X_{toluene} \times P^{o}_{toluene})$$

$$= (0.5 \times 95.1 \text{ torr}) + (0.5 \times 28.4 \text{ torr}) = 47.5 \text{ torr} + 14.2 \text{ torr}$$

$$= 61.7 \text{ torr}$$

Note: Toluene is a benzene ring with a CH_3 group attached to one carbon atom. The similarity of the molecules causes the solution to be very close to ideal. Notice that the mole fraction of benzene in the vapor is higher (47.5 torr vs. 14.2 torr) because it is the more volatile component.

Boiling Point Elevation and Freezing Point Depression

We add salt to water to increase the boiling point of the water. It takes more heat to raise the vapor pressure of a salt water solution to atmospheric pressure (its boiling point) than to raise the vapor pressure of the pure solvent to atmospheric pressure. The more solute there is in the solution, the higher the boiling point elevation:

$$\Delta T = K_b m_{solute}$$

ΔT is the difference between the boiling point of the solution and the pure solvent, K_b is the molal boiling-point elevation constant, and m_{solute} is the molality of the solute in the solution.

Boiling point elevation

What is the boiling point of an aqueous solution containing 10. g of glucose ($C_6H_{12}O_6$) in 100. mL of water? $K_b = 0.52°C$ kg/mol.

solution:

First we calculate the molality of the solution:

$$\text{Moles solute} = \frac{10.\ \text{g}}{180.\ \text{g/mol}} = 0.056\ \text{mol}$$

Mass of solvent = 0.100 kg (the density of water is 1 g/mL)

$$\text{Molality, } m = \frac{0.056\ \text{mol}}{0.100\ \text{kg}} = 0.56\ \text{mol/kg}$$

Substitute the molality into the equation for boiling point elevation and solve for ΔT:

$$\Delta T = K_b \times m_{solute}$$

$$\Delta T = 0.52°\text{C kg/mol} \times 0.56\ \text{mol/kg} = 0.29°\text{C}$$

$$\Delta T = T_{final} - T_{initial}$$

$$T_{final} = 0.29°\text{C} + 100.00°\text{C} = 100.29°\text{C}$$

We use salt on streets to lower the temperature at which water freezes. We can picture this phenomenon as follows:

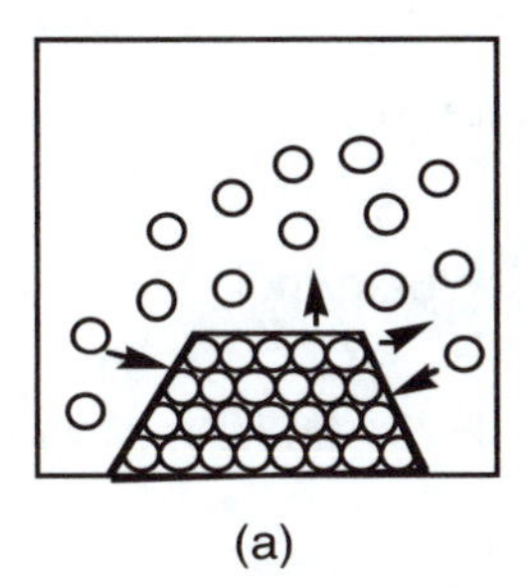

(a)

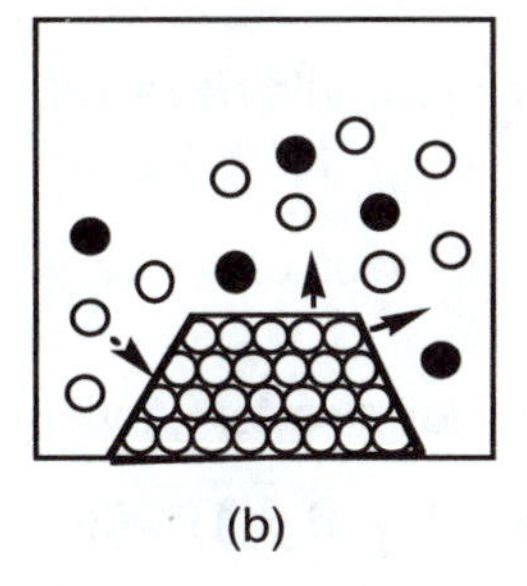

(b)

Diagram (a) shows ice in equilibrium with pure water (water molecules enter and leave the solid phase at the same rate). Diagram (b) shows ice in equilibrium with salt water. At equilibrium, the rate water molecules enter and leave the solid phase is equal. The solute in the liquid water slows the rate that water molecules enter the solid phase. It requires a lower temperature to slow the rate that water molecules leave the solid. (Note that ions from the salt do not "freeze" into the ice.)

The equation for freezing point depression is analogous to that for boiling point elevation:

$$\Delta T = K_f m_{solute}$$

Freezing point depression

Ethylene glycol ($C_2H_6O_2$) is a common antifreeze. Tetrahydrofuran (C_4H_8O) can also act as an antifreeze. How many grams of tetrahydrofuran would have to be added to water to produce the same freezing point lowering as one gram of ethylene glycol?

solution:

Freezing point lowering depends on the molality of the solution, so we need to determine how many grams of tetrahydrofuran contain the same number of moles as one gram of ethylene glycol:

$$\text{Moles of ethylene glycol} = \frac{1.0\text{ g}}{62\text{ g/mol}} = 0.016\text{ moles}$$

$$\text{Grams of tetrahydrofuran in 0.016 moles} = 0.016\text{ mol} \times \frac{72\text{ g}}{\text{mol}} = 1.2\text{ g}$$

You would need to add a little more tetrahydrofuran to obtain the same freezing point lowering as ethylene glycol. Tetrahydrofuran is more expensive than ethylene glycol, so ethylene glycol is economically the preferable choice.

Molar mass determination

7.8 kg of the antifreeze ethylene glycol is needed to lower the freezing point of 10.0 L of water to –10.0°F (–23.3°C). Calculate the molar mass of ethylene glycol. (K_f = 1.86°C kg/mol)

solution:

We know:

ΔT = freezing point lowering = 23.3°C

10.0 L water = 10.0 kg, if we assume the density of water = 1.00 g/mL

We can calculate the molality of the ethylene glycol solution:

$$m_{solute} = \frac{\Delta T}{K_f} = \frac{23.3^\circ\text{C}}{1.86^\circ\text{C kg/mol}} = 12.53\text{ mol/kg}$$

In 10.0 kg of solvent, we would have 125.3 moles solute, so:

$$\text{Molar mass of solute} = \frac{7800\text{ g}}{125.3\text{ mol}} = 62\text{ g/mol}$$

Check: The chemical formula for ethylene glycol is $C_2H_6O_2$, so its molar mass is:

$$(2 \times 12) + (6 \times 1) + (2 \times 16) = 62\text{ g/mol}$$

Note: Freezing point lowering is more commonly used to measure molar masses than boiling point elevation because the temperature change is generally larger (compare K_f and K_b values for water: 1.86 and 0.52, respectively). Organic solvents are generally preferred to water because their K_f values tend to be larger.

Osmotic Pressure

If we separate a pure solvent and a solution by a **semipermeable membrane** (a membrane which allows only solvent molecules through), solute particles block solvent molecules in the solution from passing through the membrane. The rate of solvent moving from the solution into the pure solvent is slower than in the reverse direction, so there is a net transfer of solvent into the solution:

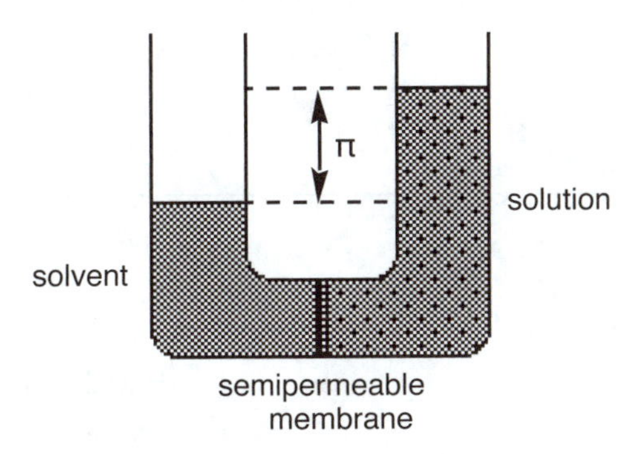

Osmotic pressure is the pressure that just stops osmosis. The dependence of the osmotic pressure on solution concentration is:

$$\pi = MRT$$

where π is the osmotic pressure (atm), M is the molarity of the solution, R is the gas law constant, and T is the temperature (Kelvin). Osmotic pressure is useful for determining solute molar masses because a small concentration of solute causes a relatively large osmotic pressure.

Determining molar mass from osmotic pressure

The osmotic pressure of a solution prepared by dissolving 10.0 g of insulin in 1000. g of water is 32.4 mm Hg at 25°C. What is the molar mass of insulin?

solution:

$$\pi = 32.4 \text{ mm Hg} \times \frac{1 \text{ atm}}{760 \text{ mm Hg}} = 0.0426 \text{ atm}$$

$$T = 25°\text{C} + 273 = 298 \text{ K}$$

$$M = \frac{\pi}{RT} = \frac{0.0426 \text{ atm}}{(0.0821 \text{ L atm/mol K}) \times (298 \text{ K})} = 0.00174 \text{ mol/L}$$

We can now determine the moles of insulin in 1.00 L of solution (we neglect the minor change in the volume of 1000. g of water when 10.0 g of insulin is added):

$$1.00 \text{ L soln} \times \frac{.00174 \text{ mol insulin}}{\text{L soln}} = 0.00174 \text{ mol insulin}$$

The molar mass of insulin is therefore:

$$\frac{10.0 \text{ g}}{0.00174 \text{ mol}} = 5750 \text{ g/mol}$$

Electrolyte Solutions

For electrolyte solutions, each ion counts as a particle. We incorporate the van't Hoff factor, *i*, into the equations for boiling point elevation, freezing point depression, and osmotic pressure:

$$\Delta T = iK_b m_{solute}$$

$$\Delta T = iK_f m_{solute}$$

$$\pi = iMRT$$

van't Hoff factor

What is the approximate value for "*i*" for the following compounds?

a) $Al(NO_3)_3$
b) $MgCl_2$
c) $Fe(NH_4)_2(SO_4)_2$
d) $C_{12}H_{22}O_{11}$

answers:

a-4, b-3, c-5, d-1(nonelectrolyte).

In the ideal case, "*i*" equals the number of ions per molecule. Most electrolyte solutions are not ideal, and "*i*" is typically somewhat lower than expected from the number of ions. Ions associate with one another and therefore do not behave as completely independent particles. Generally, the larger the charge on the ion, the greater the deviation from the ideal value.

Osmotic pressure of an electrolyte solution

1) What is the osmotic pressure for a 0.10 *M* solution of a nonelectrolyte at 25°C?

solution:

We can solve for the osmotic pressure, π, using the equation:

$$\pi = MRT$$

$$\pi = (0.1 \text{ moles/L}) \times (0.0821 \text{ L atm/mol K}) \times (298 \text{ K})$$

$$\pi = 2.45 \text{ atm.}$$

2) The osmotic pressure of a 0.10 *M* solution of an electrolyte is 10.8 atm. What is the value for *i*?

solution:

$$i = \frac{\text{measured value for electrolyte}}{\text{expected value for nonelectrolyte}}$$

$$= \frac{10.8 \text{ atm}}{2.45 \text{ atm}} = 4.41$$

The electrolyte in the above example is $Fe(NH_4)_2(SO_4)_2$, a compound which dissociates in water to produce 5 ions. The actual value of "*i*" is less than five presumably because of ion associations.

13.7 COLLOIDS

Colloids

Choose words from the list below to complete the following paragraphs:

(a) aggregate
(b) coagulate
(c) colloid
(d) scatter
(e) small
(f) repel

A suspension of tiny particles in some medium is a ____________(1). We can distinguish a colloid from a pure solution by shining light on it. The tiny particles (1 - 1000 nm) in a colloid ____________(2) the light (the **Tyndall effect)** and make it visible if you look through the solution from the side. Ions and molecules are too ____________(3) to scatter light, so the light going through a pure solution is invisible from the side.

Colloidal particles do not settle out because they attract from the medium ions of all like charge which cause the colloidal particles to ____________(4) one another. An electrolyte can neutralize the adsorbed ion layers and cause the colloid to destruct, or ____________(5). Heating a colloid can also destroy it by causing particles to ____________(6) and settle out.

answers: 1-c, 2-d, 3-e, 4-f, 5-b, 6-a.

CHAPTER CHECK

Make sure you can...

- List the major types of intermolecular forces in solution and their relative strengths.
- Predict if liquids would be miscible with each other based on intermolecular forces.
- Predict the relative solubilities of solids in liquids.
- Explain why most gases have a relatively low solubility in water.
- Explain the importance of hydrogen bonding in determining the structures of proteins, cell membranes, DNA, and cellulose.
- Explain why entropy always favors solution formation and why enthalpy may hinder solution formation.
- Predict the effect of temperature on the solubility of a substance.
- Use Henry's law to calculate the solubility of a gas in a liquid.
- Define molarity, molality, and mass, volume, and mole percent.
- Use Raoult's law to calculate the vapor pressure of mixtures.
- Calculate boiling point elevations and freezing point depressions due to solute particles.
- Determine molar mass from freezing point depression or osmotic pressure.
- Describe the dependence of osmotic pressure on solution concentration and temperature.
- Explain why we incorporate the van't Hoff factor, *i*, into equations for electrolyte solutions.
- Explain how the Tyndall effect distinguishes colloids from pure solutions.

Chapter 13 Exercises

13.1) What is the relationship between the solubility of a substance and the concentration of a saturated solution of the substance at the same temperature?

13.2) a. In any group of ions, the most strongly solvated ion is the ion with:

a. the greatest charge and the largest radius
b. the smallest charge and the largest radius
c. the greatest charge and the smallest radius
d. the smallest charge and the smallest radius

b. Which one of the following ions is most strongly solvated by water in solution?

Na^{+}	SO_4^{2-}	K^{+}	Mg^{2+}
Sr^{2+}	Cs^{+}	Ba^{2+}	NO_3^{-}

13.3) a. As crystals of potassium nitrate are dissolved in water, the resulting solution feels colder. How can the solubility of KNO_3 be increased?

b. When sodium hydroxide (NaOH) dissolves in water the overall process is strongly exothermic. However, when NaOH dissolves in a nearly saturated solution of NaOH, the process is endothermic. How can the solubility of NaOH be increased?

13.4) a. How does the solution process for a gas dissolving in water differ from the solution process for a salt dissolving in water?

b. How is it possible to increase the concentration of a gas in water?

13.5) An ideal solution of 3.00 moles of carbon disulfide in 10.0 moles of ethyl acetate has a vapor pressure of 170. torr at 28°C. If the vapor pressure of pure carbon disulfide at 28°C is 390. torr, what is the vapor pressure of pure ethyl acetate at the same temperature?

13.6) An ideal solution is made up of a 1:4 molar mixture of ethanol and methanol. At room temperature, the vapor pressure of ethanol is 44 torr and the vapor pressure of methanol is 88 torr. What is the partial pressure of methanol in the vapor above the solution? What is the mole fraction of ethanol in the vapor above the solution?

13.7) a. What is a colligative property?

b. Why does the freezing point of water decrease when salt is added to the water?

13.8) Two beakers under a bell jar, isolated from the surroundings, contain solutions of sugar in water. The first beaker initially contains 1.0 mol of sugar in 72 mL of water. The second beaker initially contains 0.50 mol of sugar in 72 mL of water. What is the mol fraction of sugar in the first beaker after the system has reached equilibrium?

13.9) a. How many grams of sodium hydroxide must be dissolved in 200. grams of water to make a 3.0 molal solution?

b. What is the molality of a solution composed of 6.4 grams of naphthalene ($C_{10}H_8$) dissolved in 250 grams of benzene (C_6H_6)?

13.10) How would you prepare the following aqueous solutions?

a. 120. mL of a 0.200 *M* $AgNO_3$ solution

b. 800. mL of a 0.010 *M* $CuSO_4$ solution from crystalline $CuSO_4 \cdot 5H_2O$

13.11) A solution is prepared by dissolving 50.0 g of moth balls, $C_6H_4Cl_2$, in 100. g of toluene, $C_6H_5CH_3$. What is the mole fraction of $C_6H_4Cl_2$ in the solution?

13.12) Jasmine perfume can be prepared by adding 5.00 g of methyl jasmonate, $C_{13}H_{18}O_3$, to 45.0 g of ethanol, C_2H_5OH. What is the molality of the resulting methyl jasmonate solution?

13.13) A dextrose intravenous solution is 5.50% glucose ($C_6H_{12}O_6$) by mass or 0.310 *M*. What is the density of this solution?

13.14) If a 3.0 *m* solution of naphthalene ($C_{10}H_8$) in cyclohexane boils at 89.4°C, what is the boiling point of pure cyclohexane? Assume that the vapor pressure of naphthalene is zero. K_b for cyclohexane = 2.80.

13.15) A by-product of the production of bromobenzene, C_6H_5Br, is biphenyl, $C_{12}H_{10}$. A quality control technician analyzed the product and found it to contain 32.0 mg $C_{12}H_{10}$/gram C_6H_5Br. What would be the boiling point of the mixture? Bromobenzene: b.p. = 155.8°C; K_b = 6.20°C/m.

13.16) 1.00 g of an unknown aromatic organic compound is dissolved in 10.0 g of biphenyl, $C_{12}H_{10}$. The freezing point of the resulting solution is 65.8°C. Elemental analysis of the unknown indicated the empirical formula to be C_5H_4. For biphenyl, the freezing point is 72.0°C and the freezing point constant, K_f, is 8.00°C/m .

a. What is the molar mass of the unknown compound?
b. What is the molecular formula?

13.17) Methanol, CH_3OH (b.p. 65°C) was used at one time as an antifreeze in radiators.

a. What is the freezing point of a 20% w/w methanol/water solution?
b. Explain why you would not want this solution in the radiator during the summer.

13.18) The osmotic pressure of blood is approximately 7.7 atm at 37°C (body temperature). What is the concentration of glucose ($C_6H_{12}O_6$) in an aqueous solution that is isotonic (same osmotic pressure) at this temperature?

Chapter 13 Answers

13.1) The solubility of a substance is the amount of solute that can be dissolved in a particular solvent. In other words, the solubility of a solute is equal to the concentration of a saturated solution of that solute at the same temperature.

If the concentration is less than the solubility at the same temperature, then the solution is unsaturated. If the concentration exceeds the solubility at the same temperature then the solution is supersaturated.

13.2) a. A major type of intermolecular force in solution is the attraction between an ion and the dipole of the solvent molecule. The strength of this ion-dipole attraction depends upon the charge density on the ion. The charge density of an ion is the ratio of the charge on the ion to the size of the ion. The larger the charge and the smaller the size, the greater is the attraction. The charge is usually more important than size. Hence, the correct answer is (c).

b. The doubly positive and doubly negative ions attract the solvent molecules more strongly than the corresponding singly charged ions. The smaller cations are more strongly solvated than the larger polyatomic anions. Of the cations, the smaller ions have a higher charge density, and are therefore more strongly solvated. Of the ions listed, Mg^{2+} will be the one most strongly solvated.

13.3) a. Endothermic processes are favored by an increase in temperature; addition of heat to the equilibrium system drives the process toward products, so an increase in temperature will increase the solubility of the potassium nitrate.

b. Even though sodium hydroxide dissolves in water with the liberation of energy initially, as the concentration of the solution increases, the heat released per mole of solute decreases. Eventually, the solution process becomes endothermic and an increase in temperature increases the solubility of sodium hydroxide.

13.4) a. Intermolecular attractions in the gaseous state are negligible, so energy is not required to separate gas molecules. Heat is liberated when a gas dissolves in water because hydration of the gas molecules releases more energy than is required to break apart attractions between solvent water molecules. The solution process for a gas is exothermic. Salts, on the other hand, often dissolve endothermically because considerable energy has to be supplied to break up the solid ionic lattice (the lattice energy) and insufficient energy is retrieved by solvation.

The other principal difference between the two cases is in the change in the disorder of the system (the entropy). For a gas dissolving in water, the entropy change is negative (unfavorable) because the gas is totally disordered and the solution is at least slightly ordered. For a salt dissolving in water, the entropy change is positive because the lattice structure of the solid salt is very ordered and the solution in comparison is disordered. A gas dissolving in water is an enthalpy-driven process. A salt dissolving in water is often an entropy-driven process.

b. Decrease the temperature, or increase the pressure of the gas above the solution (Henry's law).

13.5) Raoult's law relates the vapor pressure above a solution to the vapor pressures of the components in the solution and the composition of the solution. The total vapor pressure:

$$P = p_1 + p_2 = x_1 P_1^o + x_2 P_2^o$$

$$170. = \frac{3.00}{13.0} \times 390. + \frac{10.0}{13.0} \times P_2^o$$

$$P_2^o = 104 \text{ torr}$$

13.6) In a distillation process, the vapor above a solution always becomes richer in the more volatile component. In other words, the mole fraction of the more volatile component is greater in the vapor than it is in the liquid because that component vaporizes more easily than the other component.

Raoult's law: $p_1(\text{ethanol}) = x_1 P_1^o = (1/5) \times 44 = 8.8$ torr

Vapor pressure due to methanol above the solution = $(4/5) \times 88$ torr = 70.4 torr

Dalton's law: mole fraction (ethanol) = $p_1/P = 8.8/79.2 = 0.11$

13.7) a. A colligative property is a property that depends upon the concentration of solute particles and not upon the nature of the solute particles. The particles may be ions, they may be molecules, they may be large, or they may be small. A salt that dissociates into two ions per formula unit has approximately twice the effect as an equimolar concentration of a molecule that does not ionize because the concentration of solute particles is effectively twice as great.

Examples of colligative properties are the depression of the freezing point, the elevation of the boiling point, the lowering of vapor pressure, and osmotic pressure. In each of these examples there is an equilibrium across a phase boundary. The intrinsic cause for the colligative property is an increase or decrease in the disorder (entropy) of one of the two phases.

b. The freezing point of water decreases when salt is added because the addition of the salt increases the entropy (disorder) of the liquid phase. The change in the entropy of fusion (melting) is therefore greater and, in order to compensate, the melting point T_f must be lower since $T\Delta S$ = a constant for melting.

13.8) The system inside the bell jar will eventually reach equilibrium. Water will leave and re-enter the solutions in the beakers until both solutions have equal concentrations and the vapor pressures above the solutions therefore become equal.

Initial vapor pressure above the first beaker = $x_1 P_1^o = (4/5)P_1^o$

Initial vapor pressure above the second beaker = $x_1 P_1^o = (8/9)P_1^o$

Water will leave the second beaker at a higher rate than it will leave the first beaker. The solution in the first beaker will get more dilute and the solution in the second beaker will get more concentrated. The amount of sugar in each beaker remains the same—only water is transferred through the vapor phase.

Total amount of water = 144 mL = 8 mol

At equilibrium, 2/3 (16/3 mol) will be in beaker 1 and 1/3 (8/3 mol) will be in beaker 2.

Mole fraction of sugar in first beaker at equilibrium = 1/(19/3) = 0.16

The mole fraction of sugar in the second beaker will be the same.

13.9) a. A 3.0 molal solution of sodium hydroxide contains 3.0 moles NaOH in 1 kg of water. This is the same as 3/5 mol in 200 grams of water.

3/5 mol = 3/5 mol × 40. g/mol = 24 grams

b. 0.20

13.10) a. Dissolve 4.08 g $AgNO_3$ in about 60 mL of distilled water, dilute to 120. mL and mix.

b. Dissolve 2.0 g of copper(II) sulfate pentahydrate to about 400 mL of distilled water, dilute to 800. mL, and mix.

13.11) $X = 0.238$ C_6H_4Cl

13.12) 0.500 m $C_{13}H_{18}O_3$

13.13) 1.02 g/mL

13.14) Naphthalene is a molecular solute; the molecule does not ionize or break up. The factor i = 1.

$\Delta T_b = 2.80 \times 3.0 \times 1 = 8.4°C$

The boiling point of pure cyclohexane = 89.4 – 8.4 = 81.0°C.

13.15) new b.p. = 157.3°C

13.16) a. 129 g/mole

b. $C_{10}H_8$

13.17) a. –14.5°C

b. The boiling point of methanol is only 65°C, so the alcohol would boil off during the warm months.

13.18) molarity = 0.30 mol/L

14

Periodic Patterns in the Main-Group Elements

14.1 TRENDS ACROSS THE PERIODIC TABLE

Review of periodic trends

Label the arrows on the periodic table diagram to indicate which of the following trends increase down, and right to left across the table (arrows "A"), and which trends increase up, and left to right across the table (arrows "B"):

a) atomic radius
b) electronegativity
c) ionization energy
d) metallic character

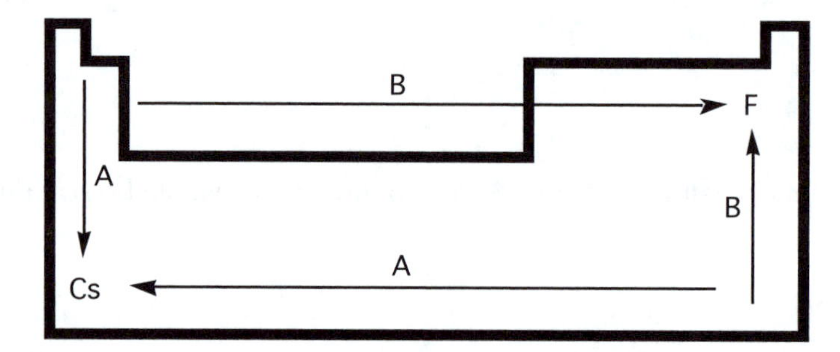

(continued on next page)

answers:

A: (a), (d). Cesium is the largest and most metallic element (excluding rare and radioactive francium).

B: (b), (c). Fluorine is the smallest (except for helium and hydrogen), and most electronegative element with a high ionization energy. Small atoms hold their electrons more strongly than large atoms.

As we go across the periodic table, bonding character in the elements changes from:

metallic →	**network covalent** →	**individual molecules** →	**separate atoms**
(metals)	(boron and carbon)	(N, O, F)	(noble gases)

The acid-base behavior of the oxides changes from basic to amphoteric (acts as an acid and a base) to acidic as the electronegativity of the atom becomes closer to that of oxygen, and the bond between the element and oxygen becomes more covalent.

Consider the chemical and physical properties of the period 2 elements, lithium through neon. The solids (lithium, beryllium, boron, and carbon) contain metallic or network covalent bonding. The gases (nitrogen, oxygen, fluorine, and neon) exist as molecules or individual atoms.

14.2 HYDROGEN

What can we learn about elemental hydrogen with a tank of hydrogen gas, a balloon, and a match?

- We can spray hydrogen into the air and observe that it is an odorless and colorless gas which is unreactive in air and in the body (i.e., nontoxic). In fact, the H–H bond energy is larger than almost any other single bond, and so is difficult to break.
- We can fill a balloon with hydrogen gas, let go of the balloon, and see as the balloon rises, that hydrogen is less dense than air.
- If we ignite the balloon, we cause a spectacular fire ball and see that hydrogen is a flammable gas. It reacts with oxygen to produce water as its only combustion product:

$$H_2(g) + 1/2O_2(g) \rightarrow H_2O(g)$$

  Hydrogen is a potentially useful, clean burning fuel.

Atomic hydrogen has a ns^1 configuration and, like the alkali metals, it can obtain an empty shell by losing its electron. Because hydrogen is so small, its nucleus holds the electron more tightly than the ns^1 electron of the alkali metals, and it therefore has a higher ionization energy. Like the halogens, hydrogen can gain an electron to obtain a filled shell. However, hydrogen has a lower electronegativity than the halogens; H^- is reactive. The three types of hydrides that hydrogen forms are:

- **Ionic hydrides**—salt-like hydrides (white, crystalline solids) that hydrogen forms with reactive metals, such as those in group 1A and the larger members of group 2A.
- **Covalent hydrides**—molecular hydrides that hydrogen forms with nonmetals. Most of the small molecule covalent hydrides are gases (such as methane and ammonia), but many of the hydrides of boron and carbon are larger molecules that are liquids or solids.

- **Interstitial hydrides**—metallic hydrides that hydrogen forms with many transition elements. H_2 molecules and H atoms occupy the holes in the metal's crystal structure and since the metal can incorporate variable amounts of hydrogen depending on the temperature and pressure of the gas, the interstitial hydrides typically do not have a single formula.

Hydrogen chemistry

1) Classify each compound as 1) an ionic hydride, 2) a covalent hydride, or 3) an interstitial hydride:

a) CaH_2	c) HF	e) LiH	g) NH_3
b) CH_4	d) H_2O	f) MgH_2	h) $TiH_{1.7}$

answers: a-1, b-2, c-2, d-2, e-1, f-1, g-2, 3-h.

2) Fill in the blanks:

Ionic hydrides form between hydrogen and _____________(1). Covalent hydrides form between hydrogen and _____________(2). Interstitial hydrides form between hydrogen and _____________(3). In interstitial hydrides, hydrogen atoms or molecules occupy _____________(4) in the metal's crystal structure.

answers: 1-metals, 2-nonmetals, 3-transition metals, 4-holes.

3) Fill in the blanks:

a) The reaction of lithium hydride with water:

$$LiH(s) + H_2O(l) \rightarrow Li^+(aq) + OH^-(aq) + H_2(g)$$

demonstrates that ionic hydrides are strong _________________.

b) The reaction of LiH with $TiCl_4$:

$$TiCl_4(l) + 4LiH(s) \rightarrow Ti(s) + 4LiCl(s) + 2H_2(g)$$

demonstrates that ionic hydrides are strong _________________.

answers: a-bases and reducing agents, b-reducing agents.

14.3 THE ALKALI METALS (GROUP 1A)

What can we learn about elemental sodium from a jar of sodium from the stockroom, some tweezers, and a beaker of water?

- We observe that sodium is a silvery-white metal that is likely reactive with oxygen since it is stored beneath a liquid. (Sodium is lustrous when freshly cut, but tarnishes on exposure to air becoming dull and gray.)

- If we swirl the jar, we can see that it is likely reactive with water since it is stored beneath unreactive mineral oil.

- When we poke it with tweezers, we see that sodium is soft. Because of its softness, we might guess that it would have a low melting point (sodium melts below the boiling point of water, at 97.8°C).

- If we put sodium into water, we see that it has a low density (it floats on the water) and it violently decomposes water. Sodium reacts with water to form a basic solution:

  $$2Na(s) + 2H_2O(l) \rightarrow 2NaOH(aq) + H_2(g)$$

  It reduces the H in H_2O from the +1 to the zero oxidation state.

The characteristics of alkali metals can be explained by their ns^1 electron configuration. Since there is only one valence electron, the attraction between the delocalized electrons and the cationic cores is weak and the crystals deform easily (are "soft"). The low ionization energy of the outer *s* electron means that alkali metals are powerful reducing agents that are always found in nature as 1+ cations. They reduce halogens to form ionic solids in exothermic reactions:

$$2Na(s) + Cl_2(g) \rightarrow 2NaCl(s) + \text{heat}$$

The ionic salts (NaCl in this case) are water soluble, as the high charge density cation attracts water molecules strongly to create a highly exothermic heat of hydration.

Reducing power of the alkali halides

For each reaction below involving an alkali metal, determine the element that is reduced:

1) $2Na(s) + 2H_2O(l) \rightarrow 2NaOH(aq) + H_2(g)$
2) $2Na(s) + Cl_2(g) \rightarrow 2NaCl(s)$
3) $4Li(s) + O_2(g) \rightarrow 2Li_2O(s)$
4) $2Li(s) + H_2(g) \rightarrow 2LiH(s)$

answers:

1) H: from +1 in H_2O to 0 oxidation state in H_2
2) Cl: from 0 in Cl_2 to –1 oxidation state in NaCl
3) O: from 0 in O_2 to –2 oxidation state in Li_2O
4) H: from 0 in H_2 to –1 oxidation state in LiH

Alkali halides can reduce hydrogen in water, elemental hydrogen, oxygen, and halides.

The Special Case of Lithium

Lithium is the only alkali metal that forms salts that exhibit covalent bond character. LiCl and LiBr are more soluble in the polar solvent ethanol than the other halide salts because the high charge density of the small lithium cation (Li^+), deforms the polarizable electron cloud of the anion.

14.4 THE ALKALINE EARTH METALS (GROUP 2A)

Magnesium is one of the most common elements in the earth's crust. What can we learn about elemental magnesium from a ribbon of magnesium metal, a flame, a beaker of water, and some solid CO_2 (dry ice)?

- We see that magnesium is a silvery-white metal that is apparently not reactive in air (it only slowly oxidizes in moist air). It is in the form of a ribbon that is bendable (ductile).

- If we put it in water, we see no apparent reaction (magnesium reacts slowly with water at ordinary temperatures).

- If we hold magnesium in the flame, it burns with an intensely bright light in air and leaves a fine white odorless powder which is practically insoluble in water. (Magnesium oxide, the combustion product, has a melting point of 2800°C!) If we put magnesium oxide powder in water, it makes the solution slightly alkaline (it forms magnesium hydroxide).

- If we ignite magnesium between two blocks of dry ice, we see that magnesium burns in carbon dioxide! Its affinity for oxygen is so great that it pulls the oxygen away from carbon to form MgO and carbon (soot):

$$2Mg(s) + CO_2(g, \text{from sublimation of dry ice}) \rightarrow 2MgO(s) + C(s)$$

The alkaline earth metals are similar in behavior to the alkali metals. They are smaller, and therefore have higher ionization energies than the alkali metals. The attraction between the delocalized ns^2 electrons and the cationic cores is stronger than it is in the alkali metals, and the crystals are therefore harder with higher melting and boiling points. Compared with the transition metals, however, alkaline earth metals are still soft and lightweight.

The alkaline earth metals are strong reducing agents:

- They reduce oxygen in air to form strongly basic oxides (see discussion of magnesium oxide above).
- The larger metals reduce water at room temperature to form H_2.
- They reduce the halogens to form ionic compounds.
- Most reduce hydrogen to form ionic hydrides, and most reduce nitrogen to form ionic nitrides.

The Special Case of Beryllium

All beryllium compounds exhibit a degree of covalent bonding. The small size and high charge density of the Be^{2+} ion polarizes nearby electron clouds. The most ionic beryllium compound, BeF_2, has a low electrical conductivity when melted.

Recall from Chapter 10 that beryllium is often electron deficient in its compounds. Gaseous BeF_2, is a linear molecule with 4 valence electrons around beryllium:

F–Be–F

When the gas condenses, the molecules bond together into chains and each beryllium atom obtains an octet:

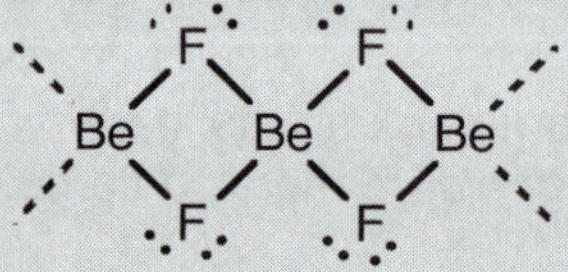

F F
Be Be Be
F F

(continued on next page)

When beryllium bonds to hydrogen, hydrogen does not have enough electrons to donate two to each beryllium atom. Solid BeH_2 consists of three atom groups (Be-H-Be) bonded by only two electrons in a three center, two electron bond:

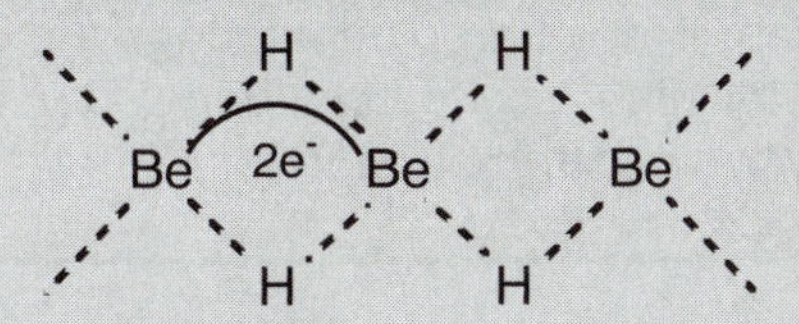

14.5 THE BORON FAMILY (GROUP 3A)

What can we infer about the element aluminum from our aluminum cookware?

- Aluminum is a shiny metal that is machinable, and relatively soft.
- It is fairly inexpensive, and so must be readily available. (Aluminum is one of the most abundant metals in the earth's crust.)
- It is not toxic to the body (studies that link aluminum to Alzheimer's disease are not definitive).
- It is a good conductor of heat (we assume, since it is used for cookware).
- It is lightweight, and unreactive with air and water at ordinary temperatures. (In moist air, an oxide film forms which protects the metal from corrosion.)

Compare aluminum to gallium. Aluminum, with an atomic number of 13, contains 13 protons in its nucleus and has an electron configuration of [Ne]$3s^23p$. Gallium, with an atomic number of 31, contains 31 protons in its nucleus and has an electron configuration of [Ar]$3d^{10}4s^24p$. The 10 $3d$ electrons do not effectively shield gallium's three valence electrons from the nuclear charge of +31, and gallium is smaller than aluminum. Many of the trends in group 3A reflect the existence of the *d* and *f* inner shell electrons of gallium, indium, and thallium.

Boron, like other first elements in a group, is unique. It is a network covalent solid: hard (its crystals are almost as hard as diamond), black, and high melting (≈2200°C). In contrast, aluminum melts at 660°, and gallium melts at 30°C. (Note: Gallium boils at 2400°C and has the largest liquid range of any element.)

Reactivity of group 3A elements:

- React sluggishly, if at all, with water
- Form oxides when heated in pure O_2. B_2O_3 is acidic; Al_2O_3 is amphoteric; Tl_2O is strongly basic
- Reduce halogens

Oxides of group 3A

Match the oxide with its acid/base behavior:

1) B_2O_3	a) acidic
2) Al_2O_3	b) amphoteric
3) Tl_2O	c) basic

answers: 1-a, 2-b, 3-c. Basicity of the oxides increases down the group.

The Special Case of Boron

All boron compounds are covalent, often in the form of network covalent compounds or large molecules. Many boron compounds are electron deficient. Boron obtains an octet of electrons by either accepting a bonding pair from electron-rich atoms, or by forming a three center, two electron bond.

Boron compounds

1) What is the Lewis structure for gaseous boron trifluoride (BF_3)?

answer:

The number of valence electrons = 3 + (3 × 7) = 24
Each fluorine atom is bonded to boron and has 3 lone pairs. Boron has 6 electrons around it in one of its resonance structures and so is electron deficient:

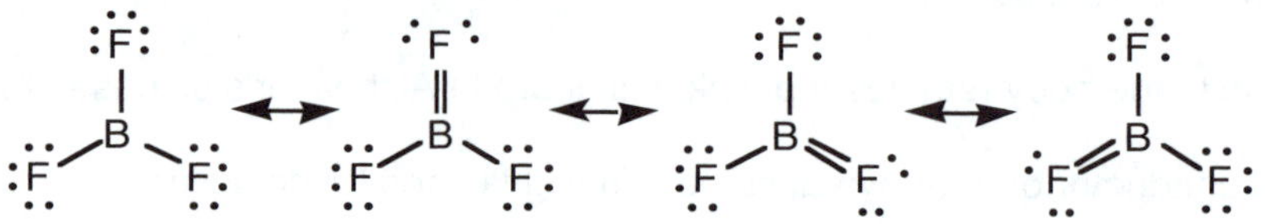

Although the last three resonance structures contain formal charges (on the boron and the double bonded fluorine atom), the B–F bonds in BF_3 have some double bond character.
Recall from Chapter 11 (page 216 in the study guide) that boron is sp^2 hybridized, and the interaction between the third *p* orbital and a lone pair of electrons on fluorine causes the B–F bond distance to be shorter than expected for a purely single B–F bond.

2) Draw the Lewis structure for ammonia, and predict how it would react with BF_3 to give boron an octet of electrons.

answer:

Lewis structure of ammonia:

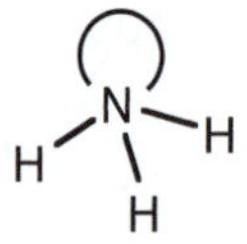

Ammonia has a lone pair that it can "donate" to boron to form the compound $F_3B–NH_3$. Sometimes you may see this type of bond, where both electrons come from the same atom, drawn with an arrow from nitrogen to boron:

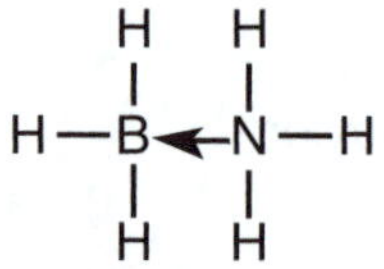

3) Compare the structure of ethane (C_2H_6) with amine-borane (BNH_6).

(continued on next page)

answer:

Boron has one less valence electron than carbon, and nitrogen has one more valence electron than carbon, so C–C and B–N bonds have the same number of valence electrons. Ethane and amine-borane have the same Lewis structures:

```
    H   H                H   H
    |   |                |   |
H — C — C — H        H — B ◄— N — H
    |   |                |   |
    H   H                H   H

   ethane             amine-borane
```

Each carbon atom in ethane donates one valence electron to the C–C bond. Nitrogen donates two valence electrons to form the B–N bond.

14.6 THE CARBON FAMILY (GROUP 4A)

The five members of the carbon family are carbon, silicon, germanium, tin, and lead. All of these elements have four electrons in their outermost energy level and form four bonds. Carbon is a nonmetal and, as discussed in Chapter 12 (Section 12.6), exists in three forms: as a network covalent solid (as either diamond or graphite), and as a molecular solid in ball or tube-shapes. Silicon and germanium have some properties of both metals and nonmetals, and are metalloids; tin and lead are metals.

Melting points of group 4A elements

Match the melting points (°C) with its element:

1) carbon (diamond)	a) 232
2) silicon	b) 327
3) germanium	c) 945
4) tin	d) 1410
5) lead	e) 4100

answers: 1-e, 2-d, 3-c, 4-a, 5-b.

Carbon and silicon are network covalent solids and have high melting points. (Silicon has longer, weaker bonds than carbon which is reflected by silicon's lower melting point.) The large decrease in the melting point of tin reflects the change to metallic bonding. Notice that the melting points of tin and lead are similar, and that tin has the lowest melting point in the group.

An unknown

Name the element that is described by each list below:

A	B
hardest substance known	soft and "greasy"
colorless	black
poor conductor	conducts electricity
network covalent solid	sheets of extended 6-membered rings

(continued on next page)

answer:

List "A"-carbon; List "B"-carbon

Carbon can exist in different molecular forms **(allotropes).** Its most stable form at ordinary temperature and pressure is graphite (list B). At high pressure, diamond is the most stable form of carbon (list A).

The Special Case of Carbon

Carbon has an intermediate electronegativity of 2.5, and it forms covalent bonds with elements from all parts of the periodic table. It displays oxidation states from +4 (in CO_2) to –4 (in CH_4).

Carbon's small size allows carbon atoms to approach each other closely enough for 2*p* orbitals to overlap side-to-side and form π bonds. Much of the chemistry of carbon centers on its ability to form double and triple bonds. The chemistry of carbon will be discussed in more detail in the next chapter on organic chemistry.

Silicon is larger than carbon, and Si–Si bonds are weaker than C–C bonds. Silicon's chemistry centers around the silicon-oxygen bond, an unusually strong bond (368 kJ/mol). Silicon is the second most abundant element on earth (behind oxygen), and much of the silicon is found in silicates, minerals containing –Si–O– repeating units.

14.7 THE NITROGEN FAMILY (GROUP 5A)

Group 5A contains the nonmetals nitrogen and phosphorus, the metalloids, arsenic and antimony, and the metal, bismuth. Nitrogen makes up more than 70% of the air we breathe. Liquid nitrogen can exist at –196°C at ordinary pressures, and is often used in laboratories to condense gases. Its triple bond is so strong that all of its six oxides have a positive heat of formation (see Table 14.3 on page 592 in your textbook). Elemental phosphorus exists as P_4 molecules and exists in several allotropes.

Reactivity of group 5A members

Consider the following reactions involving group 5A members:

a) $P_4(s) + 6Cl_2(g) \rightarrow 4PCl_3(l)$

b) $P_4(s) + 3O_2(g) \rightarrow P_4O_6(s)$ (limited oxygen)

c) $P_4(s) + 5O_2(g) \rightarrow P_4O_{10}(s)$ (excess oxygen)

d) $P_4O_{10}(s) + 6H_2O(l) \rightarrow 4H_3PO_4(l)$

e) $PCl_3(l) + Cl_2(g) \rightarrow PCl_5(s)$

f) $PCl_5(s) + 4H_2O(l) \rightarrow H_3PO_4(l) + 5HCl(g)$

g) $N_2(g) + 3H_2(g) \rightarrow 2NH_3(g)$ (high *P* and *T*)

h) $3NO(g)$ + heat $\rightarrow N_2O(g) + NO_2(g)$

i) $3NO_2(g) + H_2O(l) \rightarrow 2HNO_3(l) + NO(g)$

(continued on next page)

1) Which equation shows the **disproportionation** of an oxide to form two new oxides?
2) Which equation shows how to produce ammonia from the elements?
3) Which equation shows that the group 5A halides form by direct combination of the elements?
4) Which equations require phosphorus to expand its octet to form the product?
5) Which equations show methods for producing a weak acid used in fertilizer production, as a polishing agent for aluminum, and as an additive to soft drinks?
6) Which equations show the formation of the two important oxides of phosphorus?
7) Which equation shows the formation of an acid that is a strong oxidizing acid?

answers: 1-h; 2-g; 3-a; 4-c,d,e; 5-d,f; 6-b,c; 7-i.

Note: NO disproportionates to form NO_2 and N_2O, compounds with oxidation states higher and lower than in the original compound. The oxidation state of nitrogen in NO is +2. In NO_2 it is +4, and in N_2O, it is +1 (on the average).

14.8 THE OXYGEN FAMILY (GROUP 6A)

6A elements

Identify the element(s) from group 6A that each statement below describes:

1) The most abundant element on the earth's surface
2) Supports combustion
3) Is formed by plants during photosynthesis
4) Exists as a low boiling diatomic gas, or as a pungent gas that decomposes in heat
5) Exists in more than 10 forms; at room temperature it is an 8 membered crown-shaped ring
6) Nonmetals, which bond covalently to almost every other nonmetal
7) Behaves like a metal in its salt-like compounds
8) A strong oxidizer, second only in strength to fluorine
9) Forms a poisonous gaseous hydride (H_2E) that gives coal pits and swamps their "rotten egg" odor
10) Its hydride is the most abundant compound on earth
11) Its oxide in water produces sulfuric acid, a major contributor to acid rain

answers:

1-oxygen; 2-oxygen; 3-oxygen; 4-oxygen (O_2 and O_3); 5-sulfur; 6-sulfur and oxygen; 7-polonium; 8-oxygen; 9-sulfur (H_2S); 10-oxygen (H_2O); 11-sulfur.

Sulfur fluorides

1) Draw Lewis structures for sulfur tetrafluoride (SF_4) and sulfur hexafluoride (SF_6):

(continued on next page)

2) Which of the compounds on the previous page would you predict would be more reactive?

answer:

The lone pair on SF_4 makes it more reactive than SF_6. Sulfur hexafluoride is almost as inert as a noble gas. It has no odor or taste, is nontoxic, insoluble, and nonflammable.

Hydrides of oxygen and sulfur

Consider the Lewis structures for the hydrides of sulfur (H_2S) and oxygen (H_2O):

Do these two molecules have similar properties?

answer:

Hardly!

H_2O is a high boiling liquid essential for life. H_2S is a vile-smelling, toxic gas formed during anaerobic decomposition of plant and animal matter. The bond angles in water (104.5°) reflect sp^3 hybridization. Bond angles in hydrogen disulfide (≈90°) imply sulfur's use of unhybridized *p* orbitals.

The first element in a group always has different behavior than the rest of the group. Water, due to its relatively small and electronegative oxygen atom, has the ability to hydrogen bond and is therefore a unique substance.

14.9 THE HALOGENS (GROUP 7A)

The halogens are reactive because they require only one electron to obtain a noble gas configuration. They can either gain an electron from a metal, or share an electron pair with a nonmetal. Since halogens are readily reduced, they act as oxidizing agents in most of their reactions. Fluorine is such a powerful oxidizing agent that it oxidizes oxygen in water to O_2 and some O_3. The other halogens disproportionate in water:

$$Br_2(l) + H_2O(l) \rightleftharpoons HBr(aq) + HBrO(aq)$$

The oxidation state of Br in HBr is –1; that of Br in HBrO is +1.

The halogens

Match each halogen on the left with the properties that describe it:

halogen	description	b.p. (°C)	reactivity
1) F_2	a) faint yellow gas	e) –188.1	i) forms a weak acid with hydrogen

(continued on next page)

2) Cl_2	b) yellow-green gas	f) –34.6	j) the most reactive element
3) Br_2	c) dark red liquid	g) 58.8	k) the most electronegative element
4) I_2	d) purple-black solid	h) 184.3	l) the smallest halogen

answers:

1-a,e,i,j,k,l; 2-b,f; 3-c,g; 4-d,h.

The halogens are all diatomic molecules whose intermolecular dispersion forces increase in strength as the molecules become larger. As we move down the group, we change from gases (F_2, Cl_2), to a liquid (Br_2), to a solid (I_2). Fluorine is the most electronegative, and the most reactive element. Fluorine, with its short strong H–F bond, is the only compound to form a weak acid with hydrogen. HCl, HBr, and HI are all strong acids that dissociate completely in water.

Halogen oxoacids

Reaction of halogens and their oxides with water produce halogen oxoacids. Arrange each group of oxoacids in order of decreasing acid strength:

a) HOCl	a) $HOClO_2$
b) HOClO	b) $HOBrO_2$
c) $HOClO_2$	c) $HOIO_2$
d) $HOClO_3$	

answer: 1st column: d > c > b > a; 2nd column: a > b > c.

The strength of an acid depends on how easily the proton dissociates. The less electron density associated with the O–H bond, the easier it is to remove the proton. Another way to look at this is to consider the stability of the resulting anion once the proton is removed. For oxoacids in the first column, $HOClO_3$ contains three oxygen atoms, which stabilize the anion by giving it three electronegative sites to absorb the negative charge. $HOClO_3$ loses its proton most readily (is the strongest acid) because its anion is the most stable.

$HOClO_2$ is the strongest acid in the second column because the electronegative chlorine atom removes electron density from the O–H bond to a greater degree than the less electronegative bromine and iodine atoms, and so the proton more easily dissociates from $HOClO_2$.

14.10 THE NOBLE GASES (GROUP 8A)

The noble gases are individual atoms with extremely weak intermolecular forces. The filled valence shells in noble gases (ns^2np^6) explain their unreactivity. Helium is notably different from the other noble gases in that it contains no "*p*" electrons, and in fact, its electron configuration with a filled "*s*" subshell, would logically place it in the beryllium group.

The noble gases

Match each noble gas on the left (following page) with the properties that describe it:

(continued on next page)

noble gas		b.p. (°C)		reactivity	
1)	He	a)	–268.9	g)	smallest atom
2)	Ne	b)	–246.1	h)	cannot be frozen at ordinary pressure
3)	Ar	c)	–185.9	i)	used in balloons to make them lighter than air
4)	Kr	d)	–153.4	j)	used in light tubes to produce colored light
5)	Xe	e)	–108.1	k)	makes up ≈1% of the atmosphere
6)	Rn	f)	–62	l)	colorless, odorless, inert gas
				m)	radioactive element enters houses from rock

answers: 1-a,g,h,i,l; 2-b,j,l; 3-c,k,l; 4-d,l; 5-e,l; 6-f,l,m.

The Chemistry of Xenon

Xenon difluoride, XeF_2

prepared from the elements; colorless crystals; solubility in water at 0°: 25 g/L

Xenon tetrafluoride, XeF_4

prepared from the elements at 6 atm and 400°; colorless crystals

Xenon hexafluoride, XeF_6

colorless solid with greenish-yellow vapor;
more powerful oxidizing and fluorinating agent than XeF_2 and XeF_4

Xenon trioxide, XeO_3

colorless, hygroscopic solid; powerful explosive

Diagonal Relationships

We have seen in this chapter how the same valence electron configuration of atoms in a group imparts similar valency to the group members, and how trends based on atomic size and charge exist within a group. We have also seen unique behavior exhibited by some elements, particularly those in the 2nd period: Li, Be, B, C, N, O, F, and He. Lithium, beryllium, and boron, the first three members of the 2nd period, exhibit behavior very similar to elements in the 3rd period immediately below and to the right. These **diagonal relationships** are due to their similar atomic and ionic sizes.

CHAPTER CHECK

Make sure you can...

- Assign periodic trends for atomic radius, ionization energy, electronegativity, and metallicity.
- Name and describe the three types of hydrides hydrogen forms.
- Explain why alkali halides are strong reducing agents.
- Give the products formed when alkaline earth metals reduce oxygen, water, halogens, hydrogen, and nitrogen.
- Describe the bonding in BeH_2.
- Explain why Ga (below aluminum in the periodic table) is smaller than Al.

- Describe the reactivity of group 3A elements with water, O_2, and halogens.
- Explain why boron has a melting point of ≈2200°C and is almost as hard as diamond.
- Explain why the chemistry of carbon is so prevalent.
- Compare physical and chemical properties of H_2S and H_2O.
- Compare the physical properties of the halides, F_2, Cl_2, Br_2, and I_2.
- Name the most electronegative and most reactive element.
- Predict relative acid strengths of halogen oxoacids.
- Explain, with examples, the concept of diagonal relationships in the periodic table.

Chapter 14 Exercises

14.1) a. According to a dictionary, the word "halogen" means what?

b. Sodium is a soft, shiny, low-melting m _ _ _ _.

c. Chlorine is a toxic, pale green g _ _.

d. But NaCl is a white, high-melting s _ _ _.

14.2) a. What does the word "noble" mean in chemistry?

b. Name several metals that are sometimes found "free," uncombined in nature.

c. Cite several metals that are used in making jewelry and coins.

d. In ancient times, what were known as the metals of kings and queens—and nobles?

14.3) a. From atoms of what element is it most difficult to remove electrons (i.e., what element has the highest "ionization energy")?

b. What is the most "noble" of the gases?

c. What family of elements is found in nature never combined with other elements, or with itself, but as monatomic "molecules"?

14.4) a. What are the members of the next to the last family in the 10-membered "*d* -block"?

b. What letters of the alphabet are used to identify "blocks" of elements in periodic tables?

c. If a new block of super-heavy elements were created, what would it be called?

14.5) a. How many families are there in the *s*-, *p*-, *d*-, and *f*- blocks?

b. What series of integers is obtained when the "lengths" of the *s*-, *p*-, *d*-, and *f*- blocks are divided by 2?

c. How many families of elements would you expect there to be in a complete *g*- block?

d. What series of numbers is generated by the rule: $2\ell + 1$ with $\ell = 0,1,2,3,4...$?

e. What does the "quantum number" ℓ represent in the modern electronic model of atoms?

f. Why is the number of families in a block twice the block's value of $2\ell + 1$?

14.6) a. Excepting hydrogen, in which block(s) of the periodic table are all the elements metals?

b. Metallicity increases in what direction in a family?

c. For which families are the first members nonmetals and the last member, at least, a metal? (Name the families by their first members.)

d. For which families are the first members so nonmetallic that, in fact, all family members are nonmetallic?

14.7) a. What are the simplest chemical formulas for the oxides of the elements in sodium's period when the oxidation state of the element is the same as its group number?

b. What are the maximum "valencies" of the elements in sodium's row?

c. How many electrons are there in neutral atoms of the elements in sodium's row?

d. What are the charges of the "cores" of atoms of elements in sodium's row? (An atomic core = atom – valence shell electrons.)

e. What are the number of the groups that the following elements are in: Na, Mg, Al, Si, P, S, Cl?

14.8) In what way, electronically, is a hydrogen atom like:

a. an atom of sodium? b. an atom of carbon? c. an atom of chlorine?

14.9) True of False? At ordinary conditions:

a. Most compounds of metals are solids.
b. Compounds of nonmetals with nonmetals are often gaseous, or low boiling liquids.
c. Few compounds of metals are gaseous.

14.10) Identify the following substances as Ionic, Covalent, or Metallic.

a. Conducts in the solid state without occurrence of chemical changes at the electrodes.
b. Is an insulator in the solid state, but conducts fairly well in the molten state with occurrence of chemical changes at the electrodes.
c. Is an insulator (poor conductor) in both the solid and liquid states.

14.11) Indicate with the symbols M, I, and C (Exercise 14.10) the type of substances formed by atoms of Na, C, Cl, and H reacting with Na, C, Cl, and H:

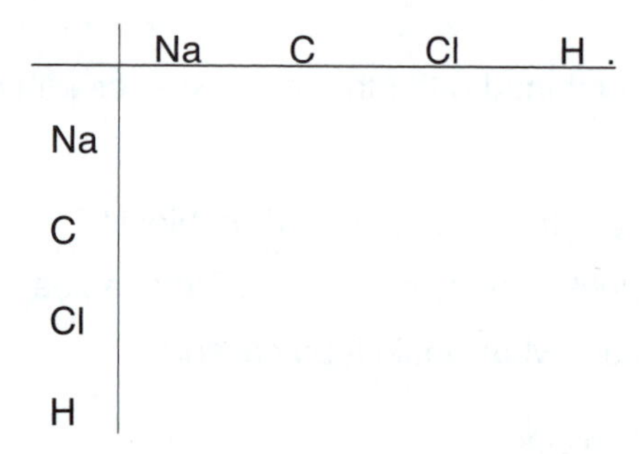

	Na	C	Cl	H .
Na				
C				
Cl				
H				

14.12) From a chemical point of view (Exercise 14.11), hydrogen is most like which element(s)?

14.13) What is sulfur's oxidation number in the following compounds?

a. SF_2, SF_4, SF_6

b. SO, SO_2, SO_3

c. SOF_2, SOF_4

d. SO_2F_2

14.14) a. Li^+ is about the same size as what ion in group 2?

b. Before atomic weights were well established, Be was often placed in the same family as the element _____.

c. Exercises a and b cite examples of the D _ _ _ _ _ _ _ Relationship.

Chapter 14 Answers

14.1)
a. "Salt-generator" (the best known example is NaCl)
b. metal
c. gas
d. salt

14.2)
a. relatively unreactive
b. gold and silver; also copper
c. copper, silver, and gold; also platinum
d. the "noble metals"

14.3) a. He; b. He; c. the noble gases

14.4)
a. Cu, Ag, Au
b. *s*, *p*, *d*, *f*
c. The "*g*-block" (Sober physicists don't find giraffes hiding in kitchens.)

14.5
a. 2, 6, 10, 14
b. 1, 3, 5, 7
c. $2 \times 9 = 18$
d. 1, 3, 5, 7, 9,...
e. an electron's angular momentum
f. Because of electrons' classically nondescribable two-valuedness called "spin," each orbital with distinct values of n, ℓ, and m_ℓ can have up to 2 electrons.

14.6)
a. *s*, *d*, *f*
b. downward
c. the boron, carbon, nitrogen, and oxygen families
d. the halogen and noble gas families

14.7)
a. Na_2O, MgO, Al_2O_3, SiO_2, P_2O_5, SO_3, Cl_2O_7
b. Na 1, Mg 2, Al 3, Si 4, P 5, S 6, Cl 7
c. Na 1, Mg 2, Al 3, Si 4, P 5, S 6, Cl 7, Ar 8
d. Na^{+1}, Mg^{+2}, Al^{+3}, Si^{+4}, P^{+5}, S^{+6}, Cl^{+7}
e. 1, 2, 3, 4, 5, 6, 7

14.8)
a. Each atom has one outer electron (which hydrogen, unlike sodium, never loses in chemical compounds).
b. Each atom has a half-filled shell (often filled by sharing, as in H_2, diamond, and CH_4).
c. Each atom is one electron short of a filled shell (often filled by electron capture from a good electron donor such as sodium, as in Na^+Cl^- (sodium chlor<u>ide</u>) and Na^+H^- (sodium hydride) – both high melting salts).

14.9) a. T; b. T; c. T.

14.10) a. M; b. I; c. C

14.11)

	Na	C	Cl	H .
Na	M	I	I	I
C	I	C	C	C
Cl	I	C	C	C
H	I	C	C	C

14.12) C, Cl (Compare four rows in the above exercise; compare to your answers in Exercise 14.8)

14.13) a. 2, 4, 6; b. 2, 4, 6; c. 4, 6; d. 6

14.14) a. Mg^{+2}; b. Al; c. Diagonal

15

Organic Compounds and the Atomic Properties of Carbon

15.1 THE SPECIAL NATURE OF CARBON

Carbon chemistry is so diverse that there is a whole branch of chemistry named for it. Carbon is a small atom with an intermediate electronegativity (2.5), which forms millions of covalent compounds. It almost always has four bonds in its compounds and makes chains, rings, branched compounds, and compounds with single, double, and triple bonds. Most reactions of organic compounds take place at a functional group, an oxygen, nitrogen, or halogen atom, for example. C–C and C–H bonds are nonpolar, and so are not reactive groups. C–C single bonds are shielded by C–H bonds, and hence are not accessible to reagents. C=C double bonds, in contrast, are exposed and readily attacked, e.g., by proton donors and halogens. Figure 15.2 on page 619 in your textbook shows examples of organic compounds with different bonding arrangements. Notice that in all the examples, each carbon atom has four bonds.

Characteristics of carbon and its compounds

Six characteristics of carbon and the bonds it forms are listed below. Match each characteristic with the corresponding observed behavior or reactivity of carbon's compounds listed a-f on the following page.

1) Carbon has an intermediate electronegativity.
2) Carbon is a small atom.
3) Carbon has four sp^3 hybridized orbitals each containing one electron.
4) The C–C and C–H bonds are both essentially nonpolar.
5) The C–O bond is polar.
6) The C–S and C–Br bonds are long and weak.

(continued on next page)

a) Carbon forms covalent compounds.
b) Carbon forms short, strong bonds.
c) Carbon forms four bonds in its compounds.
d) C–C and C–H bonds are unreactive.
e) C–O bonds are reactive.
f) C–S and C–Br bonds are reactive.

answers: 1-a; 2-b; 3-c; 4-d; 5-e; 6-f.

15.2 THE STRUCTURES AND CLASSES OF HYDROCARBONS

Hydrocarbons contain hydrogen and carbon. **Alkanes** are hydrocarbons with only single bonds (saturated hydrocarbons), **alkenes** contain one or more double bonds, and **alkynes** contain triple bonds. Alkenes and alkynes are referred to as **unsaturated hydrocarbons.**

Paraffins

Paraffin wax is a mixture of solid hydrocarbons obtained from petroleum. "Par-affin" means "without affinity." The hydrocarbons have little affinity for other reagents because their C–C and C–H bonds are unreactive.

Naming Alkanes

Carbon-carbon single bonded skeletons often survive reaction conditions, hence we name alkanes by their number of carbon atoms.

How Chemists Count Carbons

meth-, (1)	eth-, (2)	prop-, (3)	but-, (4)	pent-, (5)
hex-, (6)	hept-, (7)	oct-, (8)	non-, (9)	dec-, (10)

We name the position of substituents on the hydrocarbon chain by numbering the carbon atoms along the main chain starting from the end that will put the substituents on the lowest number carbon atom.

Naming alkanes

1) Name the following hydrocarbons:

1) CH_4
2) CH_3CH_3
3) $CH_3CH_2CH_3$
4) $CH_3CH_2CH_2CH_3$
5) $CH_3CH_2CH_2CH_2CH_3$
6) C_5H_{12}

7) $H_3C-\overset{H_2}{C}-CH_2-CH_2-CH_3$ (drawn with the chain bent: CH_3 above CH_2 above CH_2)

8) $H_3C-\underset{H}{\overset{CH_3}{C}}-\overset{H_2}{C}-CH_3$

9) cyclic: $\overset{H_2}{C}$, H_2C, CH_2, H_2C-CH_2 (five-membered ring)

10) $H_3C-\underset{CH_3}{\overset{CH_3}{C}}-CH_3$

(continued on next page)

answers:

1) methane
2) ethane
3) propane
4) butane
5) pentane
6) ?
7) pentane
8) 2-methylbutane
9) cyclopentane
10) 2,2-dimethylpropane

C_5H_{12} (Problem 6) does not tell us the specific compound. C_5H_{12} could be pentane, 2-methylbutane, or 2,2-dimethylpropane, all of which have five carbon atoms and 12 hydrogen atoms. These compounds, which have the same formula but different structures, are called **structural isomers.**

Note: 2-methylbutane is also called isopentane (common name); 2,2-dimethylpropane is also called neopentane (common name). The common names are based on the total number of carbon atoms: 5 carbon atoms is a pentane. The IUPAC name uses the longest continuous carbon chain in the molecule as the root.

Linear alkanes

How does the number of hydrogen atoms relate to the number of carbon atoms for linear alkanes?

Let's consider the first 4 members of the series:

```
    H           H  H          H  H  H           H  H  H  H
    |           |  |          |  |  |           |  |  |  |
 H— C —H     H— C— C —H    H— C— C— C —H     H— C— C— C— C —H
    |           |  |          |  |  |           |  |  |  |
    H           H  H          H  H  H           H  H  H  H
```

methane CH_4 — ethane C_2H_6 — propane C_3H_8 — butane C_4H_{10}

Each successive member adds a CH_2 group to the previous member:

```
   H |H| H                              H |H  H| H
   | ||| |                              | ||  || |
H— C—|C|—C —H   + "CH2"  ——→        H— C—|C— C|—C —H
   | ||| |                              | ||  || |
   H |H| H                              H |H  H| H
```

propane → butane

What is the formula for pentane?

```
   H |H  H| H                              H |H  H  H| H
   | ||  || |                              | ||  |  || |
H— C—|C— C|—C —H   + "CH2"  ——→        H— C—|C— C— C|—C —H
   | ||  || |                              | ||  |  || |
   H |H  H| H                              H |H  H  H| H
```

butane → pentane

answer: $C_4H_{10} + CH_2 = C_5H_{12}$.

(continued on next page)

We can see from these examples that the general formula for linear alkanes is:

$$H_3C\text{-}(CH_2)_n\text{-}CH_3$$

Linear alkanes therefore have "n" carbon atoms from the internal CH_2 groups, and 2 additional carbon atoms from the terminal CH_3 groups.

The number of hydrogen atoms is "$2n$" from the internal CH_2 groups, and 6 additional hydrogen atoms from the terminal CH_3 groups:

$$C_{n+2}H_{2n+6}$$

We can simplify the formula by letting $n + 2 = m$. Then, $n = m - 2$, and we have:

$$C_{(m-2)+2}H_{2(m-2)+6} = \mathbf{C_mH_{2m+2}}$$

Boiling points and molar mass of alkanes

1) Arrange the following hydrocarbons in order of increasing molar mass:
   1) butane
   2) cyclopentane
   3) 2,2-dimethylpropane
   4) ethane
   5) methane
   6) 2-methylbutane
   7) pentane
   8) propane

   answer: see part "2."

2) Match the hydrocarbons (listed below in order of increasing molar mass) to their boiling points (°C):

1) methane (16.04)	a) –164
2) ethane (30.07)	b) –88.6
3) propane (44.11)	c) –42.1
4) butane (58.12)	d) 0.5
5) cyclopentane (70.14)	e) 9.5
6) pentane (72.15)	f) 27.8
7) 2-methylbutane (72.15)	g) 36.1
8) 2,2-dimethylpropane (72.15)	h) 49.2

   answers: 1-a; 2-b; 3-c; 4-d; 5-h; 6-g; 7-f; 8-e.

   The boiling points of linear hydrocarbons increase with their molar mass (see Figure 15.7 on page 625 in your textbook). Cyclopentane boils at a higher temperature than the other five-carbon hydrocarbons because its disc shape allows its molecules to have more intermolecular contact. The branched 5 carbon molecules have lower boiling points than *n*-pentane because they make less intermolecular contact.

3) Which of the alkanes above are gases at room temperature (25°C)?

   answer:

   methane, ethane, propane, butane, and 2,2-dimethylpropane.

   Linear hydrocarbons with five or more carbon atoms are liquids at room temperature. The primary component of gasoline, octane, is a liquid at ambient temperature. Substances such as margarine, grease, and wax contain long hydrocarbon chains and are solids at room temperature.

Naming Alkenes

We name alkenes the same way we name alkanes, except the suffix is *-ene* instead of *-ane*. We often refer to alkenes by their common names. Ethylene is the common name for ethene, C_2H_4. Propylene is the common name for propene, C_3H_6. Polyethylene and polypropylene are polymers composed of ethylene and propylene chains, respectively.

Exercise

What is wrong with the names for the following alkenes?

1) 3-butene 2) 3-methyl-5-hexene 3) 4-ethyl-1-pentene 4) 4-ethene-heptane

answers:

1) We indicate the position of the double bond by the number of the first C atom in it. The correct name is 2-butene.

2) We number the chain from the end closest to the double bond. The correct name is 4-methyl-1-hexene.

3) The longest root chain that contains both carbon atoms of the double bond is the 6 carbon chain. The correct name is 4-methyl-1-hexene.

4) The root chain must contain the carbons of the double bond, even though it is not the longest chain. The correct name is 3-propyl-1-hexene.

Naming Alkynes

We name alkynes the same way we name alkenes, except the suffix is *-yne* instead of *-ene*. Alkynes are often called by their common names.

Exercise

Name the following alkynes:

1) $HC{\equiv}CH$

2) $H_3C-C{\equiv}CH$

3) $BrH_2C-C{\equiv}CBr$

4) $H_3C-C{\equiv}C-C_6H_5$ (benzene ring)

answers:

1) ethyne, or acetylene

2) propyne, or methyl acetylene

3) 1,3-dibromopropyne

4) 1-phenylpropyne

Ethyne is almost always called by its common name, acetylene. A benzene ring attached to a hydrocarbon chain is called a **phenyl** group.

Isomerism

Structural isomers are compounds that have the same molecular formula, but a different sequence of bonded atoms.

Structural isomers

Draw 3 structural isomers for the formula C_5H_{12}.

answer:

pentane, 2-methylbutane, and 2,2-dimethylpropane each has 5 carbon atoms and 12 hydrogen atoms, and so are isomers:

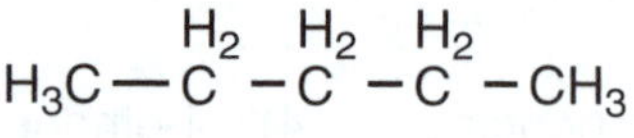

pentane

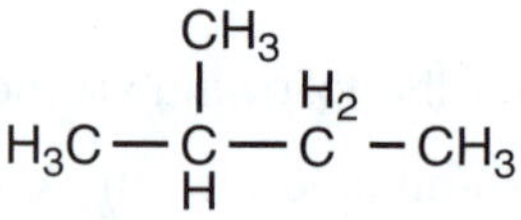

2-methylbutane

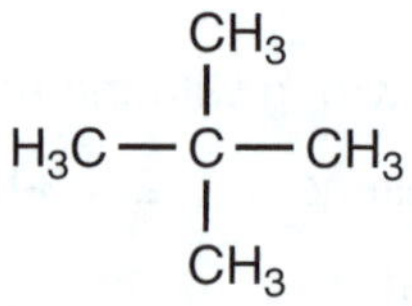

2,2-dimethylpropane

We saw in the exercise on naming alkanes that the structural isomers pentane, 2-methylbutane, and 2,2-dimethylpropane, all with the molecular formula C_5H_{12}, have different physical properties:

	b.p.
pentane	36.1°C
2-methylbutane	27.8°C
2,2-dimethylpropane	9.5°C

Stereoisomers have the same molecular formula and the same sequence of bonded atoms, but different orientations of groups in space. There are two types of stereoisomers.

- **Geometric isomers** (or *cis-trans* isomers) have atoms which take different positions around a rigid ring or bond.
- **Optical isomers** are mirror images which cannot be superimposed on each other (a good example of mirror images which cannot be superimposed is your hands). Optical isomers differ only in the direction that each isomer rotates the plane of polarized light.

Geometric isomers

1) Why can 2-butene have *cis-trans* isomers, while butane cannot?

answer:

2-butene contains a double bond, which does not allow rotation of the groups at room temperature. Carbon-carbon single bonds allow free rotation of the groups about them at room temperature.

2) Draw the *cis* and *trans* isomers for 2-butene.

solution:

The *trans* isomer has the methyl groups on opposite sides of the carbon-carbon double bond; the *cis* isomer has the methyl groups on the same side of the C=C double bond:

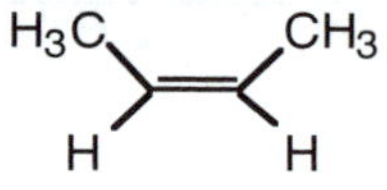

cis-2-butene

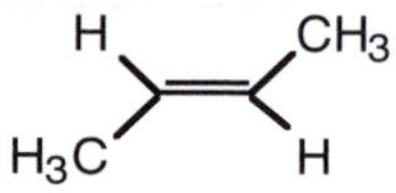

trans-2-butene

Optical isomers

An organic molecule is asymmetric (not superimposable on its mirror image) if it contains a carbon atom that is bonded to four different groups.

Which of the following compounds contain an asymmetric carbon atom?

1) Cl–C(Br)(H)–H
bromo-chloromethane

2) Cl–C(Br)(F)–H
bromo-chloro-fluoromethane

3) $H_3C-CH(CH_3)-CH_2-CH_3$
2-methylbutane

4) $H_3C-C(CH_3)_2-CH_2-CH_3$
2,2-dimethylbutane

5) $H_3C-CH(Br)-CH_2-CH_3$
2-bromobutane

answers:

Compounds #2 and #5 are asymmetric:

- Compound #2: There are four different groups attached to the carbon atom (a bromine, a chlorine, a fluorine, and a hydrogen atom), so bromo-chloro-fluoromethane is asymmetric.

- Compound #5: The second carbon atom contains four different groups (a bromine, a methyl group, an ethyl group, and a hydrogen atom), so 2-bromobutane is asymmetric.

Drugs are often asymmetric molecules, which, in order to bind to sites in our body, must have the correct "handedness." The wrong isomer may not only be ineffective, but it could be detrimental. Drug companies therefore must be able to synthesize these drugs so that they are optically pure.

Aromatic hydrocarbons contain carbon atoms with delocalized electrons (often benzene rings). Remember from Chapter 11 that the benzene molecule (C_6H_6) requires resonance structures, since all C–C bonds are known to be equivalent:

We consider the sigma bonds between carbon atoms to be localized electrons in sp^2 hybrid orbitals giving the benzene ring its planar structure. The remaining *p* orbital on each carbon atom is perpendicular to the plane of the benzene ring and contains the remaining electrons. These 6 *p* orbitals can combine to form π molecular orbitals. The electrons in these π molecular orbitals are delocalized above and below the ring. You will often see the structure of benzene written with a circle inside the carbon six-membered ring to indicate its delocalized π bonding.

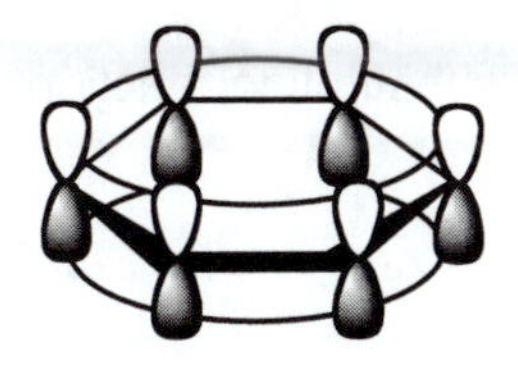

Benzene is usually the parent compound, and attached groups are named as prefixes.

Naming aromatic compounds

Name the following compounds:

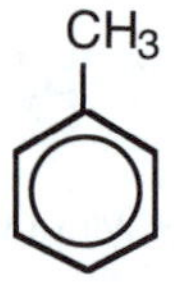

CH_3 (1) CH_3, CH_3 (2) CH_3, CH_3 (3) CH_3, CH_3  (4)

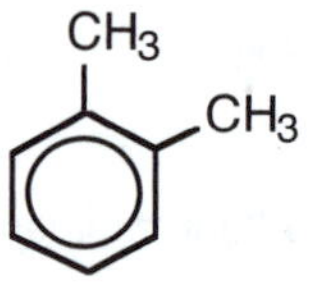

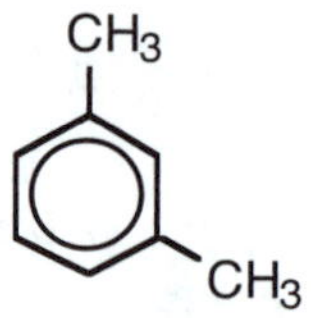

answers:

1) methylbenzene (toluene)

2) 1,2-dimethylbenzene (*o*-xylene)

3) 1,3-dimethylbenzene (*m*-xylene)

4) 1,4-dimethylbenzene (*p*-xylene)

Xylene is the common name for dimethylbenzene. We indicate the positions of the methyl groups by:

o- (ortho) for adjacent groups

m- (meta) for groups separated by one ring carbon atom

p- (para) for groups on opposite ring carbon atoms

15.3 CLASSES OF ORGANIC REACTIONS

There are three general classes of organic reactions into which many reactions fall. Two of the reaction types (addition reactions and elimination reactions) involve carbon-carbon double (or triple) bonds, bonds which are exposed and readily attacked.

- **Addition reactions** occur when the π bond of an alkene or alkyne breaks, and new sigma bonds form to the atoms which "add" to the molecule.

- **Elimination reactions** are the opposite of addition reactions. A molecule is eliminated, and a double or triple bond forms in the hydrocarbon product. The driving force for this type of reaction is usually the loss of a small, stable molecule such as HBr or water.

- **Substitution reactions** occur in alkanes when an atom from an added reagent substitutes for an atom in a hydrocarbon molecule. Usually in these reactions, halogen atoms replace hydrogen atoms.

Classes of reactions

Classify the following as addition, elimination, or substitution reactions:

(continued on next page)

1) Br_2 + $H_2C{=}CH_2$ ⟶ $BrH_2C{-}CH_2Br$

2) $CH_3CH_2CH_2Br$ ⟶ $H_3C{-}\underset{H}{C}{=}CH_2$ + HBr

3) CH_3CH_2Cl + CH_3ONa ⟶ $H_3CCH_2OCH_3$ + $NaCl$

answers:

1) Addition reaction: bromine atoms add across the double bond in ethylene.

2) Elimination reaction: HBr is eliminated and a double bond forms in the hydrocarbon product.

3) Substitution reaction: a methoxy group substitutes for the chlorine atom in chloroethane.

Oxidation and Reduction of Organic Groups

When we assign an oxidation number to a carbon atom in an organic compound, we must decide if the attached groups are more or less electronegative than carbon. For example, the oxidation number of the carbon atom in carbon tetrachloride, CCl_4, is +4 because chlorine is more electronegative than carbon, and is assigned an oxidation number of –1. The oxidation number of carbon in methane, CH_4, is –4 because hydrogen is less electronegative than carbon and is assigned an oxidation number of +1.

When hydrogen atoms add to a carbon atom, the carbon atom is reduced. Conversely, the loss of hydrogen atoms from a carbon atom results in its oxidation. Compare the oxidation number of the carbon atoms in methane, CH_4, and ethylene, $H_2C{=}CH_2$. The oxidation number of carbon in ethylene is –2, the oxidation number of carbon in methane is –4. Carbon is reduced on addition of hydrogen atoms.

The addition of oxygen to carbon results in oxidation of the carbon atom since oxygen is more electronegative than carbon.

Redox reactions

For each reaction, on the following page, determine if the carbon atom is oxidized or reduced:

1) $CH_4 + 2O_2 \rightarrow CO_2 + 2H_2O$
2) $CH_3CH(OH)CH_3 \rightarrow CH_3C(O)CH_3$
3) $CH_2CH_2 + Br_2 \rightarrow BrCH_2CH_2Br$
4) $CH_2CH_2 + HCl \rightarrow ClCH_2CH_3$

answers:

1) Addition of oxygen and loss of hydrogen to carbon results in its oxidation. Carbon is oxidized from an oxidation state of –4 in CH_4 to an oxidation state of +4 in CO_2. (Hydrogen atoms are assigned an oxidation number of +1 in hydrocarbons because they are less electronegative than carbon.)

2) Loss of a hydrogen atom from the central carbon results in its oxidation.

3) Addition of bromine atoms to carbon results in its oxidation; the bromine atoms are reduced.

4) The carbon that gains a hydrogen atom is reduced, and the carbon that gains the chlorine atom is oxidized.

15.4 REACTIVITY OF COMMON FUNCTIONAL GROUPS

A **functional group** is some group other than a C–H single bond on a hydrocarbon. We have already discussed alkenes and alkynes with double and triple bonds, respectively. Other functional groups involve atoms such as oxygen, nitrogen, and halogen atoms.

Name Each of the Following Functional Groups

1) $H_2C{=}CH_2$

2) $HC{\equiv}CH$

3) H_3C-OH (wood alcohol)

4) H_3C-Br

5) H_3C-NH_2

6) $H_3C-\overset{\overset{\displaystyle O}{\|}}{C}H$

7) $H_3C-\overset{\overset{\displaystyle O}{\|}}{C}-CH_3$ (acetone)

8) $H_3C-\overset{\overset{\displaystyle O}{\|}}{C}-OH$ (vinegar)

9) $H_3C-\overset{\overset{\displaystyle O}{\|}}{C}-OCH_3$

10) $H_3C-\overset{\overset{\displaystyle O}{\|}}{C}-NH_2$

11) $CH_3C{\equiv}N$

answers:

1) alkene (R=R)
2) alkyne (R≡R)
3) alcohol (R–OH)
4) haloalkane (R–X)
5) amine (R–N)
6) aldehyde (CH=O)
7) ketone (C=O bonded to 2 carbons)
8) carboxylic acid (C=O bonded to OH)
9) ester (C=O bonded to –OR)
10) amide (C=O bonded to N)
11) nitrile (C≡N)

The word aldehyde comes from alcohol dehydrogenated.

Functional groups

Determine what class of compound is described in each case below and on the next page:

1) They are named by adding the suffix *-ol* to the parent hydrocarbon name (example: methanol). Their –OH group allows them to hydrogen bond and gives them similar properties to water. They lose H and –OH groups (dehydration) to form alkenes. With hydrohalic acids, they react to give haloalkanes:

$$R-OH + HBr \longrightarrow R-Br + H_2O$$

2) They form when two alcohol molecules combine eliminating water. They have no –OH group, and so melt and boil at lower temperatures than alcohols.

3) They consist of a carbon atom connected to a halogen. In the presence of base, alcohols form with the elimination of the halogen:

(continued on next page)

$$R{-}Br + OH^- \longrightarrow R{-}OH$$

4) They are derivatives of ammonia that cause fish to smell "fishy." They are weakly basic and can be neutralized by the acid from a lemon (citric acid). They form hydrogen bonds, and so have fairly high melting and boiling points. They can undergo substitution reactions to form larger ammonia derivatives:

$$2\,R{-}NH_2 + R''{-}Cl \longrightarrow R{-}NH{-}R'' + RNH_3Cl$$

5) Hydrocarbons that typically undergo addition reactions due to their electron rich multiple bonds.

6) They have delocalized electrons, and therefore are less reactive than compounds with localized multiple bonds.

7) They contain an electron rich **carbonyl group** at the end of a hydrocarbon chain. They are named by adding *-al* to the parent hydrocarbon name (example: propanal). They are reduced to alcohols.

8) They contain an electron rich carbonyl group within a hydrocarbon chain. They are named by adding *-one* to the parent hydrocarbon name (example: propanone). They can be reduced to alcohols:

$$R{-}C(=O){-}R \xrightarrow{CH_3Li} \xrightarrow{H_2O} R{-}C(OH)(CH_3){-}R$$

9) They are named by adding *-oic* acid to the parent alkane (example: propanoic acid). They are weak acids in water which are neutralized completely by a base to give a salt (on evaporation) and water:

$$R{-}C(=O){-}OH + OH^- \longrightarrow R{-}C(=O){-}O^-Na^+ + H_2O$$

10) Long chains of these are **fatty acids.**

11) They form from an alcohol and a carboxylic acid with loss of water (dehydration):

$$R{-}C(=O){-}\boxed{OH \quad H}{-}O{-}R \overset{H^+}{\rightleftharpoons} R{-}C(=O){-}O{-}R + H_2O$$

12) They form from a substitution reaction of an ester and an amine:

$$\underset{\text{ester}}{R{-}C(=O){-}OR'} + \underset{\text{amine}}{H_2N{-}R''} \longrightarrow R{-}C(=O){-}NHR'' + R'{-}OH$$

13) The most important example of this group for life is the **peptide bond** which links amino acids in a protein.

answers:

1-alcohols; 2-ethers; 3-haloalkanes; 4-amines; 5-alkenes, alkynes; 6-aromatic compounds; 7-aldehydes; 8-ketones; 9,10-carboxylic acids; 11-ester; 12,13-amides.

15.5 POLYMERS: SYNTHETIC MACROMOLECULES

Small **monomer** units combine to form **polymers** (poly-mer = many molecules). Synthetic polymers form from addition reactions, or condensation reactions. **Addition polymers** form when monomers add together. Polyethylene (used for plastic bags, bottles, and toys) forms from the addition of ethylene molecules to each other. A **free radical** (species with an unpaired electron, usually from a peroxide) initiates the reaction. The polymerization reaction stops when two free radicals form a covalent bond. Figure 15.27 on page 652 in your textbook shows the steps in the free radical polymerization of ethylene. Polymerization can also be initiated by formation of an ion instead of a free radical. **Catalysts** are often used to allow polymerization reactions to occur under mild conditions. Table 15.6 on page 653 in your textbook shows the monomers used for the most important addition reactions. Notice that all the monomers are based on ethylene ($R_1 = R_2 = R_3 = R_4 = H$):

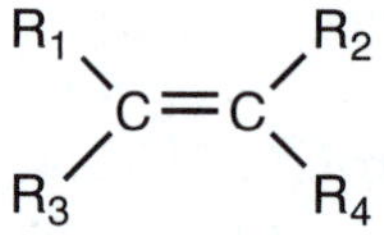

$$(R_1)(R_3)C=C(R_2)(R_4)$$

Addition polymers based on ethylene

Consider an ethylene-type molecule: $(R_1)(R_3)C=C(R_2)(R_4)$

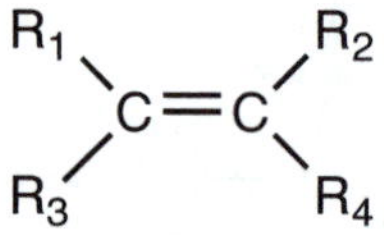

Name the polymer that forms from the monomer with:

1) $R_1 = R_2 = R_3 = R_4 = H$
2) $R_1 = R_2 = R_3 = H$ and $R_4 = Cl$
3) $R_1 = R_2 = R_3 = R_4 = F$ (see shaded box on next page)
4) $R_1 = R_2 = R_3 = H$ and R_4 = benzene ring
5) $R_1 = R_2 = R_3 = H$ and $R_4 = CH_3$
6) $R_1 = R_2 = R_3 = H$ and $R_4 = CN$

<u>answers</u>:

1-polyethylene; 2-poly(vinyl chloride); 3-polytetrafluoroethylene (Teflon); 4-polystyrene; 5-polypropylene; 6-polyacrylonitrile.

Condensation polymers form when two functional groups undergo a dehydration-condensation reaction. Condensation of carboxylic acid and amines form polyamides (nylon); condensation of carboxylic acid and alcohols form polyesters:

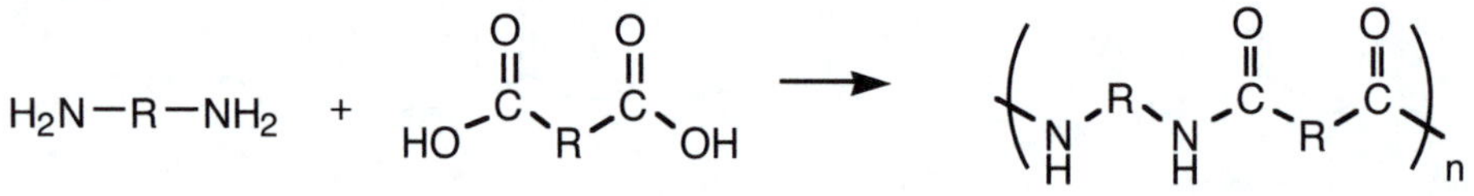

$$H_2N{-}R{-}NH_2 + HO{-}C(=O){-}R{-}C(=O){-}OH \longrightarrow \left[{-}NH{-}R{-}NH{-}C(=O){-}R{-}C(=O){-}\right]_n$$

polyamide

$$HO{-}R{-}OH + HO{-}C(=O){-}R{-}C(=O){-}OH \longrightarrow \left[{-}O{-}R{-}O{-}C(=O){-}R{-}C(=O){-}\right]_n$$

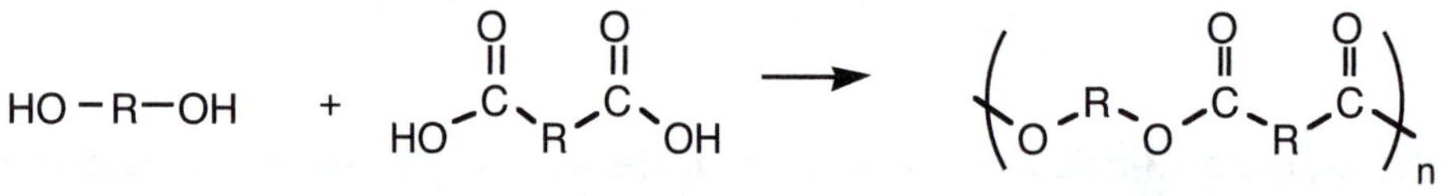

polyester

Don't Throw That Gunk Out!

The discovery of nylon and Teflon:

A dicarboxylic acid and a diamine react to form a sticky material with little apparent use. However, one day a chemist in the Carothers research group in the DuPont Company put a small ball of the sticky stuff on the end of a stirring rod and formed a string as he pulled it away from the sticky mass. He noticed the strength and silky appearance of the thread and realized that it could be a useful fiber.

In 1938, a DuPont chemist named Roy Plunkett synthesized about 100 pounds of gaseous perfluoroethylene and stored it in steel cylinders:

$$F_2C{=}CF_2$$

Unexpectedly, a cylinder produced no perfluoroethylene gas when the valve was opened. Apparently, a technician said, it was empty. Strange, thought Plunkett, we filled it recently and haven't used much. How about taking the tank over to the machine shop and having them cut it open with their large hack saw? The cylinder was cut open to reveal a white powder. The powder was a polymer of perfluoroethylene, which was developed into Teflon.

Synthetic polymers are ever present in our lives: Teflon is used for cookware and insulation; polypropylene is used for carpeting and bottles; polyvinyl chloride is used for plastic wrap and indoor (lead-free) plumbing; polystyrene is used for insulation and furniture; polyacrylonitrile is used for yarns, and the fabrics Orlon and Acrilon; polyvinyl acetate is used for adhesives, paints, textile coatings, and computer disks; polyvinylidine chloride is used in Saran wrap; polymethyl methacrylate is used as a glass substitute (e.g., Plexiglas), and bowling balls. Nylon, lycra, and all synthetic yarns are polymers.

15.6 POLYMERS: BIOLOGICAL MACROMOLECULES

Biopolymers make up the most abundant organic chemicals on earth. **Polysaccharides** form from individual sugar molecules **(monosaccharides).** Cellulose (found in plants), starch (an energy store for plants), and glycogen (the energy storage molecule for animals) all form from repeated glucose units. Figure 15.29 on page 655 in your textbook shows the structure of glucose in aqueous solution. Become familiar with the appearance of glucose, and notice how it undergoes a dehydration condensation with another monosaccharide (fructose, in the case of the example) to form a disaccharide (sucrose) with the loss of a water molecule.

Proteins are polyamides made up of amino acids, monomers that contain a carboxyl and an amine group on the same molecule. Two amino acids combine to form a peptide (amide) bond with the elimination of a water molecule (dehydration):

$$H_2N{-}CH(R){-}COOH + H_2N{-}CH(R'){-}COOH \longrightarrow H_2N{-}CH(R){-}\boxed{C({=}O){-}NH}{-}CH(R'){-}COOH + H_2O$$

peptide bond

The common amino acids are shown in Figure 15.30 on page 656 in your textbook. The approximately 20 amino acids combine in chains of 50 to several thousand to form proteins. The sequence of amino acids determines the protein's shape and function. The two broad classes of proteins are summarized below.

Fibrous proteins

- relatively simple amino acid compositions
- small number of R groups in a repeating sequence
- shaped like extended helices or sheets
- key components of materials that require strength and flexibility (hair, wool, skin, and connective tissue)

Globular proteins

- more complex amino acid compositions
- contain varying proportions of all 20 different R groups
- rounded and compact shapes
- wide variety of functions (catalysts, messengers, bacterial defenders, membrane gatekeepers)

Nucleic acids are unbranched polymers that contain the chemical information that guides the designing of proteins. The monomers (mononucleotides) that form nucleic acids have three parts: an N-containing base, a sugar, and a phosphate group. Your text shows these groups on page 659. The two types of nucleic acids, **ribonucleic acid** (RNA) and **deoxyribonucleic acid** (DNA) differ in the sugar portions of their mononucleotides. Linear segments of the DNA molecule act as **genes,** chemical blueprints for constructing an organism's proteins.

DNA exists as a double helix: two chains wrapped around each other. The sugar and phosphate groups form the twisted "tracks" of DNA. The bases hydrogen bond to form the "ties" between the tracks. There are four bases and they always hydrogen bond with the same partner: adenine (A) with thymine (T), and guanine (G) with cytosine (C). Thymine and cytosine are pyrimidines; adenine and guanine are purines:

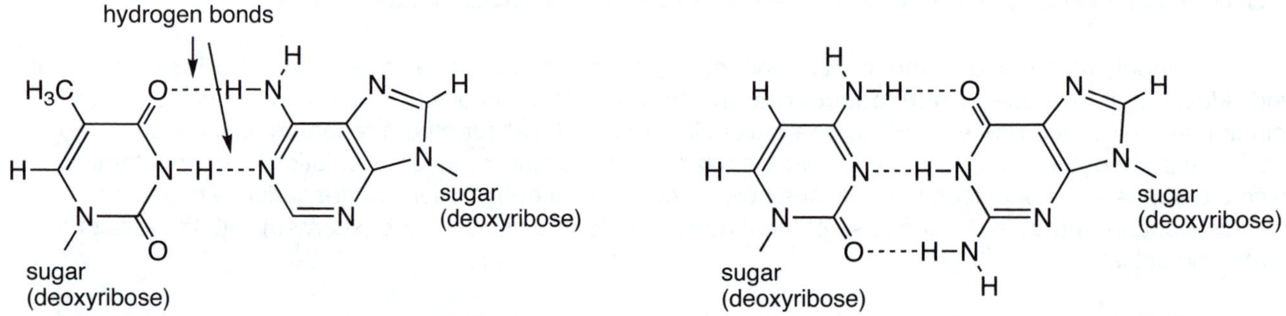

hydrogen bonds

Thymine (T) Adenine (A) Cytosine (C) Guanine (G)

A sequence of three bases acts as a code for a specific amino acid. This code is translated (with the help of messenger RNA and transfer RNA) into a sequence of amino acids that are linked to make a protein. The key to protein synthesis is the _ _ _ _ _ _ _ _ _ _ _ _ _ (hydrogen bond).

CHAPTER CHECK

Make sure you can...

- Explain why carbon forms a huge number and variety of compounds.
- Name and draw structures for alkanes, alkenes, and alkynes.
- Distinguish between structural, geometric, and optical isomers.
- Identify chiral centers of molecules.
- Name aromatic hydrocarbons; know the isomers of xylene.
- Recognize addition, elimination, and substitution reactions.
- Identify the functional groups shown on study guide page 302, and know how they react (pages 302, 303).
- Determine the reactants and products of reactions of: alcohols, alkyl halides, amines, aldehydes, ketones, and carboxylic acids.
- Recognize and name functional groups in organic molecules.
- Explain how addition and condensation polymers form.
- Draw a polymer structure based on monomer structures.
- List the three types of biopolymers and their monomers.
- Explain how base-pairing controls protein synthesis and DNA replication.
- Explain how the DNA sequence determines the amino acid sequence in proteins.

Chapter 15 Exercises

15.1) How many structural isomers are there for C_6H_{14}? Draw them and give their names.

15.2) Name the following compounds:

a. $H_3C-C(CH_3)(CH_3)-$ attached to a cyclohexane ring

b. CH_3 attached to a cyclopentene ring

c. benzene ring with two CH_3 groups (meta positions)

d. $H_3C-CH(C_6H_5)-CH_2-CHCl-CH_2-CH_3$

15.3) Classify these compounds as alkanes or alkenes (assume there are no rings):

a. C_2H_4 b. C_4H_8 c. C_8H_{18}

15.4) Write a molecular formula of a hydrocarbon containing six carbons which is an:

a. alkane b. alkene c. alkyne d. aromatic

15.5) Draw *cis-* and *trans-*2-pentene.

15.6) Circle the functional group(s) on the compounds on the following page:

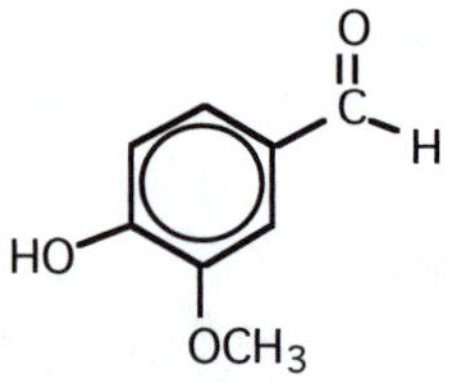

O
C
H
HO
OCH_3

Vanillin
(vanilla beans)

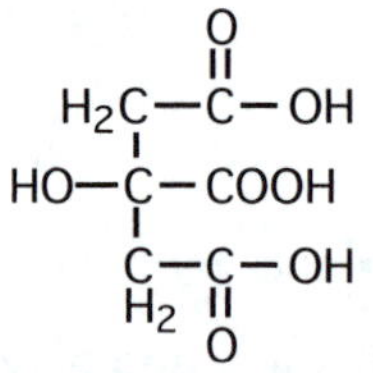

O
$H_2C-C-OH$
$HO-C-COOH$
$C-C-OH$
H_2 O

Citric acid
(citrus juice)

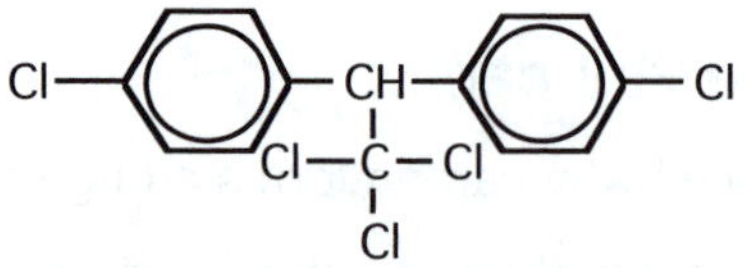

Cl — CH — Cl
Cl—C—Cl
Cl

Dichlorodiphenyltrichloroethane
(DDT insecticide)

15.7) Name the following compounds:

O C CH_3

O
C
H_3C CH_3

COOH

$CH_3-CH-CH_2-CH_2-COOH$
Cl

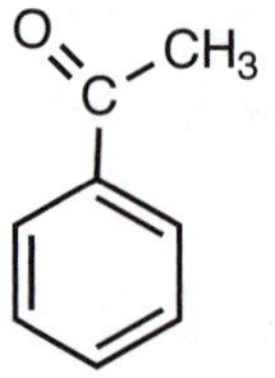

a. 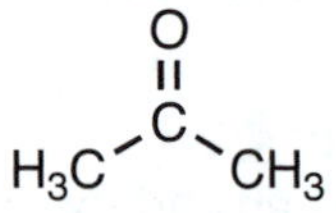 b. 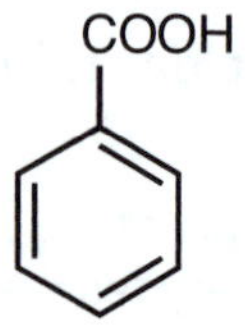 c. 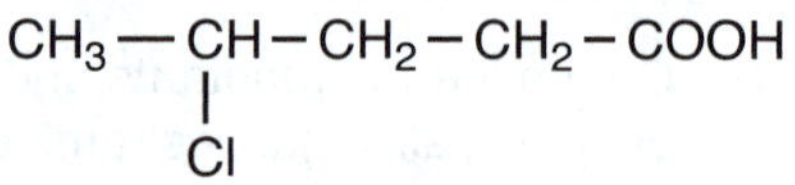 d.

15.8) Circle the functional group(s) on the following compounds:

CH_3 OH
$C\equiv CH$
O

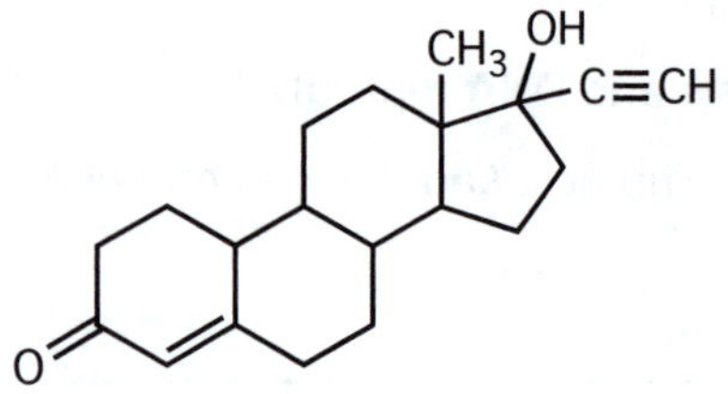

Northeindrone (the pill)

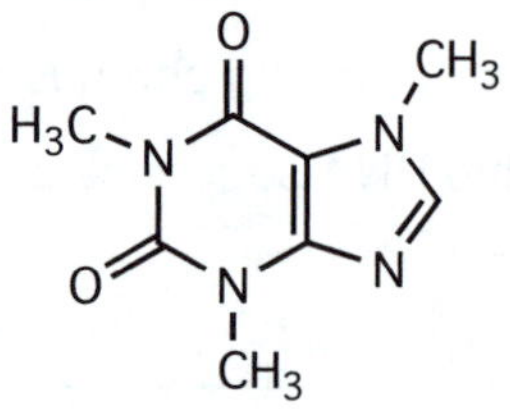

O
CH_3
H_3C N N
O N N
CH_3

Caffeine

15.9) Identify the following compounds as an alcohol, ether, aldehyde, ketone, acid, or ester:

a. CH_2O b. CH_4O c. C_2H_4O d. C_2H_6O e. C_3H_6O f. $C_2H_4O_2$

15.10) Fill in the needed reactant(s) to complete the reactions:

a. (OH on cyclohexane) ⟶ (O on cyclohexane)

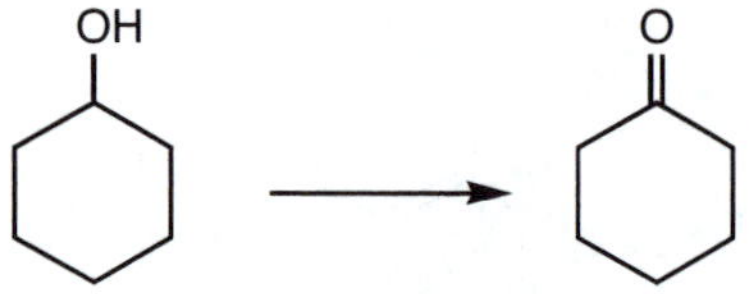

b. $H_2C{=}\underset{H}{C}{-}\underset{H}{C}{=}CH_2 \longrightarrow H_3C{-}\overset{H_2}{C}{-}\overset{H_2}{C}{-}CH_3$

c. $H_3C{-}\underset{H}{\overset{CH_3}{C}}{-}\overset{H_2}{C}{-}Br \longrightarrow H_3C{-}\overset{CH_3}{C}{=}CH_2$

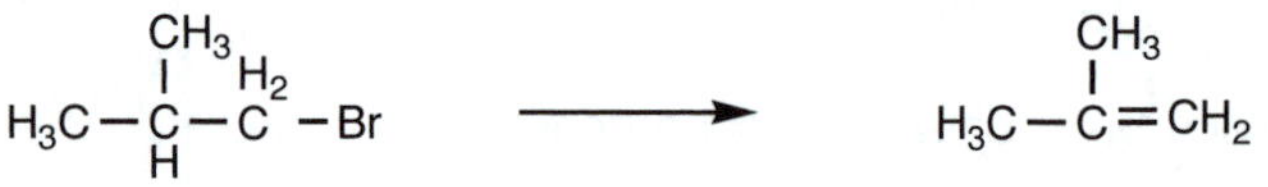

d. CH_2Cl (on benzene) ⟶ CH_2OH (on benzene)

15.11) Complete the following reactions:

a. 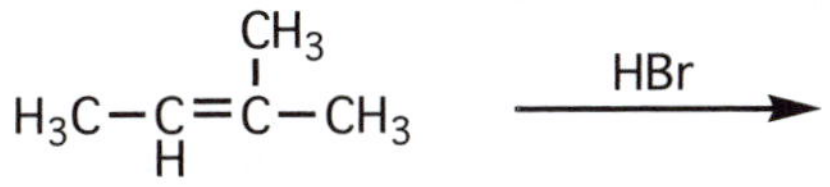 $H_3C{-}CH{=}C(CH_3){-}CH_3 \xrightarrow{\text{HBr}}$

b. 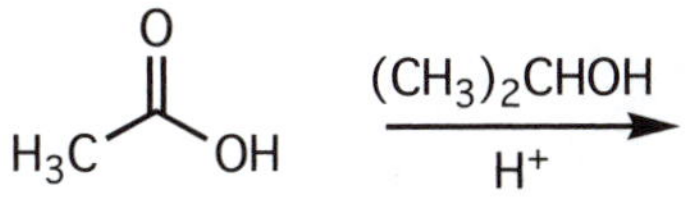 $H_3C{-}C(=O){-}OH \xrightarrow[H^+]{(CH_3)_2CHOH}$

15.12) a. How many monomers make up this segment of polyvinyl chloride (PVC)?

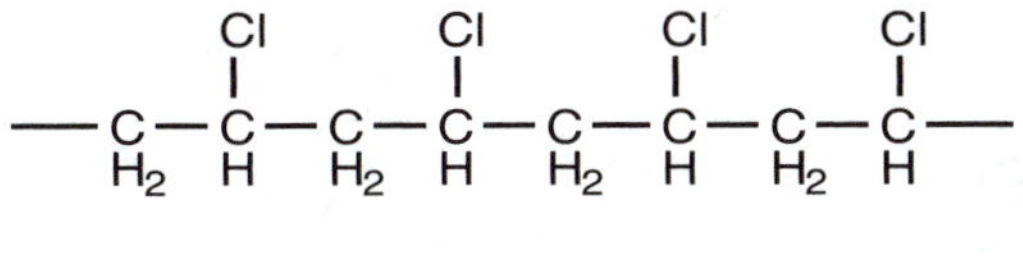

$$-CH_2-CH(Cl)-CH_2-CH(Cl)-CH_2-CH(Cl)-CH_2-CH(Cl)-$$

b. Give the structure of the monomer.

15.13) The structure below is a peptide:

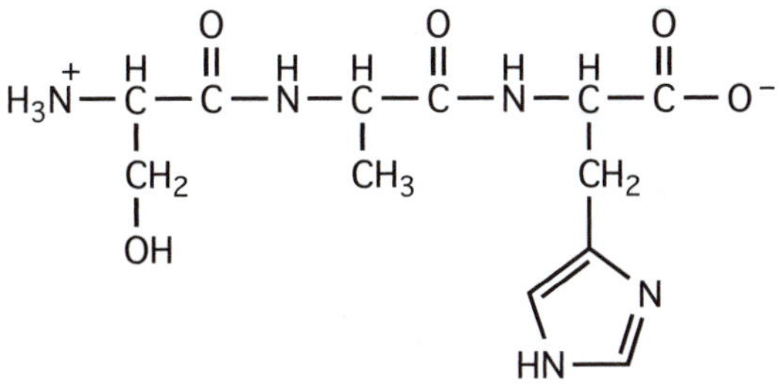

$$H_3\overset{+}{N}-CH(CH_2OH)-C(=O)-NH-CH(CH_3)-C(=O)-NH-CH(CH_2\text{-imidazolyl})-C(=O)-O^-$$

a. Circle the peptide bonds.

b. How many amino acids make up the peptide?

Chapter 15 Answers

15.1) There are 5 structural isomers of C_6H_{14}:

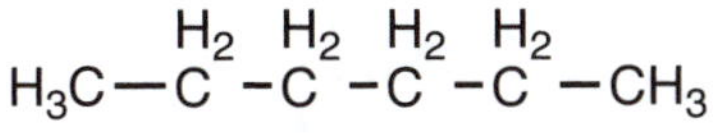

$H_3C-CH_2-CH_2-CH_2-CH_2-CH_3$

hexane

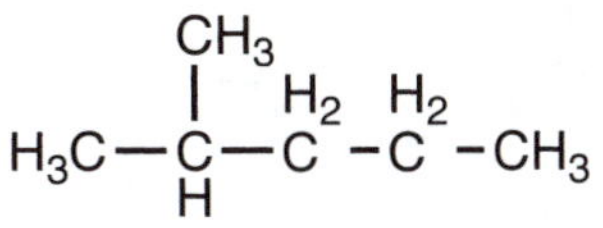

$H_3C-CH(CH_3)-CH_2-CH_2-CH_3$

2-methylpentane

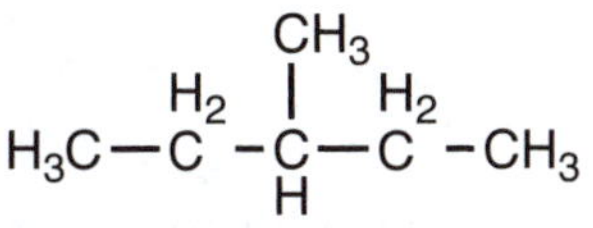

$H_3C-CH_2-CH(CH_3)-CH_2-CH_3$

3-methylpentane

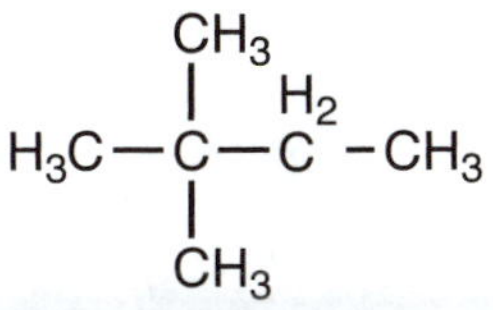

$H_3C-C(CH_3)_2-CH_2-CH_3$

2,2-dimethylbutane

$H_3C-CH(CH_3)-CH(CH_3)-CH_3$

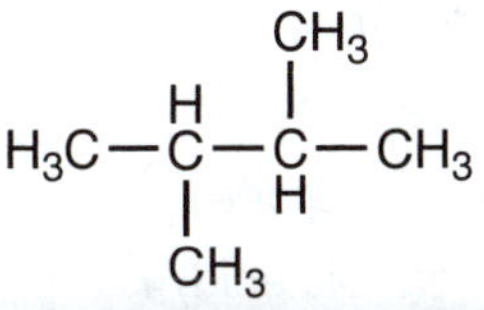

2,3-dimethylbutane

15.2) a. 1-*tert*-butylcyclohexane; b. 3-methyl-cyclopentene; c. 1,3-dimethylbenzene (*m*-xylene); d. 4-chloro-2-phenylhexane

15.3) If m = number of carbon atoms, straight chain alkanes have the formula, C_mH_{2m+2}; straight chain alkenes have the formula C_mH_{2m}. C_8H_{18} is a linear alkane; C_2H_4 and C_4H_8 are alkenes.

15.4) a. C_6H_{14}; b. C_6H_{12}; c. C_6H_{10}; d. C_6H_6

15.5)

cis-2-pentene

trans-2-pentene

15.6)

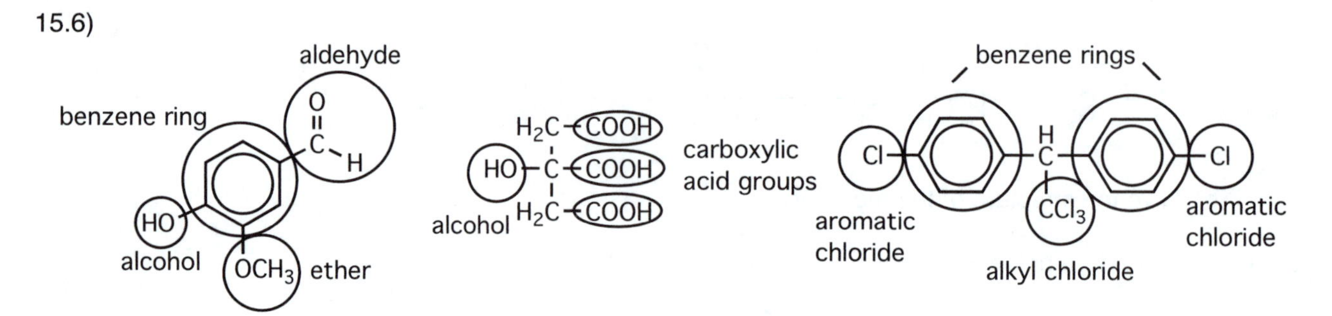

15.7) a. methyl-phenyl-ketone
b. 2-propanone (acetone)
c. benzoic acid
d. 4-chloropentanoic acid

15.8)

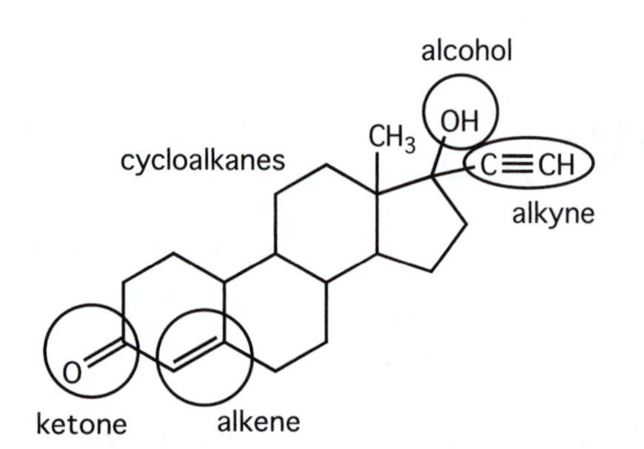

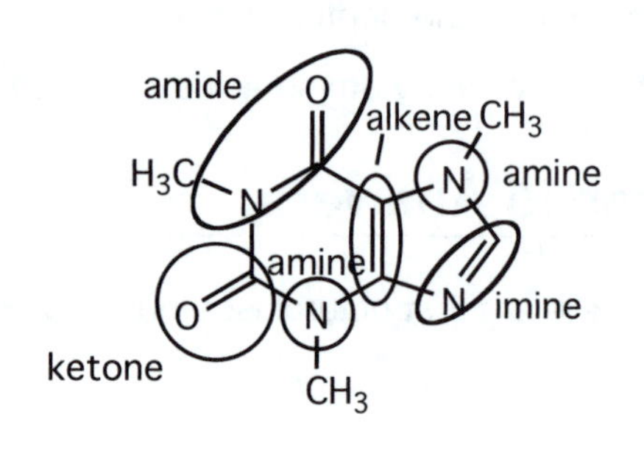

15.9) a. aldehyde (methanal or formaldehyde)
b. alcohol (methanol)
c. aldehyde (acetaldehyde)
d. ether (dimethyl ether)
e. ketone (propanone, acetone)
f. carboxylic acid (ethanoic acid)

15.10) a. CrO_3/H_2SO_4; b. H_2/Pt; c. KOH; d. NaOH

15.11) a. $H_3C{-}CH_2{-}C(CH_3)(Br){-}CH_3$ b. $H_3C{-}C(=O){-}O{-}CH(CH_3){-}CH_3$

15.12) a. four monomer units; b. $H_2C{=}CH{-}Cl$

15.13) a. $H_3\overset{+}{N}{-}CH(CH_2OH){-}C(=O){-}NH{-}CH(CH_3){-}C(=O){-}NH{-}CH(CH_2{-}\text{imidazolyl}){-}C(=O){-}O^-$ b. three amino acids

16

Kinetics: Rates and Mechanisms of Chemical Reactions "It's About Time"

16.1 REACTION RATE

Chemical kinetics is the study of reaction rates. Diamond spontaneously changes to graphite, but the reaction is extremely slow, so it seems to us as though "diamonds are forever." We saw in section 6.1 of the study guide (problem on conversion of energy into heat, page 120) that the combustion of sugar in our bodies occurs at a much slower rate (it is metabolized) than the combustion of gasoline in our cars. We slow down reactions that cause food to spoil by keeping food in the refrigerator. We slow these reactions further by placing the food in the freezer. We study kinetics to determine a possible "reaction mechanism"—what might be happening at the molecular level.

The Factors We Can Control That Affect Reaction Rates

Concentration of reactants

Since molecules must collide in order to react, reactant rate is proportional to the concentration of reactants.

Temperature

Molecules must collide with enough energy to react. Raising the temperature increases both the number of collisions between molecules and the energy of the collisions.

Physical state of reactants

If a liquid or solid reacts with another phase, the reaction occurs at the interface between phases. Therefore, the more surface area, the faster the reaction.

Catalysts

A catalyst speeds up a reaction by providing a different reaction mechanism with a lower activation energy.

Reaction rates

Molecules must physically collide with each other with enough energy in order for them to react. Which of the following would you expect might change the rate of a chemical reaction?

1) concentration of reactants
2) concentration of products
3) temperature
4) mixing speed
5) surface area of solid particles reacting with a liquid or gas

answer:

All of the above.

16.2 EXPRESSING THE REACTION RATE

The rate of a chemical reaction is a positive quantity that tells us how the concentration of a reactant or product changes with time. Let's consider the reaction between nitrogen dioxide and ozone, which destroys ozone in our atmosphere:

$$2NO_2(g) + O_3(g) \rightarrow N_2O_5(g) + O_2(g)$$

The rate of this reaction can be taken to be the disappearance of one of the reactants with time, or the appearance of one of the products with time:

$$\text{Rate} = \frac{\text{decrease in conc. of } O_3}{\text{time interval}} = \frac{-\Delta\text{conc. } O_3}{\Delta t} = \frac{-\Delta[O_3]}{\Delta t}$$

$$\text{Rate} = \frac{\text{increase in conc. of } O_2}{\text{time interval}} = \frac{\Delta\text{conc. } O_2}{\Delta t} = \frac{\Delta[O_2]}{\Delta t}$$

Since reactants disappear over time (conc. of O_3 at time 2 < conc. of O_3 at time 1), the rate equation contains a negative sign to make the reaction rate a positive number. We use square brackets to express concentration in moles per liter. The units of reaction rate depend on the unit of time:

mol/L s, mol/L min, mol/L day

Rate of reaction

Refer to the rates of reaction for the disappearance of O_3 and the appearance of O_2 (above) to write the rate of reaction in terms of the disappearance of NO_2 and the appearance of N_2O_5.

solution:

For every molecule of ozone that disappears, two molecules of nitrogen dioxide disappear (nitrogen dioxide disappears at twice the rate of ozone). The rate of the reaction written in terms of NO_2 is therefore 1/2 the rate of the reaction written in terms of O_3:

$$\text{Rate} = \frac{-\Delta\text{conc. } O_3}{\Delta t} = -\frac{1}{2}\frac{\Delta\text{conc. } NO_2}{\Delta t}$$

(continued on next page)

For every molecule of oxygen that appears, one molecule of N_2O_5 appears. Therefore, the rate of the reaction written in terms of N_2O_5 = the rate of the reaction written in terms of O_2:

$$\text{Rate} = \frac{\Delta\text{conc. } O_2}{\Delta t} = \frac{\Delta\text{conc. } N_2O_5}{\Delta t}$$

The value for the rate of reaction depends on which reaction component we monitor. The convention is to use a reactant or product with a coefficient of 1 as the reference.

Example

1) Express the reaction rate in terms of a change in concentration of each substance for the reaction of NO(g) with oxygen to produce $N_2O_3(g)$.

solution:

The balanced equation is:

$$4NO(g) + O_2(g) \rightarrow 2N_2O_3(g)$$

First, determine the change in moles of NO, O_2, and N_2O_3 between reactants and products:

$$\Delta n_{NO} = -4 \text{ mol}$$
$$\Delta n_{O_2} = -1 \text{ mol} \quad \text{(reference)}$$
$$\Delta n_{N_2O_3} = +2 \text{ mol}$$

Based on the above values, complete the equations:

$$-\Delta n_{O_2} = -(?)\,\Delta n_{NO} \quad \text{and,}$$

$$-\Delta n_{O_2} = (?)\,\Delta n_{N_2O_3}$$

answers: 1/4, 1/2.

Thus, $$\frac{-\Delta n_{O_2}}{\Delta t} = \frac{-(?)\,\Delta n_{NO}}{\Delta t} \quad \text{and,}$$

$$\frac{-\Delta n_{O_2}}{\Delta t} = \frac{(?)\Delta n_{N_2O_3}}{\Delta t}$$

answers: 1/4, 1/2.

We can now express the rate of the reaction in terms of each substance. O_2 is the reference since its coefficient is 1:

$$\text{Rate} = -\frac{[O_2]}{\Delta t} = -\frac{1}{4}\frac{[NO]}{\Delta t} = \frac{1}{2}\frac{[N_2O_3]}{\Delta t}$$

2) How fast does the concentration of oxygen decrease when N_2O_3 forms at 8.0×10^{-5} mol/L s?

solution:

Oxygen disappears at 1/2 the rate that N_2O_3 forms:

$$\text{Rate} = 1/2 \times (8.0 \times 10^{-5} \text{ mol/L s})$$
$$= 4.0 \times 10^{-5} \text{ mol/L s}$$

Reaction rates generally slow down during the course of a reaction as reactants are used up. (The higher the concentration of reactants, the more frequently they will collide and be converted to products.) The **instantaneous rate** is the rate of a reaction at a particular instant during the reaction. Often we measure the rate of a reaction when the reactants are mixed before any product accumulates so that the reverse reaction is negligible. This is the **initial rate** of a reaction, the rate of the reaction when $t = 0$.

Instantaneous reaction rate

If you measure the average rate of a reaction over the first 5 second period to be 0.006 mol/L s, and in the period between 5 and 10 seconds to be 0.002 mol/L s, what would you estimate would be the rate of the reaction at $t = 5$ seconds?

answer:

We might guess that the rate of the reaction at $t = 5$ seconds would be halfway between the two rates, or:

$$\frac{0.006 + 0.002}{2} = 0.004 \text{ mol/L s}$$

The more accurate way to estimate the reaction rate at this point would be to draw a curve of the change in reaction concentration with the change in time, and find the slope of the line tangent to the curve at $t = 5$ s:

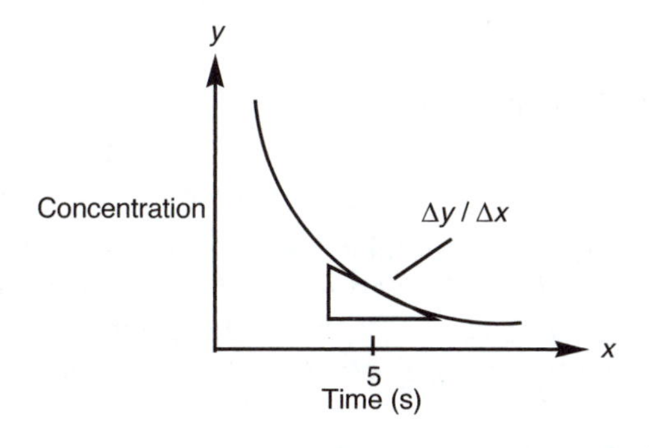

16.3 RATE LAWS

The **rate law** for a reaction tells us how reaction rate changes with the concentrations of the reactants and products. Consider the reaction between carbon monoxide and nitrogen dioxide to form carbon dioxide and nitrogen monoxide:

$$CO(g) + NO_2(g) \rightarrow CO_2(g) + NO(g)$$

We could determine how the rate of the reaction changes with, say, the concentration of carbon monoxide by measuring the initial rate of the reaction at several different concentrations of CO, holding the concentration of nitrogen dioxide constant. Likewise, we could study the dependence of the reaction rate on the concentration of NO_2 by changing its concentration and keeping the concentration of CO constant.

Rate law

1) Study the data from two sets of experiments on the next page to determine how the rate depends on the concentration of CO and NO_2. The data was taken at 400°C, and the rates have units of mol/L s:

(continued on next page)

Experiment #1			Experiment #2		
[CO]	$[NO_2]$	rate	[CO]	$[NO_2]$	rate
0.10	0.10	0.005	0.10	0.20	0.010
0.20	0.10	0.010	0.20	0.20	0.020
0.30	0.10	0.015	0.30	0.20	0.030
0.40	0.10	0.020	0.40	0.20	0.040

solution:

We see from the first experiment that when the concentration of CO doubles, the rate doubles: when [CO] increases from 0.10 to 0.20, the rate doubles from 0.005 to 0.010 mol/L s; when [CO] increases from 0.20 to 0.40, the rate doubles from 0.010 to 0.020 mol/L s.

We see from comparing experiments 1 and 2 that, in each case, when the concentration of NO_2 is doubled in experiment 2 compared to experiment 1, the rate also doubles.

We can conclude that the rate of the reaction is directly proportional to the concentrations of both CO and NO_2:

$$\text{Rate} = k[\text{CO}][\text{NO}_2]$$

We say the reaction is first order with respect to CO and NO_2. The overall reaction order is the sum of the individual orders, so the reaction is second order overall. k is the **rate constant** for the reaction and is a function of the temperature.

2) What is the value for k at 400°C for this reaction?

solution:

We calculate k by substituting experimental values into the rate equation. We could choose any set of conditions from experiments #1 and #2 above, and it is a good check on your calculated value of k to use more than one set of conditions. We'll use the values from the first set of conditions of experiment #2 (rate = 0.010 mol/L s; [CO] = 0.10 mol/L; $[NO_2]$ = 0.20 mol/L):

$$\text{Rate} = k[\text{CO}][\text{NO}_2]$$

$$0.010 \text{ mol/L s} = k(0.10 \text{ mol/L}) \times (0.20 \text{ mol/L})$$

$$k = \frac{0.010 \text{ mol/L s}}{(0.20 \text{ mol/L}) \times (0.10 \text{ mol/L})} = 0.50 \text{ L/mol s} \;\; (\text{or } 0.50\ M^{-1}\ \text{s}^{-1})$$

For the general reaction,

$$a\text{A}(g) + b\text{B}(g) \rightarrow \text{products}$$

the rate law takes the form:

$$\text{Rate} = k[\text{A}]^m[\text{B}]^n$$

Rate laws are determined by experiment; we cannot predict the rate law from the balanced equation. The exponents m and n, called reaction orders, are not necessarily related to the coefficients of the reaction a and b.

How can we measure the concentration of the reactants at various times while the reaction is in progress? We could remove a sample of the reaction mixture and titrate it, or, if one of the substances has a color, we could monitor the change in color over time. If either the reactants or the products are ionic, we could monitor the change in conductivity of the reaction mixture with electrodes. If a reaction involves a change in the number of moles of a gas, the rate could be determined from the change in pressure with time.

Following reactions

What method would you choose to study the change in concentration of the reactants or products for each of the following reactions?

1) $CH_3CHO(g) \rightarrow CH_4(g) + CO(g)$

2) $NO(g) + O_3(g) \rightarrow O_2(g) + NO_2(g)$

3) $CH_3COCH_3(l) + Br_2(l) \rightarrow CH_3COCH_3Br(l) + H^+(aq) + Br^-(aq)$

answers:

1) We could follow the rate of this reaction by measuring the change in pressure with time, since 1 mole of gas reacts to form 2 moles of gas.

2) We could follow the rate of this reaction spectrometrically by measuring the formation of brown $NO_2(g)$.

3) We could monitor this reaction rate by measuring the increase in conductivity as H^+ and Br^- ions form. More likely, this reaction would be monitored by measuring the disappearance of the dark red color from liquid bromine.

Determining rate laws

The Williamson synthesis is a major means to producing organic ethers from alkyl halides. The following Williamson synthesis at 25°C produced the rate data below:

$$CH_3I + NaOC_2H_5 \rightarrow CH_3OC_2H_5 + NaI$$

Use the data to determine the rate law for the reaction and the rate constant (k):

	Initial concentrations		
Exptl. Trial	$[CH_3I]$ (M)	$[NaOC_2H_5]$ (M)	Initial rate (M/s)
1	0.20	0.10	2.0×10^{-5}
2	0.40	0.10	4.0×10^{-5}
3	0.40	0.20	8.0×10^{-5}
4	0.10	0.20	2.0×10^{-5}

solution:

The general form of the rate equation is:

$$\text{Rate} = k[CH_3I]^m[NaOC_2H_5]^n$$

We can determine the value for m by comparing the rates from experiments 1 and 2, in which only the concentration of CH_3I changes:

(continued on next page)

$$\frac{\text{Rate}_2}{\text{Rate}_1} = \frac{4.0 \times 10^{-5}\ M\ s^{-1}}{2.0 \times 10^{-5}\ M\ s^{-1}} = \frac{k(0.40\ M)^m(0.10\ M)^n}{k(0.20\ M)^m(0.10\ M)^n}$$

$$2 = \frac{(0.40\ M)^m}{(0.20\ M)^m} = 2^m$$

$$m = 1$$

We can determine the value for n by comparing the rates from experiments 2 and 3, in which only the concentration of $NaOC_2H_5$ changes:

$$\frac{\text{Rate}_3}{\text{Rate}_2} = \frac{8.0 \times 10^{-5}\ M\ s^{-1}}{4.0 \times 10^{-5}\ M\ s^{-1}} = \frac{k(0.40\ M)^m(0.20\ M)^n}{k(0.40\ M)^m(0.10\ M)^n}$$

$$2 = \frac{(0.20\ M)^n}{(0.10\ M)^n} = 2^n$$

$$n = 1$$

The rate of this reaction is first order in both CH_3I and $NaOC_2H_5$.
The overall reaction order is $n + m = 2$. The rate law is:

$$\text{Rate} = k[CH_3I][NaOC_2H_5]$$

We can calculate the rate constant k from the results of any of the four experiments. Using the results from experiment 1 we have:

$$2 \times 10^{-5}\ M/s = k\,(0.20\ M)(0.10\ M)$$

$$2 \times 10^{-5}\ M/s = k\,(0.020\ M^2)$$

$$k = \frac{2 \times 10^{-5}\ M\ s^{-1}}{0.020\ M^2}$$

$$k = 1.0 \times 10^{-3}\ M^{-1}\ s^{-1}$$

It is a good idea to check the value for k with data from the other experiments. Note that the units for k depend on the order of the reaction and the units of time.

Often the exponents for rate equations are integers (0,1,2); in many reactions, however, they are fractions. When the product influences the reaction rate, the exponents for the rate equation may be negative meaning that an increase in that component decreases the reaction rate.

16.4 INTEGRATED RATE LAWS

The rate equation in the example above tells us how the rate of the reaction changes with concentration, but it does not tell us how the rate and concentrations change with time. Usually this question is the one of most practical interest: how much carbon monoxide is left after several minutes? several days?

The form of the equations relating concentration to time depends on the reaction order. We can integrate the rate laws over time to obtain the integrated rate laws.

Integrated Rate Laws

Zero order (rate = k):

$$[A]_0 - [A]_t = kt$$

First order (rate = k [A]):

$$\ln [A]_0 - \ln [A]_t = kt$$

Second order (rate = $k\,[A]^2$):

$$\frac{1}{[A]_t} - \frac{1}{[A]_0} = kt$$

Integrated rate laws

Rearrange the integrated rate laws to put them in the form of an equation for a straight line, $y = mx + b$. For each equation give the slope and the y-axis intercept.

solutions:

Zero order: $[A]_t = -kt + [A]_0$

A plot of $[A]_t$ vs. t will have a slope of $-k$ and a y-axis intercept of $[A]_0$.

First order: $\ln [A]_t = -kt + \ln [A]_0$

A plot of $\ln [A]_t$ vs. t will have a slope of $-k$ and a y-axis intercept of $\ln [A]_0$.

Second order: $\dfrac{1}{[A]_t} = kt + \dfrac{1}{[A]_0}$

A plot of $\dfrac{1}{[A]_t}$ vs. t will have a slope of $+k$ and a y-axis intercept of $\dfrac{1}{[A]_0}$.

If you wish to determine the reaction order from concentration-time data, it may take several plots. If you plot reactant vs. time and obtain a straight line, the reaction is zero order with respect to that reactant. If you plot ln [reactant] vs. time and obtain a straight line, the reaction is first order with respect to that reactant. If you plot 1/[reactant] vs. time, the reaction is second order with respect to that reactant.

A quick math review: The logarithm to the base e is called the natural logarithm (ln). It is the most convenient logarithm for use in calculus. The number "e" is an irrational number and equals 2.71828.... If you are working with the integrated first-order rate law ($\ln [A] = -kt + \ln [A]_0$) and wish to solve for the concentration of A, you raise both sides of the equation to the power of "e":

$$e^{\ln [A]} = [A] = e^{(-kt + \ln [A]_0)}$$

The decomposition of hydrogen peroxide

The decomposition of hydrogen peroxide is a useful way to obtain oxygen:

$$2H_2O_2(l) \rightarrow 2H_2O(l) + O_2(g)$$

It is a first-order reaction with a rate constant of 0.0410 min^{-1}.

1) If we start with a 3.00% hydrogen peroxide solution (a typical solution sold in grocery stores), what will be its concentration after 30.0 minutes?
2) How long will it take for its concentration to drop to 2.00%?
3) How long will it take for one-half of the sample to decompose?

solutions:

1) Since it is a first-order reaction, the integrated rate law is:

$$\ln [A]_0 - \ln [A]_t = 0.0410\ min^{-1} \times t\text{, or:}$$

$$\ln [A]_t = \ln [A]_0 - (0.0410\ min^{-1} \times t)$$

The concentration in mol/L of a 3.00% H_2O_2 solution is:

$$\frac{3.00\ g\ H_2O_2}{100.\ g\ water} \times \frac{mol}{34.0\ g\ H_2O_2} \times \frac{1\ g}{1\ mL} \times \frac{1000\ mL}{L} = 0.882\ mol/L$$

Its concentration after 30.0 min would be:

$$\ln [A]_{30.0\ min} = \ln (0.882\ mol/L) - (0.0410\ min^{-1} \times 30.0\ min)$$

$$[A]_{30.0\ min} = e^{[\ln (0.882) - (0.0410)(30.0)]} = 0.258\ mol/L$$

2) The concentration of a 2.00% H_2O_2 solution is 2/3 the concentration of a 3.00% solution:

$$0.882\ mol/L \times \frac{2}{3} = 0.588\ mol/L$$

The time it would take for its concentration to drop from 3.00% to 2.00% is:

$$t = \frac{\ln (0.882\ mol/L) - \ln (0.588\ mol/L)}{0.0410\ min^{-1}}$$

$t = 9.89$ min

3) The time it takes for half the sample (0.441 mol/L) to decompose is:

$$t = \frac{\ln (0.882\ mol/L) - \ln (0.441\ mol/L)}{0.0410\ min^{-1}}$$

$t = 16.9$ min

How useful is hydrogen peroxide as a way to obtain oxygen? It doesn't appear to be useful since the reaction is slow. However, the rate of decomposition of hydrogen peroxide can be increased greatly with yeast which catalyzes the reaction. Try it.

The time it takes for half of the hydrogen peroxide to decompose is its **half-life.** The half-life of a first-order reaction is independent of reactant concentration. We can see this from the integrated first-order rate law:

$$\ln[A]_0 - \ln[A]_t = kt$$

$$\ln[A]_0 - \ln\frac{1}{2}[A]_0 = kt_{\frac{1}{2}}$$

$$\ln\frac{[A]_0}{\frac{1}{2}[A]_0} = kt_{\frac{1}{2}}$$

$$\ln 2 = kt_{\frac{1}{2}}$$

$$t_{\frac{1}{2}} = \frac{\ln 2}{k} = \frac{0.693}{k} \quad \text{(first-order process)}$$

Half-life of hydrogen peroxide solution

1) How long does it take for one-half of a 0.882 *M* hydrogen peroxide solution to decompose? (The decomposition of hydrogen peroxide is a first-order reaction with $k = 0.0410\ \text{min}^{-1}$.)
2) How long does it take for one-half of a 0.441 *M* hydrogen peroxide solution to decompose?

solution:

1) Since the decomposition of hydrogen peroxide is a first-order reaction, we use the equation:

$$t_{\frac{1}{2}} = \frac{0.693}{k}$$

$$t_{\frac{1}{2}} = \frac{0.693}{0.0410\ \text{min}^{-1}} = 16.9\ \text{min}$$

2) The half-life is independent of starting concentration, so the amount of time it takes for half of any concentration of hydrogen peroxide to decompose is 16.9 min.

Half-Life Equations

Zero order (rate = k):

$$t_{\frac{1}{2}} = \frac{1}{2}\frac{[A]_0}{k}$$

First order (rate = $k[A]$):

$$t_{\frac{1}{2}} = \frac{0.693}{k}$$

Second order (rate = $k[A]^2$):

$$t_{\frac{1}{2}} = \frac{1}{k[A]_0}$$

16.5 THEORIES OF CHEMICAL KINETICS

Chemists use two models—collision theory and transition state theory—to explain the effects of concentration and temperature on reaction rate.

Most chemical reactions proceed faster as temperature increases. We store food in the refrigerator to slow down chemical reactions that spoil food. We use pressure cookers to increase the temperature at which food cooks. An approximate rule is that an increase of 10°C doubles the reaction rate.

Recall from Chapter 5 that an increase in temperature increases the average kinetic energy of molecules. At higher temperatures, a greater fraction of collisions between molecules have enough energy to overcome the energy barrier to reaction. We can think of this energy barrier as the amount of energy required to weaken the bonds holding reactant molecules together allowing them to rearrange to form products. This energy barrier is called **activation energy, E_a.**

Not all collisions that have enough energy to overcome the energy barrier result in reaction. The molecules must have the correct orientation to react. The Arrhenius equation relates temperature, activation energy, and molecular orientation to the rate constant, k:

$$\ln k = \frac{-E_a}{R} \times \frac{1}{T} + \ln A$$

where: k is the rate constant

T is the absolute temperature

R is the universal gas constant

E_a is the activation energy

A is a term which takes into account the frequency of collisions, and the fraction of collisions with an effective orientation.

If we measure the rate constant of a reaction at different temperatures, we can determine the energy of activation by plotting $\ln k$ versus $1/T$. The slope of this line is $-E_a/R$.

Exercise

The reaction: $2NOCl(g) \rightarrow 2NO(g) + Cl_2(g)$

has an activation energy, E_a, of 1.00×10^2 kJ/mol, and a rate constant of 0.286 s^{-1} at 500. K. What is the rate constant at 400. K?

solution:

We know: $T_1 = 500.$ K

$T_2 = 400.$ K

$k_1 = 0.286\ s^{-1}$

$E_a = 100$ kJ/mol $= 1.00 \times 10^5$ J/mol

Use the Arrhenius equation to write equations for k_1 and k_2:

$$\ln k_2 = \frac{-E_a}{R} \times \frac{1}{T_2} + \ln A, \text{ and } \ln k_1 = \frac{-E_a}{R} \times \frac{1}{T_1} + \ln A$$

If we subtract k_1 from k_2, we obtain the expression:

(continued on next page)

$$\ln \frac{k_2}{k_1} = \frac{-E_a}{R}\left(\frac{1}{T_2} - \frac{1}{T_1}\right)$$

Now we can substitute values into the equation and solve for k_2:
(Make sure you feel comfortable with the math in these problems.)

$$\ln \frac{k_2}{0.286\ s^{-1}} = \frac{-1.00 \times 10^5\ J/mol}{8.3148\ J/K\ mol}\left(\frac{1}{400.\ K} - \frac{1}{500.\ K}\right)$$

$$\ln \frac{k_2}{0.286\ s^{-1}} = -6.01$$

$$\frac{k_2}{0.286\ s^{-1}} = 0.00245$$

$$k_2 = 6.99 \times 10^{-4}\ s^{-1}$$

Let's see how well the approximation holds that a 10°C increase in temperature doubles the rate of the reaction. The temperature change between 500 K and 400 K is 100 K, or ten, 10 degree temperature changes. If we double the rate constant k_2 (6.99×10^{-4}) 10 times, we obtain 0.715. This number does not equal k_1 (0.286), but it is the same order of magnitude.

Collision theory and transition state theory

Choose words from the list below to complete the following paragraphs:

(a) activated complex	(d) energy	(g) products
(b) activation energy	(e) one	(h) reversible
(c) collide	(f) orientation	(i) speed

Molecules must _____________(1) in order to react. Most collisions do not result in the formation of _____________(2). The number of collisions is usually much, much larger than the rate of the reaction. Molecules must collide with enough _____________(3) and with the proper _____________(4) to react. An increase in temperature increases the average _____________(5) of particles, and therefore the frequency which they _____________(6). The effect of temperature on reaction rate, however, is primarily due to the increase in the fraction of collisions with enough energy to exceed the _____________(7) threshold. In the Arrhenius equation, the factor "A" is the product of the collision frequency (Z) and an orientation probability factor (p) which is specific for each reaction. Collisions between individual atoms have p values near _____________(8) since almost no matter how they collide, they react.

When molecules collide in an effective orientation and have enough energy to overcome the activation energy, they form a high energy species with partial bonds called an _____________(9), or transition state. Every reaction goes through a transition state from which point it can either continue on to form product, or fall back apart into reactants. This theory implies that all reactions are _____________(10).

answers: 1-c, 2-g, 3-d, 4-f, 5-i, 6-c, 7-b, 8-e, 9-a, 10-h.

Figure 16.20 on page 704 in your textbook shows the relative energy levels of the reactants, the transition state, and the products as the reaction of methyl bromide with hydroxide ion progresses. The **activation energy** for the forward reaction is the difference in energy between the reactants and the transition state. The **activated complex** formed during the transition from reactants to products cannot be isolated, but its form can be predicted based on knowledge of the reaction mechanism.

16.6 REACTION MECHANISMS

A reaction mechanism describes a sequence of elementary steps by which a reaction occurs. Usually the elementary steps are **unimolecular** (one molecule decomposing), or **bimolecular** (two molecules combining). A third less likely event is a **termolecular** reaction (three particles colliding). For elementary reactions, the reaction order equals the number of reactants. The slowest elementary step limits how fast the overall reaction proceeds and is called the **rate-limiting step.**

Atmospheric chemistry

Nitrogen oxide, a pollutant from automobile gas, reacts with oxygen to produce the colorless gas, N_2O_4, which is in equilibrium with the brownish gas, NO_2. The reaction mechanism is:

$NO(g) + O_2(g) \rightleftharpoons NO_3(g)$ pre-equilibrium

$NO(g) + NO_3(g) \rightarrow N_2O_4(g)$ rate determining step

$N_2O_4(g) \rightleftharpoons 2NO_2(g)$ equilibrium

What is the rate law for this reaction mechanism?

solution:

We use the rate-limiting step to write an expression for the rate:

$$\text{Rate} = k_2[NO][NO_3]$$

The rate law should not contain intermediate species (such as NO_3), so we use the pre-equilibrium to write the concentration of NO_3 in terms of reactants. In an equilibrium, the forward and reverse reaction rates are equal:

$$k_1[NO][O_2] = k_{-1}[NO_3]$$

$$[NO_3] = \frac{k_1}{k_{-1}}[NO][O_2]$$

Now substitute this expression for $[NO_3]$ into the rate law for the slow step:

$$\text{Rate} = \frac{k_2 k_1}{k_{-1}}[NO]^2[O_2]$$

Often all the constants are grouped together into one constant: $k = \frac{k_2 k_1}{k_{-1}}$, and

$$\text{Rate} = k[NO]^2[O_2]$$

The rate law is second order in NO. In the lab, if you squirt NO into the air, the brown color of NO_2 appears almost immediately. The reaction is much slower in the atmosphere owing to the low concentration of NO, and the second order dependence of reaction.

Reaction pathway

The reaction of hydrogen with iodine produces hydrogen iodide or hydroiodic acid:

$$H_2(g) + I_2(g) \rightarrow 2HI(g)$$

The experimentally determined rate law is: rate = $k[H_2][I_2]$. Which of the following mechanisms is consistent with this rate law?

1) proposed pathway #1:

$$H_2 + I_2 \rightarrow 2HI$$

2) proposed pathway #2:

$$I_2 \rightarrow 2I \text{ (slow)}$$
$$H_2 + 2I \rightarrow 2HI \text{ (fast)}$$

3) proposed pathway #3:

$$I_2 \rightleftharpoons 2I \text{ (fast)}$$
$$H_2 + 2I \rightarrow 2HI \text{ (slow)}$$

4) proposed pathway #4:

$$I_2 \rightleftharpoons 2I \text{ (fast)}$$
$$H_2 + I \rightleftharpoons H_2I \text{ (fast)}$$
$$H_2I + I \rightarrow 2HI \text{ (slow)}$$

answer:

The rate laws for the four proposed pathways are:

1) Rate = $k[H_2][I_2]$
2) Rate = $k[I_2]$
3) The rate law for the slow step is: Rate = $k_2[H_2][I]^2$
   For the first step, the forward and reverse reaction rates are equal at equilibrium:

$$k_1[I_2] = k_{-1}[I]^2$$

$$[I]^2 = \frac{k_1}{k_{-1}}[I_2]$$

   Substituting the expression for $[I]^2$ into the rate law for the slow step gives:

$$\text{Rate} = k_2[H_2]\frac{k_1}{k_{-1}}[I_2]$$

$$\text{Rate} = k[H_2][I_2] \text{ where } k = \frac{k_2 k_1}{k_{-1}}$$

(continued on next page)

4) The rate law for the slow step is: Rate = $k_3[H_2I][I]$

The first step is reversible, so we have:

$$k_1[I_2] = k_{-1}[I]^2$$

$$[I]^2 = \frac{k_1}{k_{-1}}[I_2] \quad \text{or} \quad [I] = \left(\frac{k_1}{k_{-1}}[I_2]\right)^{\frac{1}{2}}$$

Using the same approach with the second step we have:

$$[H_2I] = \frac{k_2}{k_{-2}}[H_2][I]$$

and substituting the relationship for [I]:

$$[H_2I] = \frac{k_2}{k_{-2}}[H_2]\left(\frac{k_1}{k_{-1}}[I_2]\right)^{\frac{1}{2}}$$

Substituting the expressions for the intermediates into the rate law for the slow step gives:

$$\text{Rate} = \frac{k_3k_2}{k_{-2}}[H_2]\left(\frac{k_1}{k_{-1}}[I_2]\right)^{\frac{1}{2}} \times \left(\frac{k_1}{k_{-1}}[I_2]\right)^{\frac{1}{2}}$$

$$\text{Rate} = k[H_2][I_2] \text{ where } k = \frac{k_3\,k_2\,k_1}{k_{-1}\,k_{-2}}$$

Proposed pathways #1, #3, and #4 are consistent with the experimentally determined rate law. Which pathways are consistent with the spectroscopic detection of free iodine atoms?

answer:

Mechanisms #3 and #4 both propose the existence of free iodine atoms. Mechanism #4 is a more likely reaction pathway than #3 because pathway #3 includes a less likely termolecular reaction between a hydrogen molecule and two free iodine atoms. We cannot prove that a particular mechanism is the one actually taken during the reaction, we simply use mechanisms which have no data to refute them.

- The overall rate law includes only the species up to and including those in the rate determining step. For reaction pathway #2, in the example above, the species in the fast step after the slow step are not included in the rate law.
- Reaction intermediates do not appear in the overall rate law.

16.7 CATALYSIS: SPEEDING UP A REACTION

A **catalyst** is a substance that increases the rate of reaction without being consumed in the reaction. A catalyst speeds up a reaction by providing a different reaction mechanism with lower activation energy:

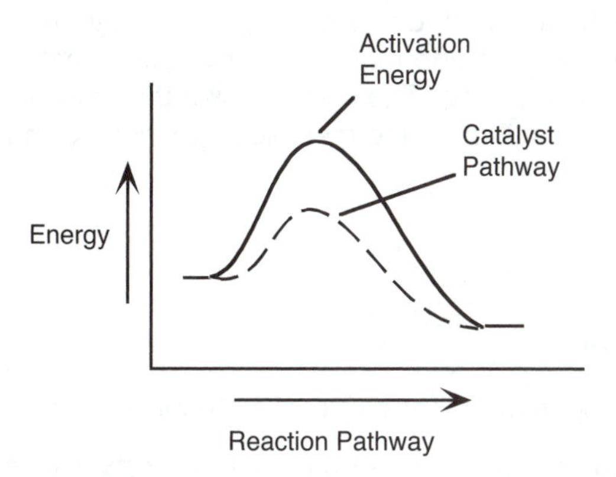

Catalysts speed up both the forward and reverse reactions, so using a catalyst does not give more product, but gives the product more quickly. **Homogeneous catalysts** exist in solution with the reaction mixture. Figure 16.23 on page 713 in your textbook shows a reaction energy diagram for a catalyzed and an uncatalyzed process. **Heterogeneous catalysts** exist as a separate phase from the reaction. Often the catalyst is a solid interacting with gases or liquids. **Enzymes,** which have features of both homogeneous and heterogeneous catalysts, are incredibly efficient biological catalysts.

One method to study the mechanisms of catalytic reactions is to carry out the reactions on single-crystal surfaces under vacuum conditions (to keep the surface of the crystal clean). Reactants introduced into the vacuum chamber are studied spectroscopically to determine how they adsorb to the metal surface, and the reaction steps they undergo to form products. (See, for example, Figure 16.21 on page 532 in your text depicting the metal-catalyzed hydrogenation of ethylene.) It is hoped that the reaction mechanisms on these crystal surfaces under vacuum conditions mimic the mechanisms of heterogeneous catalysis occurring at ambient pressures.

Enzymes are both efficient and specific catalysts. An enzyme is a protein with an **active site,** a small region of the protein with a specific shape, which binds to a reactant molecule, or **substrate.** The active site can be thought of as a **lock** and the substrate a **key,** or, because there is evidence the enzyme may change shape when the substrate lands at its active site, the interaction can be thought of as an **induced-fit model.** In binding to the substrate, the enzyme stabilizes the transition state species, lowering the reaction's activation energy, and increasing the reaction rate. Virtually every cell reaction is catalyzed by its own specific enzyme.

Whereas catalysis by enzymes is essential to life, catalysis in nature can also cause problems. The breakdown of ozone in the stratosphere by chlorine atoms from chlorofluorocarbons (CFC's) has depleted the protective ozone layer that absorbs UV radiation from the sun. The chlorine atoms, which form when UV radiation in the upper atmosphere cleaves chlorofluorocarbons, act as homogeneous catalysts to the ozone breakdown reaction reaction. Clouds and dust provide a surface that speeds formation of chlorine atoms, so heterogeneous catalysis also plays a role. CFC's have been phased out, but full recovery of the ozone layer will be slow.

Let's return to our statement at the chapter outset that even though graphite is more energetically favorable than diamond, it seems to us on our human timescale that, "diamonds are forever." It turns out that on a geologic timescale, diamonds practically are forever. Scientists recently unearthed diamonds, entrapped in crystals of zircon, more than 4 billion years old—nearly as old as the earth, itself. Although graphite is the most stable form of carbon under normal conditions, at high pressures (as those found in the interior of the earth), diamond becomes the more stable form. The carbon atoms in diamond, on average, are closer together than the carbon atoms in graphite, making diamond denser than graphite, and more stable at extremely high pressures. At lower pressures, diamonds spontaneously change to graphite, but the high activation energy of this process causes the kinetics to be extremely slow.

Synthetic diamonds, which can be made by heating graphite to temperatures >1500°C and at about 60,000 atmospheres of pressure, can be made more easily by using small amounts of metallic elements, such as iron or nickel, which act as catalysts. The metals lower the activation energy, speeding up the reaction. As the diamond lattice forms, it traps the metallic impurities, giving many synthetic diamonds a color.

CHAPTER CHECK

Make sure you can...

- List factors that might change the rate of a chemical reaction.
- Express the rate of a reaction in terms of the disappearance of reactants and the appearance of products.
- Write a rate law from experimental data and find *k*, the rate constant, at a specified temperature.
- Suggest experimental methods for following the rate of a reaction.
- Use integrated rate laws to determine how reaction rates and concentrations change with time.
- Use the half-life equations for zero-, first-, or second-order reactions, to calculate how much time it takes for half of a reactant to disappear.
- Use the Arrhenius equation to calculate the rate constant, *k*, at different temperatures.
- Write a rate law from a proposed reaction mechanism.
- Determine if a proposed reaction pathway is consistent with an experimentally determined rate law.
- Define homogeneous and heterogeneous catalysts.
- Explain, using collision theory and transition state theory, how catalysts speed up forward and reverse reaction rates.

Chapter 16 Exercises

16.1) The portable breathalyzers used to check suspects for drunk driving use the oxidation of ethyl alcohol by dichromate ion in the test:

$$3C_2H_5OH + Cr_2O_7^{2-} + 8H^+ \rightarrow 3CH_3CHO + 2Cr^{3+} + 7H_2O$$

The results of the rate law study, at 25°C are given below:

Exptl. Trial	Initial Concentration (M) $[C_2H_5OH]$	$[Cr_2O_7^{2-}]$	$[H^+]$	Initial Rate (M min^{-1})
1	0.15	0.050	0.010	5.0×10^{-4}
2	0.30	0.050	0.010	1.0×10^{-3}
3	0.30	0.100	0.010	2.0×10^{-3}
4	0.30	0.050	0.100	1.0×10^{-3}

a. Determine the rate law, including the rate constant (*k*) for the reaction.

b. What is the overall order of the reaction?

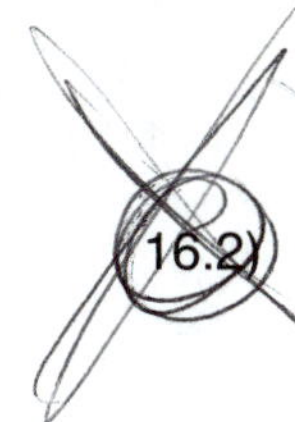

16.2) The hydrolysis reaction:

$$C_6H_5C(CH_3)_2Cl + H_2O \longrightarrow C_6H_5C(CH_3)_2OH + HCl$$

produces the experimental rate law data:

Exptl. Trial	Initial Concentration (M) $[C_9H_{11}Cl]$	$[H_2O]$	Initial Rate (M min^{-1})
1	0.12	0.20	1.4×10^{-3}
2	0.36	0.20	4.3×10^{-3}
3	0.36	0.30	4.3×10^{-3}
4	0.24	0.10	2.9×10^{-3}

a. Determine the rate law, including the rate constant, k, for the reaction.
b. What is the overall order of the reaction?

16.3) Under high temperature conditions, the following rate law data was obtained for the reaction below:

Exptl. Trial	Initial Concentration (M) $[NO_2]$	$[CO]$	Initial Rate (M min^{-1})
1	0.050	0.050	0.031
2	0.075	0.050	0.070
3	0.075	0.10	0.070

$$NO_2(g) + CO(g) \rightarrow NO(g) + CO_2(g)$$

a. Determine the rate law, including the rate constant, k, for the reaction.
b. What is the overall order of the reaction?

16.4) The rate of gas phase dehydration of t-butyl alcohol is temperature dependent:

$$C_4H_{10}O(g) \rightarrow C_4H_8(g) + H_2O(g)$$

Temperature (K)	k (1/sec)
600.	4.5×10^{-9}
800.	4.1×10^{-4}
1000.	0.39
1200.	38

Calculate the activation energy, E_a, and the Arrhenius constant, A, for the reaction.

16.5) The rate of decarboxylation of ethyl chloroformate is temperature dependent:

$$ClCOOC_2H_5 \rightarrow CO_2 + C_2H_5Cl$$

Temperature (K)	k (1/sec)
473	0.0013
523	0.026
573	0.31
623	2.43

Calculate the activation energy, E_a, and the Arrhenius constant, A, for the reaction.

16.6) Radioactive isotopes undergo a first-order rate of decay. Iodine-131 has a half-life of 8.0 days.

a. What is the rate constant for the decay?
b. Calculate the time required for a 200. mg sample of ^{131}I to decay to 10. mg.

16.7) The dimerization of tetrafluoroethylene at 403 K proceeds according to the rate law:
Rate = $1.6 \times 10^{-3}\ M^{-1}\ s^{-1}[C_2F_4]^2$

a. 0.80 mole of C_2F_4 is injected into a 1.0 L reaction chamber at 403 K. How much C_2F_4 is left after 1 hour?
b. What is the half-life of the reaction?

16.8) Oxygen difluoride undergoes decomposition at elevated temperatures. The concentration vs. time data at 523 K is given in the table below:

Time (min)	0	2	4	6	8	10	12
$[OF_2]$ *M*	0.200	0.150	0.120	0.100	0.085	0.075	0.066

Graph the data to determine the order of reaction and the rate constant, *k*.

16.9) Sulfuryl chloride undergoes decomposition at elevated temperatures. The concentration vs. time data at 593 K is given in the table below:

Time (h)	0	2	4	6	8	10	12
$[SO_2Cl_2]$ *M*	0.50	0.43	0.36	0.31	0.27	0.23	0.19

Graph the data to determine the order of reaction and the rate constant, *k*.

16.10) For many exothermic reactions, a 10°C increase in temperature will double the rate of the reaction. Explain using the collision theory and transition rate theory.

16.11) The addition of a catalyst will increase the rate of a chemical reaction. Use the collision theory and transition state theory to explain why this occurs.

Chapter 16 Answers

16.1) a. When the concentration of C_2H_5OH doubles from 0.15 *M* to 0.30 *M*, the rate doubles from $5.0 \times 10^{-4}\ M\ min^{-1}$ to $1.0 \times 10^{-3}\ M\ min^{-1}$ (experiments 1 & 2; first-order dependence).

When the concentration of $Cr_2O_7^{2-}$ doubles from 0.050 *M* to 0.10 *M*, the rate doubles from 1.0×10^{-3} to $2.0 \times 10^{-3}\ M\ min^{-1}$ (experiments 2 & 3; first-order dependence).

When the concentration of H^+ increases 10-fold from 0.010 *M* to 0.100 *M*, the rate stays the same at $1.0 \times 10^{-3}\ M\ min^{-1}$ (experiments 2 & 4; no dependence).

The rate law expression is: rate = $k[C_2H_5OH][Cr_2O_7^{2-}]$

We'll choose the data from experiment 1 to solve for *k*:

$$5.0 \times 10^{-4}\ M\ min^{-1} = k \times 0.15\ M \times 0.050\ M$$
$$k = 0.067\ M^{-1}\ min^{-1}$$

(You can check this answer by using data from experiments 2, 3, or 4.)

Rate = $6.7 \times 10^{-2}\ M^{-1}\ min^{-1}[C_2H_5OH][Cr_2O_7^{2-}]$

b. The reaction is second order.

16.2) a. When the concentration of $C_9H_{11}Cl$ triples from 0.12 to 0.36 M, the rate triples from 1.4×10^{-3} to 4.3×10^{-3} M min^{-1} (experiments 1 & 2; first-order dependence).

When the concentration of H_2O increases from 0.2 to 0.3 M, the rate remains unchanged at 4.3×10^{-3} M min^{-1} (experiments 2 & 3; no dependence).

Since the reaction rate is independent of the water concentration, and first-order with respect to $C_9H_{11}Cl$, we would expect that when the concentration of $C_9H_{11}Cl$ doubles from 0.12 to 0.24 M, as it does in experiments 1 & 4, the rate should double, as it does, from 1.4×10^{-3} to 2.9×10^{-3} M min^{-1}.

The rate expression is: rate = $k[C_9H_{11}Cl]$

We'll use the data from experiment 1 to solve for k:

$$1.4 \times 10^{-3}\ M\ \text{min}^{-1} = k \times 0.12\ M$$
$$k = 0.012\ \text{min}^{-1}$$

Rate = 1.2×10^{-2} min^{-1} $[C_9H_{11}Cl]$

b. The reaction is first order.

16.3) a. The general form of the rate equation is:

$$\text{Rate} = k[NO_2]^m[CO]^n$$

We can determine the value of m by comparing the rates from experiments 1 and 2, in which only the concentration of NO_2 changes:

$$\frac{\text{Rate}_2}{\text{Rate}_1} = \frac{0.070\ M\ \text{min}^{-1}}{0.031\ M\ \text{min}^{-1}} = \frac{k(.075\ M)^m(.050\ M)^n}{k(.050\ M)^m(.050\ M)^n}$$

$$2.26 = 1.5^m$$

$$m = 2$$

We can determine the value of n by comparing the rates from experiments 2 and 3, in which only the concentration of CO changes:

$$\frac{\text{Rate}_3}{\text{Rate}_2} = \frac{0.070\ M\ \text{min}^{-1}}{0.070\ M\ \text{min}^{-1}} = \frac{k(.075\ M)^m(0.10\ M)^n}{k(.075\ M)^m(.050\ M)^n}$$

$$1 = 2^n$$

$$n = 0$$

We'll use data from experiment 1 to solve for k :

$$\text{rate} = k[NO_2]^2$$
$$0.031\ M\ \text{min}^{-1} = k \times (0.050\ M)^2$$
$$k = 12\ M^{-1}\ \text{min}^{-1}$$

Rate = $12\ M^{-1}$ min$^{-1}[NO_2]^2$

b. The reaction is second order.

16.4) Use the Arrhenius equation to write equations for k_1 and k_2:

$$\ln k_2 = \frac{-E_a}{R} \times \frac{1}{T_2} + \ln A, \text{ and } \ln k_1 = \frac{-E_a}{R} \times \frac{1}{T_1} + \ln A$$

If we subtract k_1 from k_2, we obtain the expression:

$$\ln \frac{k_2}{k_1} = \frac{-E_a}{R}\left(\frac{1}{T_2} - \frac{1}{T_1}\right)$$

Substitute values for k_2, k_1 and T_2 and T_1 and solve for E_a:

$$\ln \frac{4.1 \times 10^{-4}\,s^{-1}}{4.5 \times 10^{-9}\,s^{-1}} = \frac{-E_a}{8.3148\ J/K\ mol}\left(\frac{1}{800\ K} - \frac{1}{600\ K}\right)$$

$E_a = 2.3 \times 10^5$ J or 230 kJ

Use this value of E_a in the Arrhenius equation to solve for A:

$$\ln 4.5 \times 10^{-9}\,s^{-1} = \frac{-2.3 \times 10^5\,J}{8.3148\ J/K\ mol} \times \frac{1}{600\ K} + \ln A$$

$A = 4.7 \times 10^{11}$

16.5) $E_a = 120$ kJ; $A = 5.0 \times 10^{10}$

16.6) a. For a first-order rate of decay: $t_{1/2} = 0.693/k$; $k = 8.7 \times 10^{-2}\,day^{-1}$

b. 34 days

Check: One half-life (8 days) leaves 100 mg of ^{131}I; 2 half-lives (16 days) leaves 50 mg; 3 half-lives (24 days) leaves 25 mg; 4 half-lives (32 days) leaves 12.5 mg, so 34 days appears to be correct for leaving 10 mg of ^{131}I.

16.7) a. The reaction is second order, so the integrated rate law is:

$$\frac{1}{[A]_t} - \frac{1}{[A]_0} = kt$$

Solving for $[A]_{1h}$ gives:

$$\frac{1}{[A]_{1h}} - \frac{1}{0.8\ M} = (1.6 \times 10^{-3} M^{-1}\,s^{-1}) \times (3600\ s)$$

$$\frac{1}{[A]_{1h}} = 5.76\ M^{-1} + \frac{1}{0.8\ M} = 7.01\ M^{-1}$$

$[A]_{1h} = 0.14$ mol

b. Use the second-order half-life equation:

$$t_{\frac{1}{2}} = \frac{1}{k[A]_0} = \frac{1}{1.6 \times 10^{-3} M^{-1} s^{-1} \times 0.8\ M}$$

$t_{1/2} = 780$ s

16.8) Plot of $1/[OF_2]$ vs. time (min) gives slope = $k = 0.843\ M^{-1}\,min^{-1}$

16.9) Plot of $\ln [SO_2Cl_2]$ vs. time (h) gives slope = $-k = -0.079\ h^{-1}$

16.10) a. An increase in temperature will increase the kinetic energy thus increase molecular speed. The increase in molecular speed will increase the frequency of collision which is one of the factors in the collision theory of reaction rates.

b. An increase in temperature will increase the fraction of molecules with sufficient energy to break and/or make bonds (E_a).

16.11) A catalyst will increase the rate of a reaction by changing the reaction pathway, enhancing the proper orientation of the reactants and/or lowering the activation energy by weakening existing bonds.

17

Equilibrium: The Extent of Chemical Reactions

17.1 THE EQUILIBRIUM STATE AND THE EQUILIBRIUM CONSTANT

We saw in Chapters 12 and 13 that a closed container of water is at **equilibrium** when the rate that water molecules escape from the liquid equals the rate that water molecules in the vapor state return to the liquid water. Although it is a dynamic system (water molecules continually move between the liquid and gaseous states), we see the system as static since the water level does not change. The chemical equation for the equilibrium is:

$$H_2O(l) \rightleftharpoons H_2O(g)$$

The position of this equilibrium (the relative amounts of the liquid and gas) depends on the temperature of the system. At a particular temperature, the number of gaseous water molecules is fixed.

Vapor pressure of water

1) What is the equilibrium vapor pressure in each 1 L beaker shown below at 50°C? at 100°C? What is the equilibrium concentration of gaseous water in each beaker at 50°C? at 100°C?

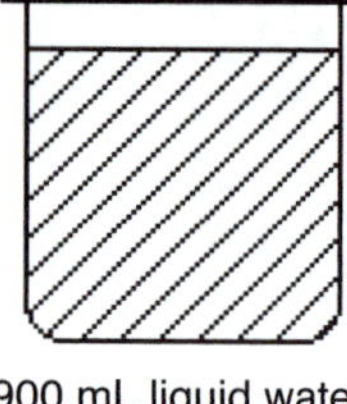

900 mL liquid water

(a)

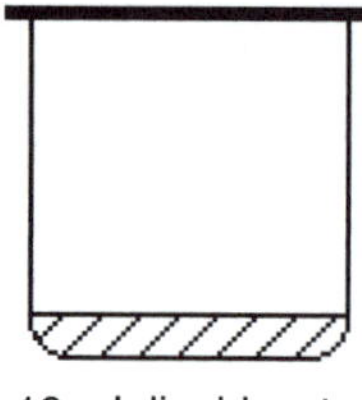

10 mL liquid water

(b)

1 mL liquid water

(c)

(continued on next page)

solutions:

The vapor pressure in each beaker at 50°C is 92.51 mm Hg (0.122 atm).

The vapor pressure in each beaker at 100°C is 760 mm Hg (1 atm).
(You can look these values up in the Handbook of Chemistry and Physics.)

We can use the ideal gas law to calculate the equilibrium concentration of $H_2O(g)$ at 50°C and 100°C:

$PV = nRT$ so $n = PV/RT$

at 50°C, $$n = \frac{(0.122\ \text{atm})\ (1\ \text{L})}{(0.0821\ \text{L atm/mol K})\ (323\ \text{K})}$$

$n = 0.00460$ mol

The concentration of $H_2O(g)$ is therefore 0.0046 mol/L.

at 100°C, $$n = \frac{(1\ \text{atm})\ (1\ \text{L})}{(0.0821\ \text{L atm/mol K})\ (373\ \text{K})}$$

$n = 0.0327$ mol and concentration = 0.0327 mol/L

2) Does 1 mL of water in a 1 liter container at 100°C maintain an equilibrium between the liquid and gaseous states, or does all the liquid water evaporate?

solution:

Moles of water molecules in 1 mL of water are:

$$1\ \text{mL} \times \frac{1\ \text{g}}{\text{mL}} \times \frac{1\ \text{mol}}{18\ \text{g}} = 0.055\ \text{mol}$$

Since the equilibrium concentration of gaseous water at 100°C is 0.0327 mol/L (Problem #1 above), one mL of liquid water contains more than enough water molecules to provide this concentration, and not all of the water evaporates.

3) What is the vapor pressure of 1/2 mL of water at 100°C in a 1 liter flask?

solution:

Moles of water molecules in 0.5 mL water are:

$$0.5\ \text{mL} \times \frac{1\ \text{g}}{\text{mL}} \times \frac{1\ \text{mol}}{18\ \text{g}} = 0.0278\ \text{mol}$$

We use the ideal gas law to solve for pressure, P:

$$P = \frac{nRT}{V}$$

$$P = \frac{(0.0278\ \text{mol})\ (0.0821\ \text{L atm/mol K})\ (373\ \text{K})}{1\ \text{L}}$$

$P = 0.851$ atm

The vapor pressure falls below 1 atm because all the water in the flask evaporates, and there is not enough to reach a pressure of 1 atm. (There must be about 0.6 mL of water in a 1 L flask to establish an equilibrium between the liquid and gaseous phases.)

For a given temperature, the equilibrium position does not depend on the amount of liquid water present, and we need to know only the concentration of gaseous water molecules to give us the equilibrium position.

Many chemical systems have more than one gaseous species present. In these systems, we cannot express the equilibrium concentration of one species independent from the others. The equilibrium between N_2O_4 (a colorless gas), and NO_2 (a brown gas) is described in your text beginning on page 732. The system reaches equilibrium when the forward and reverse reaction rates become equal and therefore the concentrations of reactants and products stop changing:

$$N_2O_4(g) \rightleftharpoons 2NO_2(g)$$

At a given temperature, we could obtain an infinite number of equilibrium concentrations of N_2O_4 and NO_2 depending on how we prepare the system. It turns out that there is a mathematical relationship between all sets of concentrations we could obtain. The quotient,

$$\frac{[NO_2]^2}{[N_2O_4]}$$

equals 0.36 at 100°C regardless of the amounts of reaction species we start with, the volume of the container, or the total pressure. We call this constant $\boldsymbol{K_c}$**,** the **equilibrium constant.** At 150°C, the equilibrium constant for this reaction is 3.2. The brackets indicate concentrations are in moles per liter (*M*), and the subscript "c" on the equilibrium constant tells us the equilibrium is expressed in terms of concentrations. Pages 732 and 733 in your text derive the above quotient using the forward and reverse rate laws.

The numerator of the quotient contains the concentration of the product raised to the power of its coefficient in the balanced equation (2), and the denominator of the quotient contains the concentration of the reactant raised to the power of its coefficient in the balanced equation (1). For every gaseous system, we can write an analogous expression for the equilibrium constant.

Equilibrium constant expressions

1) Write the equilibrium constant expression for each of the following reactions:

1) $N_2(g) + 3H_2(g) \rightleftharpoons 2NH_3(g)$
2) $C_2H_6(g) + 7/2O_2(g) \rightleftharpoons 2CO_2(g) + 3H_2O(g)$
3) $2C_2H_6(g) + 7O_2(g) \rightleftharpoons 4CO_2(g) + 6H_2O(g)$
4) $2HI(g) \rightleftharpoons H_2(g) + I_2(g)$
5) $H_2(g) + I_2(g) \rightleftharpoons 2HI(g)$
6) $CaCO_3(s) \rightleftharpoons CaO(s) + CO_2(g)$

answers:

1) $K_c = \dfrac{[NH_3]^2}{[N_2][H_2]^3}$

2) $K_c = \dfrac{[CO_2]^2[H_2O]^3}{[C_2H_6][O_2]^{\frac{7}{2}}}$

3) $K_c = \dfrac{[CO_2]^4[H_2O]^6}{[C_2H_6]^2[O_2]^7}$

(continued on next page)

4) $K_c = \frac{[H_2][I_2]}{[HI]^2}$ 5) $K_c = \frac{[HI]^2}{[H_2][I_2]}$ 6) $K_c = [CO_2]$

2) What is the relationship between the equilibrium expressions for equations 2 and 3?

solution:

K_c (eqn. 3) = K_c^2 (eqn. 2)

3) What is the relationship between the equilibrium expressions for equations 4 and 5?

solution:

K_c (eqn. 4) = $1/K_c$ (eqn. 5)

Two Comments about K_c

1) The concentration of a pure liquid or solid does not appear in the expression of K_c.
2) The numerical value of K_c is meaningful only when it is associated with a particular chemical equation.

If the equilibrium constant, K, is very small, the reaction forms only a tiny amount of product before reaching equilibrium. If K is very large, the reaction proceeds until almost all the reactants are gone and we say it "goes to completion." When significant amounts of both reactants and products are present at equilibrium, K has an intermediate value.

17.2 THE REACTION QUOTIENT

As the concentrations of reactants and products are changing before equilibrium is reached, we refer to the **reaction quotient, Q.** When the reaction system reaches equilibrium and reactant and product concentrations no longer change:

$$Q = K \text{ (at equilibrium)}$$

Q is the ratio of product concentration terms multiplied together divided by reactant concentration terms multiplied together, with each term raised to the power of its stoichiometric coefficient. If an overall reaction is the sum of two or more reactions, the overall reaction quotient is the product of the reaction quotients for each of the steps:

$$Q_{overall} = Q_1 \times Q_2 \times Q_2 \times \ldots$$

A reaction quotient for a forward reaction is the reciprocal of the reaction quotient for the reverse reaction:

$$Q_{c(fwd)} = \frac{1}{Q_{c\,(rev)}}$$

Because we are concerned only with concentrations that change as they approach equilibrium, we eliminate the terms for pure liquids and solids from the reaction quotient (their constant concentrations are incorporated into the reaction quotient, Q_c).

17.3 EXPRESSING EQUILIBRIA WITH PRESSURE UNITS

Relationship Between K_c and K_p

Often it is easier to measure the pressure of a gas than its concentration. K_p is the equilibrium constant obtained when the concentrations of all species are expressed as their partial pressures in atmospheres. The ideal gas law can be used to relate K_c and K_p:

$$K_p = K_c(RT)^{\Delta n(\text{gas})}$$

where Δn(gas) is moles of gaseous products – moles of gaseous reactants.

Calculating K_p from K_c

Calculate K_p for the following reaction:

$$PCl_3(g) + Cl_2(g) \rightleftharpoons PCl_5(g); \quad K_c = 1.67 \text{ (at 500. K)}$$

solution:

There is one mole of gaseous products and 2 moles of gaseous reactants, so $\Delta n_{gas} = 1 - 2 = -1$:

$$K_p = K_c(RT)^{\Delta n(\text{gas})} = 1.67[(0.0821)(500.)]^{-1} = 0.0407$$

K_c for an overall reaction

Consider the following reactions:

$$SO_2(g) + 1/2O_2(g) \rightleftharpoons SO_3(g)$$

$$NO_2(g) \rightleftharpoons NO(g) + 1/2O_2(g)$$

1) What is the expression for K_c for each reaction?

solution:

$$K_c = \frac{[SO_3]}{[SO_2][O_2]^{\frac{1}{2}}} \qquad K_c = \frac{[NO][O_2]^{\frac{1}{2}}}{[NO_2]}$$

2) Show the equation which is the sum of the above two equations and the K_c expression for it.

solution:

$$SO_2(g) + NO_2(g) \rightleftharpoons SO_3(g) + NO(g)$$

$$K_c = \frac{[SO_3][NO]}{[SO_2][NO_2]}$$

3) What is the relationship between the K_c for the overall reaction and the K_c's for the two reactions in step 1 that add to give the overall reaction?

solution:

If we multiply the K_c's for the two reactions in step 1, we obtain K_c for the overall reaction in step 2:

$$\frac{[SO_3]}{[SO_2][O_2]^{\frac{1}{2}}} \times \frac{[NO][O_2]^{\frac{1}{2}}}{[NO_2]} = \frac{[SO_3][NO]}{[SO_2][NO_2]}$$

If an overall reaction is the sum of two or more reactions, *K* for the overall reaction is the product of the equilibrium constants of the individual reactions.

17.4 REACTION DIRECTION: COMPARING Q AND *K*

How can we tell if a reaction mixture is at equilibrium, and if it is not at equilibrium, how can we tell which direction the reaction proceeds in order to reach equilibrium? We use the **mass-action expression (*Q*),** referred to earlier as the **reaction quotient.** *Q* has the same form as the expression for *K*, but it contains actual concentrations in place of equilibrium concentrations. For the reaction:

$$aA + bB \rightleftharpoons cC + dD$$

the **reaction quotient (*Q*)** is:

$$Q_c = \frac{[C]^c[D]^d}{[A]^a[B]^b}$$

When the system is at equilibrium, $Q_c = K_c$. If Q_c is less than K_c, the reaction proceeds to the right to consume reactants (decreasing the denominator in the above quotient), and produce products (increasing the numerator). If Q_c is greater than K_c, the reverse reaction occurs to produce reactants and consume products.

Predicting the direction of reaction

We saw earlier that a reaction that takes place in automobiles is:

$$N_2(g) + O_2(g) \rightleftharpoons 2NO(g);\quad K_c = 0.10\ (2000°C)$$

Predict the direction the system will move to attain equilibrium at 2000°C if we add:

a) 1 mole of nitrogen and 1 mole of oxygen to a 1 L container.
b) 1 mole of nitrogen, 1 mole of oxygen, and 1 mole of NO to a 10 L container.

solutions:

The Q_c expression for the reaction is:

$$Q_c = \frac{[NO]^2}{[N_2]\,[O_2]}$$

For situation a), we have:

initial concentration of N_2 = 1 mol/L
initial concentration of O_2 = 1 mol/L
initial concentration of NO = 0
$Q_c = 0$, since [NO] = 0. The reaction will proceed to the right to produce NO.

For situation b), we have:

initial concentration of N_2 = 1 mol/10 L = 0.1 mol/L
initial concentration of O_2 = 1 mol/10 L = 0.1 mol/L
initial concentration of NO = 1 mol/10 L = 0.1 mol/L

$$Q_c = \frac{(0.1)^2}{(0.1)\,(0.1)} = 1$$

$Q_c > K_c$ so the reverse reaction will take place. The concentration of NO decreases and the concentration of N_2 and O_2 increase until the quotient equals 0.10, and the system attains equilibrium.

17.5 HOW TO SOLVE EQUILIBRIUM PROBLEMS

If you are given the equilibrium quantities (concentrations or partial pressures), you can solve for K by substituting the values into the reaction quotient (Problem #1 below). If some of the equilibrium quantities are not given, use a reaction table to show *initial* quantities, the *changes* in these quantities during reaction, and the *equilibrium* quantities, and solve for the unknown, x (Problem #2 below).

Equilibrium calculations

1) For the reaction of nitrogen with hydrogen to produce ammonia:

$$N_2(g) + 3H_2(g) \rightleftharpoons 2NH_3(g)$$

the equilibrium concentrations of N_2, H_2, and NH_3 are 0.15, 0.80, and 0.20 mol/L, respectively at 400°C. What is the value of K_c?

solution:

$$K_c = \frac{[NH_3]^2}{[N_2][H_2]^3} = \frac{(0.20)^2}{(0.15)(0.80)^3} = 0.52$$

2) For the same reaction at a different temperature, if one starts with 0.18 mol/L of both hydrogen and nitrogen, the equilibrium concentration of ammonia is 0.040 mol/L. Calculate K_c.

solution:

We know the equilibrium concentration of ammonia, but we know only the starting concentrations of hydrogen and nitrogen. In cases such as these, it can be helpful to make a table:

	N_2	+	$3H_2$	=	$2NH_3$
Initial conc.	0.18		0.18		0
Change	$-x$		$-3x$		$+2x$
Equilibrium conc.	$0.18 - x$		$0.18 - 3x$		$0 + 2x$

We can solve for x since we know the equilibrium concentration of NH_3:

$$2x = 0.040, \text{ so } x = 0.020,$$

and solve for moles of N_2 and H_2 at equilibrium:

$$\text{conc. } N_2 = 0.18 - x = 0.18 - 0.020 = 0.16$$

$$\text{conc. } H_2 = 0.18 - 3x = 0.18 - 0.060 = 0.12$$

Now calculate the value of K_c at the new temperature:

$$K_c = \frac{[NH_3]^2}{[N_2][H_2]^3} = \frac{(0.040)^2}{(0.16)(0.12)^3} = 5.8$$

Sometimes you will be given K and initial quantities and asked to solve for the equilibrium quantities. In this type of problem, you use a reaction table as in Problem #2 above, and solve again for x. You may need to use the quadratic formula to solve for x, but first look to see if there is a perfect square involved (Problem #1 on the following page), or if you can simplify the math by making an assumption about the concentrations of some of the species being very small relative to others (Problem #2, next page).

Determining equilibrium concentrations

1) When gasoline burns in an automobile, the temperature is high enough that nitric oxide forms from the elements nitrogen and oxygen:

$$N_2(g) + O_2(g) \rightleftharpoons 2NO(g)$$

At 2000°C, K_c for this reaction is 0.10. What are the equilibrium concentrations of NO, N_2 and O_2, when the starting concentration of each reactant is 1.0 mol/L?

solution:

We can use the table as earlier to express initial and equilibrium concentrations for all species:

	N_2	+	O_2	=	2NO
Initial conc.	1		1		0
Change	$-x$		$-x$		$+2x$
Equilibrium conc.	$1 - x$		$1 - x$		$2x$

In this case, we do not know any of the equilibrium concentrations, but we know the value for the equilibrium constant, K_c:

$$K_c = 0.10 = \frac{[NO]^2}{[N_2][O_2]} = \frac{(2x)^2}{(1 - x)(1 - x)}$$

We could solve this equation using the "quadratic formula," but often there is a way out of that. In this case, the right-hand side of the equation is a perfect square, so we can simply take the square root of both sides to solve for x:

$$(0.10)^{\frac{1}{2}} = 0.32 = \frac{2x}{1 - x}$$

$$0.32 - 0.32x = 2x$$

$$x = 0.14$$

Now we can use values from our table to solve for the concentrations of all species:

$$[N_2] = [O_2] = 1 - x = 0.86 \text{ mol/L}$$

$$[NO] = 2x = 0.28 \text{ mol/L}$$

2) At a certain temperature, K_c for the reaction between N_2 and O_2 to form NO is 3.2×10^{-4}. What is the equilibrium concentration of NO when 0.150 moles of N_2 and 0.30 moles of O_2 are added to a 1.50 L container?

solution:

$$N_2(g) + O_2(g) \rightleftharpoons 2NO(g)$$

	N_2	+	O_2	=	2NO
Initial conc.	0.10		0.20		0
Change	$-x$		$-x$		$+2x$
Equilibrium conc.	$0.10 - x$		$0.20 - x$		$2x$

$$K_c = 3.2 \times 10^{-4} = \frac{[NO]^2}{[N_2][O_2]} = \frac{(2x)^2}{(0.10 - x)(0.20 - x)}$$

Instead of using the quadratic formula to solve this equation, let's assume that since K_c is small, very little NO forms and x is very small compared to 0.10 or 0.20. We will assume that $0.10 - x \approx 0.10$, and $0.20 - x \approx 0.20$. Now the equation to solve becomes:

(continued on next page)

$$K_c = 3.2 \times 10^{-4} = \frac{(2x)^2}{(0.10)(0.20)}$$

Solving for x gives us:

$$4x^2 = 6.4 \times 10^{-6}$$

$$x = 1.3 \times 10^{-3}$$

Our assumption seems to be OK because x is small compared to 0.1 and 0.2 (1.3×10^{-3} is less than 5% of 0.1):

$$[NO] = 2x = 2 \times (1.3 \times 10^{-3}) = 2.6 \times 10^{-3}$$

17.6 LE CHATELIER'S PRINCIPLE

Le Chatelier's principle states that if we impose a change on a system at equilibrium, the system shifts to minimize the effect of the change. We will discuss how a system at equilibrium shifts to minimize the effects of:

- changing the concentration of reactants or products
- changing the volume of the system
- changing the temperature

Shifts in equilibrium

Use Le Chatelier's principle to work through the following exercise. Consider nitrogen and hydrogen in equilibrium with ammonia:

$$N_2(g) + 3H_2(g) \rightleftharpoons 2NH_3(g) + \text{heat}$$

What would be the effect of:

1) adding more nitrogen to the system?
2) removing ammonia from the system?
3) decreasing the volume of the reaction vessel?
4) increasing the temperature?

answers:

1 & 2) The reaction shifts to restore the concentration ratio to the value required by K_c. Reaction would occur in the forward direction to remove reactants and produce products. Equilibrium would be established with new concentrations of nitrogen, oxygen, and nitric oxide.

3) Decreasing volume increases pressure. The equilibrium shifts to decrease pressure. There are 4 moles of gaseous reactants and 2 moles of gaseous products, so reaction occurs in the forward direction, the direction which produces fewer moles of gaseous species. For reactions that have no net changes in the number of moles of gas, a volume change has no effect on the position of the equilibrium.

4) Heat is produced as ammonia forms (the reaction is exothermic). If heat is added to increase the temperature, the reverse reaction occurs to consume some of the added energy. The equilibrium constant does change with a change in temperature.

The **van't Hoff equation** describes how changes in temperature affect the equilibrium constant:

$$\ln \frac{K_2}{K_1} = \frac{-\Delta H^{\circ}_{rxn}}{R}\left(\frac{1}{T_2} - \frac{1}{T_1}\right)$$

Calculating the change in *K* with a change in temperature

For the Haber process in which nitrogen is converted to ammonia:

$$N_2(g) + 3H_2(g) \rightleftharpoons 2NH_3(g)$$

$\Delta H = -92.0$ kJ. If the equilibrium constant at room temperature is 5.0×10^8, at 200°C, will K_2 be smaller or larger than K_1?

solution:

The reaction is exothermic (gives off heat), so heat is a product. If we add heat to increase the temperature, the reaction will shift to the left to consume some of the added energy, and K_2 will be smaller than K_1 as the concentration of products decrease and reactants increase. It turns out we can calculate the new K_2 value by using an equation called the **van't Hoff** equation:

$$\ln \frac{K_2}{K_1} = \frac{-\Delta H^{\circ}_{rxn}}{R}\left(\frac{1}{T_2} - \frac{1}{T_1}\right)$$

We know:

$T_2 = (200 + 273) = 473$ K

$T_1 = (25 + 273) = 298$ K

$\Delta H = -92.0$ kJ

If we substitute these values into the van't Hoff equation, we obtain:

$$\ln \frac{K_2}{K_1} = \frac{-92{,}000 \text{ J}}{8.3148 \text{ J/K mol}}\left(\frac{1}{473 \text{ K}} - \frac{1}{298 \text{ K}}\right)$$

$$\ln \frac{K_2}{K_1} = -13.737$$

$$\frac{K_2}{K_1} = 1.0815 \times 10^{-6}$$

$$K_2 = \left(1.0815 \times 10^{-6}\right)\left(5.0 \times 10^{8}\right) = 540$$

Although one could obtain higher yields of ammonia by decreasing temperature, the reaction rate decreases with temperature to a point where it is no longer practical. In practice, the temperature is raised to a moderate level, and a catalyst is used to increase the reaction rate. Catalysts affect the time to reach equilibrium, but not the equilibrium position.

Adding a catalyst to a reacting system speeds up the reaction, but it has no effect on the equilibrium position. A catalyst may optimize the yield of a reaction system, as your textbook discusses for the industrial production of ammonia.

CHAPTER CHECK

Make sure you can...

- Distinguish between the rate and the extent of a reaction.
- Write an equilibrium expression from a balanced equation.
- Calculate K_c from initial reactant concentrations and equilibrium product concentrations.
- Use the ideal gas law to relate K_c and K_p.
- Find K_c for an overall reaction that is the sum of two or more reactions.
- Predict the direction of a reaction will shift to attain equilibrium using *Q*.
- Use Le Chatelier's principle to predict which direction a reaction will shift when a change is imposed on a system at equilibrium.
- Use the van't Hoff equation to calculate the change in *K* with a change in temperature.

Chapter 17 Exercises

17.1) Methyl alcohol can be prepared commercially by the reaction below:

$$CO(g) + 2H_2(g) \rightleftharpoons CH_3OH(g)$$

For the reaction at 200°C, the equilibrium concentrations are $[CO] = 0.60\ M$, $[H_2] = 0.13\ M$, and $[CH_3OH] = 0.43\ M$. What is K_c for the reaction?

17.2) For the reaction below at 25°C, the equilibrium concentrations for P_4, H_2, and PH_3 are 0.933 *M*, 0.600 *M*, and 0.267 *M*, respectively:

$$P_4(g) + 6H_2(g) \rightleftharpoons 4PH_3(g)$$

What is K_c for the reaction? What is K_p?

17.3) The following reaction occurs within an erupting volcano:

$$2H_2S(g) + SO_2(g) \rightleftharpoons 3S(s) + 2H_2O(g)$$

In an effort to understand the equilibrium of the reaction, a researcher analyzed the equilibrium mixture in a reaction vessel. At 650 K, the partial pressures of H_2S, SO_2, and H_2O are 1.17 atm, 1.07 atm, and 0.83 atm, respectively. Calculate K_p for the reaction.

17.4) For the reaction below at 50°C, the equilibrium partial pressures for N_2O, O_2, and NO_2 are 0.904 atm, 1.36 atm, and 2.19 atm, respectively:

$$2N_2O(g) + 3O_2(g) \rightleftharpoons 4NO_2(g)$$

Calculate K_p for the reaction. What is K_c?

17.5) 34.3 g of phosphorous pentachloride is placed in a 500. mL reaction vessel at 300°C:

$$PCl_5(g) \rightleftharpoons PCl_3(g) + Cl_2(g)$$

Chemical analysis of the equilibrium mixture indicates 52.0% of the reactant decomposes. Calculate K_c for the reaction.

17.6) 1.00 mole of N_2O_4 is placed in a 1.00 L reaction vessel at 298 K:

$$N_2O_4(g) \rightleftharpoons 2NO_2(g)$$

Chemical analysis of the equilibrium mixture indicates only 3.5% of the reactant decomposes. Calculate K_c for the reaction.

17.7) K_c for the reaction below at 100°C is 1.41:

$$CH_3CH{=}CH_2(g) + H_2O(g) \rightleftharpoons C_3H_8O(g)$$

The equilibrium concentrations are $[H_2O] = 0.35\ M$ and $[C_3H_8O] = 0.65\ M$. What is the equilibrium concentration of $CH_3CH{=}CH_2$?

17.8) Toluene, $C_6H_5CH_3$, can be prepared at an oil refinery from heptane, C_7H_{16}, and a catalyst at high temperature:

$$C_7H_{16}(g) \rightleftharpoons C_6H_5CH_3(g) + 4H_2(g)$$

K_p for the reaction at 600 K is 11.9. At equilibrium the partial pressures of C_7H_{16} and $C_6H_5CH_3$ are 4.28 atm and 0.723 atm, respectively. What is the equilibrium partial pressure of H_2?

17.9) Urea, $CO(NH_2)_2$, is a major nitrogen fertilizer produced in the U.S.:

$$2NH_3(g) + CO_2(g) \rightleftharpoons CO(NH_2)_2(s) + H_2O(g) \qquad \Delta H_{rxn} = -21.4\ kJ$$

What direction, if any, will the following changes have on the equilibrium?

a. addition of CO_2
b. removal of $CO(NH_2)_2$
c. addition of H_2O
d. increase in the temperature
e. increase in the pressure
f. addition of a catalyst

17.10) 0.50 mole of CH_3Br and 0.75 mole of H_2S are injected into a 1.00 L reaction vessel at 500 K:

$$CH_3Br(g) + H_2S(g) \rightleftharpoons CH_3SH(g) + HBr(g)$$

K_c for the reaction at 500 K is 0.33. What are the equilibrium concentrations for each of the species in the reaction?

17.11) Ethyl chloride, C_2H_5Cl, undergoes thermal decomposition at 250°C:

$$C_2H_5Cl(g) \rightleftharpoons CH_2{=}CH_2(g) + HCl(g)$$

K_p at 250°C is 2.09. If the initial partial pressure of C_2H_5Cl is 1.00 atm, what is the partial pressure of C_2H_5Cl, $CH_2{=}CH_2$, and HCl at equilibrium?

17.12) Dimethyl ether can be produced by the dehydration of methyl alcohol:

$$2CH_3OH(g) \rightleftharpoons CH_3OCH_3(g) + H_2O(g)$$

K_c for the reaction at 600 K is 12.9. 1.00 mole each of CH_3OH, CH_3OCH_3, and H_2O are placed in a 2.00 L reaction vessel. What are the equilibrium concentrations of each species?

17.13) K_c for the following reaction is 0.068 at 273 K:

$$CO(g) + H_2(g) \rightleftharpoons CH_2O(g)$$

The initial concentrations of CO, H_2, and CH_2O are 1.25 *M*, 2.00 *M*, and 1.00 *M*, respectively. What are the equilibrium concentrations of each species?

Chapter 17 Answers

17.1) Write the expression for K_c, and substitute in the equilibrium concentrations:

$$K_c = \frac{[CH_3OH]}{[H_2]^2[CO]} = \frac{0.43}{(0.13)^2(0.60)} = 42$$

17.2) Write the expression for K_c, and substitute in the equilibrium concentrations:

$$K_c = \frac{[PH_3]^4}{[P_4][H_2]^6} = \frac{(0.267)^4}{(0.933)(0.600)^6} = 0.117$$

$K_p = K_c(RT)^{\Delta n\,(gas)} = 0.117[(0.0821)(298)]^{-3} = 7.99 \times 10^{-6}$

17.3) Write the expression for K_p, and substitute in the equilibrium concentrations:

$$K_p = \frac{[H_2O]^2}{[H_2S]^2[SO_2]} = \frac{(0.83)^2}{(1.17)^2(1.07)} = 0.47$$

17.4) Write the expression for K_p, and substitute in the equilibrium concentrations:

$$K_p = \frac{[NO_2]^4}{[N_2O]^2[O_2]^3} = \frac{(2.19)^4}{(.904)^2(1.36)^3} = 11.2$$

$$K_c = \frac{K_p}{(RT)^{\Delta n\,(gas)}} = \frac{11.2}{[(0.0821)(323)]^{-1}} = 297$$

17.5) Each mole of PCl_5 that decomposes forms 1 mole of PCl_3 and 1 mole of Cl_2; 52% of the PCl_5 decomposes leaving 48% as PCl_5.

The initial concentration of PCl_5 is: 34.3 g × 1 mol/208.24 g × 0.5 L = 0.329 *M*

Final concentration of PCl_5 = 0.48 × 0.329 *M* = 0.158 *M*

Final concentrations of PCl_3 and Cl_2 = 0.52 × 0.329 *M* = 0.171 *M*

$$K_c = \frac{[Cl_2][PCl_3]}{[PCl_5]} = \frac{(0.171)(0.171)}{(0.158)} = 0.185$$

17.6) $K_c = 5.1 \times 10^{-3}$

17.7) Write the expression for K_c, and solve for $[CH_3CH{=}CH_2]$:

$$K_c = \frac{[C_3H_8O]}{[CH_3CH=CH_2][H_2O]} \Rightarrow [CH_3CH=CH_2] = \frac{[C_3H_8O]}{K_c[H_2O]} = \frac{0.65}{(1.41)(0.35)} = 1.3\ M$$

17.8) $[H_2] = 2.90\ M$

17.9)
a. shifts towards product
b. no effect; $CO(NH_2)_2$ is a solid
c. shifts towards reactant
d. shifts towards reactant; reaction is exothermic
e. shifts towards product
f. no effect; catalysts change the reaction rate

17.10) The K_c expression is:

$$K_c = \frac{[HBr][CH_3SH]}{[CH_3Br][H_2S]} = 0.33$$

The initial and final concentrations of the reactants and products are:

	CH_3Br	H_2S	CH_3SH	HBr
Initial	0.5	0.75	0	0
Final	$0.5 - x$	$0.75 - x$	x	x

Substituting into the K_c expression gives:

$$K_c = \frac{x^2}{(0.5 - x)(0.75 - x)} = 0.33$$

$x = [CH_3SH] = [HBr] = 0.22\ M$

$[CH_3Br] = 0.5 - 0.22 = 0.28\ M$

$[H_2S] = 0.75 - 0.22 = 0.53\ M$

17.11) The partial pressures at equilibrium are:

$C_2H_5Cl = 0.26$ atm
$C_2H_4 = 0.74$ atm
$HCl = 0.74$ atm

17.12) The equilibrium concentrations are:

$[CH_3OH] = 0.18\ M$
$[CH_3OCH_3] = 0.66\ M$
$[H_2O] = 0.66\ M$

17.13) The equilibrium concentrations are:

$[CH_2O] = 0.35\ M$
$[CO] = 1.9\ M$
$[H_2] = 2.7\ M$

18

Acid-Base Equilibria

18.1 ACIDS AND BASES IN WATER

The simplest (and earliest) definition of acids and bases is the **Arrhenius acid-base definition:**

Acids contain hydrogen and dissociate in water to give H^+:

$$HCl(aq) \rightarrow H^+(aq) + Cl^-(aq)$$

Bases contain the hydroxyl group and dissociate in water to give OH^-:

$$NaOH(aq) \rightarrow Na^+(aq) + OH^-(aq)$$

An acid, such as HCl, and a base, such as NaOH, react to form water and a salt. We say the acid and base are **neutralized.** The salt that forms (in this case, sodium chloride) exists as hydrated ions:

$$HCl(aq) + NaOH(aq) \rightarrow Na^+(aq) + Cl^-(aq) + H_2O(l)$$

All strong acids and bases (ones that dissociate completely in water) react together to form water:

$$H^+(aq) + OH^-(aq) \rightarrow H_2O(l)$$

The heat of reaction for this reaction is about –56 kJ per mole of water formed. The salts that form during strong acid / strong base reactions do not affect this heat of reaction.

Acids and bases may be strong or weak depending on the extent to which they dissociate into ions in water. Strong acids and bases dissociate completely into ions; weak acids and bases dissociate partially. For a weak acid, HA, we write its dissociation as:

$$HA(aq) + H_2O(l) \rightleftharpoons H_3O^+(aq) + A^-(aq)$$

The equilibrium constant for this reaction is:

$$K_c = \frac{[H_3O^+][A^-]}{[HA][H_2O]}$$

The concentration of water is so large that its concentration changes negligibly during this reaction and we therefore simplify the K_c expression by including the $[H_2O]$ term with the value of K_c:

$$K_c[H_2O] = \frac{[H_3O^+][A^-]}{[HA]} = K_a$$

The **acid dissociation constant, K_a,** is a specific equilibrium constant used for acid dissociation. It tells us how strong an acid is by telling us how much of the acid exists in its dissociated form (H_3O^+ and A^-). The larger the K_a value is, the stronger is the acid. We generally do not write K_a expressions for the strong acids because strong acids dissociate completely, and virtually no undissociated HA molecules exist.

Strong acids and bases

The most common strong acids are:	HCl, HBr, HI, HNO_3, $HClO_4$, and H_2SO_4
The most common strong bases are:	LiOH, NaOH, KOH, RbOH, CsOH, $Mg(OH)_2$, $Ca(OH)_2$, $Sr(OH)_2$, $Ba(OH)_2$

18.2 WATER AND pH

We saw in the last chapter how to write the equilibrium expression, K_c, for a chemical equilibrium. Water molecules are in equilibrium with hydronium ion (H_3O^+) and hydroxide ion (OH^-):

$$H_2O(l) + H_2O(l) \rightleftharpoons H_3O^+(aq) + OH^-(aq)$$

Water is **amphoteric** because it may act as both an acid (donates a proton), and a base (accepts a proton). Chemists sometimes abbreviate this reaction as:

$$H_2O(l) \rightleftharpoons H^+(aq) + OH^-(aq)$$

which focuses on the **dissociation** of water. Water dissociates to a very small extent: only about one in every 555 million water molecules exists in its dissociated form.

The equilibrium expression for water (K_w)

1) Write the equilibrium expression for the dissociation of water shown on the previous page.

   solution:

   When the species dissociating is an acid, we call the equilibrium constant K_a:

$$K_a = \frac{[H_3O^+][OH^-]}{[H_2O]^2}$$

2) What is the concentration of water ($[H_2O]$) in mol/L, assuming the concentration of dissociated ions is negligible?

   solution:

   One liter of water contains 1000 g of water (density of water = 1.00 g/mL); we also know the molar mass of water is 18.0 g/mole:

$$1000 \text{ g} \times \frac{1 \text{ mol}}{18.0 \text{ g}} = 55.5 \text{ mol}$$

   The molarity of water is therefore 55.5 mol/L.

Since the molarity of water is large, and remains essentially constant for dilute solutions, we include its concentration with the constant, K_a:

$$\mathbf{K_a[H_2O]^2 = K_w = [H_3O^+] \times [OH^-] = 1.0 \times 10^{-14}} \quad (25°C)$$

The **dissociation constant of water** has the symbol $\mathbf{K_w}$. Any water solution contains H_3O^+ and OH^- at concentrations such that their product is 1.0×10^{-14}. Although we sometimes refer to protons in solution as H^+, this notation is an abbreviation. The small proton with its high charge density bonds covalently to one of the lone electron pairs on an oxygen atom of water to form the hydronium ion, H_3O^+.

The dissociation constant of water

1) Given that $K_w = [H^+] \times [OH^-] = 1.0 \times 10^{-14}$, what is the concentration of H_3O^+ and OH^- ions in pure water?

   solution:

   In pure water, H_3O^+ ions and OH^- ions form in equal amounts:

$$[H_3O^+] = [OH^-] = \sqrt{1.0 \times 10^{-14}} = 1.0 \times 10^{-7}\ M$$

2) What percent of water molecules exist in the dissociated form?

   solution:

   We calculated in the problem above that the concentration of water is 55.5 mol/L. The concentration of H_3O^+ and OH^- ions in pure water is 1×10^{-7} mol/L. Therefore, the percent of water molecules in the dissociated form is:

(continued on next page)

$$\frac{1.0 \times 10^{-7} \text{ mol/L}}{55.5 \text{ mol/L}} \times 100\% = 1.8 \times 10^{-7}\%$$

or one out of every approximately 555 million water molecules exists in the dissociated form.

When the concentration of H_3O^+ ions equals the concentration of OH^- ions, the solution is **neutral.** In most cases, water solutions are not neutral. Drinking water is not neutral because carbon dioxide from the air dissolves in water to form carbonic acid, which dissociates in water:

$$H_2O(l) + CO_2(g) \rightarrow H_2CO_3(aq) \overset{H_2O}{\rightleftharpoons} H_3O^+(aq) + HCO_3^-(aq)$$

Solutions that have higher H_3O^+ concentrations than OH^- concentrations are **acidic.** Solutions that have higher OH^- concentrations than H_3O^+ concentrations are **basic.** Are drinking water solutions typically acidic or basic? (answer: acidic.)

pH

We saw in the earlier problem that the concentration of H_3O^+ and OH^- ions in pure (neutral) water is 1×10^{-7}. The concentration of H_3O^+ ions is small and is expressed using negative exponents. We can use a logarithmic scale to express hydronium ion concentration without negative exponents because a logarithm is an exponent:

$$\log 10^5 = 5$$

$$\log 10 = 1$$

$$\log 1 = \log 10^0 = 0$$

$$\log 10^{-7} = -7$$

$$\log 10^{-8.3} = -8.3$$

For pure water, the log of the hydronium ion concentration (1.0×10^{-7}) is -7. In order to work with positive numbers, we take minus the log of the hydronium ion concentration to obtain **pH.** The "p" in pH means "minus the log of." pH = "minus the log of the hydronium ion concentration":

$$\mathbf{pH = -\log [H_3O^+]}$$

The pH of neutral water = $-\log (1.0 \times 10^{-7}) = 7.0$

pH and pOH

We use pH to express the acidity of a solution, but we could just as easily express the acidity as pOH, or "minus the log of the hydroxide ion concentration." Use the expression for K_w to derive the relationship between pH and pOH.

solution:

The expression for K_w is:

$$K_w = [H_3O^+] \times [OH^-] = 1.0 \times 10^{-14} \text{ (25°C)}$$

(continued on next page)

We take the negative log of each side to obtain:

$$-\log K_w = -\log [H_3O^+] \times -\log [OH^-] = -\log (1.0 \times 10^{-14})$$

$$\mathbf{pK_w = pH + pOH = 14.00 \ (25°C)}$$

pH problems

1) What is the pH and the pOH of neutral water?

solution:

In neutral water, $\left[H_3O^+\right] = \left[OH^-\right] = \sqrt{1.0 \times 10^{-14}} = 1.0 \times 10^{-7}\ M$

$$pH = pOH = -\log (1.0 \times 10^{-7}) = 7.00$$

The pH (and the pOH) of a neutral solution is 7.00

2) If the pH of a sample of drinking water is 6.50, what is the hydrogen ion concentration? What is the concentration of OH^- ions? What is the pOH?

solution:

We need to take the antilog of the negative of the pH to solve for $[H^+]$:

$$[H_3O^+] = 10^{-6.50} = 3.2 \times 10^{-7}$$

We can solve for the concentration of OH^- using either relationship:

$$[H_3O^+] \times [OH^-] = 1.0 \times 10^{-14}$$

or, pH + pOH = 14

Using the first relationship: $[H_3O^+] = 3.2 \times 10^{-7}$, and

$$\left[OH^-\right] = \frac{1.0 \times 10^{-14}}{3.16 \times 10^{-7}} = 3.2 \times 10^{-8}\ M$$

Using the second equation, pOH = 14 – 6.5 = 7.5, and $[OH^-] = 10^{-7.5} = 3.2 \times 10^{-8}$

(The answers are slightly different due to rounding off the number for $[H_3O^+]$.)

3) Human blood has a pH of about 7.4. What is the concentration of H_3O^+ ions in blood?

solution:

We take the antilog of the negative of the pH to solve for $[H_3O^+]$:

$$[H_3O^+] = 10^{-7.4} = 4 \times 10^{-8}\ M$$

As hydrogen ion concentration increases, pH decreases.

When pH < 7, the solution is acidic.

When pH > 7, the solution is basic.

Basicity and acidity

Which of the following solutions is acidic, and which is basic?

1) beer (pH = 4-5)
2) blood (pH = 7.4)
3) household ammonia (pH = 11.9)
4) milk (pH = 6.4)
5) milk of magnesia (pH = 10.5)
6) 1 *M* HCl (pH = 0.0)
7) 1 *M* NaOH (pH = 14.0)
8) seawater (pH = 8.3)
9) stomach acid (pH = 2.2–2.4)
10) vinegar

answers: acidic (pH < 7): 1,4,6,9,10; basic (pH > 7): 2,3,5,7,8.

Acids are generally sour, react with active metals to produce hydrogen gas, and cause "burns." Bases are usually bitter and feel slippery (because they dissolve a layer of your skin).

The pH Scale

Many aqueous solutions have pH values that fall between 1 and 14 (hydrogen ion concentrations between 0.1 and 1×10^{-14}). As you saw in the above problem, the pH of 1 *M* HCl is 0.0. Can you have a negative pH? Yes, for example: concentrated hydrochloric acid is about 12 *M*. Hydrochloric acid dissociates completely in water. What is the pH of concentrated HCl? (answer: The concentration of hydronium ions in concentrated hydrochloric acid is 12 *M*. Therefore its pH = –log (12) = –1.08.)

Strong Acids and Bases

Hydrochloric acid dissociates in water to produce hydronium ions and chloride ions:

$$HCl(aq) + H_2O(l) \rightarrow H_3O^+(aq) + Cl^-(aq)$$

The reaction goes essentially 100% to completion. No HCl molecules exist because they all dissociate into ions. We do not write an equilibrium expression for this reaction because we assume K_a is infinite. Acids such as hydrochloric acid are called **strong acids.** Sometimes you may see the dissociation of hydrochloric acid written as:

$$HCl(aq) \rightarrow H^+(aq) + Cl^-(aq)$$

This equation does not indicate the association of the proton to an oxygen atom of water to form the hydronium ion, H_3O^+.

pH of solutions containing a strong acid

1) What is the pH of an aqueous solution containing 0.050 *M* HCl?

solution:

Hydrochloric acid is a strong acid that dissociates completely in water:

(continued on next page)

$$HCl(aq) + H_2O(l) \rightarrow H_3O^+(aq) + Cl^-(aq)$$

Since all the HCl forms hydronium ions and chloride ions, $[H_3O^+] = 0.050$

$$pH = -\log(0.050) = 1.30$$

2) What is the pH of an aqueous solution containing 1.00×10^{-8} *M* HCl?

solution:

If we try to solve this problem analogous to the one above, we run into a problem:

$$[H_3O^+] = 1.00 \times 10^{-8}, \text{ and pH} = -\log(1.00 \times 10^{-8}) = 8.00$$

We know an acid solution will not have a pH greater than 7, so what is going on? It turns out there is such a small amount of acid, that even though it is a strong acid, there is not enough of it for it to be the major source of hydrogen ions. Water itself is the major source of hydrogen ions: $[H_3O^+]$ from water = 1.00×10^{-7}. We find the pH of this solution by adding together the hydrogen ion concentration from the dissociation of water and the hydrogen ion concentration from the strong acid, HCl:

$$pH = -\log(1.00 \times 10^{-7} + 1.00 \times 10^{-8}) = 6.959$$

In aqueous solutions, water always contributes a small number of hydronium ions. Often another source (such as an acid) will swamp out the contribution from water, but keep in mind that the contribution from water will be significant whenever $[H_3O^+]$ is close to, or less than, 1.0×10^{-7}.

Analogous to the strong acids are the **strong bases.** The most common strong base is sodium hydroxide:

$$NaOH(aq) + H_2O(l) \rightarrow Na^+(aq) + OH^-(aq)$$

Sodium hydroxide is 100% ionized in solution to sodium and hydroxide ions. The concentration of undissociated NaOH is virtually zero. Strong acids and bases behave in solution as strong electrolytes. Compare a solution of 0.1 *M* sodium hydroxide and 0.1 *M* sodium chloride. Both conduct electric current well and behave as completely ionized substances in their colligative properties. One has a pH of 13, and the other has a pH of close to 7.

We saw in Chapter 4 that there are not very many strong acids and bases. The most important ones are listed in the table below.

The Strong Acids and Bases

Acids		Bases
Hydrochloric acid	HCl	Hydroxides of the 1A and 2A metals:
Hydrobromic acid	HBr	LiOH, NaOH*, KOH*, RbOH, CsOH,
Hydroiodic acid	HI	$Mg(OH)_2$, $Ca(OH)_2$, $Sr(OH)_2$,
Nitric acid	HNO_3	$Ba(OH)_2$
Perchloric acid	$HClO_4$	Metal oxides
Sulfuric acid	H_2SO_4 (first H^+)	* most commonly used in the lab

Concentrated hydrochloric acid (HCl) is about 12 *M*; concentrated nitric acid (HNO_3) is about 16 *M*; sulfuric acid is provided as essentially pure H_2SO_4. Reaction of a strong acid (such as hydrochloric acid) with a strong base (such as sodium hydroxide) quickly produces water as its product in an exothermic neutralization reaction:

Molecular equation:

$$HCl(aq) + NaOH(aq) \rightarrow H_2O(l) + NaCl(aq)$$

Total ionic equation:

$$H_3O^+(aq) + Cl^-(aq) + Na^+(aq) + OH^-(aq) \rightarrow 2H_2O(l) + Na^+(aq) + Cl^-(aq)$$

Net ionic equation:

$$H_3O^+(aq) + OH^-(aq) \rightarrow 2H_2O(l)$$

Reaction of a strong acid with a strong base

1) What is the pH of a 2.0 L solution containing 50. mL of 0.10 *M* HCl and 10. mL of 1.0 *M* NaOH?

solution:

First we calculate initial concentrations of HCl and NaOH:

$$[HCl] = [H_3O^+] = \frac{0.050\text{ L} \times 0.10\text{ mol/L}}{2.0\text{ L}} = 0.0025\ M$$

$$[NaOH] = [OH^-] = \frac{0.010\text{ L} \times 1.0\text{ mol/L}}{2.0\text{ L}} = 0.005\ M$$

Next we determine the equilibrium concentrations of H_3O^+ and OH^-:

	$H_3O^+(aq)$	+ $OH^-(aq)$	$\rightarrow$ $2H_2O(l)$
Start conc.	0.0025	0.005	
Change	–0.0025	–0.0025	
Equil. conc.	0	0.0025	

Calculate pH from the pOH:

$$pOH = -\log(.0025) = 2.60$$

$$pH = 14 - 2.6 = 11.40$$

2) How would you prepare 2 L of a solution with a pH of 3 from a 1 *M* HCl stock solution?

solution:

A solution with a pH of 3 has $[H_3O^+] = 1 \times 10^{-3}\ M$. 2 L of that solution contains:

$$2\text{ L} \times (1 \times 10^{-3}\text{ mol/L}) = 2 \times 10^{-3}\text{ mol } H_3O^+$$

Since HCl completely dissociates, we need 2×10^{-3} mol HCl from our 1 *M* stock solution. The volume of stock solution that contains 2×10^{-3} mol HCl is:

$$\frac{2 \times 10^{-3}\text{ mol HCl}}{1\text{ mol/L}} = 2 \times 10^{-3}\text{ L stock solution}$$

We would dilute 2×10^{-3} L (2 mL) of the stock solution to 2 L to obtain a solution with pH = 3.

18.3 THE BRØNSTED-LOWRY ACID-BASE DEFINITION

The **Brønsted-Lowry** definition of acids and bases is broader than the Arrhenius definition of acids and bases in two ways:

- it includes bases that do not contain OH in their formulas, and
- it does not specify that the acids and bases be in aqueous solutions.

Brønsted and Lowry define acid-base reactions as a proton transfer process where:

- the acid is the proton donor, and
  (An acid must contain H in its formula.)
- the base is the proton acceptor.
  (A base must contain a lone pair of electrons to bind the proton.)

Ammonia (NH_3) is a Brønsted-Lowry base that is not an Arrhenius base. Nitrogen contains a lone pair of electrons that bonds with a proton from an acid to form NH_4^+. If the acid is hydrogen sulfide, H_2S, we have the following acid-base reaction:

$$H_2S + NH_3 \rightleftharpoons HS^- + NH_4^+$$

H_2S and HS^- are a **conjugate acid-base pair.** H_2S is the acid for the reaction in the forward direction, and HS^- acts as a base for the reaction in the reverse direction. What is the other conjugate acid-base pair in this reaction? (answer: NH_3 and NH_4^+.) Use Figure 18.8 on page 790 in your textbook, which lists the relative strengths of conjugate acid-base pairs, to determine if the equilibrium above lies predominately to the right or predominately to the left. (answer: The reaction between hydrogen sulfide and ammonia proceeds to the right because ammonia, NH_3, is a stronger base than HS^- and wins the competition for a proton.)

Conjugate acid-base pairs

Table 18.4 on page 788 in your textbook shows six Brønsted-Lowry acid-base reactions. Notice that water and HPO_4^{2-} can both donate or accept protons acting as either Brønsted-Lowry acids or bases. Consider reactions 1, 3, 4, and 6 in the table. Which of these reactions proceeds to the right?

answer:

Reactions 3 and 4 proceed to the right. In reaction 3, CO_3^{2-} is a stronger base than NH_3 and it therefore wins the competition for a proton; in reaction 4, OH^- is a stronger base than HPO_4^{2-} so OH^- wins the competition for a proton and shifts the equilibrium to the right.

18.4 WEAK ACIDS

Most acids do not dissociate completely in water. The acid that makes vinegar acidic, acetic acid (CH_3COOH), dissociates only a small extent in water:

$$CH_3COOH(aq) + H_2O(l) \rightleftharpoons CH_3COO^-(aq) + H_3O^+(aq)$$

$$K_a = \frac{[CH_3COO^-][H_3O^+]}{[CH_3COOH]} = 1.8 \times 10^{-5}$$

Analogous to expressing H^+ concentrations as pH, we can express K_a as pK_a, or "minus the log of K_a." The pK_a for acetic acid is $-\log(1.8 \times 10^{-5}) = 4.74$. Most of the acetic acid molecules are in the undissociated form, CH_3COOH. In a 0.10 *M* solution of acetic acid for example, only about 1.3% of the acetic acid molecules exist in ionized form.

We can think of this reaction as a competition between acetate ions (CH_3COO^-) and water molecules for a proton. $K_a = 1.8 \times 10^{-5}$, so the equilibrium favors the reactants. Acetate wins the competition for the proton, and therefore acetate is a stronger base than water.

Acetic acid

1) Calculate the pH of a 0.1 *M* acetic acid solution. K_a for acetic acid = 1.8×10^{-5}.

solution:

It may be helpful to make a table:

	$CH_3COOH(aq) + H_2O$	$\rightleftharpoons$	$CH_3COO^-(aq)$	$+ H_3O^+(aq)$
Start conc.:	0.1		0	0
Equil. conc.:	$0.1 - x$		x	x

$$K_a = \frac{[CH_3COO^-][H_3O^+]}{[CH_3COOH]} = 1.8 \times 10^{-5} = \frac{x^2}{0.1 - x}$$

We could now solve for x by rearranging the equation to the form $x^2 + bx + c = 0$ and using the quadratic formula. However, since acetic acid dissociates to only a small extent, we know that the concentration of H_3O^+ will be small, and the concentration of CH_3COOH will be close to its original concentration. We make the approximation that x is much less than 0.1 and therefore $0.1 - x \approx 0.1$. Now we can solve for x easily:

$$\frac{x^2}{0.1} = 1.8 \times 10^{-5}$$

$$x = 1.34 \times 10^{-3} = [H_3O^+]$$

Notice that x is only 1.3% of 0.1, so our approximation was OK.
(*see note on following page)

$$\text{pH} = -\log(1.34 \times 10^{-3}) = 2.9$$

2) Calculate the pH of a 1 *M* acetic acid solution.

solution:

As above:

	$CH_3COOH(aq) + H_2O$	$\rightleftharpoons$	$CH_3COO^-(aq)$	$+ H_3O^+(aq)$
Start conc.:	1		0	0
Equil. conc.:	$1 - x$		x	x

$$K_a = \frac{[CH_3COO^-][H_3O^+]}{[CH_3COOH]} = 1.8 \times 10^{-5} = \frac{x^2}{1.0 - x}$$

(continued on next page)

We make the approximation that x is << 1.0 and therefore $1.0 - x \approx 1.0$. Now we can solve for x easily:

$$\frac{x^2}{1.0} = 1.8 \times 10^{-5}$$

$$x = 4.24 \times 10^{-3} = [H^+]$$

(Notice that x is only 0.42% of 1.0, so our approximation was OK.)

$$pH = -\log(4.24 \times 10^{-3}) = 2.4$$

*Note from the problem on the previous page: You are justified in setting molarity – x = molarity as long as this approximation does not introduce an error of more than about 5%. It is a reasonable start to make this approximation for all weak acid and weak base problems, and then determine if your calculated x is less than 5% of the original molarity. If it is more than 5% of the original molarity, you can repeat the calculation subtracting the value just calculated for x from the original molarity. If the calculated value for x is still more than 5% of the original molarity, you could repeat the procedure. This method of successive approximations is an alternative to solving the quadratic formula.

The previous example illustrates that the percent of dissociated molecules of a weak acid increases as the acid becomes more dilute. Acetic acid is about 1.3% dissociated in a 0.1 *M* solution, but only 0.42% dissociated in a 1 *M* solution.

Relative strengths of acids

Arrange the following acids in order of decreasing strength:

1) NH_4^+, $K_a = 5.6 \times 10^{-10}$
2) $HClO_2$, $K_a = 1.12 \times 10^{-2}$
3) HF, $K_a = 6.8 \times 10^{-4}$
4) CH_3COOH, $K_a = 1.8 \times 10^{-5}$
5) HCN, $K_a = 6.2 \times 10^{-10}$
6) HIO, $K_a = 2.3 \times 10^{-11}$

solution:

The strongest acid has the largest K_a, the weakest acid has the smallest K_a:
$HClO_2 > HF > CH_3COOH > HCN > NH_4^+ > HIO$

Strong vs. weak acids

Assign each of the following properties to:

a) a 0.1 *M* HCl solution, or
b) a 0.1 *M* acetic acid solution.

1) has a conductivity greater than a 0.1 *M* NaCl solution (see *note on the following page)
2) is a relatively poor conductor
3) has a pH of 2.9
4) has a pH of 1.0
5) has a freezing point almost the same as a 0.10 *M* NaCl solution
6) has a freezing point comparable to a 0.10 *M* sugar solution

answers: 1-a, 2-b, 3-b, 4-a, 5-a, 6-b.

Strong acids and bases, because they completely dissociate into ions, behave in solution as strong electrolytes. They are good electrical conductors and lower the freezing point of water to about the same extent as a strong electrolyte. Weak acids are poor conductors because there are few ions present, and they lower the freezing point of water to about the same extent as a nonelectrolyte such as sugar.

*Note: The very high conductivity of an HCl solution is due to the H^+ ions, which are able to jump from one water molecule to the next, transferring charge quickly. This process of proton transfer results in a more rapid transfer of positive charge than would be possible if the H_3O^+ ion had to push its way through the solution as other ions (such as Na^+) must.

Polyprotic Acids

Polyprotic acids have more than one ionizable proton. Acetic acid (CH_3COOH) is not a polyprotic acid because it has only one acidic hydrogen atom. (The hydrogen atoms bonded to carbon are not acidic.) Sulfuric acid (H_2SO_4), sulfurous acid (H_2SO_3), phosphoric acid (H_3PO_4), phosphorous acid (H_3PO_3), carbonic acid (H_2CO_3), and hydrosulfuric acid (H_2S) are all polyprotic acids. It is always more difficult to pull off each successive proton: successive dissociation constants usually differ by several orders of magnitude. In calculations, we can usually assume that the first ionization produces almost all the hydrogen ion. Consider phosphoric acid:

$$H_3PO_4(aq) + H_2O(l) \rightleftharpoons H_2PO_4^-(aq) + H_3O^+(aq); \qquad K_{a_1} = 7.2 \times 10^{-3}$$

$$H_2PO_4^-(aq) + H_2O(l) \rightleftharpoons HPO_4^{2-}(aq) + H_3O^+(aq); \qquad K_{a_2} = 6.3 \times 10^{-8}$$

$$HPO_4^{2-}(aq) + H_2(l) \rightleftharpoons PO_4^{3-}(aq) + H_3O^+(aq); \qquad K_{a_3} = 4.2 \times 10^{-13}$$

There are 5 orders of magnitude difference between the first and second dissociation constants, and another 5 orders of magnitude difference between the second and third dissociation constants.

A polyprotic acid

Calculate the hydrogen ion concentration in a 0.050 *M* phosphoric acid solution.

solution:

We will begin by assuming that the only significant source of hydrogen ions comes from the first dissociation of phosphoric acid. We will check this assumption after our calculation.

$$H_3PO_4(aq) + H_2O(l) \rightleftharpoons H_2PO_4^-(aq) + H_3O^+(aq)$$

	H_3PO_4	$H_2PO_4^-$	H_3O^+
Initial conc.	0.050	0	0
Change	$-x$	x	x
Equil. conc.	$0.050 - x$	x	x

$$K_a = \frac{[H_2PO_4^-][H_3O^+]}{[H_3PO_4]} = \frac{x^2}{0.050 - x} = 7.2 \times 10^{-3}$$

We'll assume that x is small compared to 0.050 ($0.050 - x \approx 0.050$) and solve for x:

$$\frac{x^2}{0.050} = 7.2 \times 10^{-3}$$

$$x = [H_3O^+] = 0.019\,M$$

Was our assumption that x is small compared to 0.050 valid?

(continued on next page)

$$\frac{0.019\ M}{0.05\ M} \times 100\% = 38\%$$

Drat! x is not less than 5% of 0.050 M, and so is not small enough to ignore. We could either go back and solve the quadratic equation, or we can use this value for x to make a new estimate for the concentration of H_3PO_4:

$$[H_3PO_4] = 0.050 - 0.019 = .031$$

Our estimate of x is:

$$\frac{x^2}{0.031} = 7.2 \times 10^{-3}$$

$$x = [H_3O^+] = .015$$

If we repeat this process, our new estimate of $[H_3PO_4] = 0.050 - 0.015 = .035$

$$\frac{x^2}{0.035} = 7.2 \times 10^{-3}$$

$$x = \left[H_3O^+\right] = 0.016\ M$$

x is no longer changing much, so we will take 0.016 as the concentration of hydrogen ions from the first dissociation of H_3PO_4.

Is the concentration of hydrogen ions from the second dissociation of phosphoric acid significant?

$$H_2PO_4^-(aq) + H_2O(l) \rightleftharpoons HPO_4^{2-}(aq) + H_3O^+(aq)$$

	$H_2PO_4^-$	HPO_4^{2-}	H_3O^+
Initial conc.	0.016	0	0
Change	$-x$	x	x
Equil. conc.	$0.016 - x$	x	x

$$\frac{\left[HPO_4^{2-}\right]\left[H_3O^+\right]}{\left[H_2PO_4^-\right]} = \frac{x^2}{0.016 - x} = 6.3 \times 10^{-8}$$

We'll assume that x is small compared to 0.016:

$$\frac{x^2}{0.016} = 6.3 \times 10^{-8}$$

$$x = \left[H_3O^+\right] = 3.17 \times 10^{-5}\ M$$

Was our assumption that x is small compared to 0.016 valid?

$$\frac{3.17 \times 10^{-5}\ M}{0.016\ M} \times 100\% = 0.2\%$$

Good! We're OK this time. The hydrogen ion concentration from the second dissociation of phosphoric acid (3.17×10^{-5}) is insignificant compared to the concentration from the first dissociation (0.016), so, as is usually the case with polyprotic acids, we can assume that the first ionization produces almost all the hydrogen ion.

18.5 MOLECULAR PROPERTIES AND ACID STRENGTH

Consider the dissociation of an acid, HA:

$$HA + H_2O \rightleftharpoons H_3O^+ + A^-$$

As the electronegativity of A increases, acidity of the acid HA increases:

$$HF > H_2O > NH_3$$

The electronegative fluorine atoms in hydrofluoric acid displace electron density away from the hydrogen atoms allowing them to dissociate into H^+ ions. For oxoacids with the same number of oxygens around a central nonmetal, acid strength increases with the electronegativity of the central nonmetal:

$$HOCl > HOBr > HOI$$

$$H_2SO_4 > H_2SeO_4$$

$$H_3PO_4 > H_3AsO_4$$

For oxoacids with different numbers of oxygen atoms around a central nonmetal, acid strength increases with the number of oxygen atoms. The greater number of electronegative oxygen atoms can share the negative charge from the loss of a proton and therefore make a more stable conjugate base:

$$HOClO_3 > HOClO_2 > HClO_2 > HOCl$$

$$HNO_3 > HNO_2$$

$$H_2SO_4 > H_2SO_3$$

As the atomic radius of A increases, for example down a column in the periodic table, the HA bond becomes weaker, and acidity increases:

$$H_2Se > H_2S > H_2O$$

$$HI > HBr > HCl >> HF$$

(In water, HI, HBr, and HCl are equally strong because strong acids dissociate completely in water to form H_3O^+, and since the strong acid is no longer present, we observe the acid strength of H_3O^+.)

Hydrated metal ions can act as Brønsted-Lowry acids if the metal ion is small and highly charged. A small, highly charged metal ion withdraws electron density from the O–H bonds of the bound water molecules causing a proton to be released:

$$Al(H_2O)_6^{3+}(aq) + H_2O(l) \rightleftharpoons Al(H_2O)_5OH^{2+}(aq) + H_3O^+(aq)$$

Acidity of hydrated metal ions increases as the size of the metal ion decreases and the charge of the metal ion increases.

18.6 WEAK BASES

Weak bases produce hydroxide ions in water, but to a much smaller extent than do strong bases. One of the most common weak bases is ammonia, NH_3. Ammonia produces hydroxide ions in water:

$$NH_3(aq) + H_2O(l) \rightleftharpoons NH_4^+(aq) + OH^-(aq)$$

Most of the ammonia molecules are undissociated, but a small percent of them dissociate to produce ammonium ion and hydroxide ion. For weak bases, we can write a K_b expression that is analogous to the K_a expression for weak acids:

$$K_b = \frac{[NH_4^+][OH^-]}{[NH_3]} = 1.8 \times 10^{-5}$$

It is only coincidence that K_a for the common weak acid, acetic acid, is the same as K_b for the common weak base, ammonia! Without working out the math, we know from our exercises with acetic acid, that a 0.1 *M* ammonia solution is 1.3% dissociated, and a 1 *M* NH_3 solution is 0.42% dissociated. It may be confusing to see ammonia solutions labeled as NH_4OH. There is no evidence that this species exists. Think of these solutions as mainly ammonia (NH_3) and water (H_2O) molecules, with a small percent of NH_4^+ and OH^- ions.

Any molecule that removes a proton from water to produce hydroxide ions is a base. A salt that contains the anion of a weak acid is a weak base. $NaCH_3COO$ is a weak base because the acetate anion removes protons from water to form acetic acid:

$$CH_3COO^-(aq) + H_2O(l) \rightleftharpoons CH_3COOH(aq) + OH^-(aq)$$

The acetate anion is the **conjugate base** of acetic acid. Acetic acid is the **conjugate acid** of the acetate anion. Remember that all weak acid/weak base pairs can be related this way. What is the other conjugate acid/conjugate base pair in this equilibrium? (Answer: OH^- is the conjugate base of the conjugate acid, H_2O.)

Relation between K_a and K_b

Consider the dissociation reactions of ammonia (NH_3) and its conjugate acid, the ammonium ion (NH_4^+):

$$NH_3(aq) + H_2O(l) \rightleftharpoons NH_4^+(aq) + OH^-(aq); \quad K_b = 1.8 \times 10^{-5}$$

$$NH_4^+(aq) + H_2O(l) \rightleftharpoons NH_3(aq) + H_3O^+(aq) \quad K_a = ?$$

1) Write K_a and K_b expressions for these dissociation reactions.

solution:

$$K_b = \frac{[NH_4^+][OH^-]}{[NH_3]} = 1.8 \times 10^{-5}$$

$$K_a = \frac{[NH_3][H_3O^+]}{[NH_4^+]} = ?$$

2) What is $K_a \times K_b$?

solution:

$$K_a \times K_b = \frac{[NH_3][H_3O^+]}{[NH_4^+]} \times \frac{[NH_4^+][OH^-]}{[NH_3]} = [H_3O^+][OH^-] = K_w$$

$$\mathbf{K_a \times K_b = K_w}$$

3) What is K_a for the ammonium ion?

(continued on next page)

$K_a \times K_b = K_w$, or

$$K_a = \frac{K_w}{K_b} = \frac{1.0 \times 10^{-14}}{1.8 \times 10^{-5}} = 5.6 \times 10^{-10}$$

4) Consider the reaction between ammonium, NH_4^+, and base:

$$NH_4^+(aq) + OH^-(aq) \rightleftharpoons NH_3(aq) + H_2O(l)$$

a) Identify the conjugate acid/base pairs for this reaction.

b) Is water (H_2O) or NH_4^+ a stronger acid?

c) Would you expect this equilibrium to lie farther to the right or to the left?

solutions:

a) NH_4^+ = conj. acid; NH_3 = conj. base

H_2O = conj. acid; OH^- = conj. base

b) The dissociation constant for water is 1×10^{-14}. The dissociation constant for NH_4^+ is 5.6×10^{-10}. The ammonium ion dissociates more than water, and so the ammonium ion is a stronger acid than water.

c) Since NH_4^+ is a stronger acid than water, the equilibrium lies to the right.

The relationship $K_a \times K_b = K_w$ tells us that the stronger the conjugate acid, the weaker the conjugate base and vice versa.

Reaction of a weak acid and a strong base

What is the pH of 100. mL of a 0.10 *M* acetic acid solution to which 50. mL of 0.10 *M* NaOH has been added?

solution:

The initial concentrations of acetic acid and NaOH are:

$$[CH_3COOH] = \frac{0.100 \text{ L} \times 0.10 \text{ mol/L}}{0.150 \text{ L}} = 0.067\ M$$

$$[NaOH] = \frac{0.050 \text{ L} \times 0.10 \text{ mol/L}}{0.150 \text{ L}} = 0.033\ M$$

The strong base (hydroxide ion) deprotonates the weak acid until the base is used up. The reaction goes essentially to completion:

$$CH_3COOH(aq) + OH^-(aq) \rightarrow CH_3COO^-(aq) + H_2O(l)$$

(continued on next page)

Initial conc.	0.067	0.033	0
Change	–0.033	–0.033	+0.033
End conc.	0.033	0	0.033

Now we can solve for the pH of the acetic acid/acetate ion solution:

$$CH_3COOH(aq) + H_2O(l) \rightleftharpoons CH_3COO^-(aq) + H_3O^+(aq)$$

$$K_a = \frac{[CH_3COO^-][H_3O^+]}{[CH_3COOH]} \quad \text{or} \quad [H_3O^+] = \frac{K_a\,[CH_3COOH]}{[CH_3COO^-]}$$

$$[H_3O^+] = \frac{(1.8 \times 10^{-5})(0.033)}{(0.033)} = 1.8 \times 10^{-5}$$

pH = 4.74

We calculated in the **acetic acid** exercise that a 0.1 *M* acetic acid solution has a pH of 2.9. The pH of the acetic acid solution to which base has been added is higher (less acidic).

pH of a conjugate acid/base pair

To find the pH of weak acid, HA, it is helpful to set up a table that shows the initial concentration of the acid HA (say 0.1 *M*), the change in the concentration when some HA dissociates into H_3O^+ and A^-, and the equilibrium concentrations of HA, H_3O^+, and A^-:

	$HA(aq)$ $\rightleftharpoons$	$H_3O^+(aq)$ +	$A^-(aq)$
Start conc.:	0.1	0	0
Change:	$-x$	x	x
Equil. conc.:	$0.1 - x$	x	x

The K_a expression allows us to solve for the concentration of H_3O^+ ions, and hence to find the pH of the weak acid:

$$K_a = \frac{[H_3O^+][A^-]}{[HA]} = \frac{(x)(x)}{(0.1 - x)}$$

We usually can make the approximation that $0.1 - x \approx 0.1$, and solve for x, which is the concentration of H_3O^+ ions.

In problems such as the one on the previous page where a strong base reacts with a weak acid (CH_3COOH), there is a source of the conjugate base (CH_3COO^-) that does not come from the dissociation of acetic acid in water, but from deprotonation of acetic acid by OH^-. The amount of CH_3COO^- formed in this manner (0.033 *M* in the earlier exercise), is much, much more than the amount that dissociates from acetic acid, and so we disregard the dissociated amount. Another example of an "independent source" of CH_3COO^- would be the addition of a salt such as $NaCH_3COO$ to an acetic acid solution.

Caution

Don't confuse conjugate acid-base pairs with the K_a and K_b expressions for amphoteric substances, substances that behave as either acids or bases. The sodium bicarbonate anion may act as either an acid or a base:

$$HCO_3^-(aq) + H_2O(l) \rightleftharpoons CO_3^{2-}(aq) + H_3O^+(aq); \quad K_a = 4.8 \times 10^{-11}$$

$$HCO_3^-(aq) + H_2O(l) \rightleftharpoons H_2CO_3(aq) + OH^-(aq); \quad K_b = 2.4 \times 10^{-8}$$

$$K_a = \frac{[CO_3^{2-}][H_3O^+]}{[HCO_3^-]} \qquad K_b = \frac{[H_2CO_3^-][OH^-]}{[HCO_3^-]}$$

In these two equations, K_a and K_b are not related by $K_a \times K_b = K_w$ since they are not expressions for a conjugate acid/conjugate base pair.

Will a solution containing HCO_3^- be acidic or basic?
(answer: The solution will be slightly basic since $K_b > K_a$.)

18.7 ACID-BASE PROPERTIES OF SALT SOLUTIONS

When salts (strong electrolytes) dissolve in water, they dissociate into ions. The ions may be acidic, basic, or neutral. Neutral anions are derived from strong acids; neutral cations are derived from strong bases. To see why this is the case, consider the strong acid, HCl:

$$HCl(aq) + H_2O(l) \rightarrow H_3O^+(aq) + Cl^-(aq)$$

The anion, Cl^-, has no tendency to combine with H^+ to form HCl since HCl is a strong acid and completely dissociates in water. Cl^- does not affect the pH of a solution and is a neutral ion. Likewise, the sodium cation, Na^+, has no tendency to combine with OH^- since NaOH is a strong base, and so Na^+ is a neutral cation. Conjugate bases of weak acids are basic anions. The acetate ion, CH_3COO^-, is the conjugate base of acetic acid, and is an example of a basic anion. Examples of acidic cations are the ammonium ion, NH_4^+, Al^{3+}, and the transition metal ions.

Acidity and basicity of salt solutions

Would you expect each of the following solutions to lower the pH, raise the pH, or have no effect on the pH of a neutral solution?

1) NaCl
2) NH_4Cl
3) NH_4NO_3
4) $NaCH_3COO$
5) $KClO_2$
6) KCN
7) $AlCl_3$
8) Na_2CO_3

answers:

1) neutral cation and neutral anion; no pH effect
2) acidic cation and neutral anion; lowers pH

(continued on next page)

3) acidic cation and neutral anion; lowers pH
4) neutral cation and basic anion; raises pH
5) neutral cation and basic anion; raises pH
6) neutral cation and basic anion; raises pH
7) acidic cation and neutral anion; lowers pH
8) neutral cation and basic anion; raises pH

18.8 THE LEVELING EFFECT

Why are the strong acids HCl, HBr, HI, HNO_3, $HClO_4$, and H_2SO_4 all equally strong in water? The answer is that all strong acids dissociate completely in water to form H_3O^+. Since the strong acid is no longer present, we observe the acid strength of H_3O^+. Similarly, strong bases dissociate completely in water to form OH^-, and we observe the base strength of OH^-. We say that water exerts a **leveling effect** on strong acids and bases.

If we wish to rank the strong acids in terms of relative strength, we must dissolve them in a solvent that accepts their protons less readily than water. Acetic acid is an acid in water, but when it is used as a solvent for strong acids, it acts as a base. HI protonates acetic acid to a greater extent than HBr which protonates acetic acid to a greater extent than HCl, so although HI, HBr, and HCl are equally strong acids in water, their acid strength in acetic acid is HI > HBr > HCl, as you would predict based on the sizes of their atomic radii.

18.9 LEWIS ACID-BASE DEFINITION

The Lewis acid-base definition focuses on the role of the electron pair.

An acid is any species that accepts an electron pair.
A base is any species that donates an electron pair.

This definition greatly increases the classes of acids. Electron deficient atoms act as a Lewis acid by accepting an electron pair from a base to form a covalent bond. The product of a Lewis acid-base reaction is called an **adduct.**

Some examples of Lewis acids (electron deficient central atoms) are:

- BF_3, $AlCl_3$
  (central atom does not have an octet)
- SO_2, CO_2
  (polar multiple bonds)
- Any metal ion dissolved in water:

$$\underset{\text{acid}}{M^{2+}(aq)} + \underset{\text{base}}{4H_2O(l)} \rightleftharpoons \underset{\text{adduct}}{M(H_2O)_4{}^{2+}(aq)}$$

Review of acid-base definitions

Choose words from the list below to complete the following paragraphs:

(a) acid
(b) base
(c) electron pair
(d) hydrogen
(e) hydroxide
(f) proton
(g) proton transfer
(h) water

Arrhenius proposed the earliest definition for acids and bases, and it is the most restrictive definition. In the Arrhenius model, an acid is a substance that in water produces an excess of _______________(1) ions; a base forms an excess of _______________(2) ions.

The Brønsted-Lowry theory extends this concept of acids and bases. A Brønsted-Lowry acid-base reaction involves a _______________(3) from one species to another. The species that donates the proton is the _______________(4), and the species which accepts the proton is the _______________(5). Often the proton acceptor is _______________(6) in which case the hydronium ion forms, and water acts as a _______________(7). Water can act as both an acid and a base according to the Brønsted-Lowry approach. Every acid has a conjugate _______________(8), and vice versa. In order to be a Brønsted-Lowry acid, the species must contain an ionizable _______________(9).

The Lewis acid-base concept, extends the definition of an acid further. An acid must be a species that can accept an _______________(10). Examples of Lewis acids are protons, metal ions, and molecules with an incomplete octet of electrons, such as boron trifluoride, BF_3. Generally "Lewis acids" refer to a species such as BF_3 which is a powerful electron pair acceptor.

answers: 1-d, 2-e, 3-g, 4-a, 5-b, 6-h, 7-b, 8-b, 9-f, 10-c.

CHAPTER CHECK

Make sure you can...

- Use the Arrhenius definition to classify acids and bases.
- Show the relationship between K_c and K_a.
- Use the dissociation constant for water, K_w, to find the percent of dissociated water molecules.
- Use the definition of K_w to derive the relationship between pH and pOH.
- Identify acidic and basic solutions from their pH.
- Find the pH of strong acid and strong base solutions.
- List the most common strong acids and strong bases.
- Classify compounds as acids or bases using the Brønsted-Lowry definition.
- Identify conjugate acid and conjugate base pairs.
- Calculate the pH of weak acid solutions and polyprotic acid solutions.

- Calculate K_a if you know K_b and vice versa.
- Calculate the pH of a solution containing a weak acid and a strong base.
- Predict acid strength based on periodic trends and molecular properties.
- Explain the leveling effect.
- Predict if a salt solution would be neutral, acidic, or basic.
- Define a "Lewis acid" and give examples.

Chapter 18 Exercises

18.1) Fill in the empty blocks in the table below:

$[H^+]$	$[OH^-]$	pH	pOH
4.3×10^{-4} *M*			
		8.90	
	1.5×10^{-11} *M*		
			4.22

18.2) The freezing point depression of meta-hydroxybenzoic acid, $HC_7H_5O_3$, in water indicates it is a weak electrolyte that ionizes 4.0% in a 0.050 *M* solution. Calculate K_a for the acid.

18.3) The pH of a 0.10 *M* adipic acid solution is 5.43. Calculate K_a for the weak acid.

18.4) The pH of a 0.010 *M* strychnine solution is 6.00. Calculate K_b for the conjugate base.

18.5) Calculate pK_a for the following weak acids:

a. trans-cinnamic acid, $HC_9H_7O_2$; $K_a = 3.7 \times 10^{-5}$
b. diethylbarbituric acid, $HC_8H_{11}O_3N_2$; $K_a = 3.7 \times 10^{-8}$
c. saccharin, $HC_7H_4O_3NS$; $K_a = 2.5 \times 10^{-2}$

18.6) Calculate pK_b for the following weak bases:

a. Novocain, $C_{13}H_{20}O_2N_2$; $K_b = 7 \times 10^{-6}$
b. piperdine, $C_5H_{11}N$; $K_b = 1.6 \times 10^{-3}$
c. methyl red, $(CH_3)_2N{\cdot}C_{13}H_9O_2N_2$; $K_b = 3 \times 10^{-12}$

18.7) Calculate the pH and the percent ionization for each of the following weak acid solutions:

a. 0.10 *M* HF
b. 0.10 *M* H_3BO_3
c. 0.10 *M* HCN

18.8) What is the relationship, if any, between the percent ionization of a weak acid and the size of K_a for the acid?

18.9) Calculate the pH and percent ionization for each of the following solutions of acetic acid:

a. 1.00 *M* CH_3COOH
b. 0.10 *M* CH_3COOH
c. 0.010 *M* CH_3COOH

18.10) What is the relationship, if any, between the percent ionization of a weak acid and the concentration of the acid solution?

18.11) Calculate the pH and the percent ionization for each of the following weak base solutions:

a. 0.10 *M* CH_3NH_2; $pK_a = 10.657$
b. 0.020 *M* $C_6H_5NH_2$; $pK_a = 4.63$

18.12) Calculate K_b and pK_b for the salts of the following weak acids:

a. sodium pelargonate, $NaC_9H_{17}O_2$; $K_a = 1.1 \times 10^{-5}$
b. potassium hydrogen phosphate, K_2HPO_4; $K_a = 6.2 \times 10^{-8}$

18.13) Calculate the pH of the solutions below:

a. 0.020 *M* potassium acid phthalate ($KHC_8H_4O_4$) solution; $K_a = 1.3 \times 10^{-3}$
b. 0.15 *M* potassium glutarate ($KHC_5H_8O_4$) solution; $K_a = 4.5 \times 10^{-5}$
c. 0.050 *M* morphine hydrochloride ($C_{17}H_{19}O_3N{\cdot}HCl$) solution; $K_b = 7.4 \times 10^{-7}$

18.14) Fumaric acid, $H_2C_4H_2O_4$, is one of the intermediates of the citric acid cycle of glucose and fatty acid metabolism. It may be represented as H_2Fum. Calculate the pH and the concentrations of the [$HFum^-$] and [Fum^{2-}] ions in a 0.040 *M* H_2Fum solution ($pK_{a1} = 3.03$; $pK_{a2} = 4.44$).

18.15) Citric acid, $H_3C_6H_5O_7$, can be expressed as H_3Cit ($pK_{a1} = 3.14$; $pK_{a2} = 4.77$; $pK_{a3} = 6.39$). For a 0.10 *M* citric acid solution, determine the concentration of each of the following species: [H^+], [H_3Cit], [H_2Cit^-], [$HCit^{2-}$], and [Cit^{3-}].

18.16) For each of the aqueous solutions of the following substances, indicate whether it is acidic, neutral or basic, and indicate whether the substance is a strong acid, weak acid, strong base, weak base, acidic salt, basic salt, or neutral salt:

a. NaOCl
b. HNO_3
c. $CaCO_3$
d. $Mg(OH)_2$
e. K_2SO_4
f. NH_2OH
g. $HC_7H_5O_2$
h. $(CH_3)_2NH$
i. H_3AsO_4
j. $(NH_4)_2SO_4$
k. $AgNO_3$
l. NH_3

Chapter 18 Answers

18.1)

[H^+]	[OH^-]	pH	pOH
4.3×10^{-4} *M*	2.3×10^{-11} *M*	3.37	10.63
1.3×10^{-9} *M*	7.9×10^{-6} *M*	8.90	5.10
6.7×10^{-4} *M*	1.5×10^{-11} *M*	3.18	10.82
1.7×10^{-10} *M*	6.0×10^{-5} *M*	9.78	4.22

18.2) 4% ionization means .048 *M* remains undissociated as $HC_7H_5O_3$, and .002 *M* dissociates:

$$K_a = \frac{[H^+][C_7H_5O_3^-]}{[HC_7H_5O_3]} = \frac{(.002)^2}{0.048} = 8.3 \times 10^{-5}$$

18.3) $K_a = 1.4 \times 10^{-10}$

18.4) If the acid, HA, dissociates into H^+ and A^-, [H^+] = [A^-] = 10^{-6} (from the pH).

The concentration of undissociated HA is: $0.010\ M - 10^{-6} \approx 0.010\ M$:

$$K_a = \frac{(10^{-6})^2}{.010} = 10^{-10}$$

$$K_b = K_w / K_a = 1.0 \times 10^{-14}/1.0 \times 10^{-10} = 1.0 \times 10^{-4}$$

18.5) a. $K_a = 3.7 \times 10^{-5}$; $pK_a = -\log(3.7 \times 10^{-5}) = 4.43$; b. 7.43; c. 1.60

18.6) a. $K_b = 7 \times 10^{-6}$; $pK_b = -\log(7 \times 10^{-6}) = 5.2$; b. 2.80; c. 11.5

18.7) a. K_a of HF = 6.8×10^{-4}

let $[H^+] = [F^-] = x$; then $[HA] = 0.1 - x$ and:

$$K_a = \frac{x^2}{0.1 - x} = 6.8 \times 10^{-4}$$

$x = 7.9 \times 10^{-3}$; $pH = -\log(7.9 \times 10^{-3}) = 2.10$;

% ionized = $7.9 \times 10^{-3}/(0.1 - 7.9 \times 10^{-3}) \times 100\% = 8.6\%$

b. pH = 5.12; 7.6×10^{-3}% ionized; c. pH = 5.10; 7.9×10^{-3}% ionized

18.8) For a given concentration, the smaller the K_a of the acid, the smaller the percent ionization.

18.9) a. pH = 2.37; 0.42% ionized; b. pH = 2.87; 1.3% ionized; c. pH = 3.37; 4.290% ionized.

18.10) The smaller the concentration, the greater the percent ionization of the acid.

18.11) a. pH = 11.81; 7% ionized; b. pH = 8.46; 1.5×10^{-2} % ionized

18.12) a. $K_a = 1.1 \times 10^{-5}$

$$K_b = \frac{K_w}{K_a} = \frac{1.0 \times 10^{-14}}{1.1 \times 10^{-5}} = 9.1 \times 10^{-10}$$

$pK_b = -\log(9.1 \times 10^{-10}) = 9.04$

b. $K_b = 1.6 \times 10^{-7}$; $pK_b = 6.79$

18.13) a. $HC_8H_4O_4^- \rightleftharpoons H^+ + C_8H_4O_4^{2-}$

$[HC_8H_4O_4^-] = 0.020\ M - x$ and $[H^+] = [C_8H_4O_4^{2-}] = x$

$$K_a = \frac{x^2}{0.02 - x} = 1.3 \times 10^{-3}$$

$x = 5.1 \times 10^{-3}$; $pH = -\log(5.1 \times 10^{-3}) = 7.59$

b. pH = 8.76; c. pH = 4.59

18.14) pH = 2.21; $[HFum^-] = 5.7 \times 10^{-3}\ M$; $[Fum^{2-}] = 4.4 \times 10^{-5}\ M$

18.15) $[H^+] = 8.53 \times 10^{-3}\ M$; $[H_3Cit] = 9.2 \times 10^{-2}\ M$; $[H_2Cit^-] = 8.2 \times 10^{-3}\ M$; $[HCit^{2-}] = 3.7 \times 10^{-4}\ M$; $[Cit^{3-}] = 1.2 \times 10^{-5}\ M$

18.16)
a. basic; basic salt
b. acidic; strong acid
c. basic; basic salt
d. basic; strong base
e. neutral; salt
f. basic; weak base
g. acidic; weak acid
h. basic; weak base
i. acidic; weak acid
j. acidic; acidic salt
k. acidic; acidic salt
l. basic; weak base

19

Ionic Equilibria in Aqueous Systems

19.1 ACID-BASE BUFFERS

Buffers temper the change in pH of a solution. A lake has buffering capacity if acid rain does not significantly change its pH. A buffer usually contains a weak acid, which reacts with OH^- ions, and the conjugate base of the weak acid, which reacts with H^+ ions. A buffer has maximum buffering capacity when the acid/conjugate base ratio is 1:1.

Acetic acid (CH_3COOH) mixed with acetate ion (CH_3COO^-) is a buffer. The buffer in blood is carbonic acid (H_2CO_3) and bicarbonate ion (HCO_3^-). Lactic acid, produced by exercise, could lower pH to a point where it would kill body cells if blood were not buffered.

pH change of a buffered solution

A liter of a buffered solution contains 0.10 mole CH_3COOH and 0.10 mole $NaCH_3COO$:

a) What is the pH of this solution?

b) What is the pH after addition of 100 mL of 0.10 *M* NaOH?

c) What is the pH after addition of 100 mL of 0.10 *M* HCl?

d) What is the pH of 100 mL of 0.10 *M* NaOH added to 1.0 L of water?

e) What is the pH of 100 mL of 0.10 *M* HCl added to 1.0 L of water?

<u>solutions</u>:

a) CH_3COOH (acetic acid, a weak acid) and CH_3COO^- (acetate ion, its conjugate base) form a buffer solution. To find pH, we must determine the concentration of H^+. The equilibrium is:

$$CH_3COOH + H_2O \rightleftharpoons CH_3COO^- + H_3O^+$$

(continued on next page)

Initial conc.	0.10	0.10	0
Change	$-x$	$+x$	$+x$
Equil. conc.	$0.10 - x$	$0.10 + x$	x

$$K_a = \frac{[CH_3COO^-][H_3O^+]}{[CH_3COOH]} = \frac{(0.1 + x)(x)}{(0.1 - x)} = 1.8 \times 10^{-5}$$

We look for approximations to make equations such as these easier to solve. Since acetic acid is a weak acid, we can assume that not much dissociates, and $0.1 - x \approx 0.1$. Then:

$$\frac{(0.1)(x)}{0.1} = 1.8 \times 10^{-5}$$

$$x = 1.8 \times 10^{-5}\ M = [H_3O^+]$$

(Notice that 1.8×10^{-5} is much smaller than 0.1, so our assumption was OK.)

$$pH = -\log(1.8 \times 10^{-5}) = 4.74$$

b) NaOH dissociates completely in water producing sodium and hydroxide ions. The initial concentrations are:

$$[OH^-] = \frac{(0.1\ L)(0.10\ mol/L)}{1.1\ L} = 0.0091\ M$$

$$[CH_3COOH] = \frac{0.10\ mol/L}{1.1\ L} = 0.091\ M$$

$$[CH_3COO^-] = \frac{0.10\ mol/L}{1.1\ L} = 0.091\ M$$

Hydroxide ion, a strong base, deprotonates acetic acid to form acetate anion and water. The reaction goes essentially to completion:

$$CH_3COOH + OH^- \rightarrow CH_3COO^- + H_2O$$

	CH_3COOH	OH^-	CH_3COO^-
Initial conc.	0.091	0.0091	0.091
Change	−0.0091	−0.0091	+0.0091
Equil. conc.	0.0819	0	0.1001

We find pH from the K_a expression for the equilibrium:

$$CH_3COOH + H_2O \rightleftharpoons CH_3COO^- + H_3O^+$$

$$K_a = \frac{[CH_3COO^-][H_3O^+]}{[CH_3COOH]} \quad \text{or} \quad [H_3O^+] = \frac{K_a[CH_3COOH]}{[CH_3COO^-]}$$

$$[H_3O^+] = \frac{(1.8 \times 10^{-5})(0.0819)}{(0.1001)} = 1.47 \times 10^{-5}\ M$$

$$pH = 4.83$$

The pH change of the buffered solution after addition of a strong base (NaOH) is 0.09 pH unit.

(continued on next page)

c) HCl is a strong acid that completely dissociates in water to give H_3O^+ and Cl^- ions. H_3O^+ protonates acetate ion to form acetic acid and water. The reaction goes essentially to completion:

$$CH_3COO^- + H_3O^+ \rightarrow CH_3COOH + H_2O$$

	CH_3COO^-	H_3O^+	CH_3COOH
Initial conc.	0.091	0.0091	0.091
Change	–0.0091	–0.0091	+0.0091
Equil. conc.	0.0819	0	0.1001

We find pH from the K_a expression for the equilibrium:

$$CH_3COOH + H_2O \rightleftharpoons CH_3COO^- + H_3O^+$$

$$\left[H_3O^+\right] = \frac{\left(1.8 \times 10^{-5}\right)(0.1001)}{(0.0819)} = 2.20 \times 10^{-5}\ M$$

pH = 4.65

The pH change of the buffered solution due to addition of a strong acid (HCl) is 0.08 pH unit.

d) The pH of pure water is 7.0. To find the pH of 100 mL of 0.10 *M* NaOH added to 1.0 L of water, calculate the concentration of NaOH, and use the relationship that pH = 14 – pOH:

$$[OH^-] = \frac{(0.10\ L)(0.10\ M)}{1.1\ L} = 0.0091\ M$$

pOH = –log (0.0091) = 2.04

pH = 14 – pOH = 14 – 2.04 = 11.96

The pH change from pure water is 12 – 7 = 5 pH units.

e) Analogous to part "d," we calculate the concentration of HCl to find pH:

$$[H_3O^+] = \frac{(0.10\ L)(0.10\ M)}{1.1\ L} = 0.0091\ M$$

pH = –log (0.0091) = 2.04

In the buffered solution, the strong acid and base were consumed by acetate ion and acetic acid, respectively. The pH change in these solutions was less than 0.1 pH unit. The same amount of acid and base added to pure water changed the pH by 5 pH units.

When we solved for the pH of acetic acid in the above problem, we used the expression for K_a and then solved for H_3O^+ to find the pH:

$$CH_3COOH + H_2O \rightleftharpoons CH_3COO^- + H_3O^+$$

$$K_a = \frac{[CH_3COO^-][H_3O^+]}{[CH_3COOH]} \quad \text{or} \quad [H_3O^+] = \frac{K_a[CH_3COOH]}{[CH_3COO^-]}$$

We can solve directly for pH by taking the negative logarithm (base 10) of both sides of the equation:

$$-\log [H_3O^+] = -\log K_a - \log \left[\frac{CH_3COOH}{CH_3COO^-}\right]$$

$$\mathbf{pH = p}\boldsymbol{K}_\mathbf{a} \mathbf{+ \log \frac{[CH_3COO^-]}{[CH_3COOH]}}$$

(the **Henderson-Hasselbalch** equation).

pH of buffer solutions

1) What is the pH of a solution containing 0.1 M CH_3COOH and 0.1 M CH_3COO^-?
2) What is the pH of a solution containing 1.0 M CH_3COOH and 1.0 M CH_3COO^-?
3) What is the pH of a solution containing 0.1 M CH_3COOH?
4) What is the pH of one liter of a 0.20 M CH_3COOH solution to which 0.10 mol OH^- has been added?
5) Which of the above solutions has the greatest buffer capacity?

solutions:

1) We can use the Henderson-Hasselbalch equation to solve directly for pH:

$$pH = pK_a + \log \frac{[CH_3COO^-]}{[CH_3COOH]}$$

$$pH = pK_a + \log \left(\frac{0.1}{0.1}\right)$$

$$pH = pK_a = 4.74$$

2) Analogous to the solution for "1":

$$pH = pK_a + \log \left(\frac{1.0}{1.0}\right)$$

$$pH = pK_a = 4.74$$

Problems 1 and 2 illustrate that when **HA = A⁻, pH = p*K*a.**

3) In this case, we need to consider the equilibrium:

	CH_3COOH	$+ H_2O$	$\rightleftharpoons$	CH_3COO^-	$+ H_3O^+$
Initial conc.	0.10			0	0
Change	$-x$			$+x$	$+x$
Equil. conc.	$0.10 - x$			x	x

(continued on next page)

$$K_a = \frac{[CH_3COO^-][H_3O^+]}{[CH_3COOH]} = \frac{(x)(x)}{(0.1 - x)} = 1.8 \times 10^{-5}$$

Since acetic acid is a weak acid, we can assume that only a small amount dissociates, and $0.1 - x \approx 0.1$. Therefore:

$$K_a = \frac{x^2}{0.1}$$

$$x^2 = (1.8 \times 10^{-5}) \times 0.1$$

$$x = \sqrt{1.8 \times 10^{-6}} = 1.34 \times 10^{-3} M = [H_3O^+]$$

$$pH = -\log(1.34 \times 10^{-3}) = 2.87$$

4) OH^-, a strong base deprotonates CH_3COOH, a weak acid. The reaction goes essentially to completion:

	CH_3COOH	+ OH^-	→ CH_3COO^- + H_2O
Initial conc.	0.20	0.10	0
Change	−0.10	−0.10	+0.10
Equil. conc.	0.10	0	0.10

We have made a buffer with equal amounts of acetic acid and acetate ions—the same buffer solution as in Problem "1" when equal amounts of acetic acid and sodium acetate were mixed. Again, the Henderson-Hasselbalch equation allows us to solve directly for pH:

$$pH = pK_a + \log\left(\frac{0.1}{0.1}\right)$$

$$pH = pK_a = 4.74$$

5) Solutions 1, 2, and 4 all have the same pH, since for these solutions, $[CH_3COOH] = [CH_3COO^-]$. Solution 2 has the greatest buffer capacity since it has the highest concentration of the weak acid and its conjugate base.

A buffer is generally effective if it is within ±1 pH unit of the pK_a. The most effective buffer is one in which its pH = pK_a. A buffer may be prepared by mixing an acid and the salt of its conjugate base, or by partially deprotonating a weak acid to form its conjugate base (or vice versa).

Buffer range

Consider an acetic acid/acetate ion buffer.

a) If pH = pK_a, what are the relative concentrations of $[CH_3COOH]$ and $[CH_3COO^-]$?

b) If pH = $pK_a + 1$, what are the relative concentrations of $[CH_3COOH]$ and $[CH_3COO^-]$?

c) If pH = $pK_a - 1$, what are the relative concentrations of $[CH_3COOH]$ and $[CH_3COO^-]$?

(continued on next page)

solutions:

a) We know from the Henderson-Hasselbalch equation:

$$pH = pK_a + \log \frac{[CH_3COO^-]}{[CH_3COOH]}$$

that pH = pK_a when $\log \frac{[CH_3COO^-]}{[CH_3COOH]} = 0$

The log of 1 = 0, so pH = pK_a when $[CH_3COOH] = [CH_3COO^-]$.

b) The log of 10 = 1, so pH = pK_a + 1 when the concentration of acetate ion is 10 times the concentration of acetic acid.

c) The log of 1/10 = −1, so pH = pK_a − 1 when the concentration of acetic acid is 10 times the concentration of acetate ion.

In general, a buffer is effective if a conjugate acid is less than 10 times as concentrated as its conjugate base (or vice versa).

19.2 ACID-BASE TITRATION CURVES

We can determine the concentration of an acid by titrating with a base or vice versa. We use pH meters or indicators to determine the end point of a titration.

Indicators are weak organic acids whose acid form (HIn) is a different color than its base form (In^-). The indicator bromthymol blue is yellow when protonated, and blue when deprotonated:

$$HIn(aq) + H_2O(l) \rightleftharpoons In^-(aq) + H_3O^+(aq) \qquad K_a = \frac{[In^-][H_3O^+]}{[HIn]}$$

yellow — blue

The acid and/or base forms are intensely colored, so only tiny amounts are required to color a solution, and they therefore do not affect the pH of the test solution. An indicator changes color over a pH range of about two units. You may be wondering how a system which changes color over two pH units (or a 100-fold change in $[H^+]$) can be used to accurately determine the end point of a titration. As you will see, the pH changes more than two pH units with addition of only a tiny amount of titrant at the equivalence point of many acid-base titrations.

Strong Acid-Strong Base Titrations

The **equivalence point** of a titration is the point when the number of moles of added OH^- equals the number moles of H_3O^+ originally present or vice versa. The **end point** of a titration occurs when the indicator changes color. We choose an indicator so that the end point is close to the equivalence point.

Equivalence point

What is the pH at the equivalence point for each of the following:

a) the titration of a strong acid with a strong base.
b) the titration of a weak acid with a strong base.

answers:

a) The reaction of a strong acid with a strong base is:

$$H_3O^+ + OH^- \rightarrow 2H_2O$$

At the equivalence point, all the acid is neutralized, and the pH = 7.

b) The reaction of a weak acid with a strong base is:

$$HA + OH^- \rightarrow A^- + H_2O$$

At the equivalence point, the pH is greater than 7 because the solution contains the weak base, A^-. The pH is determined by the pK_b and the concentration of A^-.

Titration of HCl with NaOH

40.00 mL of 0.1000 *M* HCl is titrated with 0.1000 *M* NaOH. What is the pH of the solution after the following volumes of 0.1000 *M* NaOH have been added?

a) 10.00 mL NaOH
b) 39.99 mL NaOH
c) 40.00 mL NaOH
d) 40.01 mL NaOH
e) 50.00 mL NaOH

solutions:

a) The reaction is: $H_3O^+ + OH^- \rightarrow 2H_2O$

initial moles H_3O^+ = 0.04000 L × 0.1000 mol/L = 0.004000 mol

moles added OH^- = 0.01000 L × 0.1000 mol/L = 0.001000 mol

moles H_3O^+ not neutralized = 0.004000 mol – 0.001000 mol = 0.003000 mol

$$[H_3O^+] = \frac{0.003000 \text{ mol}}{(0.04000\text{L} + 0.01000\text{L})} = 0.06000\ M$$

pH = –log (0.06) = 1.2218

b) initial moles H_3O^+ = 0.04000 L × 0.1000 mol/L = 0.004000 mol

moles added OH^- = 0.03999 L × 0.1000 mol/L = 0.003999 mol

moles H_3O^+ not neutralized = 0.004000 mol – 0.003999 mol = 1×10^{-6} mol

(continued on next page)

$$[H_3O^+] = \frac{1 \times 10^{-6} \text{ mol}}{(0.04000 \text{ L} + 0.03999 \text{ L})} = 1.429 \times 10^{-5}\ M$$

pH = 4.9

c) The 40.00 mL of 0.1 *M* NaOH exactly neutralizes the 40.00 mL of 0.1 *M* HCl. The pH is that of pure water, 7. (The presence of the neutral species Na^+ and Cl^- does not affect the pH.)

d) The additional 0.01 mL of base takes us past the equivalence point, and base is in excess:

moles added OH^- = 0.04001 L × 0.1000 mol/L = 0.004001 mol

moles of OH^- not neutralized = 0.004001 mol – 0.004000 mol = 0.000001 mol

$$[OH^-] = \frac{0.000001 \text{ mol}}{(0.04000 \text{ L} + 0.04001 \text{ L})} = 1.250 \times 10^{-5}\ M$$

pOH = –log (1.250 × 10^{-5}) = 4.903

pH = 14 – pOH = 14 – 4.903 = 9.1

e) We are farther past the equivalence point, and the amount of excess base determines the pH:

moles added OH^- = 0.05000 L × 0.1000 mol/L = 0.005000 mol

end moles OH^- = 0.005000 mol – 0.004000 mol = 0.001000 mol

$$[OH^-] = \frac{0.001000 \text{ mol}}{(0.04000 \text{ L} + 0.05000 \text{ L})} = 1.111 \times 10^{-2} M$$

pOH = –log (1.111 × 10^{-2}) = 1.9542

pH = 14 – pOH = 14 – 1.9542 = 12.0458

Near the equivalence point, addition of only 0.02 mL of base (less than half a drop) changes the pH from 4.845 to 11.10. Phenolphthalein is often used as an indicator for these titrations. It changes from colorless at pH 8.3 to pink at pH 10.0. Although the color change does not occur at the equivalence point, it changes within a partial drop of the equivalence point.

The titration curve for this titration is shown on page 838 of your text. What other indicators would be appropriate for this titration? (answer: Any that change color between pH 4 and 10; see the list of indicators on page 837 of your text, showing the colors the indicators assume at different pH values.)

Weak Acid-Strong Base Titrations

When we titrate a weak acid with a strong base, we create a buffered solution during the middle of the titration as the weak acid is deprotoned and exists in solution with its conjugate base.

Titration of acetic acid with sodium hydroxide

Consider the titration of acetic acid (a weak acid) with sodium hydroxide (a strong base):

$$CH_3COOH(aq) + OH^-(aq) \rightleftharpoons CH_3COO^-(aq) + H_2O$$

Choose words from the list below to complete the following paragraph:

(a) buffered
(b) equals
(c) higher
(d) methyl red (end point pH 5)
(e) phenolphthalein (end point pH 9)
(f) rapidly
(g) shorter
(h) slowly
(i) acetate ion

If we compare the titration of acetic acid (a weak acid) to a strong acid, the pH at the beginning of the titration of acetic acid will be _______________(1) than that of a strong acid of the same molarity. Initially the pH will rise more _______________(2) than it did with a strong acid. When enough base is added to make the acetate concentration 1/10th of the acetic acid concentration, the solution is _______________(3). In the buffered region, the pH changes _______________(4). At the midpoint of the buffer region, the concentration of acetic acid _______________(5) the concentration of acetate ion, and pH = pK_a. At the equivalence point, the solution contains only _______________(6), which makes the solution basic. The steep portion of the curve is _______________(7) than when a strong acid is titrated, so the choice of indicator is more crucial. The pH of a dilute sodium acetate solution is 9, so an appropriate indicator for the titration would be _______________(8). An indicator that would change color too early, and therefore be inappropriate would be _______________(9).

answers: 1-c, 2-f, 3-a, 4-h, 5-b, 6-i, 7-g, 8-e, 9-d.

Weak Base-Strong Acid Titrations

Consider the titration of the weak base, ammonia with HCl:

$$NH_3(aq) + H_3O^+(aq) \rightleftharpoons NH_4^+(aq) + H_2O$$

When we titrate ammonia (a weak base) with a hydrochloric acid (a strong acid), the shape of the titration curve is inverted from the titration curve for acetic acid (a weak acid) and sodium hydroxide (a strong base). The pH starts out high, drops quickly initially, and then drops slowly as a buffered solution forms. At the midpoint of the buffer region, the pH equals the pK_b of ammonia. The pH at the equivalence point is below 7 because the solution contains ammonium ions (NH_4^+), which are weak acids. The indicator must change color in an acidic pH range.

19.3 EQUILIBRIA OF SLIGHTLY SOLUBLE IONIC COMPOUNDS

In Chapter 13, we saw that most solutes do not dissolve completely, and that in a saturated solution, equilibrium exists between the undissolved solid and the dissolved solute:

$$CaF_2(s) \rightleftharpoons Ca^{2+}(aq) + 2F^-(aq)$$

We will assume in this chapter that any salt that dissolves, dissociates completely into ions. In reality, some of the dissolved calcium is in the form of CaF^+. When we assume that all the dissolved calcium exists as the Ca^{2+} ion (as we will do in this chapter), the calculated value for the solubility of calcium fluoride will be lower than the observed solubility of calcium fluoride. For this equilibrium we write:

$$K_c = \frac{[Ca^{2+}][F^-]^2}{[CaF_2]}$$

We assume the concentration of a solid is constant, and so its concentration is combined with the constant K_c to obtain K_{sp} (the solubility-product constant):

$$K_c[CaF_2] = [Ca^{2+}][F^-]^2 = K_{sp}$$

K_{sp} for CaF_2 at 25°C is 3.2×10^{-11}. K_{sp} depends on temperature.

Writing K_{sp} for slightly soluble ionic compounds

Write the solubility-product constant expression for:

a) calcium fluoride
b) barium carbonate
c) lead(II) iodate
d) mercury(II) sulfide

Which of these compounds is the least soluble in water?

solutions:

a) The equilibrium expression is:

$$CaF_2(s) \rightleftharpoons Ca^{2+}(aq) + 2F^-(aq)$$

$$K_{sp} = [Ca^{2+}][F^-]^2$$

b) The equilibrium expression is:

$$BaCO_3(s) \rightleftharpoons Ba^{2+}(aq) + CO_3^{2-}(aq)$$

$$K_{sp} = [Ba^{2+}][CO_3^{2-}]$$

c) The equilibrium expression is:

$$Pb(IO_3)_2(s) \rightleftharpoons Pb^{2+}(aq) + 2IO_3^-(aq)$$

$$K_{sp} = [Pb^{2+}][IO_3^-]^2$$

d) The equilibrium expression is:

$$HgS(s) + H_2O(l) \rightleftharpoons Hg^{2+}(aq) + HS^-(aq) + OH^-(aq)$$

$$K_{sp} = [Hg^{2+}][HS^-][OH^-]$$

S^{2-} is so basic that it reacts completely with water to form hydrogen sulfide ion and hydroxide ion. The least soluble compound is HgS with the smallest K_{sp} value (2×10^{-53}).

Before working problems, let's review from Chapter 4 solubility rules for ionic compounds in water.

Solubility Rules for Ionic Compounds in Water

1. **Anions of strong acids are soluble.**
   Most nitrate (NO_3^-), sulfate (SO_4^{2-}) and perchlorate (ClO_4^-) salts are soluble; exceptions are $CaSO_4$, $BaSO_4$, and $PbSO_4$ which are insoluble.

2. **Anions of weak acids are only slightly soluble.**
   Most sulfide (S^{2-}), carbonate (CO_3^{2-}), and phosphate (PO_4^{3-}) salts are only slightly soluble.

3. **Most hydroxide salts are only slightly soluble.**
   Exceptions are group 1 and heavy group 2 metals;
   NaOH, KOH, and $Ca(OH)_2$ are soluble.

4. **Most salts of Na^+, K^+, and NH_4^+ are soluble.**

5. **Most chloride, bromide, and iodide salts are soluble.**
   Exceptions are salts of Ag^+, Pb^{2+}, Cu^+, and Hg_2^{2+}, which are insoluble.

Determining K_{sp} from solubility

CaF_2 is used in the manufacture of steel and glass. When powdered fluorite is shaken with water at 18°C, 1.5×10^{-4} g dissolves in 10.0 mL of solution. Calculate K_{sp} of CaF_2 at 18°C.

solution:

First write the equilibrium and K_{sp} expressions:

$$CaF_2(s) \rightleftharpoons Ca^{2+}(aq) + 2F^-(aq); \quad K_{sp} = [Ca^{2+}][F^-]^2$$

Calculate the molar solubility of CaF_2:

$$\frac{1.5 \times 10^{-4}\ \text{g}}{0.010\ \text{L}} \times \frac{1\ \text{mol}}{78.08\ \text{g}} = 1.92 \times 10^{-4}\ \text{mol/L}$$

When one mole of CaF_2 dissolves, 1 mol of Ca^{2+} and 2 mol of F^- form:

$$[Ca^{2+}] = 1.92 \times 10^{-4}\ \text{mol/L}$$

$$[F^-] = 2 \times (1.92 \times 10^{-4}\ \text{mol/L}) = 3.84 \times 10^{-4}\ \text{mol/L}$$

Now we can calculate K_{sp}:

$$K_{sp} = [Ca^{2+}][F^-]^2 = (1.92 \times 10^{-4}\ \text{mol/L}) \times (3.84 \times 10^{-4}\ \text{mol/L})^2$$

$$= 2.8 \times 10^{-11}$$

Compare this value for K_{sp} to the one at 25°C in Table 19.3 of your text on page 850 (3.2×10^{-11}). K_{sp} is smaller at 18°C; less CaF_2 dissolves at the lower temperature.

Determining solubility from K_{sp}

K_{sp} for CaF_2 at 25°C is 3.2×10^{-11}. Estimate the solubility, in mol/L of this mineral.

solution:

We can see from the equation:

$$CaF_2(s) \rightleftharpoons Ca^{2+}(aq) + 2F^-(aq)$$

that 1 mol of Ca^{2+} and 2 mol of F^- form for every mol of CaF_2 that dissolves.
If S is the solubility of CaF_2 in mol/L, then:

$$[Ca^{2+}] = S, \text{ and } [F^-] = 2S$$

We can substitute into the expression for K_{sp}:

$$K_{sp} = [Ca^{2+}][F^-]^2 = S \times (2S)^2 = 3.2 \times 10^{-11}$$

$$4S^2 = 3.2 \times 10^{-11}$$

$$S = 2.0 \times 10^{-4}\ M$$

The solubility of calcium fluoride changes if a **common ion** is present. Would the solubility of CaF_2 in 0.10 M $Ca(NO_3)_2$ (a soluble salt) be greater or lower than in water? (answer: The dissolved $Ca(NO_3)_2$ increases the concentration of Ca^{2+} ions, so the concentration of F^- must decrease to maintain a constant K_{sp} value. The equilibrium shifts to the left, making calcium fluoride less soluble in the 0.10 M $Ca(NO_3)_2$ solution.)

Solubility of CaF_2 in the presence of a common ion

What is the solubility of CaF_2 in 0.10 M $Ca(NO_3)_2$? K_{sp} of $CaF_2 = 3.2 \times 10^{-11}$.

solution:

The main thing to keep in mind when working this problem is that most of the solid calcium fluoride does not go into solution. The tiny amount that *does* dissolve produces a concentration of calcium ion, Ca^{2+}, which is insignificant compared to the amount which comes from the soluble salt, $Ca(NO_3)_2$.

	$CaF_2(s)$	$\rightleftharpoons$	$Ca^{2+}(aq)$	+	$2F^-(aq)$
Initial conc.	—		0.10		0
Change	—		$+S$		$+2S$
Equilibrium conc.	—		$0.1 + S$		$2S$

$$K_{sp} = [Ca^{2+}][F^-]^2 = (0.1 + S) \times (2S)^2 = 3.2 \times 10^{-11}$$

We assume that $0.1 >> S$, so $(0.1 + S) \approx S$. Then:

$$(0.10)(2S)^2 = 3.2 \times 10^{-11}$$

$$4S^2 = 3.2 \times 10^{-10}$$

$$S = 8.9 \times 10^{-6}$$

Compare this solubility to that in pure water worked in the earlier example problem ($S = 2.0 \times 10^{-4}$). The presence of the common ion, Ca^{2+}, decreased solubility, as predicted.

Predicting whether a precipitate will form

Does a precipitate form when 0.500 L of 0.50 M $Ca(NO_3)_2$ is mixed with 0.500 L of 0.050 M NaF?

solution:

Nitrate and sodium salts are soluble, so the ions present in solution are:

Ca^{2+}, NO_3^-, Na^+, and F^-

$NaNO_3$ is soluble, so the only possibility for a precipitate is CaF_2.
Calculate the initial concentrations of calcium and fluoride ions:

$$[Ca^{2+}] = \frac{0.500\text{ L} \times 0.50\text{ mol/L}}{1\text{L}} = 0.025\ M$$

$$[F^-] = \frac{0.500\text{ L} \times 0.050\text{ mol/L}}{1\text{L}} = 0.025\ M$$

Substituting into Q_{sp} gives: $Q_{sp} = [Ca^{2+}][F^-]^2 = (0.025) \times (0.025)^2 = 1.6 \times 10^{-5}$.
Since $Q_{sp} > K_{sp}$, CaF_2 will precipitate until $Q_{sp} = 3.2 \times 10^{-11}$.

Separating Ions by Selective Precipitation

Ions in mixtures may be separated by selectively precipitating them. The less soluble compounds (ones with the smallest K_{sp} values) precipitate first.

Separating ions

100. mL of a 0.50 M $Sn(NO_3)_2$ solution is mixed with 100. mL of a 0.20 M $Cd(NO_3)_2$ solution. If aqueous NaOH is added to the mixture, which ion precipitates first? (K_{sp} of $Sn(OH)_2 = 5.4 \times 10^{-27}$ and K_{sp} of $Cd(OH)_2 = 8.1 \times 10^{-13}$.) Calculate the concentration of OH^- that will separate Sn from Cd.

solution:

$Sn(NO_3)_2$ and $Cd(NO_3)_2$ are soluble, so the ions in solution are Sn^{2+}, Cd^{2+}, and NO_3^-.
Hydroxide ion added to the solution will precipitate Sn^{2+} and Cd^{2+}:

$$Sn(OH)_2(s) \rightleftharpoons Sn^{2+}(aq) + 2OH^-(aq) \qquad K_{sp} = [Sn^{2+}][OH^-]^2 = 5.4 \times 10^{-27}$$

$$Cd(OH)_2(s) \rightleftharpoons Cd^{2+}(aq) + 2OH^-(aq) \qquad K_{sp} = [Cd^{2+}][OH^-]^2 = 8.1 \times 10^{-13}$$

We want to add enough hydroxide ion to precipitate as much of the least soluble compound as we can, but not enough to start precipitating the more soluble compound. $Sn(OH)_2$ has the smallest K_{sp} and is therefore the least soluble compound. If we add hydroxide ion until we have a saturated solution of $Cd(OH)_2$, that will be the most Sn^{2+} that can be removed from the mixture without precipitating any Cd^{2+}. We can then calculate the remaining concentration of Sn^{2+} to see if we accomplished the separation.

The initial concentration of Cd^{2+} is:

$$[Cd^{2+}] = \frac{(0.1\text{L} \times 0.20\text{ mol/L})}{0.2\text{L}} = 0.10\ M$$

(continued on next page)

The $[OH^-]$ that gives a saturated $Cd(OH)_2$ solution is:

$$[OH^-] = \sqrt{\frac{K_{sp}}{[Cd^{2+}]}} = \sqrt{\frac{8.1 \times 10^{-13}}{0.10}} = 2.8 \times 10^{-6}\,M$$

If the hydroxide concentration increases above 2.8×10^{-6} *M*, Cd^{2+} ion will begin to precipitate out as $Cd(OH)_2$. What is the concentration of Sn^{2+} ion remaining at this hydroxide ion concentration?

$$[Sn^{2+}] = \frac{K_{sp}}{[OH^-]^2} = \frac{5.4 \times 10^{-27}}{(2.84 \times 10^{-6})^2} = 6.7 \times 10^{-16}\ M$$

The initial concentration of Sn^{2+} was 0.25 *M*, so if we add hydroxide solution to make the concentration of hydroxide ion 2.8×10^{-6} *M*, essentially all the Sn^{2+} will precipitate leaving Cd^{2+} in solution.

19.4 EQUILIBRIA INVOLVING COMPLEX IONS

A complex ion is a metal ion (the **central atom**) covalently bonded to two or more anions or molecules (called **ligands**). The metal ion acts as a Lewis acid, which accepts an electron pair from each ligand. Compounds containing complex ions are called **coordination compounds.** Coordination compounds will be discussed in more detail in Chapter 23.

Coordination compounds

For each of the following coordination compounds, give the charge of the complex ion, the charge of the central atom, and list the ligands.

a) $[Pt(NH_3)_4]Cl_2$

b) $[Cu(NH_3)_4]SO_4$

c) $K_2[PtCl_4]$

d) $[Zn(H_2O)_3(OH)]Cl$

answers:

Ligands are written within the brackets; species written outside the brackets are not bonded directly to the central metal atom, and are present as free ions.

a) The NH_3 molecules are neutral and the chloride ions have a charge of –1.
charge of complex = +2; charge of metal (Pt) = +2
ligands: 4 NH_3 molecules

b) The NH_3 molecules are neutral and the sulfate ion has a charge of –2.
charge of complex = +2; charge of metal (Cu) = +2
ligands: 4 NH_3 molecules

c) The potassium ions have a charge of +1 and chloride ions have a charge of –1.
charge of complex = –2; charge of metal (Pt) = +2
ligands: 4 Cl^- ions

d) The water molecules are neutral. The hydroxide and chloride ions have –1 charges.
charge of complex = +1; charge of metal (Zn) = +2
ligands = 3 H_2O molecules and 1 OH^- ion

The **formation constant,** K_f, for a complex ion determines the extent of ligand substitution:

$$Zn(H_2O)_4^{2+}(aq) + 4NH_3(aq) \rightleftharpoons Zn(NH_3)_4^{2+}(aq) + 4H_2O(l)$$

$$K_f = \frac{[Zn(NH_3)_4^{2+}]}{[Zn(H_2O)_4^{2+}][NH_3]^4} = 7.8 \times 10^8$$

Notice that this formation constant is large, so the complex ion forms readily and is stable.

Calculating the concentration of complex ions

If 25.5 mL of 3.10×10^{-2} M $Fe(H_2O)_6^{3+}$ is mixed with 35.0 mL of 1.50 M NaCN, what is the final $[Fe(H_2O)_6^{3+}]$? K_f of $Fe(CN)_6^{3-} = 4.0 \times 10^{43}$.

solution:

The equation and K_f expressions are:

$$Fe(H_2O)_6^{3+}(aq) + 6CN^-(aq) \rightleftharpoons Fe(CN)_6^{3-}(aq) + 6H_2O(l)$$

$$K_f = \frac{[Fe(CN)_6^{3-}]}{[Fe(H_2O)_6^{3+}][CN^-]^6}$$

The initial reactant concentrations are:

$$[Fe(H_2O)_6^{3+}] = \frac{(0.0255\ L) \times (3.1 \times 10^{-2}\ mol/L)}{(0.0255\ L + 0.0350\ L)} = 0.013\ M$$

$$[CN^-] = \frac{(0.0350\ L) \times (1.5\ mol/L)}{(0.0255\ L + 0.0350\ L)} = 0.87\ M$$

We set up a reaction table with $x = [Fe(H_2O)_6^{3+}]$ at equilibrium. Since K_f is so large, we can assume that essentially all the $Fe(H_2O)_6^{3+}$ is converted to $Fe(CN)_6^{3-}$:

	$Fe(H_2O)_6^{3+}$	+ $6CN^-$	$\rightleftharpoons$ $Fe(CN)_6^{3-}$	+ $6H_2O(l)$
Initial conc.	0.013	0.87	0	
Change	$-\approx 0.013$	$-6(0.013)$	$+0.013$	
Equil. conc.	x	0.792	0.013	

Solving for x gives:

$$K_f = \frac{[Fe(CN)_6^{3-}]}{[Fe(H_2O)_6^{3+}][CN^-]^6} = \frac{(0.013)}{(x)(0.792)^6} = 4.0 \times 10^{43}$$

$$x = 1.3 \times 10^{-45}\ M$$

x, the amount of $Fe(H_2O)_6^{3+}$ left at equilibrium, is very small, as we would expect since the reaction proceeds nearly to completion (K_f is a large number).

CHAPTER CHECK

Make sure you can...

- Explain how a buffer works and what makes an effective buffer.
- Find the pH of a buffered solution using the Henderson-Hasselbalch equation.
- Explain how indicators work in acid-base titrations.
- Distinguish between a strong acid/strong base titration curve, a weak acid/strong base titration curve, and a weak base/strong acid titration curve.
- Calculate the pH at any point in an acid-base titration.
- Write the solubility-product constant expressions for slightly soluble ionic compounds.
- Determine K_{sp} from solubility data and vice versa.
- Calculate the decrease in solubility caused by the presence of a common ion.
- Use ion concentrations to calculate Q_{sp} to predict whether a precipitate forms.
- Calculate how to separate ions by selective precipitation based on their K_{sp} values.
- Calculate, using K_f values, the concentration of complex ions.

Chapter 19 Exercises

19.1) a. How would you prepare 1 L of a lactic acid/sodium lactate buffer solution with a pH of 4.00 from a 0.10 *M* $HC_3H_5O_3$ lactic acid solution and solid sodium lactate, $NaC_3H_5O_3$? (pK_a of $HC_3H_5O_3$ = 3.08)

b. How would you prepare a buffer solution with a pH of 4.50?

19.2) a. How would you prepare 500. mL of a buffer with a pH of 9.00 from 0.50 *M* aqueous NH_3 and a bottle of reagent grade NH_4Cl? A NH_3/NH_4Cl buffer containing equal molar amounts of base and salt has a pH of 9.25. (K_b of NH_3 = 1.77×10^{-5})

b. Could you prepare a buffer with a pH of 8.00 using the same salt and base?

19.3) a. What effect does the addition of the salt of the weak acid have on the pH of the acid/salt buffer?
b. If the salt of a weak base is added to a weak base/salt buffer, is the effect on the pH the same?

19.4) What volume of 0.50 *M* NaClO must be added to 0.20 *M* HClO in order to prepare 500 mL of a HClO/NaClO buffer with a pH of 7.00? (K_a HClO = 3.5×10^{-8})

19.5) The average adult has 5.0 L of blood of which 3.0 L is blood plasma with a pH of 7.4. The pH of the blood must be maintained within a very narrow region of pH, 7.35–7.45. Thus, the pH of the blood can vary only 0.05 pH units before a serious health risk results. What is the buffer capacity, with respect to excess acid, for the blood supply of an average adult?

19.6) 20.0 mL of 0.10 *M* KH_2PO_4 solution is titrated with 0.10 *M* KOH. Calculate the pH of the solution after the following amounts of KOH solution are added:

a. 0 mL b. 10.0 mL c. 19.8 mL d. 20.0 mL e. 25.0 mL

19.7) Calcium carbonate, $CaCO_3$, is an active ingredient in many antacids such as Tums. What is the molar solubility of $CaCO_3$ in g/L? How does the calcium carbonate work if it does not dissolve in our stomach? (K_{sp} = 8.7×10^{-9})

19.8) Calculate the K_{sp} for each of the salts below:

a. 0.35 g TlCl/100 mL solution
b. 1.35 g $Cu(IO_3)_2$/L solution

19.9) A primary treatment of municipal sewage involves the addition of calcium hydroxide and aluminum sulfate to a sedimentation tank to produce aluminum hydroxide, $Al(OH)_3$. As the $Al(OH)_3$ slowly settles to the bottom, it traps suspended particles and bacteria. What is the $[Al^{3+}]$ if the pH in the sedimentation tank is 8.00? (K_{sp} of $Al(OH)_3 = 2 \times 10^{-32}$ at 25°C)

19.10) Hard water produces soap scum on clothes and in bathtubs. One way to remove the Ca ions and Mg ions that causes hard water is to add hydrated lime, $Ca(OH)_2$. The Mg ions are removed in the form of $Mg(OH)_2$. If the metal ion concentration is less than 65 mg/gal, the water is considered soft. If the target $[Mg^{2+}]$ is only 30 mg/gal, what is the pH needed to cause the remaining Mg^{2+} to precipitate?

19.11) A nuclear chemist, working with a 5.0×10^{-3} M $Ra(NO_3)_2$ solution, precipitates the Ra^{2+} ion as $RaSO_4$ ($K_{sp} = 2.0 \times 10^{-11}$) in an effort to reduce the amount of radium in the waste. What concentration of Na_2SO_4 must be used in order to reduce the free $[Ra^{2+}]$ to 1.0 parts per trillion ? (1 ppt = 1 nanogram/L)

19.12) The concentrations of chloride ion and bromide ion in seawater are 0.55 M and 8.4×10^{-4} M, respectively. 1.0 M $AgNO_3$ is added to the seawater by a micropump. Which will precipitate first, AgCl or AgBr? The change in volume is negligible. What percent of the first ion has precipitated when the second ion begins to precipitate?

19.13) The ferrous ion concentration, $[Fe^{2+}]$, in the ground water of a Midwest aquifer is 3.6×10^{-4} M. What is the minimum concentration of NH_3 needed to initiate the precipitation of $Fe(OH)_2$?

Chapter 19 Answers

19.1) a. Dissolve 93 g of $NaC_3H_5O_3$ in about 500 mL of 0.10 M $HC_3H_5O_3$, dilute to 1.00 L with 0.100 M $HC_3H_5O_3$, and mix.

b. Dissolve 295 g of $NaC_3H_5O_3$ in about 500 mL of 0.10 M $HC_3H_5O_3$, dilute to 1.00 L with 0.100 M $HC_3H_5O_3$, and mix.

19.2) a. Dissolve 24 g NH_4Cl in about 250 mL of the 0.50 M NH_3 solution, dilute to 500 mL with the 0.20 M NH_3 solution, and mix:

$$0.500 \text{ L} \times 0.50 \text{ mol/L} = 0.25 \text{ mol } NH_3$$

$$9.00 = 9.25 + \log \frac{0.25 \text{ mol}}{x \text{ mol } NH_4Cl} \Rightarrow x = 0.44 \text{ mol } NH_4Cl$$

$$0.44 \text{ mol } NH_4Cl \times 53.49 \text{ g/mol} = 23.5 \text{ g } NH_4Cl$$

b. Dissolve 240 g of NH_4Cl in the 500 mL of 0.50 M NH_3 and mix:

$$8.00 = 9.25 + \log \frac{0.25 \text{ mol}}{x \text{ mol } NH_4Cl} \Rightarrow x = 4.45 \text{ mol } NH_4Cl$$

$$4.45 \text{ mol } NH_4Cl \times 53.49 \text{ g/mol} = 238 \text{ g } NH_4Cl$$

19.3) a. The addition of the salt of the weak acid will increase the pH of the buffer.
b. The addition of the salt of a weak base will decrease the pH of the buffer.

19.4) 61 mL of NaClO must be added to 439 mL of 0.20 *M* HClO.

19.5) 1.4×10^{-8} moles of H^+

19.6) a. pH = 4.10; b. pH = 7.20; c. pH = 9.20; d. pH = 9.95; e. pH = 12.18

19.7) 9.3×10^{-3} g/L
The excess acid consumes the soluble carbonate ion and by Le Chatelier's principle, the lost carbonate ion is replaced by the dissolution of more calcium carbonate.

19.8) a. $K_{sp} = [Tl^+][Cl^-]$

$[Tl^+] = [Cl^-] = 0.35 \text{ g}/239.836 \text{ g/mol}/0.100 \text{ L} = 1.459 \times 10^{-2}$

$K_{sp} = (1.459 \times 10^{-2})^2 = 2.1 \times 10^{-4}$

b. $K_{sp} = [Cu^{2+}][IO_3^-]^2$

$[Cu^{2+}] = 1.35 \text{ g}/413.35 \text{ g/mol}/1 \text{ L} = 3.266 \times 10^{-3}$; $[IO_3^-] = 2 \times (3.266 \times 10^{-3}) = 6.532 \times 10^{-3}$

$K_{sp} = (3.266 \times 10^{-3}) \times (6.532 \times 10^{-3})^2 = 1.39 \times 10^{-7}$

19.9) $[Al^{3+}] = 2 \times 10^{-14}$ *M*

19.10) The target concentration of Mg^{2+} in mol/L is:

$$\frac{30 \text{ mg}}{\text{gal}} \times \frac{1 \text{ gal}}{4 \text{ qt}} \times \frac{1.056 \text{ qt}}{\text{L}} \times \frac{1 \text{ mol}}{24.305 \text{ g}} \times \frac{1 \text{ g}}{1000 \text{ mg}} = 3.26 \times 10^{-4} M$$

$Mg(OH)_2(s) \rightleftharpoons Mg^{2+}(aq) + 2OH^-(aq)$ $\quad K_{sp} = [OH^-]^2[Mg^{2+}] = 6.3 \times 10^{-10}$

Let $x = [OH^-]$, then: $x^2 (3.26 \times 10^{-4}) = 6.3 \times 10^{-10}$

$x = 1.39 \times 10^{-3} = [OH^-]$; $1.39 \times 10^{-3} \times [H^+] = 1 \times 10^{-14}$; $[H^+] = 7.19 \times 10^{-12}$

$pH = -\log (7.19 \times 10^{-12}) = 11.14$

19.11) $[Na_2SO_4] = 4.5$ *M*

19.12) Since the K_{sp} for AgBr is smaller than that for silver chloride, we might expect AgBr to precipitate first. However, because there is such a large excess of chloride ion compared to bromide ion, silver chloride precipitates until $[Ag^+] = 5.95 \times 10^{-10}$ at which point, when more silver is added, silver bromide begins to precipitate. The chloride ion concentration remaining at this point is $[Cl^-] = 0.27$ *M* (51% of the chloride has precipitated).

19.13) pH = 8.53; $[NH_3] = 6.33 \times 10^{-7}$

20

Thermodynamics: Entropy, Free Energy, and the Direction of Chemical Reactions

20.1 THE SECOND LAW OF THERMODYNAMICS

The First Law

In Chapter 6 we saw that the total energy of the universe is constant; energy cannot be created or destroyed (the first law of thermodynamics). The first law of thermodynamics recasts thermochemical "heats of reaction" into ΔH values. The first law does not tell us why a process occurs in a given direction. The second law of thermodynamics answers the why question.

Exothermic and endothermic reactions

Determine which of the following reactions are endothermic and which are exothermic. Which reactions are spontaneous (ones that occur by themselves)?

1) ice at room temperature melting
2) water below 0°C freezing
3) water vaporizing
4) methane burning
5) sodium chloride dissolving in water
6) iron rusting
7) diamond converting to graphite

answers:

The processes that absorb energy from their surroundings (endothermic reactions) are:

1) ΔH^{o}_{rxn} = 6.02 kJ/mol
3) ΔH^{o}_{rxn} = 44.0 kJ/mol
5) ΔH^{o}_{rxn} = 3.90 kJ/mol

(continued on next page)

The processes that release energy (exothermic reactions) are:

2) $\Delta H^{o}_{rxn} = -6.02$ kJ/mol

4) $\Delta H^{o}_{rxn} = -802$ kJ/mol

6) $\Delta H^{o}_{rxn} = -826$ kJ/mol

7) $\Delta H^{o}_{rxn} = -1.90$ kJ/mol

All of the reactions are spontaneous. Spontaneous reactions may be exothermic or endothermic, and may be fast or slow. Diamond changes to graphite spontaneously, but hardly instantaneously.

We cannot use the standard heat of formation for a reaction to determine if a reaction is spontaneous.

The Second Law

Picture a ball rolling uphill; a rusty nail changing back to a shiny one; carbon dioxide and water coming together to form wood; water freezing at room temperature. These processes never happen. Experience tells us you can't go back. **All spontaneous processes increase the entropy of the universe.** This is the second law of thermodynamics. **Entropy (*S*)** is the measure of a system's disorder, or its number of microstates; entropy increases as systems become more disordered (have more microstates):

$$S_{\text{disordered system}} > S_{\text{ordered system}} \text{ or, } S_{\text{more microstates}} > S_{\text{fewer microstates}}$$

The entropy of the universe is always increasing. A spontaneous process may occur in which the system becomes more ordered, but the entropy of the system + the surroundings must be greater than zero:

$$\Delta S_{\text{universe}} = \Delta S_{\text{sys}} + \Delta S_{\text{surr}} > 0$$

We will show how we can calculate both ΔS_{sys} and ΔS_{surr}.

If entropy is always increasing, then why does water freeze?

Water freezes spontaneously when the temperature is below 0°C. Does the entropy of the water molecules increase or decrease as they change from the liquid to the solid state? How can we understand the fact that water freezing below 0°C is a spontaneous process?

answer:

The entropy of water molecules decreases as they become more ordered in a solid crystal lattice. Since water freezes spontaneously below 0°C, the increase in entropy of the surroundings must be larger than this decrease in entropy of the system, making the total entropy change of the system + surroundings > 0.

The heat released to the surroundings as water freezes increases the random motion of particles in the surroundings causing their disorder, or number of microstates (entropy), to increase.

Entropy of the System (ΔS_{sys})

Entropy is a state function, so it depends only on the initial and final states:

$$\Delta S_{sys} = S_{final} - S_{initial}$$

Unlike enthalpies, which are all relative values, we can determine absolute entropies. The **third law of thermodynamics** states that a perfect crystal has zero entropy at absolute zero. At absolute zero, the crystal has no thermal motion, and so has only one arrangement of atoms. Entropy is an extensive property; it depends on the amount of a substance. The entropy of a mole of a substance in its standard state is its **standard molar entropy** (S^o in J/K mol).

Predicting relative S^o values

Which of the following would you predict would increase the entropy of a system?

a) temperature increase
b) solid changing to a liquid
c) liquid changing to a gas
d) solid or liquid dissolving in a liquid
e) gas dissolving in a solid or liquid
f) number of atoms increases (same physical state)

answers: a,b,c,d (sometimes),f.

All of the above would increase the disorder of a system except for a gas dissolving in a liquid or solid. Gas molecules are so disordered, that their disorder decreases as they incorporate into a liquid or solid matrix. The entropy of some solids may decrease when dissolved because, although the entropy of their ions is much greater in solution than in the crystal, the entropy of the solvent may decrease as water molecules become highly ordered around hydrated ions.

Predicting relative entropy values

Choose the member of each pair with the higher entropy:

1) 1 mol liquid nitrogen; 1 mol gaseous nitrogen
2) 1 mol $O_2(g)$; 1 mol $O_3(g)$
3) water at room temperature; water at 90°C
4) 1 mol dry ice ($CO_2(s)$); 1 mol $CO_2(g)$
5) 1 mol $CO_2(g)$; 1 mol $CO_2(g)$ dissolved in a cola

solutions:

1) 1 mol gaseous nitrogen; for a given substance, entropy increases in the sequence solid < liquid < gas.

2) 1 mol O_3; both oxygen and ozone are gaseous molecules that contain oxygen atoms. Ozone has more atoms in a molecule, so there are more types of motion available to it and its entropy is greater.

(continued on next page)

3) water at 90°C; entropy increases with an increase in temperature.

4) 1 mol gaseous CO_2; see solution "a."

5) 1 mol gaseous CO_2; the entropy of a gas decreases as it dissolves into a liquid or solid matrix.

Predicting the sign of $\Delta S°$

Predict the sign of ΔS^o for each of the following reactions:

1) $4NH_3(g) + 5O_2(g) \rightarrow 4NO(g) + 6H_2O(g)$

2) $CH_4(g) + 2O_2(g) \rightarrow CO_2(g) + 2H_2O(g)$

3) $CaCO_3(s) \rightarrow CaO(s) + CO_2(g)$

answers:

1) In general, when a reaction involves gaseous molecules, the change in the number of molecules of gaseous reactants and products dominates the entropy change. In reaction (1), 9 gaseous molecules produce 10 gaseous molecules. The number of gaseous molecules increases, and entropy increases; ΔS^o is positive.

2) 3 gaseous molecules produce 3 gaseous products, so the entropy change for this reaction is approximately 0.

3) A gas is produced from a solid reactant, so entropy increases, and ΔS^o is positive.

20.2 CALCULATING THE CHANGE IN ENTROPY OF A REACTION

We can add standard molar entropies to find the standard entropy of reaction:

$$\Delta S^o_{rxn} = \sum m\, S^o_{(products)} - \sum n\, S^o_{(reactants)}$$

where *m* and *n* are the coefficients of the balanced equation.

Calculating the standard entropy of reaction (ΔS_{sys})

Calculate ΔS^o at 25°C for the reaction:

$$CH_4(g) + 2O_2(g) \rightarrow CO_2(g) + 2H_2O(g)$$

given the standard entropy values:

$CH_4(g)$	186 J/K mol	$CO_2(g)$	214 J/K mol
$O_2(g)$	205 J/K mol	$H_2O(g)$	189 J/K mol

solution:

$$\Delta S^o_{rxn} = \sum m\, S^o_{(products)} - \sum n\, S^o_{(reactants)}$$

$$= (1\text{ mol} \times 214\text{ J/K mol}) + (2\text{ mol} \times 189\text{ J/K mol}) - (1\text{ mol} \times 186\text{ J/K mol}) - (2\text{ mol} \times 205\text{ J/K mol})$$

(continued on next page)

$$\Delta S^{\circ} = 214\ \text{J/K} + 378\ \text{J/K} - 186\ \text{J/K} - 410\ \text{J/K}$$

$$\Delta S^{\circ} = -4\ \text{J/K}$$

Notice that in the previous example we predicted that the entropy change for this reaction would be approximately 0 because moles of gaseous reactants = moles of gaseous products. When we calculate ΔS°, we find that it is close to 0.

Entropy of the Surroundings (ΔS_{surr})

Knowing the entropy change of a system will not tell us if a reaction is spontaneous. We must know the entropy change of a system and the surroundings ($\Delta S_{universe} = \Delta S_{system} + \Delta S_{surroundings}$). If the surroundings add heat to the system (endothermic reaction), the entropy of the surroundings decreases; if the system gives off heat to the surroundings (exothermic reaction), the entropy of the surroundings increases. The sign of ΔS_{surr} depends on the direction of heat flow, and the magnitude of ΔS_{surr} depends on the temperature. The change in entropy of the surroundings is greater when heat is added at a lower temperature:

$$\Delta S_{surr} = -\frac{q_{sys}}{T}$$

$$\boldsymbol{\Delta S_{surr} = -\frac{\Delta H_{sys}}{T}\ \text{(const. } P\text{)}}$$

We can calculate ΔS_{surr} by measuring ΔH_{sys} and the temperature at which the change takes place.

Entropy of solution formation

We saw in Chapter 13 that the entropy of solution formation (ΔS_{sys}) is positive, and favorable, but that ΔS_{surr} depends on the heat of solution. Explain why one of the following processes is spontaneous, and the other is not:

a) oil mixing with water

b) ethanol mixing with water

solution:

a) ΔH for the mixing of polar water molecules and nonpolar oil molecules is large and positive. Since:

$$\Delta S_{surr} = -\frac{\Delta H_{sys}}{T},$$

a large positive ΔH makes ΔS_{surr} a large negative number. Although the entropy of solution formation (ΔS_{sys}) is positive, the large negative ΔS_{surr} makes $\Delta S_{universe}$ negative, and solution formation is nonspontaneous.

b) ΔH for the mixing of polar molecules is small, and the positive value for ΔS_{sys} predominates over the change in entropy of the surroundings (ΔS_{surr}). $\Delta S_{universe}$ is positive and so solution formation is spontaneous.

Determining if a reaction is spontaneous (calculation of $\Delta S_{universe}$)

Does the combustion of methane with oxygen occur spontaneously at 298 K?

$$CH_4(g) + 2O_2(g) \rightarrow CO_2(g) + 2H_2O(g)$$

$$\Delta S^o_{sys} = -4 \text{ J/K}$$

$$\Delta H^o_f\ (CH_4) = -75 \text{ kJ}$$

$$\Delta H^o_f\ (CO_2) = -393.5 \text{ kJ}$$

$$\Delta H^o_f\ (H_2O) = -242 \text{ kJ}$$

solution:

For this reaction to be spontaneous, $\Delta S_{universe}$ must be > 0.
Since $\Delta S_{universe} = \Delta S_{sys} + \Delta S_{surr}$, and since $\Delta S^o_{sys} = -4$ J/K (above), ΔS^o_{surr} must be greater than +4 J/K in order for $\Delta S_{universe}$ to be > 0.

We calculate ΔS^o_{surr} from ΔH^o_{sys}:

$$\Delta S^o_{surr} = -\frac{\Delta H^o_{sys}}{T}$$

$$\Delta H^o_{sys} = \Delta H^o_{rxn} = \sum m\Delta H^o_{f\ (products)} - \sum n\Delta H^o_{f\ (reactants)}$$

$= (2 \text{ mol } H_2O \times -242 \text{ kJ/mol}) + (1 \text{ mol } CO_2 \times -393.5 \text{ kJ/mol}) - (1 \text{ mol } CH_4 \times -75 \text{ kJ/mol})$

$= (-484 \text{ kJ/mol}) - (-393.5 \text{ kJ/mol}) + (75 \text{ kJ/mol})$

$= -802.5 \text{ kJ/mol}$

$$\Delta S^o_{surr} = -\frac{\Delta H^o_{sys}}{T} = -\frac{-802.5 \text{ kJ/mol}}{298 \text{ K}} \times \frac{1000 \text{ J}}{\text{kJ}} = 2693 \text{ J/K}$$

(Note: Don't forget that the units of ΔS^o are J/K, not kJ/K.)

$$\Delta S^o_{universe} = \Delta S^o_{surr} + \Delta S^o_{sys} = 2693 \text{ J/K} - 4 \text{ J/K} = 2689 \text{ J/K}$$

$\Delta S^o_{universe} > 0$, so the reaction is spontaneous at 298 K. The entropy change of the system is almost zero, but the heat that this reaction gives off to its surroundings greatly increases the entropy of the surroundings and makes $\Delta S^o_{universe} > 0$.

At equilibrium, neither the forward nor the reverse reaction is spontaneous:

$$\Delta S_{sys} = -\Delta S_{surr}$$

$$\Delta S_{universe} = \Delta S_{surr} + \Delta S_{surr} = 0$$

20.3 ENTROPY, FREE ENERGY, AND WORK

We can predict the spontaneity of a process from ΔS_{univ}. Another thermodynamic function that is related to spontaneity is the Gibbs free energy (G):

$$G = H - TS$$

For a process at constant temperature and pressure, the change in free energy (ΔG) is:

$$\Delta G = \Delta H - T\Delta S$$

All quantities in this equation refer to the system. (When no subscript is used, assume the quantity refers to the system.) When ΔG is negative, the process is spontaneous.

Usually the enthalpy contribution (ΔH) to the free energy change (ΔG) is much larger than the entropy contribution ($T\Delta S$), especially at lower temperatures. The entropy contribution is more likely to become critical to the spontaneity of a process if the temperature is high. Most exothermic reactions, which have a negative ΔH, are spontaneous. In order for an exothermic reaction to be nonspontaneous (ΔG positive), there would need to be an increase in the order of the system (a negative ΔS term) and a temperature that is high enough to make the $T\Delta S$ term larger than ΔH.

Temperature and spontaneity

The formation of solid sodium chloride from elemental sodium and chlorine is an exothermic reaction ($\Delta H^o = -822.2$ kJ). The increase in order of the sodium and chlorine atoms in the sodium chloride lattice makes the entropy of the system negative ($\Delta S^o = -181.7$ J/K). At what temperatures is this exothermic reaction nonspontaneous?

solution:

The reaction is not spontaneous when ΔG^o is positive. Since both ΔH^o and ΔS^o are negative, ΔG^o will be positive when the $T\Delta S^o$ term is larger than the ΔH^o term.

Calculate when $\Delta H^o = T\Delta S^o$:

$$-822.2 \text{ kJ} = T \times (-181.7 \text{ J/K})$$

$$T = \frac{-822{,}200 \text{ J}}{-181.7 \text{ J/K}} = 4525 \text{ K}$$

The reaction is not spontaneous at temperatures above 4525 K (4252°C).

Gibbs free energy (G)

A process is spontaneous if $\Delta S_{univ} > 0$. To see how ΔG relates to spontaneity, relate ΔG to ΔS_{univ}.

solution:

We know that: $\Delta S_{univ} = \Delta S_{surr} + \Delta S_{sys}$, and

(continued on next page)

$$\Delta S_{surr} = -\frac{\Delta H_{sys}}{T} \text{ (const. } P\text{)}$$

Substituting $-\Delta H_{sys}/T$ for ΔS_{surr} in the first equation gives:

$$\Delta S_{univ} = -\frac{\Delta H_{sys}}{T} + \Delta S_{sys}$$

Multiply through by $-T$:

$$-T\Delta S_{univ} = \Delta H_{sys} - T\Delta S_{sys}$$

Comparing this equation to the one for ΔG earlier ($\Delta G = \Delta H - T\Delta S$), we see that the right hand sides of the two equations are the same, and therefore:

$$\Delta G = -T\Delta S_{univ} \text{ (const. } T, P\text{)}$$

or,
$$\Delta S_{univ} = -\frac{\Delta G_{sys}}{T}$$

A process is spontaneous if $\Delta S_{univ} > 0$. Since temperature in Kelvin is always > 0, a process carried out at constant temperature and pressure will be spontaneous when ΔG is negative.

When:

$\Delta G < 0$, the process is spontaneous

$\Delta G > 0$, the process is nonspontaneous

$\Delta G = 0$, the system is at equilibrium

ΔS_{univ} and ΔG

When any process is spontaneous, the entropy of the universe increases ($\Delta S_{univ} > 0$). When a process at constant temperature and pressure is spontaneous, the change in free energy is less than 0 ($\Delta G < 0$).

Spontaneity of melting ice

Use the change in free energy (ΔG°) and the change in the entropy of the universe (ΔS_{univ}) to predict the spontaneity of ice melting at: 1) –10°C, 2) 0°C, 3) 10°C.

$H_2O(s) \rightarrow H_2O(l)$, $\Delta H^\circ = 6.03 \times 10^3$ J/mol;

$\Delta S^\circ = 22.1$ J/K mol

(The superscript zero (°) indicates all substances are in their standard states.)

solution:

(continued on next page)

1) ΔG^o :

substitute –10°C (263 K), and the values for ΔH^o and ΔS^o into the free energy equation:

$$\Delta G^o = \Delta H^o - T\Delta S^o$$

$$\Delta G^o = 6.03 \times 10^3 \text{ J/mol} - (263 \text{ K} \times 22.1 \text{ J/K mol})$$

$$\Delta G^o = 2.17 \times 10^2 \text{ J/mol}$$

ΔS_{univ}:

$$\Delta S_{univ} = \Delta S^o + \Delta S_{surr}$$

$$\Delta S^o_{surr} = -\frac{\Delta H^o_{sys}}{T} = \frac{-6.03 \times 10^3 \text{ J/mol}}{263 \text{ K}} = -22.9 \text{ J/K mol}$$

$$\Delta S_{univ} = (22.1 \text{ J/K mol}) + (-22.9 \text{ J/K mol}) = -0.8 \text{ J/K mol}$$

ΔG^o is positive and ΔS_{univ} is negative, so ice does not melt spontaneously at –10°C. (Is this the answer you expected from experience?)

2) ΔG^o :

substitute 0°C (273 K), and the values for ΔH^o and ΔS^o into the free energy equation:

$$\Delta G^o = \Delta H^o - T\Delta S^o$$

$$\Delta G^o = 6.03 \times 10^3 \text{ J/mol} - (273 \text{ K} \times 22.1 \text{ J/K mol})$$

$$\Delta G^o = 0 \text{ J/mol}$$

ΔS_{univ}:

$$\Delta S_{univ} = \Delta S^o + \Delta S_{surr}$$

$$\Delta S^o_{surr} = -\frac{\Delta H^o_{sys}}{T} = \frac{-6.03 \times 10^3 \text{ J/mol}}{273 \text{ K}} = -22.1 \text{ J/K mol}$$

$$\Delta S_{univ} = (22.1 \text{ J/K mol}) + (-22.1 \text{ J/K mol}) = 0 \text{ J/K mol}$$

ΔG^o and ΔS_{univ} equal 0, so water and ice are at equilibrium at 0°C.

3) ΔG^o :

substitute 10°C (283 K), and the values for ΔH^o and ΔS^o into the free energy equation:

$$\Delta G^o = \Delta H^o - T\Delta S^o$$

$$\Delta G^o = 6.03 \times 10^3 \text{ J/mol} - (283 \text{ K} \times 22.1 \text{ J/K mol})$$

$$\Delta G^o = -2.24 \times 10^2 \text{ J/mol}$$

ΔS_{univ}:

$$\Delta S_{univ} = \Delta S^o + \Delta S_{surr}$$

$$\Delta S^o_{surr} = -\frac{\Delta H^o_{sys}}{T} = \frac{-6.03 \times 10^3 \text{ J/mol}}{283 \text{ K}} = -21.3 \text{ J/K mol}$$

$$\Delta S_{univ} = (22.1 \text{ J/K mol}) + (-21.3 \text{ J/K mol}) = 0.793 \text{ J/K mol}$$

ΔG^o is negative and ΔS_{univ} is positive, so ice melts spontaneously at 10°C.

When Ice Melts

H_2O molecules become more disordered and their entropy increases:

ΔS_{sys} (ΔS^o) favors the melting of ice.

Heat is absorbed from the surroundings and the entropy of the surroundings decreases; ΔS_{surr} favors the freezing of ice.

Below 0°C, $\Delta S_{surr} > \Delta S_{sys}$, and ice freezes.

Above 0°C, $\Delta S_{sys} > \Delta S_{surr}$, and ice melts.

At 0°C, $\Delta S_{surr} = \Delta S_{sys}$, and ice and water coexist.

When mercury boils

Mercury has an enthalpy of vaporization of 58.51 kJ/mol, and an entropy of vaporization of 92.92 J/K mol. What is the normal boiling point of mercury?

$Hg(l) \rightarrow Hg(g)$ $\quad \Delta H^o = 58.51$ kJ/mol

$\Delta S^o = 92.92$ J/K mol

solution:

At the boiling point of mercury, liquid and gaseous mercury are in equilibrium ($\Delta G^o = 0$):

$$\Delta G^o = \Delta H^o - T\Delta S^o = 0$$

$$\Delta H^o = T\Delta S^o$$

$$T = \frac{\Delta H^o}{\Delta S^o} = \frac{58{,}510 \text{ J/mol}}{92.92 \text{ J/k mol}} = 629.7 \text{ K } (356.7^o\text{C})$$

ΔG^o_{rxn} can be calculated using values for the standard free energy of formation:

$$\mathbf{\Delta G^o_{rxn} = \sum m \Delta G^o_{f\,(products)} - \sum n \Delta G^o_{f\,(reactants)}}$$

where $\sum$ means "the sum of," ΔG^o_f is the standard free energy of formation, and *m* and *n* are the coefficients of the reactants and products from the balanced equation.

Calculation of ΔG^o_{rxn}

Calculate ΔG^o_{rxn} at 25°C for the following reactions:

1) $CH_4(g) + 2O_2(g) \rightarrow CO_2(g) + 2H_2O(g)$
2) $2H_2S(g) + SO_2(g) \rightleftharpoons 3S(s) + 2H_2O(g)$

(continued on next page)

given the values for the standard free energies of formation:

$CH_4(g)$	–51 kJ/mol	$H_2S(g)$	–33 kJ/mol
$CO_2(g)$	–394 kJ/mol	$H_2O(g)$	–229 kJ/mol
$SO_2(g)$	–300 kJ/mol		

solution:

1) $\Delta G^o_{rxn} = \sum m\,\Delta G^o_{f\ (products)} - \sum n\,\Delta G^o_{f\ (reactants)}$

$= (1 \text{ mol} \times -394 \text{ kJ/mol}) + (2 \text{ mol} \times -229 \text{ kJ/mol}) - (1 \text{ mol} \times -51 \text{ kJ/mol}) - (2 \text{ mol} \times 0)$

$= -394 \text{ kJ} - 458 \text{ kJ} + 51 \text{ kJ}$

$= -801 \text{ kJ}$

2) $\Delta G^o_{rxn} = \sum m\,\Delta G^o_{f\ (products)} - \sum n\,\Delta G^o_{f\ (reactants)}$

$= (3 \text{ mol} \times 0) + (2 \text{ mol} \times -229 \text{ kJ/mol}) - (2 \text{ mol} \times -33 \text{ kJ/mol}) - (1 \text{ mol} \times -300 \text{ kJ/mol})$

$= -458 \text{ kJ} + 68 \text{ kJ} + 300 \text{ kJ}$

$= -92 \text{ kJ}$

Both reactions are spontaneous in the forward direction at 25°C.

Free Energy and Work

For a spontaneous process, ΔG represents the energy that is free to do useful work:

$$\Delta G = w_{max}$$

For a nonspontaneous process, ΔG represents the minimum amount of work that must be done to make the process occur. The amount of work we obtain from any real process is always less than the maximum possible amount because some of the work is changed to heat the surroundings. We can determine the work potential of a process, but we can never achieve this potential. Thermodynamicist Henry Bent sums up these thoughts as follows:

First law: You can't win, you can only break even.

Second law: You can't break even.

20.4 FREE ENERGY, EQUILIBRIUM, AND REACTION DIRECTION

In Chapter 16 on kinetics, we saw that at equilibrium, the forward and reverse reaction rates are equal. From a thermodynamic point of view, equilibrium is the lowest value of free energy available to the system. The free energy is related to the equilibrium constant by:

$$\Delta G = \Delta G^o + RT \ln Q$$

At equilibrium, $\Delta G = 0$, and $Q = K$, so:

$$\Delta G^o = -RT \ln K$$

ΔG^o for the autoionization of water

What is ΔG^o for the autoionization of water at 25°C?

solution:

For the process, $H_2O(l) \rightleftharpoons H^+(aq) + OH^-(aq)$, $K = 1.0 \times 10^{-14}$.

We can use the relationship between ΔG^o and K to calculate ΔG^o:

$$\Delta G^o = -RT \ln K$$

$$\Delta G^o = -(8.3148 \text{ J/K mol} \times 298 \text{ K}) \ln (1.0 \times 10^{-14})$$

$$\Delta G^o = 80. \text{ KJ}$$

ΔG^o is positive, so the reverse reaction is favored; only a tiny percent of water molecules exist in their ionized form.

CHAPTER CHECK

Make sure you can...

- State the second law of thermodynamics mathematically.
- Predict relative S^o values for physical changes.
- Predict the sign of ΔS^o for chemical reactions.
- Calculate the change in entropy for a reaction.
- Calculate the entropy of the surroundings knowing ΔH_{sys} and the temperature.
- Determine if a reaction is spontaneous by calculation of $\Delta S_{universe}$.
- Relate Gibbs free energy (G) to H, T, and S.
- Determine if a reaction is spontaneous by calculation of ΔG.
- Calculate ΔG^o_{rxn} using values for the standard free energy of formation.
- Know the relationship between ΔG and useful work, w_{max}.
- Relate free energy to the equilibrium constant, K.

Chapter 20 Exercises

20.1) Without consulting any tables, indicate which of the two substances has a greater amount of entropy. Explain your answer:

a. $CH_4(g)$ (natural gas) or $CH_3CH_2CH_3(g)$ (propane gas)
b. $N_2(g)$ (atmospheric nitrogen) or $N_2(l)$ (liquid nitrogen)
c. $Fe(OH)_2$ (Iron(II) hydroxide) or $Fe(OH)_3$ (Iron(III) hydroxide)
d. $Br(g)$ (atomic bromine) or $Br_2(g)$ (molecular bromine)

20.2) Calculate: ΔS^o_{sys}, ΔS^o_{surr}, and ΔS^o_{univ} for the reaction below (25°C) and indicate if the reaction is spontaneous.

An iron nail placed in a solution of copper(II) sulfate will develop a copper coating on the nail:

$$Fe(s) + CuSO_4(aq) \rightleftharpoons Cu(s) + FeSO_4(aq)$$

ΔH^o_f $FeSO_4$ = –998.3 kJ/mol; S^o $FeSO_4$ = –117.6 J/mol K

ΔH^o_f $CuSO_4$ = –884.5 kJ/mol; S^o $CuSO_4$ = –79.5 J/mol K

20.3) Is the entropy change generally greater when a substance changes from a solid to a liquid, or when it changes from a liquid to a gas? Explain your answer.

20.4) Calculate ΔG^o_{rxn} for each reaction at 25°C:

a. Near the earth surface ozone is actually an air pollutant with a tolerance level of 0.12 ppm. Ozone reacts in the presence of ultraviolet light with another air pollutant, NO_2, to produce nitric acid:

$$O_3(g) + H_2O(g) + 2NO_2(g) \rightleftharpoons 2HNO_3(l) + O_2(g)$$

b. Phosgene gas, $COCl_2$, was first used as a chemical weapon during WWI. It reacts with water in the lungs:

$$COCl_2(g) + H_2O(l) \rightleftharpoons CO_2(g) + 2HCl(aq)$$

20.5) Before development of metal catalysts, SO_2 was converted to SO_3 using NO_2 as a catalyst in a method referred to as the lead chamber process:

$$2SO_2(g) + 2NO_2(g) \rightleftharpoons 2SO_3(g) + 2NO(g)$$

$$2NO(g) + O_2(g) \rightleftharpoons 2NO_2(g)$$

$$2SO_2(g) + O_2(g) \rightleftharpoons 2SO_3(g) \quad \text{(net reaction)}$$

What is ΔG^o_{rxn} for each step at 100°C?

20.6) Calculate ΔG^o_{rxn} for the reaction:

$$C_{diamond}(s) \rightarrow C_{graphite}(s)$$

Given that (25°C):

$$C_{diamond}(s) + O_2(g) \rightarrow CO_2(g) \qquad \Delta G^o = -397 \text{ kJ}$$

$$C_{graphite}(s) + O_2(g) \rightarrow CO_2(g) \qquad \Delta G^o = -394 \text{ kJ}$$

Is the conversion of diamonds to graphite a spontaneous process?

20.7) The decomposition of $KClO_3$ is a spontaneous process:

$$KClO_3(s) \rightarrow KCl(s) + 3/2O_2(g) \qquad \Delta G^o = -117 \text{ kJ}$$

How much useful work can we obtain when 1 mole of $KClO_3$ decomposes?

Could the reaction be reversed by doing 50 kJ of work? 120 kJ of work?

20.8) The free energy of formation of NO(g) and NO_2(g) from the elements nitrogen and oxygen is 86.6 and 51.9 kJ/mol, respectively (25°C, 1 atm). Under ordinary conditions, would you expect these compounds to decompose to N_2(g) and O_2(g)? How do you explain the fact that these compounds are a major source of air pollution?

20.9) At what temperatures are the following reactions spontaneous?

a. An innovative, efficient coal gasification process converts coal into methane, which is easier to transport and burns more cleanly than coal. The overall process is:

$$2C(s) + 2H_2O(g) \rightleftharpoons CH_4(g) + CO_2(g)$$

b. The use of ammonia as a fertilizer in agriculture makes it one of the most important chemicals manufactured in the world. Fritz Haber developed a method for the synthesis of ammonia directly from nitrogen and hydrogen:

$$N_2(g) + 3H_2(g) \rightleftharpoons 2NH_3(g)$$

c. The first step in the manufacturing of nitric acid by the Ostwald process is the oxidation of ammonia:

$$4NH_3(g) + 5O_2(g) \rightleftharpoons 4NO(g) + 6H_2O(g)$$

20.10) Describe the effect of temperature on reaction spontaneity at 1 atm pressure for the following conditions:

a. ΔH is negative; ΔS is positive
b. ΔH is positive; ΔS is negative
c. ΔH and ΔS are both positive
d. ΔH and ΔS are both negative

20.11) What are the equilibrium constants for the following reactions?

a. The photochemical decomposition of Freon in the stratosphere leads to the decomposition of the ozone present. The average temperature in the stratosphere is –55°C:

$$2O_3(g) \rightleftharpoons 3O_2(g)$$

b. The slightly acidic nature of unpolluted rainfall is the result of the carbon dioxide in the atmosphere reacting with water:

$$H_2O(l) + CO_2(g) \rightleftharpoons H_2CO_3(aq)$$

Chapter 20 Answers

20.1) a. Propane gas; more complex molecules contain more types of molecular motion.
b. N_2(g); gases have more random movement than liquids, less intermolecular forces of attraction.
c. $Fe(OH)_3$; more complex molecules contain more types of molecular motion.
d. Br_2(g); more complex molecules contain more types of molecular motion.

20.2) $\Delta S^o_{sys} = -32.2$ J/mol K

$$\Delta S^o_{surr} = -\frac{\Delta H^o_{sys}}{T} = -\frac{-153{,}800 \text{ J/mol}}{298 \text{ K}} = 516 \text{ J/mol K}$$

$$\Delta S^o_{univ} = \Delta S^o_{sys} + \Delta S^o_{surr} = -32.2 \text{ J/mol K} + 216 \text{ J/mol K} = 484 \text{ J/mol K}$$

Since ΔS^o_{univ} is greater than zero, the reaction is spontaneous.

20.3) The increase in randomness is usually much greater when a liquid vaporizes than when a solid melts. Recall our model of a solid, liquid, and gas first introduced on page 2 in the study guide. Disorder increases more when molecules separate in going from a liquid to a gas, than when they become disordered, but remain touching, in going from a solid to a liquid.

20.4) a. $\Delta G^o_{rxn} = -196$ kJ; spontaneous

b. $\Delta G^o_{rxn} = -214$ kJ; spontaneous

20.5) Step 1: $\Delta H^o = -84.2$ kJ/mol; $\Delta S^o = -41.3$ J/K·mol

$\Delta G^o_{rxn} = \Delta H^o - T\Delta S^o = -84.2 \text{ kJ/mol} - 373 \text{ K}(-.0413 \text{ kJ/ K·mol}) = -68.8 \text{ kJ/mol}$

Step 2: $\Delta H^o = -114.2$ kJ/mol; $\Delta S^o = -146.5$ J/ K·mol

$\Delta G^o_{rxn} = \Delta H^o - T\Delta S^o = -114.2 \text{ kJ/mol} - 373 \text{ K}(-0.1465 \text{ kJ/ K·mol}) = -59.54 \text{ kJ/mol}$

Net reaction: $\Delta G^o_{rxn} = -68.8 \text{ kJ/mol} - 59.54 \text{ kJ/mol} = -128.3 \text{ kJ/mol}$

20.6) Reverse the second equation to make graphite a product and add it to the first equation:

	$C_{diamond}(s) + O_2(g) \rightarrow CO_2(g)$	$\Delta G^o = -397$ kJ
	$CO_2(g) \rightarrow C_{graphite}(s) + O_2(g)$	$\Delta G^o = 394$ kJ
Sum:	$C_{diamond}(s) \rightarrow C_{graphite}(s)$	$\Delta G^o = -3$ kJ

As we discussed at the beginning of Chapter 16 on kinetics, diamonds spontaneously change to graphite, but the process is so slow that it seems to us as though "diamonds are forever."

20.7) For a spontaneous process, ΔG represents the energy that is free to do useful work: $\Delta G = w_{max}$. 117 kJ of useful work can be obtained when 1 mole of $KClO_3$ decomposes. The reaction could not be reversed by doing 50 kJ of work per mole of $KClO_3$, but it could by doing 120 kJ of work.

20.8) The formation of NO(g) and $NO_2(g)$ from the elements is a nonspontaneous process (ΔG^o is positive), and therefore, the reverse process, the decomposition to $N_2(g)$ and $O_2(g)$ is spontaneous. This decomposition reaction occurs slowly enough that it gives NO(g) and $NO_2(g)$ time to hang around in our atmosphere as sources of air pollution.

20.9) a. $\Delta H^o = 15.28$ kJ; $\Delta S^o = 11.0$ J/K

$\Delta G^o_{rxn} = \Delta H^o - T\Delta S^o = -15.28 \text{ kJ} - T(-0.011 \text{ kJ/K}) = 0$

Solve for T to give: $T = 1389$ K

The reaction is spontaneous when $\Delta G^o < 0$, or when the temperature is above 1389 K.

b. $\Delta H^{\circ} = -91.8$ kJ; $\Delta S^{\circ} = -197.3$ J/K
$\Delta G^{\circ}_{rxn} = \Delta H^{\circ} - T\Delta S^{\circ} = -91.8 \text{ kJ} - T(-0.1973 \text{ kJ/K}) = 0$
Solve for T to give: $T = 465$ K
The reaction is spontaneous when $\Delta G^{\circ} < 0$, or when the temperature is below 465 K.

c. $\Delta H^{\circ} = -906.2$ kJ; $\Delta S^{\circ} = 363.8$ J/K
$\Delta G^{\circ}_{rxn} = \Delta H^{\circ} - T\Delta S^{\circ} = -906.2 \text{ kJ} - T(0.3638 \text{ kJ/K}) = 0$
Solve for T to give: $T = -2490$ K
The reaction is always spontaneous since T cannot be lower than 0 K.

20.10) We evaluate the effect on ΔG° using the relationship:

$$\Delta G^{\circ} = \Delta H^{\circ} - T\Delta S^{\circ}$$

a. ΔG° is always negative; reaction is spontaneous at all temperatures.

b. ΔG° is always positive; reaction is nonspontaneous at all temperatures; the reverse reaction is spontaneous.

c. ΔG° is positive at small values of T, and becomes negative at higher values of T; the reaction is nonspontaneous at low temperatures, but becomes spontaneous as the temperature increases.

d. ΔG° is negative at small values of T, and becomes positive at higher values of T; the reaction is spontaneous at low temperatures, and the reverse reaction becomes spontaneous at higher temperatures.

20.11) a. $K_p = 1.13 \times 10^{78}$

b. $K_p = 0.034$

21

Electrochemistry: Chemical Change and Electrical Work

21.1 REDOX REACTIONS AND ELECTROCHEMICAL CELLS

Remember from Chapter 4 that electron transfer reactions are called **oxidation-reduction** (or redox) reactions. **O**xidation **i**s the **l**oss of electrons, **r**eduction **i**s the **g**ain of electrons. **"Oil rig"** is a good mnemonic device for remembering this. We will see how to harness a redox reaction to produce an electrical current, or to use an electrical current to drive a redox reaction.

Balancing a redox equation (review)

In Chapter 4, we balanced the redox reaction that occurs between chloride and permanganate ions in acidic solution, by the oxidation-number method. Here we will balance the same equation using the half-reaction method. The equation is:

$$MnO_4^-(aq) + H^+(aq) + Cl^-(aq) \rightarrow Mn^{2+}(aq) + Cl_2(g) + H_2O(l)$$

Split the overall reaction into two half-reactions containing only species which are oxidized or reduced:

$$\text{oxidation: } Cl^- \rightarrow Cl_2$$

$$\text{reduction: } MnO_4^- \rightarrow Mn^{2+}$$

Cl^- ions lose electrons (are oxidized) to obtain an oxidation number of 0 in elemental Cl_2. The oxygen atoms in MnO_4^- (oxidation number –2) contribute a negative charge of $4 \times -2 = -8$. The charge on the polyanion is –1, so manganese contributes a +7 charge. Mn gains electrons (is reduced) to obtain a +2 charge.

(continued on next page)

Balance the atoms and charges in each half-reaction:

$$\text{oxidation: } 2Cl^- \rightarrow Cl_2 + 2e^-$$

$$\text{reduction: } 5e^- + MnO_4^- + 8H^+ \rightarrow Mn^{2+} + 4H_2O$$

In the oxidation reaction, we simply balance the atoms by adding a coefficient of 2 in front of Cl^-, and balance the charge by adding two electrons to the right hand side of the equation.

In the reduction reaction, we must add 4 waters to the right hand side of the equation to make oxygen atoms balance. Then, since the reaction is in acidic solution, we balance the hydrogen atoms with protons on the left side of the equation. The charge balances by adding $5e^-$ to the left side of the equation. The $5e^-$ reduce Mn in MnO_4^- from +7 to +2. (Each side of the reaction has an overall charge of +2.)

Multiply each half-reaction by an integer to make e^- gained in reduction equal electrons lost in oxidation:

$$\text{oxidation: } 5[2Cl^- \rightarrow Cl_2 + 2e^-]$$

$$10Cl^- \rightarrow 5Cl_2 + 10e^-$$

$$\text{reduction: } 2[5e^- + MnO_4^- + 8H^+ \rightarrow Mn^{2+} + 4H_2O]$$

$$10e^- + 2MnO_4^- + 16H^+ \rightarrow 2Mn^{2+} + 8H_2O$$

This step is easy to check: the 10 e^- on the right side of the oxidation equation equal the10 e^- on the left side of the reduction equation. Make sure to multiply each member of the equation by the multiplication integer.

Add the balanced half-reactions and include states of matter:

$$10Cl^- + 10e^- + 2MnO_4^- + 16H^+ \rightarrow 2Mn^{2+} + 8H_2O + 5Cl_2 + 10e^-$$

$$10Cl^-(aq) + 2MnO_4^-(aq) + 16H^+(aq) \rightarrow 2Mn^{2+}(aq) + 8H_2O(l) + 5Cl_2(g)$$

Check that the atoms and charges balance.

For a reaction in basic solution, balance the equation as if it were in acidic solution. Add an OH^- ion to both sides of the equation for every H^+ ion present. The OH^- and H^+ ions on one side of the equation make water, and OH^- ions appear on the other side of the equation. Cancel excess water molecules that appear on both sides of the equation.

Electrochemical cells use a redox reaction to produce or utilize electrical energy. There are two types of electrochemical cells:

- **Voltaic cell** (or **galvanic cell**): uses a spontaneous reaction to produce an electrical current
- **Electrolytic cell:** uses electrical energy to drive a nonspontaneous reaction

Voltaic cells will be discussed in the next section, section 21.2. Electrolytic cells will be discussed later in section 21.6.

21.2 VOLTAIC CELLS

A spontaneous redox reaction occurs on its own. For example, a spontaneous redox reaction occurs when we add zinc to a blue copper sulfate solution. The blue color disappears, and a reddish brown deposit forms on the surface of the zinc. What is the redox reaction?

$$Zn(s) + Cu^{2+}(aq) \rightarrow Zn^{2+}(aq) + Cu(s)$$

Zinc metal reduces aquated Cu^{2+} ions (blue colored) to Cu metal (the reddish brown deposit). Zn is oxidized to Zn^{2+}. What color are aquated Zn^{2+} ions? (answer: Colorless.)

We can use this spontaneous redox reaction to do electrical work by constructing a **voltaic cell** (or **galvanic cell).** We separate the two half-reactions and connect them with a wire:

$$\text{oxidation: } Zn \rightarrow Zn^{2+} + 2e^-$$

$$\text{reduction: } Cu^{2+} + 2e^- \rightarrow Cu$$

The electrons given up by zinc atoms travel through the wire and do electrical work before they reduce the Cu^{2+} ions. Oxidation occurs at the **anode;** reduction occurs at the **cathode.** The electrode that "pumps" electrons into the external circuit (the anode) is considered to be the negative pole of the cell. The electrode that "pulls" electrons from the external circuit (the cathode) is considered to be the positive pole of the cell. Which are the negative and positive electrodes for the Zn/Cu^{2+} cell? (answer: The zinc half-reaction produces electrons that flow to the copper half-reaction. Zinc is the negative electrode; copper is the positive electrode.)

An excess of positive Zn^{2+} ions builds up around the zinc electrode, and the region around the copper electrode becomes deficient in positive ions as Cu^{2+} ions are reduced. A **salt bridge** connection allows ions to flow from one half-reaction to the other to eliminate this charge build up. A salt bridge contains (guess what?) a salt solution, which takes no part in the electrode reactions. Ions diffuse through the salt bridge from one half-reaction to the other keeping both solutions neutral.

The voltaic cell containing this reaction is shown on page 926 of your text.

Memory Devices

Oxidation occurs at the **anode** (both begin with vowels).

Reduction occurs at the **cathode** (both begin with consonants).

A way to remember more information about the cathode and anode for a galvanic cell is:

The **anode** is drawn on the **left** and is where **oxidation** takes place; it is the **negative** electrode for a **galvanic** cell.

anode = left = oxidation = negative = galvanic **(ALONG)**

Conversely then, the cathode is drawn on the right and is where reduction takes place; it is the positive electrode.

A voltaic cell

Consider the voltaic cell on page 926 of your text containing the reaction between zinc and Cu^{2+} ion. The container on the left contains a zinc rod in a 1 *M* solution of $ZnSO_4$ (a Zn^{2+} electrolyte). The container on the right contains a copper rod in a 1 *M* solution of $CuSO_4$ (a Cu^{2+} electrolyte). The electrodes are connected by a wire containing a voltmeter, and the solutions are connected by a salt bridge.

Choose words from the list below to complete the following paragraphs:

(a) active	(e) electrons	(i) negative
(b) anode	(f) inactive	(j) positive
(c) cathode	(g) increases	(k) reduce
(d) decreases	(h) left to right	(l) right to left

When zinc atoms are oxidized to Zn^{2+}, ______________(1) are generated which enter the wire. The zinc electrode, which generates electrons, is considered to have a ______________(2) charge. The cell containing the oxidation reaction is called the ______________(3).

Electrons flow through the wire to the solution containing Cu^{2+} ions where they ______________(4) Cu^{2+} to Cu^0. The electrode on the copper side of the cell has a ______________(5) charge. The cell containing the reduction reaction is called the ______________(6).

It is the convention to draw the anode on the left and the cathode on the right. In this setup, electrons move through the wire from ______________(7). In order to maintain neutrality in the two solutions, negatively charged anions (in this case, SO_4^{2-} ions) move through the salt bridge from ______________(8).

For this voltaic cell, the electrodes are components of the reaction, and so are ______________(9). As the reaction progresses, the mass of the zinc electrode ______________(10) as zinc atoms are oxidized to Zn^{2+} ions, and the mass of the copper electrode ______________(11) as Cu^{2+} ions are reduced to Cu^0 and "plate out" on the electrode. In some redox reactions, there are no reactants or products that can serve as electrodes. In these cases, electrodes are used that allow electrons to flow, but are not part of the reaction chemistry. These electrodes are called ______________(12). For the example problem showing the redox reaction between chloride and permanganate ions, a graphite or platinum electrode might be used.

<u>answers</u>: 1-e, 2-i, 3-b, 4-k, 5-j, 6-c, 7-h, 8-l, 9-a, 10-d, 11-g, 12-f.

Notation for a Voltaic Cell

A shorthand notation is useful for describing a voltaic cell. For the cell described above, and shown on page 926 of your text, the shorthand notation is:

$$Zn(s) \mid Zn^{2+}(aq) \parallel Cu^{2+}(aq) \mid Cu(s)$$

The convention is:

- The anode is written on the left.

- The double lines, ||, separate the two half-reactions.
- The single lines, |, represent a phase boundary.
- A comma separates members of the same phase.

Shorthand notation for a voltaic cell

Write the shorthand notation for a galvanic cell containing the reaction described in the first example of this section. Use graphite (inactive) electrodes:

$$MnO_4^-(aq) + H^+(aq) + Cl^-(aq) \rightarrow Mn^{2+}(aq) + Cl_2(g) + H_2O(l)$$

solution:

$$\text{Graphite} \mid Cl^-(aq) \mid Cl_2(g) \parallel H^+(aq), MnO_4^-(aq), Mn^{2+}(aq) \mid \text{Graphite}$$

Check:

- The anode (oxidation) is written on the left. Cl^- is oxidized to Cl_2.
- A double line separates the anode half-reaction and the cathode half-reaction.
- A single line indicates the phase boundaries between the solid graphite electrodes and the aqueous solutions; and between aqueous Cl^- and gaseous Cl_2.
- Commas separate aqueous members of the same phase.

21.3 CELL POTENTIAL

Which metal, zinc or copper, gives up its electrons more easily? (answer: Zinc; it gives up its electrons to reduce Cu^{2+} ions to Cu^0.) The difference in electrical potential between two electrodes is the **voltage,** or **cell potential (E_{cell}).** The Zn/Cu^{2+} cell described earlier measures 1.1 volts. This voltage is the sum of the voltage of the two half-reactions:

$$E^o_{cell} = E^o\,(Cu^{2+} \rightarrow Cu) + E^o\,(Zn \rightarrow Zn^{2+})$$

The superscript indicates the reactants are in their standard states. When a reaction is spontaneous, $E_{cell} > 0$.

We cannot measure the voltage of an electrode by itself; we can only measure the difference in voltage between two electrodes. We arbitrarily choose an electrode to have a zero voltage so that all other electrodes can be measured relative to it. The half-reaction of 1.0 *M* aqueous protons reduced to hydrogen gas (1 atm) is defined to be exactly 0.00 volts:

$$2H^+(aq, 1\,M) + 2e^- \rightarrow H_2(g, 1\text{ atm}) \qquad E^o_{reference} = 0.00\text{ V}$$

Let's set up a cell to measure the voltage for the Zn/Zn^{2+} half-reaction. The anode compartment contains a zinc metal electrode in a 1 *M* solution of Zn^{2+} and SO_4^{2-} ions. The cathode contains a platinum electrode in a 1 *M* solution of H^+ ions, and bathed with hydrogen gas at 1 atm **(the standard hydrogen electrode):**

$$Zn(s) \mid Zn^{2+}(aq) \parallel H^+(aq) \mid H_2(g) \mid Pt$$

Note: All components are in their standard states: the aqueous solutions are 1 *M*; H_2 gas is at 1 atm pressure.

- Which is the active electrode, and which is the inactive electrode?
  (answer: Zn(*s*) is the active electrode; Pt is the inactive electrode.)

- The voltage reads 0.76 volts. What is E° for the two half-reactions?
  We cannot measure the potential of the individual electrode processes, but since the H_2/H^+ half-reaction is assigned a potential of 0.00 volts, the reaction $Zn \rightarrow Zn^{2+} + 2e^-$ has a potential of 0.76 volt:

$$E^\circ_{cell} = E^\circ (H^+ \rightarrow H_2) + E^\circ (Zn \rightarrow Zn^{2+})$$

$$0.76 = 0.00 \text{ V} + E^\circ (Zn \rightarrow Zn^{2+})$$

$$0.76 = E^\circ (Zn \rightarrow Zn^{2+})$$

Voltage of Zn and Cu half-reactions

Assign E° values for the two half-reactions in the Zn/Cu^{2+} cell. (The voltmeter reads 1.10 V.)

$$Zn(s) + Cu^{2+}(aq) \rightarrow Zn^{2+}(aq) + Cu(s)$$

solution:

The half-reactions are:

oxidation: $Zn \rightarrow Zn^{2+} + 2e^-$

reduction: $Cu^{2+} + 2e^- \rightarrow Cu$

$$E^\circ_{cell} = E^\circ (Cu^{2+} \rightarrow Cu) + E^\circ (Zn \rightarrow Zn^{2+})$$

$$E^\circ (Zn \rightarrow Zn^{2+}) = 0.76 \text{ V (above)}$$

$$1.10 \text{ V} = E^\circ (Cu^{2+} \rightarrow Cu) + 0.76 \text{ V}$$

$$E^\circ (Cu^{2+} \rightarrow Cu) = 0.34 \text{ V}$$

In standard tables, all half-reactions are given as reduction processes. E° values for these half-reactions are called **standard reduction potentials.**

Standard reduction potentials for zinc and copper

What are the standard reduction potentials for zinc and copper?

solution:

In the previous exercise, we found that:

$$E^\circ (Cu^{2+} \rightarrow Cu) = 0.34 \text{ V}$$

$$E^\circ (Zn \rightarrow Zn^{2+}) = 0.76 \text{ V}$$

We must write both half-reactions as reductions:

(continued on next page)

E° ($Cu^{2+} \rightarrow Cu$) is a reduction half-reaction, so the standard reduction potential of copper is 0.34 V.

E° ($Zn \rightarrow Zn^{2+}$) is an oxidation half-reaction, so the reduction reaction equals

$-E^\circ$ ($Zn \rightarrow Zn^{2+}$) = –0.76 V.

Table 21.2 in your text (page 934) lists the standard reduction potentials at 298 K for many reactions. More values are given in Appendix D. The more positive the E° value, the more readily the reduction reaction occurs.

Relative oxidizing and reducing strengths

Consider the following reduction reactions:

$$2H^+(aq) + 2e^- \rightarrow H_2(g) \qquad E^\circ = 0.00 \text{ V}$$

$$Zn^{2+} + 2e^- \rightarrow Zn \qquad E^\circ = -0.76 \text{ V}$$

$$Cu^{2+} + 2e^- \rightarrow Cu \qquad E^\circ = +0.34 \text{ V}$$

1) Which reaction has the greatest tendency to take place? The least?
2) Which ion is the strongest oxidizing agent?
3) Which is the strongest reducing agent, H_2, Zn, or Cu?

answers:

1) Cu^{2+} reduction has the largest E° value, and so has the most tendency to take place; Zn^{2+} reduction has the smallest E° value, and so has the least tendency to take place.

2) Cu^{2+} is reduced most easily and so is the strongest oxidizing agent.

3) If we reverse the direction of the half-reactions, the signs of E° change. Zn oxidation has the largest E° value, and so Zn is the strongest reducing agent.

All this oxidizing and reducing can be confusing. Keep the following in mind:

- Standard electrode potentials are written as reductions.
- If the table shows a large E° value, the reaction has a high tendency to occur.
- An atom, ion, or molecule that is easily reduced is a strong oxidizing agent; it will take electrons from another species in order to be reduced.
- An atom, ion, or molecule that is easily oxidized is a strong reducing agent; it will donate electrons to another species in order to be oxidized.
- E° is an intensive property and does not depend on amount; *do not change an* E° *value when coefficients of a balanced equation are changed*.

The extreme

Fluoride ion is the most electronegative element. Would you predict the reduction reaction of $F_2(g)$ to $2F^-(aq)$ would have a large or small E° value? Is F_2 a strong or weak oxidizing agent? Is F^- a strong or weak reducing agent?

answers:

Fluorine molecules have a very high tendency to pick up electrons to become fluoride ions, so the reduction of fluorine to fluoride should have a large E° value. (In fact, $E^\circ = 2.87$, the largest value in the table.)

Fluorine is easily reduced so it is a strong oxidizing agent. F^- is not easily oxidized to F_2 ($E^\circ = -2.87$), so it is a weak reducing agent.

Comparing oxidizing agents

Use the data below to answer the following questions:

$$Sn^{4+}(aq) + 2e^- \rightarrow Sn^{2+}(aq) \qquad E^\circ = +0.13$$

$$Sn^{2+}(aq) + 2e^- \rightarrow Sn(s) \qquad E^\circ = -0.14$$

1) Which is the stronger oxidizing agent, $Sn^{2+}(aq)$ or $Sn^{4+}(aq)$?
2) Which is the stronger reducing agent, $Sn^{2+}(aq)$ or $Sn(s)$?
3) Is $Sn^{2+}(aq)$ a stronger oxidizing or reducing agent?

answers:

1) The strongest oxidizing agent is the most easily reduced. $Sn^{4+}(aq)$ is the stronger oxidizing agent because the reduction potential is more positive.
2) The strongest reducing agent is the most easily oxidized:

$$Sn^{2+}(aq) \rightarrow Sn^{4+}(aq) + 2e^- \qquad E^\circ = -0.13$$

$$Sn(s) \rightarrow Sn^{2+}(aq) + 2e^- \qquad E^\circ = +0.14$$

Sn(*s*) is the stronger reducing agent because the E° value is more positive.

3) Compare oxidation and reduction reactions:

$$Sn^{2+}(aq) \rightarrow Sn^{4+}(aq) + 2e^- \qquad E^\circ = -0.13$$

$$Sn^{2+}(aq) + 2e^- \rightarrow Sn(s) \qquad E^\circ = -0.14$$

It is slightly easier to oxidize $Sn^{2+}(aq)$ than to reduce it. $Sn^{2+}(aq)$ is a slightly stronger reducing agent than oxidizing agent.

Calculating cell voltages

We can use Appendix D in your text containing standard reduction potentials to calculate the standard cell voltage for a cell. Consider the cell:

$$Zn(s) \mid Zn^{2+}(aq) \parallel Fe^{3+}(aq), Fe^{2+}(aq) \mid Pt \text{ (all solns. 1 } M)$$

What voltage will the cell develop?

solution:

First write both half-reactions as reductions:

$$Zn^{2+} + 2e^- \rightarrow Zn \qquad E^\circ = -0.76 \text{ V}$$

$$Fe^{3+} + e^- \rightarrow Fe^{2+} \qquad E^\circ = +0.77 \text{ V}$$

Fe^{3+} is the most easily reduced; Fe^{3+} will be reduced and Zn will be oxidized.

Combine the two half-reactions to give the spontaneous cell reaction:

$$2(Fe^{3+} + e^- \rightarrow Fe^{2+}) \qquad E^\circ = +0.77 \text{ V}$$

$$Zn \rightarrow Zn^{2+} + 2e^- \qquad E^\circ = +0.76 \text{ V}$$

$$\text{Sum: } Zn + 2Fe^{3+} \rightarrow Zn^{2+} + 2Fe^{2+} \qquad E^\circ_{cell} = 1.53 \text{ V}$$

(Remember not to multiply the Fe^{3+}/Fe^{2+} half-reaction potential by 2 even though the half-reaction is multiplied by 2 to give an overall balanced equation with no electrons.)

Reactivity of metals with acid and water

In Chapter 14 we learned that the alkali metal, sodium, violently decomposes water to sodium hydroxide and hydrogen gas, while the alkaline earth metal, magnesium reacts only slowly with water at ordinary temperatures. Iron does not react with water at all, but reacts with acid. Explain these observations using the standard electrode potentials found in Table 21.2 and Appendix D in your text.

solution:

Sodium reacts with water to form a basic solution:

$$2Na(s) + 2H_2O(l) \rightarrow 2NaOH(aq) + H_2(g)$$

Sodium is oxidized from 0 to +1. It reduces hydrogen in H_2O from +1 to 0. The half-reactions are:

$$2H_2O + 2e^- \rightarrow 2H_2(g) + 2OH^-(aq) \qquad E^\circ = -0.83 \text{ V}$$

$$2Na(s) \rightarrow 2Na^+(aq) + 2e^- \qquad E^\circ = +2.71 \text{ V}$$

$$E^\circ_{cell} = E^\circ\,(H_2O \rightarrow H_2) + E^\circ\,(Na \rightarrow Na^+)$$

$$E^\circ_{cell} = -0.83 + 2.71 = 1.88 \text{ V}$$

(continued on next page)

A positive E^o_{cell} means the reaction is spontaneous.

The half-reaction for the oxidation of magnesium is:

$$Mg(s) \rightarrow Mg^{2+}(aq) + 2e^- \qquad E^o = +2.37 \text{ V}$$

$$E^o_{cell} = E^o (H_2O \rightarrow H_2) + E^o (Mg \rightarrow Mg^{2+})$$

$$E^o_{cell} = -0.83 + 2.37 = 1.54 \text{ V}$$

Again, the positive value for E^o_{cell} means the reaction is spontaneous. The difference in E^o_{cell} for sodium and magnesium does not account for the large difference in their reactivity in water. Magnesium quickly forms an oxide coating that prevents further reaction with water.

Metals that lie below the water half-reaction in Appendix D oxidize spontaneously as H^+ in water is reduced. Iron lies above the reduction of water, but below the standard hydrogen (reference) half-reaction. Therefore, E^o_{cell} for reaction of iron with water is negative, but for reaction with acid is positive (spontaneous).

Copper in nitric acid

A copper penny dissolves in dilute nitric acid. How can you explain this observation using the standard electrode potentials? Is a gold ring safe if you put it in acid?

solution:

The oxidation of Cu(*s*) to $Cu^{2+}(aq)$ lies above the standard hydrogen half-reaction, so reaction of copper metal with acid is nonspontaneous:

$$2H^+(aq) + 2e^- \rightarrow H_2(g) \qquad E^o = 0.00 \text{ V}$$

$$Cu(s) \rightarrow Cu^{2+}(aq) + 2e^- \qquad E^o = -0.34 \text{ V}$$

$$E^o_{cell} = E^o (H^+ \rightarrow H_2) + E^o (Cu \rightarrow Cu^{2+})$$

$$E^o_{cell} = 0 - 0.34 = -0.34 \text{ V}$$

Non-oxidizing acids (i.e., HCl) have little effect on copper metal, but copper dissolves in the oxidizing acid, HNO_3. Gold does not dissolve in nitric acid, but it dissolves in concentrated HCl if a strong oxidizing agent is present. A 3:1 mixture of conc. HCl and conc. HNO_3 is named *aqua regia* by alchemists because it dissolves gold, the king of metals. (See problem 21.4 at the end of this chapter.)

21.4 FREE ENERGY AND ELECTRICAL WORK

Charge

The charge on one electron = 1.60219×10^{-19} C. What is the charge of one mole of electrons? What is the charge on *n* moles of electrons?

(continued on next page)

solution:

The charge on one mole of electrons is:

$$1.60219 \times 10^{-19}\ \frac{\text{C}}{\text{e}^-} \times 6.02205 \times 10^{23}\ \frac{\text{e}^-}{\text{mol}} = 96{,}487\ \frac{\text{C}}{\text{mol e}^-}$$

The charge on one mole of electrons is the **faraday** (F):

$$\mathbf{1\ \mathit{F} = 96{,}487\ \frac{C}{mol\ e^-}}$$

The charge on n moles of electrons is:

$$\text{charge } (q) = 1.60219 \times 10^{-19}\ \frac{\text{C}}{\text{e}^-} \times 6.02205 \times 10^{23}\ \frac{\text{e}^-}{\text{mol}} \times n$$

$$\text{or, charge } (q) = 96{,}487\ \frac{\text{C}}{\text{e}^-} \times n$$

$$\mathbf{\mathit{q} = \mathit{nF}}$$

Electrical Work

We can harness a current flow to do work, just as we can harness a water flow to do work. The maximum amount of work we can harness depends on the difference in potential for two points in a circuit, and the amount of charge transferred:

$$w_{max} = -E_{cell} \times \text{charge}$$

where work is in joules, potential is in volts, and charge is in coulombs. How much work is done when one coulomb of charge is transferred over a potential difference of 1 volt? (answer: 1 joule.)

Charge = nF, so $w_{max} = -E_{cell} \times nF$

Units of a Joule

Thermochemistry (Chapter 6): **1 J = 1 kgm²/s²** $\mathbf{1\ J = 1\ kg\,m^2/s^2}$

Electrochemistry (Chapter 21): $\mathbf{1\ J = 1\ V \times 1\ C}$

Note: C = J/V

F = 96,487 J/V mol e^-

Remember from Chapter 20 that the free energy change, ΔG, represents the maximum work we can obtain from a spontaneous process:

$$w_{max} = -E_{cell} \times nF = \Delta G$$

When all components of a cell are in their standard states:

$$\Delta G^\circ = -nFE^\circ_{cell}$$

Dependence of Cell Voltage on Concentrations

For problems in this chapter so far, we have calculated E^{o}_{cell} for a very specific set of conditions: all constituents of the cells have been in their standard states (1 *M* concentration for aqueous solutions, and 1 atm pressure for gases). How does changing the concentration of reactants affect the cell voltage?

Recall from Chapter 20 that $\Delta G = \Delta G^{o} + RT \ln Q$. From earlier, we have that $\Delta G = -nFE$, and $\Delta G^{o} = -nFE^{o}$ therefore:

$$-nFE = -nFE^{o} + RT \ln Q$$

Dividing through by $-nF$ gives the **Nernst equation:**

$$E = E^{o} - \frac{RT}{nF} \ln Q$$

At 25°C, this equation becomes:

$$E = E^{o} - \frac{0.0592}{n} \log Q$$

At equilibrium, $Q = K$ and $E = 0$:

$$E^{o} = \frac{RT}{nF} \ln K$$

At 25°C, this equation becomes:

$$E^{o} = \frac{0.0592}{n} \log K$$

Equilibrium constant from E^{o}

1) When copper foil is placed in a silver nitrate solution, the solution turns light blue, and silver precipitates. What is the equilibrium constant for the following reaction:

$$Cu(s) + 2AgNO_3(aq) \rightarrow Cu(NO_3)_2(aq) + 2\,Ag(s)$$

solution:

First we'll write the half-reactions and calculate E^{o} for the overall reaction:

$Cu(s) \rightarrow Cu^{2+}(aq)$ $\quad E^{o} = -0.34$ V

$Ag^{+}(aq) \rightarrow Ag(s)$ $\quad E^{o} = +0.80$ V

$E^{o}_{cell} = -0.34 + 0.80 = 0.46$ V

Now we can calculate K:

$E^{o} = \frac{0.0592}{n} \log K$; $\quad n = 2$ since $2e^{-}$ are transferred in the reaction

(continued on next page)

$$0.46\text{ V} = \frac{0.0592}{2}\log K$$

$$\log K = \frac{2 \times 0.46}{0.0592} = 15.54$$

$K = 3.5 \times 10^{15}$

2) How many faradays of electricity are given off in this spontaneous reaction when 190.5 g of copper metal are used up?

solution:

$$\text{moles of copper} = 190.5\text{ g Cu} \times \frac{1\text{ mol}}{63.5\text{ g Cu}} = 3\text{ mol Cu}$$

For every mole of copper, 2 moles of electrons are used, so:

$$3\text{ mol Cu} \times \frac{2\text{ mol e}^-}{1\text{ mol Cu}} \times \frac{1\text{ faraday}}{\text{mol e}^-} = 6\text{ faradays}$$

Concentration cell

What is the cell voltage of the following cell?

$$Zn(s) \mid Zn^{2+}(aq, 0.0010\ M) \parallel Zn^{2+}(aq, 1.0\ M) \mid Zn(s)$$

solution:

$E^\circ_{cell} = 0$, since the two half-reactions are the same:

$Zn^{2+}(aq) + 2e^- \rightarrow Zn(s)$ $\quad E^\circ = -0.76$ V

$Zn(s) \rightarrow Zn^{2+}(aq) + 2e^-$ $\quad E^\circ = +0.76$ V

Q for the reaction is:

$$Q = \frac{[Zn^{2+}]_{dilute}}{[Zn^{2+}]_{conc.}} = \frac{0.0010}{1.0} = 0.0010$$

Calculate E:

$$E = E^\circ - \frac{0.0592}{n}\log Q = 0 - \frac{0.0592}{2}\log 0.0010$$

$$E = 0 - \frac{0.0592}{2} \times (-3)$$

$E = 0.089$ V

This cell is called a concentration cell. Spontaneous reaction occurs in the direction that reduces the amount of the concentrated reactant; the 1 M zinc solution is reduced to zinc metal.

21.5 ELECTROCHEMICAL PROCESSES IN BATTERIES

A **battery** is a group of voltaic cells arranged in series so that their individual voltages are added together. Many types of batteries, which fall into 3 major classes, have been developed for a variety of uses.

1) A **primary** battery cannot be recharged.

   Examples:

   Alkaline (precursor was the dry cell)
   Uses: toys, portable radios, and flashlights

   Mercury and **Silver (Button)**
   Uses: calculators (mercury), cameras, heart pacemakers, and hearing aids (silver)

   Lithium
   Uses: watches and implanted medical devices

2) A **rechargeable** (or **secondary**) battery can be recharged when it runs down by supplying electrical energy to reverse the cell reaction and form more reactant. When we recharge batteries, we convert the voltaic cells to electrolytic cells to restore nonequilibrium concentrations.

   Examples:

   Lead-Acid
   Uses: automobiles and trucks

   Nickel-Metal Hydride (replacing the nickel-cadmium battery)
   Uses: cordless razors, photo flash units, and power tools

   Lithium-Ion
   Uses: laptop computers, cell phones, and camcorders

3) A **fuel cell** (or **flow battery**) works when the reactants—usually a combustible fuel (such as hydrogen) and oxygen—generate electricity through the controlled oxidation of the fuel. The products leave the cell, so it is not a self-contained system.

Fuel Cells for Transportation

At first glance, hydrogen fuel cells would seem to be an ideal energy system for automobiles because the fuel (hydrogen) is lightweight and plentiful, and its oxidation product (water) is environmentally benign. However, the lightest gas in the universe is highly reactive and isn't easy to corral. It takes energy to produce the hydrogen (either by treatment of methane with steam, or by the electrolysis of water), so unless solar or wind energy powers these processes, the pollution only shifts from where the cars are driven to the power plants where the hydrogen is produced. It also takes energy to compress, store, and transport the hydrogen.

To make hydrogen from water by electrolysis, water is split into protons, electrons, and oxygen at the anode, and the electrons reduce protons at the cathode to form hydrogen (see the Electrolysis of Water example on page 403). The energy requirement for this process is 286 kJ/mole. Precious metal catalysts are used to reduce the voltage required to drive the electrochemical reaction.

Running a hydrogen fuel cell is the electrolysis of water in reverse: Hydrogen is split into protons and electrons, the electrons flow through a circuit to create a current that is harnessed to produce electricity, and the protons combine with oxygen to form water.

(continued on next page)

Needed before fuel cell cars are ready for primetime are: 1) energy efficient mass production of hydrogen, 2) efficient onboard storage of hydrogen, and 3) a national hydrogen delivery system. Until there is a national infrastructure of hydrogen fueling stations and an efficient way to get the hydrogen to the stations, fuel cell cars will have limited use.

Hydrogen can be stored on a vehicle in one of three ways:

- As a gas in high-pressure tanks
  Automakers store hydrogen gas in steel or carbon composite tanks filled to 10,000 psi. The energy in 1 kg of hydrogen translates into 2.5 gal of gasoline in terms of miles traveled. The goal is to store 6 kg of hydrogen to match the 300-mile per tank of driving range of gasoline-powered cars, and tanks that store this amount of hydrogen are large and heavy.

- As a liquid at extremely cold temperatures (−423°F)
  Since hydrogen is most dense as a liquid, this method allows more hydrogen storage than gaseous high-pressure storage, but it takes energy to liquefy hydrogen, it costs many times more, and a tank of liquid hydrogen on board basically creates a giant, heavy, rolling bomb.

- As a solid matrix
  Hydrogen stored inside solid materials has the potential to be a smaller and lighter-weight storage system, and may be the best long-term solution. Researchers are investigating metal hydrides, boron compounds, and metal-organic frameworks for hydrogen storage.

A full transition to fuel cell vehicles is likely still decades away. In addition, often overlooked in the energy analyses are the resources and energy required to build 3-ton vehicles. The only lightweight, practically pollution-free transportation system on the horizon is the bicycle.

21.6 ELECTROLYTIC CELLS

(Corresponds to section 21.7 in your textbook)

Galvanic cells harness spontaneous reactions to do work. An electrolytic cell uses electrical energy to drive a nonspontaneous reaction. We can convert voltaic cells into electrolytic cells by supplying an electric potential greater than the cell potential of the voltaic cell. Electrolysis has many practical applications; for example, producing aluminum metal, purifying metals, charging a battery, and chrome plating an object are all done electrolytically.

Electrolytic Cells

Oxidation occurs at the **anode** (both begin with vowels).

Reduction occurs at the **cathode** (both begin with consonants);

(same as galvanic cells).

For an electrolytic cell, the cathode, where electrons are supplied, is the negative electrode; the anode is the positive electrode.

The Stoichiometry of Electrolysis

Faraday's law of electrolysis states: The amount of substance produced at each electrode is directly proportional to the quantity of charge flowing through the cell. We cannot measure charge directly, but we can measure current, the charge flowing per unit time.

Current (i)

Current is the *rate* of flow of electric charge; it is measured in amperes (or amps).

1 ampere = 1 coulomb of charge transferred per second:

$$i = \frac{q}{t}, \text{ or}$$

$$q = i \times t$$

Metal Plating

Plating a metal means depositing a neutral metal on an electrode by reducing the metal ions. When a spoon is electroplated with silver, a silver bar is the anode where oxidation to Ag^+ occurs. A power source "pumps" the electrons to the cathode, a spoon, where silver plates out:

$$Ag^+ + e^- \rightarrow Ag$$

Silver plating

A 7.00 amp current is applied to a Ag^+ solution to produce silver metal. How long must the current be applied to produce 15.5 g of silver metal?

solution:

We need to know how many electrons are required to reduce enough Ag^+ ions to produce 15.5 g of silver metal. We know that each Ag^+ ion requires one electron to become a silver atom:

$$Ag^+ + e^- \rightarrow Ag$$

Once we know the number of electrons required, we can calculate the charge, since we know the charge on one electron = 1.60219×10^{-19} C (or the charge of a mole of electrons is 96,487 C). If we know q (charge) and i (current, 7.00 A, or 7.00 C/s), we can calculate time, t:

$$i = \frac{q}{t}, \text{ so } t = \frac{q}{i}$$

moles of e^- required to produce 15.5 g Ag:

$$15.5 \text{ g Ag} \times \frac{1 \text{ mol Ag}}{107.9 \text{ g Ag}} \times \frac{1 \text{ mol e}^-}{1 \text{ mol Ag}} = 0.144 \text{ mol e}^-$$

Charge on 0.144 mol e^-:

$$0.144 \text{ mol e}^- \times \frac{96{,}487 \text{ C}}{\text{mol e}^-} = 13{,}900 \text{ C}$$

(continued on next page)

Time that 7.00 A current must be applied to generate 13,900 C:

$$t = \frac{13{,}900\ \text{C}}{7.00\ \text{C/s}} = 1990\ \text{s, or } 33.2\ \text{minutes}$$

Tin plating

Tin cans are actually steel cans with a thin coating of tin. How much tin is plated out when a 5.00 amp current is passed through a solution containing Sn^{2+} for 40.0 minutes?

solution:

We first calculate charge from current and time:

$$q = i \times t$$

$$q = 5.00\ \text{C/s} \times 2400\ \text{s} = 12{,}000\ \text{C}$$

12,000 coulombs is how many electrons?

$$12{,}000\ \text{C} \times \frac{1\ \text{mol e}^-}{96{,}487\ \text{C}} = 0.124\ \text{mol e}^-$$

Each Sn^{2+} ion requires two electrons to become a tin atom, so the amount of tin that 0.124 mol e^- plates out is:

$$0.124\ \text{mol e}^- \times \frac{1\ \text{mol Sn}}{2\ \text{mol e}^-} \times \frac{118.7\ \text{g Sn}}{\text{mol Sn}} = 7.36\ \text{g Sn}$$

We worked examples of silver and tin plating out at a cathode. If an electrolytic cell contains both Ag^+ and Sn^{2+} ions, which order will the metals plate out onto the cathode if the voltage starts low, and is gradually increased? Compare the standard reduction potentials of Ag^+ and Sn^{2+} ions:

$$Ag^+ + e^- \rightarrow Ag \qquad E^\circ = 0.80\ \text{V}$$

$$Sn^{2+} + 2e^- \rightarrow Sn \qquad E^\circ = -0.14\ \text{V}$$

The more positive the E° value, the more the reaction has a tendency to proceed. Ag^+ is reduced more easily than Sn^{2+}, so silver will plate out first.

Why can't aluminum be plated out of an aqueous solution of Al^{3+}? The answer is that water itself is reduced before Al^{3+}:

$$2H_2O + 2e^- \rightarrow H_2 + 2OH^- \qquad E^\circ = -0.83\ \text{V}$$

$$Al^{3+} + 3e^- \rightarrow Al \qquad E^\circ = -1.66\ \text{V}$$

Until a practical electrolytic process for producing aluminum was discovered, aluminum was a more precious metal than gold or silver. Ions must be mobile for electrolysis to occur, such as when they are in a water solution, or when a salt is melted. The high melting point of aluminum oxide (2050°C), however, makes the melting of solid Al_2O_3 an impractical route to electrolysis. In 1854, it was discovered that a mixture of aluminum oxide and Na_3AlF_6 has a melting point of 1000°C, and aluminum metal could be obtained electrolytically from the molten mixture. After this discovery, the price of aluminum dropped by a factor of one thousand.

Electrolysis of Water

Electrolysis of water produces oxygen and hydrogen:

$$2H_2O \rightarrow O_2 + 2H_2$$

The anode reaction is:

$$2H_2O \rightarrow O_2 + 4H^+ + 4e^- \qquad -E^\circ = -1.23\ V$$

The cathode reaction is:

$$4H_2O + 4e^- \rightarrow 2H_2 + 4OH^- \qquad E^\circ = -0.83\ V$$

The net reaction is:

$$6H_2O \rightarrow 2H_2 + O_2 + \underbrace{4H^+ + 4OH^-}_{(4H_2O)}$$

$$2H_2O \rightarrow O_2 + 2H_2 \qquad E^\circ = -2.06\ V$$

The potentials assume 1 M H^+ and 1 M OH^-. In pure water, $[H^+] = [OH^-] = 10^{-7}$ M, and the potential for the process is –1.24 V.
($-E_{oxidation} = -0.82$ V; $E_{reduction} = -0.42$ V; $E_{cell} = -1.24$ V)

Overvoltage

When an aqueous sodium chloride solution is electrolyzed, hydrogen gas forms at the cathode, and chlorine gas forms at the anode. Can you explain these observations using the standard electrode potentials?

solution:

The possible oxidizing agents are Na^+ and H_2O:

$$2H_2O + 2e^- \rightarrow H_2 + 2OH^- \qquad E = -0.42\ V$$

$$Na^+ + e^- \rightarrow Na \qquad E^\circ = -2.71\ V$$

Water has the less negative electrode potential, so it is easier to reduce than Na^+; H_2 forms at the cathode, as observed.

The possible reducing agents are Cl^- and H_2O:

$$2H_2O \rightarrow O_2 + 4H^+ + 4e^- \qquad -E = -0.82\ V$$

$$2Cl^- \rightarrow Cl_2 + 2e^- \qquad -E^\circ = -1.36\ V$$

Water has the more positive electrode potential, so water is easier to oxidize than Cl^- and we expect to see O_2 produced at the anode. But, instead, chlorine forms. We cannot explain this observation using the standard electrode potentials. An additional 0.4 to 0.6 volts (the **overvoltage**) is required to produce oxygen at metal electrodes. The overvoltage results from kinetic factors, such as the large activation energy required for gases to form at the electrode. Overvoltage makes it possible to produce chlorine from aqueous sodium chloride (see the discussion in your textbook on page 958).

We see from the previous example that, because of overvoltage, **chlorine** gas, $Cl_2(g)$, can be prepared by oxidation of an aqueous sodium chloride solution. **Bromine** and **iodine** may also be prepared electrolytically by oxidation of aqueous solutions of their salts. Fluorine cannot be prepared this way. Remember from section 21.3 that F^- is not easily oxidized to F_2 ($E^o = -2.87$). Other anions that cannot be oxidized are the common oxoanions, SO_4^{2-}, CO_3^{2-}, NO_3^-, and PO_4^{3-} (the oxoanion nonmetal is in its highest oxidation state).

Elements that can be prepared electrolytically by reduction of aqueous solutions of their salts are **copper, silver,** and **gold** (the coinage metals), and **chromium, platinum,** and **cadmium.** More active metals such as those in groups 1A and 2A, and aluminum cannot be prepared this way.

CHAPTER CHECK

Make sure you can...

- Balance redox equations by the half-reaction method.
- Identify oxidizing and reducing agents.
- Describe the physical make-up of a voltaic cell.
- Diagram, in shorthand notation, a voltaic cell.
- Rank relative oxidizing and reducing strengths from E^o values.
- Calculate the standard cell voltage from E^o values for half-reactions.
- Use an emf series to write spontaneous redox reactions.
- Show how E_{cell}, ΔG (maximum work), and the charge flowing through the cell are related.
- Use the Nernst equation to calculate an equilibrium constant for a redox reaction, or to calculate a nonstandard cell potential, E_{cell}.
- Find the cell voltage of a concentration cell.
- State Faraday's law of electrolysis and define current, *i*.
- Calculate the current needed to produce a certain amount of product.
- Predict the products of electrolysis of aqueous salts.
- Explain how the electrolysis of water influences the products of aqueous electrolysis and explain the importance of overvoltage.

Chapter 21 Exercises

21.1) Calculate the standard cell potential, E^o_{cell}, and draw the voltaic cell for the following reactions:

a. $2Cr(s) + 3Pb^{2+}(aq) \rightarrow 3Pb(s) + 2Cr^{3+}(aq)$

b. $Cl_2(g) + Fe^{2+}(aq) \rightarrow 2Cl^-(aq) + Fe^{3+}(aq)$

21.2) Electrical energy and drinking water are generated from hydrogen-oxygen fuel cells while a space shuttle is in flight:

$$2H_2(g) + O_2(g) \rightarrow 2H_2O(l)$$

The fuel cell reaction involves a hot potassium hydroxide electrolyte solution and uses Ni/NiO embedded carbon electrodes instead of platinum. What is the cell potential produced by a hydrogen-oxygen cell?

21.3) Calculate the cell potential, E_{cell}, for the following reactions:

a. $2Al(s) + 3Ni^{2+}(aq, 0.50\ M) \rightarrow 2Al^{3+}(aq, 0.20\ M) + 3Ni(s)$

b. $2VO_2^{+}(aq) + 4H^{+}(aq) + Zn(s) \rightarrow 2VO^{2+}(aq) + 2H_2O(l) + Zn^{2+}(aq)$ $E^{o}_{cell} = 1.76$ V

$[VO_2^{+}] = 1.5\ M$; $[H^{+}] = 0.60\ M$; $[VO^{2+}] = 1.3 \times 10^{-2}\ M$; $[Zn^{2+}] = 2.0 \times 10^{-1}\ M$

21.4) Use data from Appendix D of your textbook to predict if 1 *M* HNO_3 will dissolve gold metal to form a 1 *M* Au^{3+} solution.

21.5) If you were to add iodine crystals to a sodium chloride solution, would any reaction occur? Write the balanced equation, and calculate E^{o}_{cell}, ΔG^{o}, and *K* at 25°C.

21.6) In the following reaction, the pH of the MnO_2/Mn^{2+} half-cell can be changed. Calculate the cell potential, E_{cell} for the conditions listed below:

$2Ag(s) + 2Cl^{-}(aq, 0.10\ M) + MnO_2(s) + 4H^{+}(aq) \rightarrow 2AgCl(s) + Mn^{2+}(aq, 0.10\ M), + 2H_2O(l)$

a. pH = 3.00 and pH = 7.00; b. What effect does pH have on E_{cell} for the reaction?

21.7) Aluminum is prepared by the electrolysis of aluminum oxide in the Hall process. The overall reaction is:

$$2Al_2O_3(s) \rightarrow 4Al(s) + 3O_2(g)$$

From the standard Gibbs free energy of the reaction, calculate the electromotive force for this reaction.

21.8) Chrome plated bumpers and grills on automobiles were the style during the 1950's. How long would it take to plate out 50.0 g of Cr metal by passing a 5.00 ampere current through a $Cr(NO_3)_3$ solution?

21.9) The final step in the preparation of copper is the electrolytic refining of "blister" copper to pure copper. 590. kg slabs of the crude copper are placed in a vat containing copper(II) sulfate. The pure copper plates out at the cathode on copper starter sheets. What current (amperes) is required to plate out 500. kg of copper in a 24 hour period?

21.10) A lab technician came across a bottle in the stockroom labeled only "Cobalt Chloride." In an effort to determine the oxidation number of the cobalt, she decided to carry out an analytical electrolysis of a sample of the solution. The mass of the cathode before electrolysis was 38.4124 g. After 20.00 minutes of electrolysis with a 0.500 amp current, the mass of the cathode had increased to 38.5346 g. What is the oxidation number of the cobalt ion? How should the lab technician relabel the bottle?

Chapter 21 Answers

21.1) a.

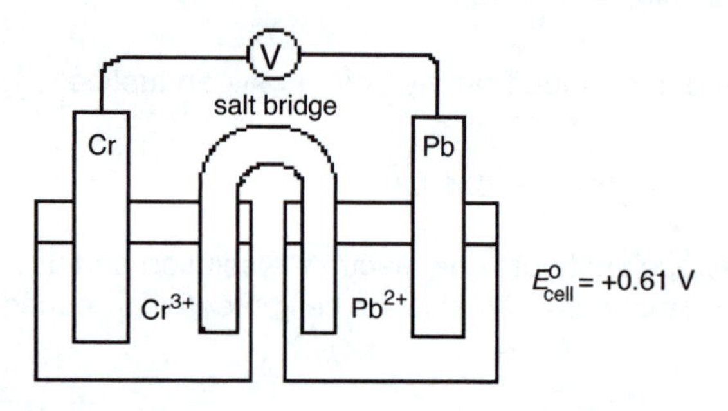

$E^{o}_{cell} = +0.61$ V

b.

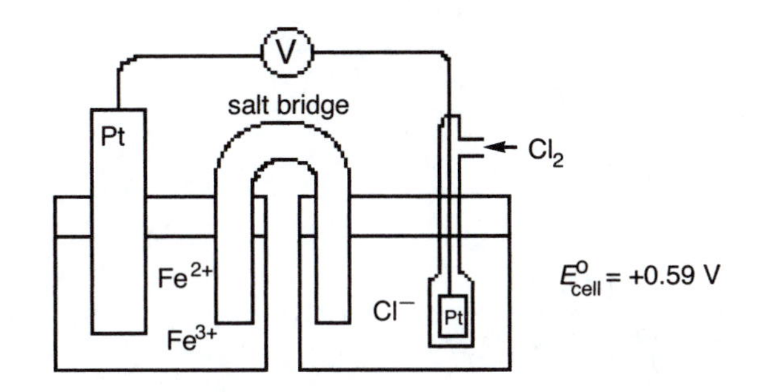

21.2) $E^{\circ}_{cell} = +1.23\ \text{V}$

21.3) a. $E_{cell} = 1.41\ \text{V} - \dfrac{0.592\ \text{V}}{6} \log \dfrac{(0.20)^2}{(0.50)^3}$

$E_{cell} = 1.41\ \text{V} - \left(\dfrac{-0.0262\ \text{V}}{6}\right)$

$E_{cell} = 1.41\ \text{V} + 0.0044\ \text{V} = 1.41\ \text{V}$

b. $E_{cell} = 1.88\ \text{V}$

21.4) The half-reaction for the reduction of HNO_3 is:

$NO_3^- + 4H^+ + 3e^- \rightarrow NO + 2H_2O \qquad E^{\circ} = 0.96\ \text{V}$

The half-reaction for the oxidation of gold is:

$Au \rightarrow Au^{3+} + 3e^- \qquad -E^{\circ} = -1.50\ \text{V}$

The sum gives:

$Au(s) + NO_3^-(aq) + 4H^+(aq) \rightarrow Au^{3+}(aq) + NO(g) + 2H_2O(l)$

$E^{\circ}_{cell} = 0.96\ \text{V} - 1.50\ \text{V} = -0.54\ \text{V}$

E°_{cell} is negative, so this process does not occur spontaneously. As mentioned in the example problem on page 414, a mixture of concentrated HCl and concentrated HNO_3 *(aqua regia)* is required to dissolve gold.

21.5 The balanced equation is:

$I_2 + 2Cl^- \rightarrow 2I^- + Cl_2$

The half-reactions are:

$I_2 + 2e^- \rightarrow 2I^- \qquad E^{\circ} = 0.53\ \text{V}$

$2Cl^- \rightarrow Cl_2 + 2e^- \qquad E^{\circ} = -1.36\ \text{V}$

The sum is: $I_2 + 2Cl^- \rightarrow 2I^- + Cl_2 \qquad E^{\circ}_{cell} = 0.53 - 1.36\ \text{V} = -0.83\ \text{V}$ (not spontaneous)

$\Delta G^{\circ} = -nFE^{\circ}_{cell} = -2\ \text{mol} \times 96{,}487\ \text{kJ/V} \times -0.83\ \text{V} = 160\ \text{kJ}$

$E^{\circ} = \dfrac{0.0592}{n} \log K;\ K = 9.1 \times 10^{-29}$

21.6) a. $MnO_2 + 2H^+ + 2e^- \rightarrow Mn^{2+} + H_2O$ $\quad E^o = 1.23$ V
$Ag(s) + Cl^- \rightarrow AgCl(s) + e^-$ $\quad E^o = -0.22$ V
$E^o_{cell} = 1.01$ V

$$E_{cell} = E^o_{cell} - \frac{0.0592}{n} \log Q \Rightarrow n = 2$$

$$pH = 3: Q = \frac{[Mn^{2+}]}{[Cl^-]^2[H^+]^4} = \frac{(0.10)}{(0.10)^2(1.0\times10^{-3})^4} = 1 \times 10^{13}$$

$$pH = 7: Q = \frac{[Mn^{2+}]}{[Cl^-]^2[H^+]^4} = \frac{(0.10)}{(0.10)^2(1.0\times10^{-7})^4} = 1 \times 10^{29}$$

$$E_{cell} = 1.01 - \frac{0.0592}{2} \log(1 \times 10^{13}) = 0.63 \text{ V @ pH} = 3.00$$

$$E_{cell} = 1.01 - \frac{0.0592}{2} \log(1 \times 10^{29}) = 0.16 \text{ V @ pH} = 7.00$$

b. Increasing the pH decreases the cell potential.

21.7) $$E^o_{cell} = -\frac{\Delta G^o}{nF} = \frac{3{,}164{,}000 \text{ J}}{12 \text{ mol e}^- \times 96{,}487} = 2.73 \text{ V}$$

21.8) The reduction is:

$$Cr^{3+} + 3e^- \rightarrow Cr^0$$

The charge on the electrons is:

$$50.0 \text{ g Cr} \times \frac{1 \text{ mol Cr}}{51.996 \text{ g Cr}} \times \frac{3 \text{ mol e}^-}{1 \text{ mol Cr}} \times \frac{96{,}487 \text{ C}}{\text{mol e}^-} = 278{,}349 \text{ C}$$

The time that a 5.00 A current must be applied to generate 278,349 C is:

$$t = \frac{278{,}349 \text{ C}}{5.00 \text{ C/s}} = 55{,}700 \text{ s, or } 15.5 \text{ h}$$

21.9) 1.76×10^4 amperes, or 18 kA

21.10) 3.0 eq/mole; cobalt(III) chloride

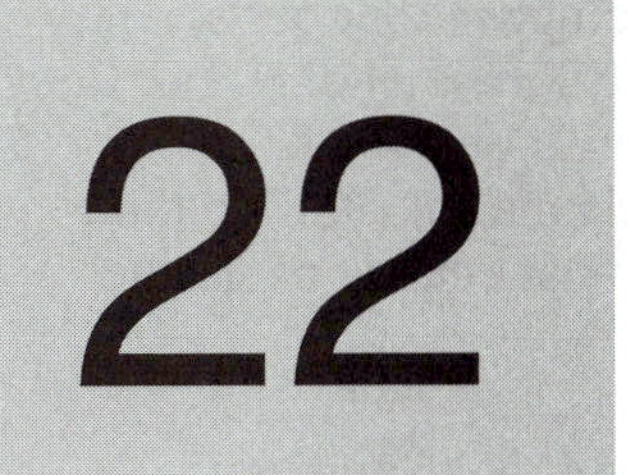

The Elements in Nature and Industry

22.1 ELEMENTS IN NATURE

Hydrogen is by far the most abundant element in the universe, accounting for almost 74% of its mass (and 90% of the atoms). Hydrogen and helium together make up nearly 99% of the mass of the universe. The stars are gigantic nuclear furnaces that produce heavier nuclei from hydrogen and helium, but clearly, in the whole universe, the synthesis of the heavier elements from hydrogen and helium has not proceeded very far. However, in the earth's crust, hydrogen makes up only 15% of the atoms.

The elements iron, oxygen, silicon, and magnesium account for over 90% of the mass of the earth. **Iron** is the most abundant element in the whole earth (oxygen is the most abundant element on the crust, followed by silicon, aluminum, and then iron). The nucleus of iron is especially stable, making iron's abundance in the universe comparatively high (0.19% by mass, from Table 22.1 in your textbook). Iron is the main constituent of the earth's core and the second most abundant metal on the earth's crust. It is also a major component of meteorites and is present in the moon's soil. Iron played a large role in man's material progress. When the process of smelting iron became widespread (around 1200 BC), the "Iron Age" began. In 1773 when an inexpensive process to convert coal to carbon (in the form of coke) was discovered, iron smelting became cheap, and large-scale iron production led to the Industrial Revolution. Iron is necessary for most forms of life because it stores and transports oxygen.

Oxygen is the most abundant element on the earth's surface. It exists as both the free element, O_2, and combined in compounds such as water and as a constituent of most rocks, minerals, and soils. It is not the major component of our atmosphere (nitrogen is), but it makes up 23% by mass of the atmosphere, and approximately 85% of the oceans. Green plants have generated essentially all our oxygen from water and carbon dioxide during the process of photosynthesis, a process that began about 2500 million years ago. It wasn't until approximately 50 million years ago that the oxygen level of the atmosphere reached present levels. We burn the products of photosynthesis (wood and fossil fuels) to produce most of the energy we use. Our fossil fuels, coal (formed by the decomposition of plants under high temperature and pressure in the absence of free O_2), and petroleum (formed by the decomposition of animals under similar conditions) are finite, and non-renewable energy sources that we consume in our everyday living. The United States, home to 5% of the world's population, consumes about 25% of the world's consumption of petroleum.

Silicon has a great affinity for oxygen, and together, these two elements comprise four out of every five atoms near the surface of the earth. In the entire universe, silicon's abundance comes after H, He, C, O, and Fe (see Table 22.1 on page 978 in your textbook). In the earth's core, silicon is essentially nonexistent, but the crust of the earth contains the lighter silicon-containing materials that "floated" to the surface as the earth formed. Our rocks and soils, clays and sands are made up almost entirely of silicate minerals and silica (SiO_2).

Aluminum is the most abundant metal on the earth's crust. The production of aluminum is exceeded only by iron and steel. Aluminum, as with most elements, is extracted from ores. Commercially, the most important mineral is bauxite, a mixed aluminum oxide hydroxide mineral.

Abundance of elements in nature

Name the following elements:

a) Most abundant element in the universe

b) Second most abundant element in the universe

c) Most abundant element in the whole earth

d) Most abundant element on the earth's crust

e) Second most abundant element on the earth's crust

f) Most abundant metal on the earth's crust

g) Second most abundant metal on the earth's crust

h) Together they comprise 4 out of 5 atoms near the earth's surface

i) Photosynthesis caused its presence in our atmosphere

answers:

a) hydrogen, b) helium, c) iron, d) oxygen, e) silicon, f) aluminum, g) iron, h) oxygen and silicon, i) oxygen.

22.2 THE CYCLING OF ELEMENTS THROUGH THE ENVIRONMENT

The Carbon Cycle

Carbon exists on earth as the free element (diamond and graphite), in compounds such as magnesium and calcium carbonates, and in the atmosphere as CO_2. Figure 22.5 in your textbook shows the sizes of these carbon sources and the interplay between them. Notice that fossil fuel is the third largest source of carbon on our planet after inorganic sediments and inorganic carbon deep in the ocean.

Fossil fuel depletion

If the typical family car produces about 1.70 metric tons of carbon in a year, how many pounds of carbon does it produce? If you assume that all the carbon is in the form of carbon dioxide, how many pounds of carbon dioxide are produced in a year from a typical family car? (1 metric ton = 1000 kg)

(continued on next page)

solution:

$$1.7 \text{ metric ton} \times \frac{1000 \text{ kg}}{1 \text{ metric ton}} \times \frac{2.2 \text{ lbs}}{\text{kg}} = 3740 \text{ lb C}$$

$$3740 \text{ lb C} \times \frac{44 \text{ lb } CO_2}{12 \text{ lb C}} = 13{,}700 \text{ lb } CO_2$$

Carbon dioxide accounts for only 0.003% of carbon in the earth's crust, but it is responsible for connecting the land and sea carbon cycles. Plant photosynthesis removes carbon dioxide from the atmosphere; plant and animal respiration and decomposition of dead organic matter returns carbon dioxide to the atmosphere. Natural fires and volcanoes also release carbon dioxide into the air.

Carbon dioxide molecules absorb infrared radiation that traps the earth's heat. Without this natural greenhouse effect, our planet would be a cold rock, unable to support life. However, we have vastly increased the amount of carbon dioxide in the atmosphere in recent years (primarily from the calcination of limestone to make cement, and by burning huge amounts of fossil fuel for energy). Automobile use is one of the largest depleters of fossil fuel. In the process of burning gasoline, our vehicles exude tons of carbon dioxide into the atmosphere (see the problem on the previous page and below), causing our average temperatures to increase (the "enhanced greenhouse effect"). Recent measurements from the top of Mauna Loa volcano in Hawaii measured the level of CO_2 in the atmosphere above 400 parts per million. The last time the Earth's atmosphere contained 400 ppm of carbon dioxide was 2.5 million year ago, during the Pleistocene Epoch. Scientists estimate that average temperatures rose as much as 18 degrees Fahrenheit during that time, and sea levels ranged between 16 and 131 feet higher than current levels. For the previous 800,000 years, CO_2 levels never exceeded 300 ppm, and there is no known geologic period in which the rate of increase has been so sharp.

Carbon dioxide emission from cars

1) When we burn 1 gallon of gasoline in our cars, how much carbon dioxide do we add to our atmosphere (assume complete combustion of the hydrocarbons to CO_2)? Use C_8H_{18} for the formula for gasoline (density = 0.70 g/mL).

solution:

The complete combustion of C_8H_{18} is:

$$2\, C_8H_{18}(l) + 25O_2(g) \rightarrow 16CO_2(g) + 18H_2O(g)$$

We convert gallons of gasoline to moles of gasoline, find moles of carbon dioxide produced, and convert to pounds:

$$1 \text{ gallon} \times \frac{4 \text{ quarts}}{\text{gallon}} \times \frac{1 \text{ liter}}{1.057 \text{ quarts}} \times \frac{1000 \text{ mL}}{1 \text{ L}} \times \frac{0.70 \text{ g}}{\text{mL}} \times \frac{1 \text{ mole}}{114 \text{ g}} = 23 \text{ moles } C_8H_{14}$$

$$23 \text{ moles } C_8H_{14} \times \frac{16 \text{ moles } CO_2}{2 \text{ moles } C_8H_{14}} \times \frac{44 \text{ g } CO_2}{\text{mole}} \times \frac{1 \text{ lb}}{454 \text{ g}} = 18 \text{ pounds } CO_2$$

2) How many gallons of gas a week does the average family car burn if it produces 13,700 pounds of CO_2 per year (from the previous problem)?

(continued on next page)

solution:

$$\frac{13{,}700 \text{ lb } CO_2}{\text{year}} \times \frac{\text{gallon gas}}{18 \text{ lb } CO_2} \times \frac{\text{year}}{52 \text{ weeks}} = 15 \text{ gallons gas/week}$$

Our automobiles produce almost 30% of the greenhouse gases, and 50% of the toxic air pollutants in our atmosphere. Before it is ever driven a mile, the manufacturing of a car adds 4 tons of carbon and 700 pounds of other pollutants to the atmosphere (The Consumer's Guide to Effective Environmental Choices, Michael Brower and Warren Leon)—and as we calculated on the previous page, each week, on average, each vehicle on the road pumps another 260 pounds (13,700/52) of carbon dioxide into the atmosphere. The desire for a renewable fuel that has less of an impact on the environment and on human health has led to the use in some areas of an ethanol/gasoline blend or "gasohol."

Ethanol as a fuel

Supporters pitch ethanol as a clean-burning, homegrown (from corn) fuel with potential CO_2 savings over gasoline. The analysis of the overall process is complex, however, and requires an understanding of the chemistry involved in ethanol's production, as well as the complex atmospheric chemistry of gasohol's combustion products. The basic chemistry of ethanol production and combustion are:

1) A fermentation process produces ethanol and CO_2 from glucose (isolated from corn):

$$C_6H_{22}O_6(aq) \rightarrow 2C_2H_6O(aq) + 2CO_2$$

2) The combustion of ethanol produces carbon dioxide and water:

$$2C_2H_6O(aq) + 6O_2(g) \rightarrow 4CO_2(g) + 6H_2O + \text{heat}$$

Energy considerations

The fermentation process (step 1 above) is slow and inefficient, but also adding to the energy costs are corn production (farm machinery, fertilizers, pesticides, and irrigation), isolation of glucose from the corn, distillation of the ethanol from the products of fermentation, and transportation of the ethanol. Ethanol is hygroscopic—it attracts moisture from the air, leading to corrosion in fuel pipelines—and therefore must be transported separately before it's blended with gasoline. Some analyses of the overall process conclude that it requires more energy to produce ethanol from corn than the resulting ethanol provides, and that greenhouse gas emissions are similar to those of gasoline.

Others studies disagree, citing the recent use of more efficient source crops, which, they argue, give ethanol production a positive net energy balance. Crops such as switchgrass and sugar cane (widely used in Brazil) are more efficient in producing ethanol than corn. The United States has begun to research technologies that convert cellulosic plant material (plant by-products, including cornstalks) into fuels. Because plant cells are difficult to dismantle, researchers are looking for new enzymes or bacteria to break down plant cell walls. Cellulosic refineries are expensive to build, and it turns out that turning wood or grass into fuel on a commercial scale is really hard to do. There is also concern that a cellulosic ethanol market could provide an incentive for landowners to clear-cut their forests. Other research involves using microbes to generate "biogasoline"—hydrocarbons that do not need to be mixed with petroleum—from plant matter. The excitement over these technologies is tempered by the reality that ethanol is likely to be only a short-term answer to the country's fuel needs. Sustainable crops—ones requiring no fertilizer or irrigation—are already difficult to achieve, and will become more so if waste biomass, normally used to replenish soils, is harvested for ethanol production.

Human health, the environment, and ethics

Mark Jacobson, civil and environmental engineer at Stanford University, uses a sophisticated air-pollution computer model to conclude that a high blend of ethanol poses an equal or greater risk to public health than gasoline, which already causes significant health damage. His model, which simulates air quality in the year 2020 when ethanol-fueled vehicles are expected to be widely available in the United States, predicts that E85 vehicles (vehicles that run on 85% ethanol and 15% gasoline) could result in more ozone, a prime ingredient of smog, resulting in higher ozone-related mortality, hospitalization, and asthma. The deleterious health effects are the same whether the ethanol is made from corn, switchgrass, or other plant products.

Using corn for cars raises ethical questions. Critics of ethanol from corn point out that the amount of corn that would supply just a single 25-gallon fill-up for a vehicle would feed an individual for a year, and the cost of corn for food will inevitably increase as the demand for ethanol production increases. There is also a supply issue—if every acre of the nation's corn were assigned to ethanol, it would supply only about 7% of what our cars use today.

Beyond ethanol?

The problems with using ethanol for fuel have spurred the search for alternative fuels with higher energy densities and more efficient syntheses than ethanol. Researchers at the University of Wisconsin have produced a synthetic fuel alternative, DMF, 2,5-dimethylfuran (*Science*, June 23, 2007), which, like ethanol, can be derived from sugars, but has a 40% higher energy density. The synthesis involves two steps using catalysts instead of fermentation:

1) Acidic catalysts strip oxygen off of glucose in the presence of salt, and the resulting HMF (5-hydroxymethylfurfural) is extracted into a hydrocarbon solvent.

2) Hydrogen, mixed with HMF in the presence of a copper-ruthenium catalyst, strips off two more oxygen atoms from HMF, producing water and DMF.

The research is promising, but there is no silver bullet: the toxicity and potential environmental impacts of DMF are unknown, and meanwhile the energy demands from our nation show no signs of slowing.

The Nitrogen Cycle

Nitrogen makes up 78% (by volume) of our atmosphere and is the most abundant uncombined element accessible to man. In the United States, the production of ammonia from nitrogen is exceeded only by the production of sulfuric acid. Nitrogen makes up 15% by weight of proteins and is thus essential to all forms of life.

Fixation of N_2 occurs through atmospheric, industrial, and biological processes.

- Atmospheric fixation:
  Lightning, fires, and car engines produce NO from nitrogen and oxygen, which oxidizes to NO_2. Rain converts NO_2 to nitric acid, which enters both sea and land and is used by plants.

- Industrial fixation:
  The Haber process produces ammonia from nitrogen and hydrogen. Most of the ammonia is used as fertilizer which is taken up by plants. Overuse of fertilizers and automobiles causes water pollution as rain washes fertilizers and nitrates into natural waters. Aquatic and animal life die from excessive algal and plant growth and decay which depletes oxygen and causes *eutrophication*.

- Biological fixation:
  Marine blue-green algae and bacteria that live on the roots of leguminous plants fix atmospheric nitrogen by reducing it to NH_3/NH_4^+ through the action of enzymes containing molybdenum at their active site. Other enzymes oxidize NH_4^+ to NO_3^-, which plants reduce to make their proteins. Animals eat the plants, and use the plant proteins to make their own proteins. When the animals die and decay, the nitrogen in the proteins converts by soil bacteria to NO_2^- and NO_3^- again.

The Phosphorus Cycle

Phosphorus exists in all living things and occurs in many crustal rocks. Unlike carbon and nitrogen, phosphorus has no volatile compound that circulates in the atmosphere and so its cycles are land- and water-based. The three phosphorus cycles are summarized below.

- The inorganic cycle:
  Weathering of phosphate rocks slowly leaches phosphates into the rivers and seas. In the ocean, some phosphate is absorbed by water organisms, but the majority is precipitated by Ca^{2+} ion and deposits on the continental shelf. Geological uplifts return the phosphate to the land.

- The land-based biological cycle:
  Plants absorb soluble dihydrogen phosphates ($H_2PO_4^-$), and animals that eat the plants excrete soluble phosphate which is used by newly growing plants. As plants and animals excrete, die, and decay, phosphate washes into rivers and oceans.

- The water-based biological cycle:
  The water-based phosphorus cycle is the most rapid of the three phosphate cycles. Inorganic phosphate is absorbed by algae and bacteria within minutes of its entering an aqueous environment. Animals eat the plants and as they excrete, die, and decay, some of the released phosphate is used by aquatic organisms, and some returns to the land when land animals eat marine animals. Some aquatic phosphate precipitates with Ca^{2+}, sinks to the seabed and returns to the mineral deposits of the inorganic cycle.

Detergents, human excreta, and industrial processes cause phosphates to enter water sources and cause unintended over-fertilization, or *eutrophication*. Overgrowth of algae and other plants depletes water of dissolved oxygen and kills fish and other aquatic life. Elimination of phosphates and more stringent requirements for sewage treatment have restored water quality in many lakes that once were "dead." Lake Erie was considered one such "success" story, recovering from its severely polluted days in the 60's and 70's. Recently, however, a toxic blue-green algae—a neurotoxic form of cyanobacteria—has been spreading in Lake Erie, caused by overflows from sewage systems and phosphorus run-off from agricultural land. A Canadian official on the International Joint Commission charged with remedying water quality concerns along the border commented, "I don't understand why people aren't marching in the streets in Toledo saying 'we want our lake back.'"

22.3 METALLURGY

The steps for extracting metals from the earth are:

1) **Mining**

2) **Pretreating the ore**
   - Crushing, grinding, or pulverizing the rock
   - Separating the mineral from attached soil, rock, and clay by:
     magnetic attraction (iron-containing minerals such as magnetite, Fe_3O_4)
     cyclone separation (density difference required)

flotation (example: copper recovery)
leaching (example: extraction of gold with a cyanide ion solution)

$$4Au(s) + O_2(g) + 8CN^-(aq) + 2H_2O(l) \rightarrow 4Au(CN)_2^-(aq) + 4OH^-(aq)$$

3) Converting the mineral to a compound (usually oxidized):

- pyrometallurgy (uses heat):

$$2ZnS(s) + 3O_2(g) \rightarrow 2ZnO(s) + 2SO_2(g)$$

- hydrometallurgy (uses aqueous solution chemistry):

$$2Cu_2S(s) + 5O_2(g) + 4H^+(aq) \rightarrow 4Cu^{2+}(aq) + 2SO_4^{2-}(aq) + 2H_2O(l)$$

4) Converting the compound to a metal through:

- chemical redox

reduction with carbon (smelting):
heating an oxide with carbon (as coke or charcoal) to obtain the metal
examples: zinc oxide, tin(IV) oxide, and iron oxide (hematite):

$$ZnO(s) + C(s) \rightarrow Zn(g) + CO(g)$$
$$2Fe_2O_3(s) + C(s) \rightarrow 4Fe(l) + 3CO_2(g)$$

reduction with hydrogen:
used for less active metals such as tungsten:

$$WO_3(s) + 3H_2(g) \rightarrow W(s) + 3H_2O(g)$$

reduction with an active metal:
used for metals that might form undesirable hydrides
example: the *thermite reaction*; aluminum powder is the reducing agent:

$$Cr_2O_3(s) + 2Al(s) \rightarrow 2Cr(l) + Al_2O_3(s)$$

oxidation with an active metal:
oxidation of a mineral is used to obtain a nonmetal (such as iodine):

$$2I^-(aq) + Cl_2(g) \rightarrow 2Cl^-(aq) + I_2(s)$$

- electrochemical redox

minerals are converted to elements in an electrochemical cell
the cation (Be^{2+}) is reduced to the metal (Be) at the cathode; the anion (Cl^-) is oxidized to the nonmetal (Cl_2) at the anode:

$$BeCl_2(l) \rightarrow Be(s) + Cl_2(g)$$

5) Refining (purifying)

There are three common methods used to purify the element:

- electrorefining
  the impure metal acts as the anode; a sample of pure metal acts as the cathode
  metal ions are reduced from the anode and deposited on the cathode
- distillation
  used for metals with relatively low boiling points such as zinc and mercury
- zone refining
  an element is recrystallized from a molten zone

6) Alloying the element

Combining metals to enhance properties such as luster, conductivity, malleability, ductility, and strength. Examples are:

- Stainless steel (Fe, Cr, and Ni)
- Brass (Cu and Zn)
- Sterling silver (Ag and Cu)
- 14-Carat gold (Au, Ag, and Cu)

From ore to element

Choose words from the list below to complete the following paragraph:

(a) recrystallize
(b) smelting
(c) cyclone separator
(d) steel
(e) electrorefining
(f) alloys
(g) hydrogen
(h) mine
(i) refined
(j) iron
(k) active metal
(l) distillation
(m) magnet
(n) oxide
(o) brass

The first step in obtaining a pure metal from the earth is to dig it out, or ______________(1) it, and then to crush and grind it. To separate an ore from the dirt, sand, or clay around it, we can use a ______________(2) (good for iron-containing ores), or a ______________(3), which blows high-pressure air through the pulverized mixture to separate the particles. Next, the mineral is usually converted to another compound, generally an ______________(4) that is easy to reduce. The metal oxide is then reduced to its metal. If carbon (in the form of coke or charcoal) is used as the reducing agent, we call this process (5). Other common reducing agents are the gas, ______________(6) and an ______________(7). Chemical oxidation may be used to obtain a nonmetal such as iodine. The element is then purified or ______________(8). If an electrolytic cell is used for this process, we call it ______________(9). Metals with low boiling points may be purified by ______________(10). Zone refining is used to ______________(11) an element from a thin molten zone. Metals are often mixed together to form ______________(12) with superior properties. ______________(13), a metal that rusts and is soft, can be alloyed with Ni and Mo to produce a harder material, ______________(14) that is corrosion resistant. Likewise, copper stiffens when zinc is added to make ______________(15).

answers:

1-h, 2-m, 3-c, 4-n, 5-b, 6-g, 7-k, 8-i, 9-e, 10-l, 11-a, 12-f, 13-j, 14-d, 15-o.

22.4 ISOLATION AND USES OF THE ELEMENTS

The Alkali Metals: Sodium and Potassium

Sodium and potassium salts are found in large salt deposits where ancient seas evaporated. This process continues today in the Great Salt Lake in Utah, for example. A **Downs cell** (Figure 22.16 on page 995 in your textbook) reduces sodium ions to sodium metal at a steel cathode, and oxidizes chloride ions to chlorine gas at a large anode. Calcium chloride mixes with the sodium chloride to form a mixture with a lower melting point than pure sodium chloride. Sodium is used to cool breeder reactors and exchange heat to the steam generator.

Potassium metal is formed by the reduction of K^+ by liquid sodium:

$$Na(l) + K^+(l) \rightleftharpoons Na^+(l) + K(g)$$

The reduction is carried out at 850°C so that the product formed is gaseous potassium. The gaseous product is removed from the reaction mixture causing the equilibrium reaction to shift towards products and produce more potassium. Potassium is used as an alloy with sodium as a heat exchanger in chemical and nuclear reactors. Its oxide, KO_2, is used in breathing masks as a source of oxygen:

$$4\mathbf{KO_2}(s) + 4CO_2(g) + 2H_2O(g) \rightarrow 4KHCO_3(s) + 3\mathbf{O_2}(g)$$

Iron, Copper, and Aluminum

Iron is recovered from ore by reduction with carbon in a blast furnace. The overall reaction is:

$$2Fe_2O_3(s) + C(s) \rightarrow 4Fe(l) + 3CO_2(g)$$

The major reactions that occur in a blast furnace are shown in Figure 22.17 on page 996 in your textbook. The process depends on a temperature range of about 2000°C between the top and the bottom of the blast furnace. Iron's major use is as the major component in a wide variety of steels. World production of steel exceeds 700 million tons.

Copper is an expensive metal today because many centuries of mining it have left ores less rich in copper. The most common copper ore is chalcopyrite, a mixed sulfide of FeS and CuS, and it usually contains less than 0.5% copper by mass. To obtain pure copper, the iron must be removed, and the resulting copper compound (Cu_2S) is partially oxidized to Cu_2O. These two copper compounds react to form copper and sulfur dioxide. The copper is purified for electrical applications by electrorefining.

From ore to copper metal

The most common copper ore, chalcopyrite, is a mixture of the sulfides FeS and CuS ($FeCuS_2$). The first step in copper extraction is pretreatment by flotation (see Figure 22.11), which concentrates the copper from 0.5% to about 15% by mass. The next step is to selectively oxidize FeS from the ore. A product of this reaction is sulfur dioxide. Write the balanced equation for this oxidation.

solution:

$$2FeCuS_2(s) + 3O_2(g) \rightarrow 2CuS(s) + 2FeO(s) + 2SO_2(g)$$

The compounds copper sulfide and iron oxide must now be separated. If the mixture is heated to 1100°C with sand (SiO_2), FeO reacts with the sand to form a molten slag, and the copper sulfide decomposes to Cu_2S which can be drawn off as a liquid. Show the reaction of FeO with sand.

solution:

$$FeO(s) + SiO_2(s) \rightarrow FeSiO_3(l) \text{ (slag)}$$

Cu_2S is separated from the mixture, but before it can be reduced to copper, some of the Cu_2S is converted to Cu_2O by roasting in air. Show this reaction (SO_2 is a product of this reaction).

solution:

$$2Cu_2S(s) + O_2(g) \rightarrow 2Cu_2O(s) + 2SO_2(g)$$

Finally, the two copper compounds Cu_2S and Cu_2O react and are reduced to copper. Show this reaction (SO_2 is a product), and indicate the oxidation states of the oxidized and reduced species.

solution:

$$Cu_2S(s) + 2Cu_2O(s) \rightarrow 6Cu(l) + SO_2(g)$$

Copper is reduced from +1 to 0 in both Cu_2S and Cu_2O; sulfur is oxidized from –2 to +4.

Aluminum is the most abundant metal in the earth's crust. The major ore of aluminum is *bauxite*. In a two-step process, Al_2O_3 is extracted from bauxite, and then is converted to the metal.

1) Extraction of Al_2O_3 from bauxite:
Boiling bauxite in 30% NaOH, dissolves Al_2O_3 which precipitates as $Al(OH)_3$ upon addition of acid. Drying at high temperature converts the hydroxide to the oxide:

$$2Al(H_2O)_3(OH)_3(s) \xrightarrow{\Delta} Al_2O_3(s) + 3H_2O(g)$$

2) Conversion of Al_2O_3 to aluminum metal (the *Hall-Heroult* process):
Al_2O_3 is dissolved in molten cryolite (Na_3AlF_6) to produce a mixture that can be electrolyzed at ~1000°C. (The melting point of pure Al_2O_3 is 2030°C). Pure aluminum forms at the cathode and the graphite anode is oxidized and forms carbon dioxide gas. The overall reaction for the process is:

$$2\mathbf{Al_2O_3} \text{ (in } Na_3AlF_6) + 3C(\text{graphite}) \rightarrow 4\mathbf{Al}(l) + 3CO_2(g)$$

Aluminum is lightweight, easy to work, and forms strong alloys. Aluminum objects are often *anodized*, coated with an oxide layer, to prevent deterioration when in contact with less active metals. Figure 22.21 on page 1001 in your textbook shows many of the familiar uses of aluminum.

Magnesium and Bromine

Magnesium and bromine are mined from the oceans as well as from land. Dissolved Mg^{2+} in sea-water converts to magnesium hydroxide on reaction with slaked lime, $Ca(OH)_2$:

$$Ca(OH)_2(aq) + Mg^{2+}(aq) \rightarrow Mg(OH)_2(s) + Ca^{2+}(aq)$$

Magnesium hydroxide converts to magnesium chloride on reaction with excess HCl. Electrolysis of magnesium chloride produces chlorine gas and molten magnesium. Pure magnesium is a strong reducing agent, and as such, is used for sacrificial anodes and in the extraction of other metals. Magnesium alloys are used in such diverse uses as aircraft bodies, luggage, and auto engine blocks.

Bromine ion, Br^-, dissolved in the oceans readily oxidizes to liquid bromine, Br_2, with aqueous chlorine:

$$2Br^-(aq) + Cl_2(aq) \rightarrow Br_2(\textit{red liquid}) + 2Cl^-(aq)$$

The bromine is removed by passing steam through the mixture and then cooling and drying the liquid bromine. Some products requiring bromine are flame retardants in rugs and textiles. Bromine is required for the synthesis of AgBr used for photographic emulsions (see next section, 22.5).

Hydrogen

Hydrogen's abundance

Choose words from the list below to complete the following paragraph:

(a) 90%
(b) H atoms
(c) hydrogen
(d) deuterium
(e) weaker
(f) 15%
(g) water
(h) H_2
(i) two
(j) no
(k) organic compounds

(continued on next page)

Recall from section 22.1 that ____________(1) is by far the most abundant element in the universe, accounting for ______________(2) of its atoms, but in the earth's crust, hydrogen accounts for only ______________(3) of the atoms. Most hydrogen in the universe exists as the molecule, ____________(4) or as free ______________(5), but on earth virtually all the hydrogen exists in combination with oxygen as ______________(6), or with carbon as ______________(7). There are three isotopes of hydrogen: protium, or ordinary hydrogen with ______________(8) neutrons, ______________(9) with one neutron, and radioactive tritium with ______________(10) neutrons. Bonds to hydrogen are ______________(11), and thus quicker to break than corresponding bonds to deuterium. This difference in the rate of bond breaking between isotopes is called the *kinetic isotope effect*.

answers: 1-c, 2-a, 3-f, 4-h, 5-b, 6-g, 7-k, 8-j, 9-d, 10-i, 11-e.

In the United States, we usually produce hydrogen from methane in a two-step thermal process. In the first step, methane is heated with water to around 1000°C over a nickel-based catalyst:

$$CH_4(g) + H_2O(g) \rightarrow CO(g) + H_2(g)$$

The product mixture is heated with steam at 400°C to increase the yield of hydrogen during the *water-gas shift* reaction:

$$H_2O(g) + CO(g) \rightleftharpoons CO_2(g) + H_2(g)$$

Carbon dioxide and hydrogen are separated either by passing the mixture through water which removes the more soluble CO_2, or by reaction with calcium oxide which removes CO_2 by conversion to $CaCO_3$.

Pure hydrogen can also be obtained through the electrolysis of water with Pt or Ni electrodes:

$$H_2O(l) \rightarrow H_2(g) + 1/2O_2(g)$$

This process, with current technology, is a very energy intensive process. Because hydrogen is such a clean-burning fuel, the race is on to find less energy intensive ways to decompose water.

Most hydrogen produced industrially is used to make ammonia or petrochemicals. Another major use of hydrogen is to hydrogenate C=C double bonds in liquid oils to form C–C single bonds in solid fats and margarine (hydrogenated oils). Hydrogen is also used in the manufacture of chemicals such as methanol, CH_3OH, a chemical with many uses.

22.5 CHEMICAL MANUFACTURING OF FILM: A CASE STUDY

Introduction

Industrial chemists often work in research groups made up of scientists with diverse backgrounds, as knowledge in chemistry, biochemistry, physics, mathematics, and engineering may all be required to invent and produce new products. The manufacture and processing of photographic film offers an example of a many-step process that requires interdisciplinary knowledge. Although silver halide film is beginning to seem outdated, its use to capture images seemed miraculous when it was discovered, and it continues to find applications today.

The invention of the camera dates back to sometime in the early 1500's when the first crude camera was simply a darkened box with a tiny hole in one wall that admitted light and formed an upside-down image on the opposite wall. It wasn't until a couple hundred of years later that it was discovered that silver salts are sensitive to light, and it was still another hundred years before a French physicist, Joseph Niepce, produced a permanent image using a light-sensitive plate. Soon after, Niepce's partner, Louis Daguerre exposed a light-sensitive metal plate, developed the image with mercury vapor, and "fixed" the image with salt. This process became known as the *daguerreotype*.

The invention of light-sensitive paper and improvements in lenses, gelatin emulsions, and light capturing abilities throughout the 1800's led to the invention of the Kodak camera, introduced by George Eastman in 1888. This introduction changed photography in that it made it accessible to amateurs. In 1924, 35-millimeter film was introduced, and soon after, the electronic flash. Next came the development of high-speed cameras, underwater cameras, infrared and ultraviolet film, and aerial photography that found uses by scientists, police, and the military. Even as film becomes increasingly scarce as electronic imaging takes its place, there is a tremendous amount of chemistry to be learned from film manufacturing and the film developing processes.

A cross section of photographic film has the general structure:

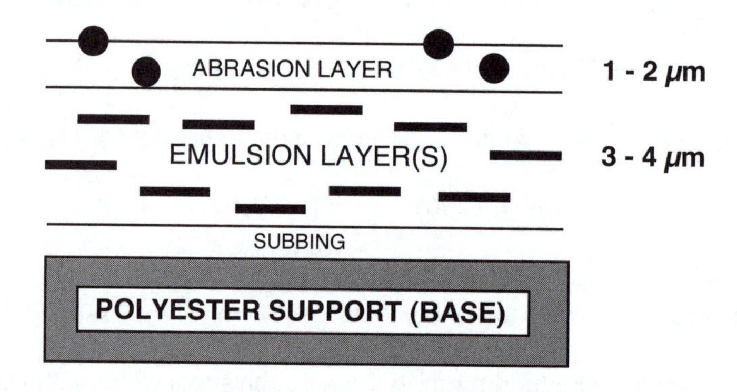

A flexible support (the polyester base) is coated with the light-sensitive silver halide crystals suspended in a gelatin-based matrix (the emulsion layer). This "layer" may have up to 15 separate layers in specialty color films. The overcoat is often a hardened gelatin-based layer that protects the emulsion layers from scratches and abrasions. It may contain polymer beads as matte particles to facilitate smooth transport of film in camera equipment.

It is an engineering accomplishment to coat emulsion and overcoat layers simultaneously at speeds of 300 ft/min, or faster, without coating defects, and at a uniform coating weight across the width of the coating. Furthermore, continuous drying of the film requires not only engineering skill, but also an understanding of the kinetics of the film matrix hardening reactions that occur during the coating and drying processes.

Coating weight reduction

The silver coating weight of your standard product is 4.2 g/m^2. You discover a way to reduce the thickness of the silver halide grains, and so reduce the coating weight of the silver to 3.8 g/m^2. If the price of silver is $72 per pound, how much money will you save your company per square foot of film?

solution:

The cost per gram of silver is:

(continued on next page)

$$\frac{\$72}{\text{lb}} \times \frac{1\text{ lb}}{454\text{ g}} = \$0.16/\text{g}$$

The initial coating weight in g/ft^2 and cost per square foot is:

$$\frac{4.2\text{ g}}{\text{m}^2} \times \left(\frac{1\text{ m}}{100\text{ cm}}\right)^2 \times \left(\frac{2.54\text{ cm}}{1\text{ in}}\right)^2 \times \left(\frac{12\text{ in}}{1\text{ ft}}\right)^2 = 0.390\text{ g/ft}^2$$

$$\frac{0.390\text{ g}}{\text{ft}^2} \times \frac{\$0.16}{\text{g}} = \$.062/\text{ft}^2$$

The reduced coating weight in g/ft^2 and cost per square foot is:

$$\frac{3.8\text{ g}}{\text{m}^2} \times \left(\frac{1\text{ m}}{100\text{ cm}}\right)^2 \times \left(\frac{2.54\text{ cm}}{1\text{ in}}\right)^2 \times \left(\frac{12\text{ in}}{1\text{ ft}}\right)^2 = 0.353\text{ g/ft}^2$$

$$\frac{0.353\text{ g}}{\text{ft}^2} \times \frac{\$0.16}{\text{g}} = \$.057/\text{ft}^2$$

The savings is:

$\$.062/\text{ft}^2 - \$.057/\text{ft}^2 = \$.005$, about half a cents per square foot.

To put this in context, for a product sales volume of 130 million sq ft per year, this represents a savings of $650,000.00/year.

Base

The polyester support (or base) on which the emulsion is coated is a polymer sheet, which is flexible (to allow the film to bend through a processor), strong (to withstand relatively high tensions used during coating), and dimensionally stable. The base may be clear, or it may be tinted with a color. Medical X-ray film base is often tinted blue to help give the final image a blue-black color that radiologists prefer.

The polyester base is a hydrophobic material. An aqueous coating solution will neither wet nor adhere to the base. Typically, a submicron layer of a saran polymer coating (the subbing) provides a bridge between the hydrophobic base and the hydrophilic binder, binding with each.

Gelatin Matrix

Gelatin, a familiar biopolymer, has been used as a binder for silver halide emulsions for over a hundred years. Gelatin is produced from collagen (a protein found in the bones and tissue of animals) by either acid or base digestion. Collagen has a helical structure of polypeptide chains that is stabilized by hydrogen bonds. Gelatin retains some of this three-dimensional structure. Since gelatin is derived from a natural product, it is a challenge to consistently produce batches of gelatin with identical properties. Gelatin must be highly purified for photographic use since the photographic response of silver halide crystals is highly dependent on the nature of the binder that surrounds the crystals.

No synthetic polymers have been found which can effectively replace gelatin. The ability of gelatin to chill set, or gel upon cooling, is crucial in film manufacture. Gelatin swells in aqueous solution to allow penetration of processing solutions. Gelatin contains reactive side groups on the polymer chains, which can be chemically crosslinked to harden the coating, and gelatin acts as a protective colloid of the silver halide microcrystals and prevents their coalescence.

Gelatin has its drawbacks. It exerts extreme pressure as it dries and can cause films to curl. Other binder components are added to counter these effects and adjust the physical properties of the gelatin binder. Latex particles can be added to the emulsion to impart flexibility and shock absorption qualities. Humectants, which prevent the film from drying out and becoming brittle, are often added. Surfactants are used to improve material compatibility and coatability.

Gelatin is hardened by chemically crosslinking reactive side groups. One type of crosslinking is amine-amine crosslinking, in which a hardening agent such as formaldehyde links amine groups (the first example on the next page, where X = an amine-amine crosslinker). Another type of crosslinking, peptide coupling, involves reaction of a gelatin amine group with a carboxyl group (the second example on the next page, where PC = peptide coupler) to form a peptide bond.

$$2\ \text{gel—NH}_2 \xrightarrow{\text{X}} \text{gel—N(H)—X—N(H)—gel}$$

$$\text{gel—N(H)—}\boxed{\text{H}\quad\text{HO}}\text{—C(=O)—gel} \xrightarrow{\text{PC}} \text{gel—N(H)—C(=O)—gel}$$

$-H_2O$

peptide bond

Crosslinking of gelatin

If there are 0.4 mmol of amine groups in 1 g of gelatin, how many grams of formaldehyde (CH_2O) would you need to add to crosslink 25% of the amine groups in 200 g of gelatin?

It requires 1 molecule of formaldehyde to crosslink 2 amine groups:

$$0.25 \times 200\text{ g gel} \times \left(\frac{0.4\text{ mmol amine}}{1\text{ g gel}}\right) \times \left(\frac{1\text{ mmol CH}_2\text{O}}{2\text{ mmol amine}}\right) \times \left(\frac{30\text{ mg CH}_2\text{O}}{\text{mmol}}\right) = 300\text{ mg CH}_2\text{O}$$

Silver Halide Grains

Silver halide microcrystals are solid-state devices; when the microcrystals are exposed to light, photoelectrons are generated. The valence and conduction bands of the microcrystal allow these photogenerated electrons to move throughout the crystal. If the silver halide grain is properly sensitized, these photogenerated electrons create catalytic centers (called "latent image" sites), which catalyze reduction of the crystal in developer solution.

Silver chloride, silver bromide, and silver iodide are all used in silver halide crystals in film. Often the best crystal for use in photographic film is comprised of a mixture of halides.

Solubility of silver halides

Given the following K_{sp} values:

AgI $K_{sp} = 8.49 \times 10^{-17}$

$AgBr$ $K_{sp} = 5.32 \times 10^{-13}$

$AgCl$ $K_{sp} = 1.76 \times 10^{-10}$

order silver iodide, silver bromide, and silver chloride from most soluble to least soluble.

solution:

The smallest K_{sp} indicates the least soluble compound:

$AgCl > AgBr > AgI$

Silver nitrate reacts with potassium bromide to produce light-sensitive silver bromide crystals (or "grains," as they are referred to in the field). The grains are produced in a kettle containing a gelatin/water mixture and excess potassium bromide at around 60°C:

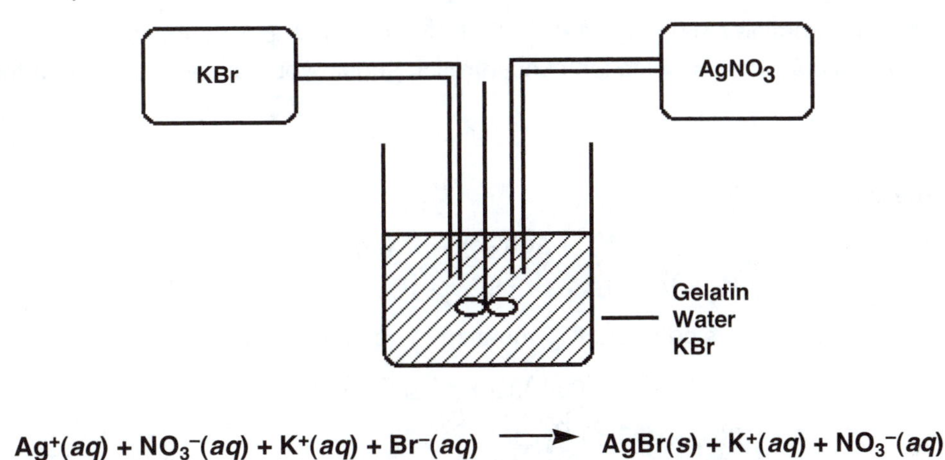

$$Ag^+(aq) + NO_3^-(aq) + K^+(aq) + Br^-(aq) \longrightarrow AgBr(s) + K^+(aq) + NO_3^-(aq)$$

The temperature, the flow rates of silver nitrate and potassium bromide, the bromide ion concentration, and the gelatin concentration all determine the size and shape of the resulting silver bromide grains. In particular, the bromide ion concentration determines the shape (or "morphology") of the grains. Silver halide precipitations are usually carried out under excess halide ion concentration. As halide ion increases, silver halide exists in solution mainly as complexes such as $AgBr_2^-$ and $AgBr_3^{2-}$. These complexes influence the crystal morphology. At lower bromide ion concentration, cubic grains form. As the bromide ion concentration increases, the morphology of the grains changes from cubic to tetradecahedral (cubes with their corners truncated), to octahedral, to tabular:

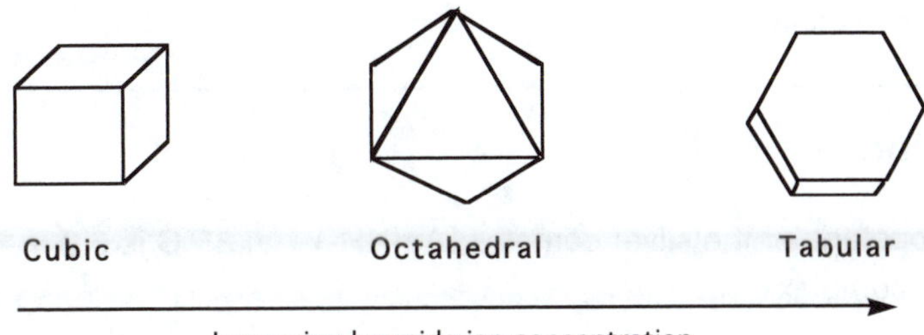

Ammonia is sometimes used during part, or all, of the precipitation to increase solubility by formation of the complex ion $Ag(NH_3)_2^+$. The increase in solubility increases the size of the silver halide crystals and may make the grain size distribution more uniform as smaller grains dissolve and redeposit onto larger grains.

pAg	Morphology
< 8.5	cubes
8.5 – 9.5	tetradecahedra
9.5 – 10.5	octahedra
> 10.5	tabulars, irregular

pAg and crystal morphology

The K_{sp} of silver bromide at 60°C is 1.58×10^{-11}. If a patent discloses that their process begins precipitation at a pAg of 8.5, what bromide ion concentration should you have in the kettle at the start of precipitation?

solution:

For the equilibrium at 60°C,

$$AgBr(s) \rightleftharpoons Ag^+(aq) + Br^-(aq)$$

$$K_{sp} = [Ag^+][Br^-] = 1.58 \times 10^{-11}$$

We can take – log of each side of the K_{sp} equation to obtain:

$$-\log K_{sp} = -\log [Ag^+] + -\log [Br^-]$$

$$pK_{sp} = pAg + pBr$$

Substituting in for pK_{sp} and pAg gives:

$$10.801 = 8.5 + pBr$$

$$pBr = 10.8 - 8.5 = 2.3$$

$$[Br^-] = 10^{-2.3} = 5 \times 10^{-3}\ M$$

Silver ions can form water-soluble complex ions with many inorganic and organic compounds. This property explains the ability of the insoluble silver salts to dissolve in some aqueous solutions. As you will see later, the formation of a silver thiocyanate complex allows silver halide to be dissolved from the film during the fixing stage of processing.

Complex ion formation

The dissociation constant for the silver complex, $Ag(NH_3)_2^+$ at 25°C is 6.2×10^{-8}. What is the formation constant for this complex?

solution:

(continued on next page)

The formation constant, or "stability" constant for $Ag(NH_3)_2^+$, is the reciprocal of the dissociation constant:

$$Ag(NH_3)_2^+ \rightleftharpoons Ag^+ + 2NH_3 \quad \text{(dissociation)}$$

$$K = \frac{[Ag^+][NH_3]^2}{[Ag(NH_3)_2^+]} = 6.2 \times 10^{-8}$$

$$Ag^+ + 2NH_3 \rightleftharpoons Ag(NH_3)_2^+ \quad \text{(formation)}$$

$$K = \frac{[Ag(NH_3)_2^+]}{[Ag^+][NH_3]^2} = \frac{1}{6.2 \times 10^{-8}} = 1.6 \times 10^{7}$$

Silver bromide crystallizes in a face-centered cubic structure with alternating silver and halide ions:

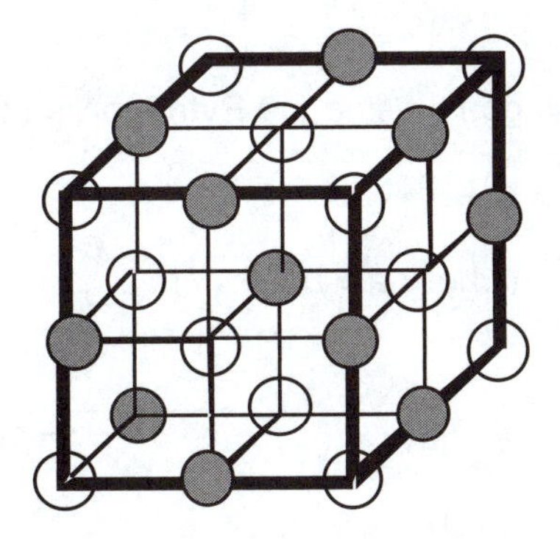

The alternating arrangement of ions on the surface of this cubic crystal is a 100 surface, or a "cubic surface." A second important type of crystal surface is the 111, or octahedral surface, which we obtain if we slice diagonally through the cubic structure shown above. 100 surfaces tend to be less reactive than 111 surfaces because the ions are bound more symmetrically. The adsorption of sensitizing materials such as sensitizing dyes, to be discussed, may be different on the two types of surfaces.

The AgBr unit cell

1) The ionic radii of silver and bromide are given by Pauling as: rAg = 1.26 Å, rBr = 1.95 Å. What would the dimension of the unit cell for silver bromide be based on these values?

We can show one face of a unit cell for a silver bromide crystal as:

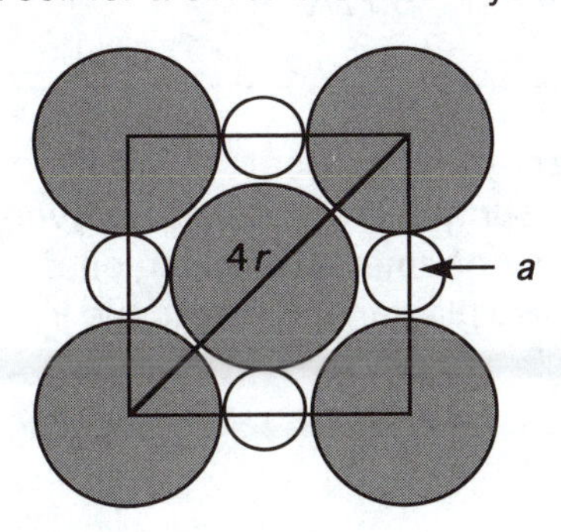

(continued on next page)

where "*a*" is the unit cell dimension. "*a*" is twice the radius of a bromide ion plus the diameter (twice the radius) of a silver ion:

$$a = (2 \times r\text{Ag}) + (2 \times r\text{Br})$$

$$a = (2 \times 1.26\ \text{Å}) + (2 \times 1.95\ \text{Å})$$

$$a = 6.42\ \text{Å}$$

2) The measured unit cell dimension for silver bromide is 5.77 Å. How can you explain the discrepancy between this measured value and our calculated value above (6.42 Å)?

solution:

The hard sphere model does not apply accurately to the silver halides because the bonding in silver halides is not strictly ionic.

3) Using 5.77 Å as the value for the unit cell dimension for silver bromide, are the bromide ions touching along the diagonal of the face of the face-centered cube (drawn on the previous page)?

solution:

We can find the length of the diagonal using the Pythagorean theorem:

$$a^2 + b^2 = c^2$$

(where $a = b = 5.77$).

Solving for c gives:

$$c = 4r = \sqrt{(5.77)^2 + (5.77)^2}$$

$$c = 4r = 8.16$$

$$r = 2.04$$

The radius of a bromide ion is 1.95 Å, so the bromide ions do not touch along the diagonal face of the unit cell.

4) If you measure a tabular silver halide grain to have a thickness of 0.15 μ, how many unit cells thick is the grain?

solution:

$$0.15\ \mu \times \frac{10^4\text{Å}}{1\ \mu} \times \frac{\text{unit cell}}{5.77\text{Å}} = 260 \text{ unit cells}$$

Crystal defects such as twin planes and dislocations, which are interruptions in the regular crystal lattice, and surface kink sites, which are surface irregularities, are very important in determining the response of the crystal to sensitization. The number of "interstitial silver ions"—ions which leave lattice positions and move interstitially through the crystal—also play an important role in the ability of a silver halide crystal to be sensitized.

Sensitization

1. Chemical sensitization

Photographic sensitivity can be increased with chemical treatment of the silver halide grains. Sulfur and gold are common sensitizing agents. Sulfur, often in the form of thiosulfate, adsorbs to silver bromide and reacts to produce a silver sulfide speck on the grain surface:

$$S_2O_3^{2-} + 2AgBr + H_2O \rightarrow Ag_2S + 2Br^- + SO_4^{2-} + 2H^+$$

The Ag_2S acts as a trap for electrons and is the site of latent image formation discussed on the next page. If sensitization is carried too far, the sulfide particles may catalyze development without exposure.

Oxidation states of sulfur

For the reaction of thiosulfate with silver bromide to produce silver sulfide shown earlier, what are the oxidation states of sulfur in each of the compounds?

solution:

- In $S_2O_3^{2-}$, three oxygen atoms with an oxidation number of –2 give a –6 charge. The charge on the anion is –2, so the sulfur atoms must contribute a +4 charge. Each sulfur atom has an oxidation number of +2.
- In Ag_2S, each silver atom has a +1 oxidation number, so sulfur has an oxidation number of –2.
- In SO_4^{2-}, four oxygen atoms with oxidation numbers of –2 give a –8 charge. The charge on the anion is –2, so the sulfur atom must contribute a +6 charge.

One sulfur atom in thiosulfate loses four electrons as it is oxidized to an oxidation state of +6; the other sulfur atom gains four electrons as it is reduced to an oxidation state of –2.

2. Spectral sensitization

Silver halide crystals absorb blue and ultraviolet light. Color films form images from light absorbed primarily by spectral sensitizers. Spectral sensitizers are organic dyes that adsorb to the surface of the grain and tune it to the incident wavelength of light. Spectral sensitization occurs by electron transfer from a dye to the silver halide grain. If the conduction band of a dye is at the right energy level in relation to the conduction band of the silver halide crystal, it can inject an electron into the conduction band:

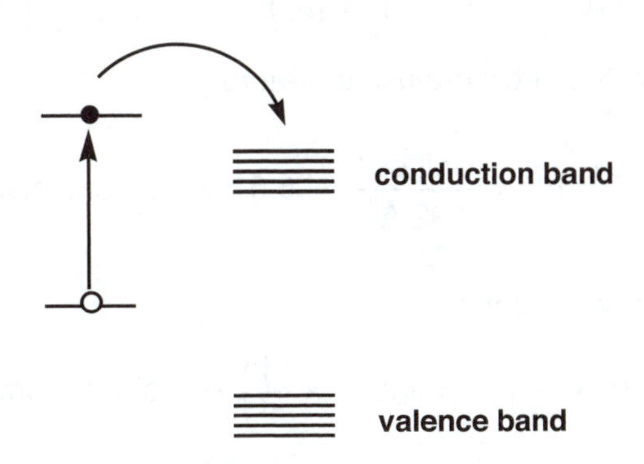

A dye must adsorb to the silver halide surface correctly to be an effective spectral sensitizer. Some dyes may be effective on a 100 (cubic) surface, but ineffective on a 111 (octahedral) surface. Many dye molecules can aggregate on the surface, which changes their adsorption wavelength.

If not all the sensitizing dye is removed during processing, dye stain results. Organic chemists work to change dye structures so that the dye molecules diffuse readily from the emulsion during processing, or so the dye molecules react with an ingredient in the processing solution to become decolorized.

Development and Fix

A photon ($h\upsilon$) of the right energy generates a mobile photoelectron in a silver halide crystal that reduces interstitial Ag^+ ions in the crystal to Ag^0. A cluster of a minimum of three silver atoms forms a "latent image." On development, the latent image catalyzes reduction of the rest of the grain to Ag^0. Development of a silver halide grain is a kinetic process: given enough time in developer, even the unexposed grains will eventually be reduced.

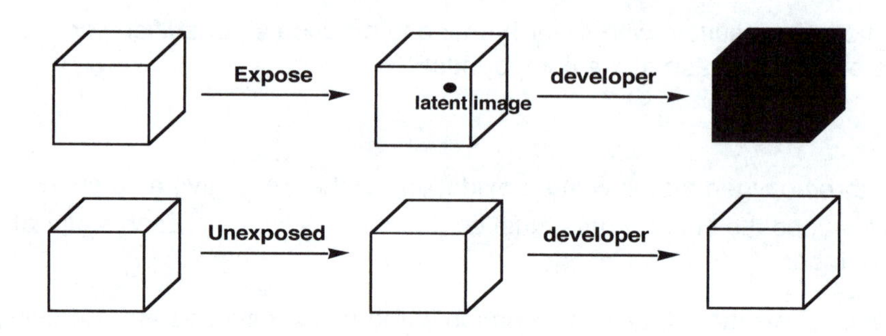

Amplification on development

If a cubic grain, volume = 0.125 μ^3, contains a latent image of three silver atoms, what is the amplification factor on development (reduction) of the grain?

solution:

The measured unit cell dimension for silver bromide is 5.77 Å (earlier exercise), so the volume of the unit cell is: $(5.77\ \text{Å})^3 = 192\ \text{Å}^3$

The volume of the grain in Å^3 is:

$$\text{volume} = \left(0.125\ \mu^3\right) \times \left(\frac{10^4\ \text{Å}}{1\ \mu}\right)^3 = 1.25 \times 10^{11}\ \text{Å}^3$$

The number of unit cells in the cubic grain is therefore:

$$1.25 \times 10^{11}\ \text{Å}^3 \times \frac{\text{unit cell}}{192\ \text{Å}^3} = 6.51 \times 10^8\ \text{unit cells}$$

Each unit cell contains 4 silver atoms:

$$\left(8\ \text{corners} \times \frac{1}{8}\right) + \left(6\ \text{faces} \times \frac{1}{2}\right) = 4\ \text{silver atoms/unit cell}$$

so the number of silver atoms in the grain is:

$$6.51 \times 10^8\ \text{unit cells} \times \frac{4\ \text{Ag atoms}}{\text{unit cell}} = 2.60 \times 10^9\ \text{Ag atoms}$$

The amplification factor on development is:

(continued on next page)

$$\frac{2.60 \times 10^9 \text{ Ag atoms after development}}{3 \text{ Ag atoms before development}} = 8.68 \times 10^8$$

This amplification factor of over 800 million is why the "relatively old" technology of silver halide photographic film remains important today. No other storage media (magnetic tape, disks, etc.) possesses this level of data amplification and data storage density.

After film is developed, it is "fixed" to wash off undeveloped silver halide. A silver halide solvent such as thiosulfate solubilizes the silver halide allowing it to wash off the film:

$$AgBr(s) + 2S_2O_3^{2-}(aq) \rightarrow Ag(S_2O_3)_2^{3-}(aq) + Br^-(aq)$$

And now, finally, you've got your image. How does this image compare to the one your friend shot with their digital camera—the one that allows them to see the photographs immediately, edit as they go, save only the best images for printing, and make their own high-quality color prints on an inexpensive digital printer? Depending on the application, you probably would not notice much difference between the two pictures. In fact, 35 mm film for professional use is practically obsolete, and digital technology has replaced film in news coverage as well as in many medical applications. However, film still has a niche. It remains the choice for landscapes and still shots that require big prints. And looking ahead, images that have been captured on film can be scanned at higher resolution as scanner technology improves; digital images, on the other hand, will always be stuck at the image quality level at which they were recorded. As long as film offers some technical advantages over digital cameras, silver halide technology, a technology that has been around for over 200 years, will likely be around, in some form, for many more.

23

Transition Elements and Their Coordination Compounds

REVIEW

In Chapter 8 we learned rules for filling orbitals with electrons, and saw how electron configurations explain periodic patterns of chemical reactivity. Transition elements are the elements that have atomic *d* or *f* orbitals being filled. The naturally occurring *d*-block metals in periods 4, 5, and 6 fill the 3*d*, 4*d*, and 5*d* shells. The lanthanides fill the 4*f* shell, and the actinides (radioactive elements) fill the 5*f* shell. Transition metals form compounds that are often colored and paramagnetic (contain unpaired electrons).

Review exercises

1) Show the electron configurations and orbital diagrams for Cu, Ag, Au, Cr, and Mo.

Cu: $[Ar]4s^1 3d^{10}$ — [Ar] ↑ (4*s*) ↑↓ ↑↓ ↑↓ ↑↓ ↑↓ (3*d*)

Ag: $[Kr]5s^1 4d^{10}$ — [Kr] ↑ (5*s*) ↑↓ ↑↓ ↑↓ ↑↓ ↑↓ (4*d*)

Au: $[Xe]6s^1 4f^{14} 5d^{10}$ — [Xe] ↑ (6*s*) ↑↓ ↑↓ ↑↓ ↑↓ ↑↓ ↑↓ ↑↓ (4*f*) ↑↓ ↑↓ ↑↓ ↑↓ ↑↓ (5*d*)

Cr: $[Ar]4s^1 3d^5$ — [Ar] ↑ (4*s*) ↑ ↑ ↑ ↑ ↑ (3*d*)

Mo: $[Kr]5s^1 4d^5$ — [Kr] ↑ (5*s*) ↑ ↑ ↑ ↑ ↑ (4*d*)

Remember: half-filled and filled shells are unusually stable.

(continued on next page)

Generally orbitals fill in the order: *ns*, $(n-1)d$, *np*. However, the energies of the *ns* and $(n-1)d$ sublevels are so close, deviations in filling patterns occur. Irregularities are found particularly in the *d* transition row, and in the lanthanides and actinides where the orbitals are very close in energy.

2) Write electron configurations for the Zn^{2+}, Cu^{2+}, and Cr^{3+}.

answers:

Zn^{2+}: $[Ar]3d^{10}$ — [Ar] $\underline{\quad}$ (4s) $\underline{\uparrow\downarrow}\ \underline{\uparrow\downarrow}\ \underline{\uparrow\downarrow}\ \underline{\uparrow\downarrow}\ \underline{\uparrow\downarrow}$ (3d)

Cu^{2+}: $[Ar]3d^{9}$ — [Ar] $\underline{\quad}$ (4s) $\underline{\uparrow\downarrow}\ \underline{\uparrow\downarrow}\ \underline{\uparrow\downarrow}\ \underline{\uparrow\downarrow}\ \underline{\uparrow}$ (3d)

Cr^{3+}: $[Ar]3d^{3}$ — [Ar] $\underline{\quad}$ (4s) $\underline{\uparrow}\ \underline{\uparrow}\ \underline{\uparrow}\ \underline{\quad}\ \underline{\quad}$ (3d)

The 4*s* orbitals are lower in energy than the 3*d* orbitals when they are filled by the aufbau process, but when they form ions, transition metals appear to lose the valence shell electrons first. Except for the lanthanides, electrons with the highest *n* value are lost first.

Zinc loses its two 4*s* electrons to form Zn^{2+} with an electron configuration of $[Ar]3d^{10}$. This filled inner 3*d* level is a **pseudo-noble gas configuration.**

23.1 PROPERTIES OF THE TRANSITION METALS

Size

Atomic size of the transition metals in period 4 decreases slightly from scandium to iron, and then remains fairly constant (see Figure 22.3, A in your text). Remember from Chapter 8 that electrons that add to the same shell (approximately the same distance from the nucleus) are not effective at shielding one another from the nuclear charge. As we move across a period then, each additional proton added to the nucleus increases the effective charge on the outer electrons and pulls them in more tightly.

Transition metals in period 5 are larger than in period 4, but transition metals in period 6 are nearly the same size as those in period 5 (see exercise below).

Lanthanide contraction

The atomic radius (in Å) of the transition metals in group 4B are:

titanium: 1.44; zirconium: 1.59; hafnium: 1.56

Use their electron configurations to explain why zirconium in period 5 and hafnium in period 6 are nearly the same size.

solution:

Ti: $[Ar]4s^23d^2$; Zr: $[Kr]5s^24d^2$; Hf: $[Xe]6s^24f^{14}5d^2$

Titanium and zirconium are the 4th elements in their periods. Hafnium is the 18th member of its period because, in addition to Cs, Ba, and Lu, the 14 elements La-Yb (the lanthanides) fall before it. The lanthanides steadily decrease in size with increasing nuclear charge as electrons fill the 4*f* sublevel. This **"lanthanide contraction"** offsets the normal increase in atomic size between periods, so transition elements in periods 5 and 6 are about the same size.

Electronegativity, Ionization Energy, and Density

Consistent with the small changes in the size of transition metals in a period, electronegativity and first ionization energies show only small differences. Density increases with atomic mass since volume across a period is approximately constant.

Densities of gold and lead

The density of lead is 11.3 g/mL; the density of gold is 18.9 g/mL. Why is gold denser than lead?

answer:

A lead atom (**M** = 207 amu/atom) weighs more than a gold atom (**M** = 197 amu/atom), so in order for the density of gold to be greater than lead, the volume of a gold atom must be smaller than the volume of a lead atom ($D = m / V$). The valence electrons in gold occupy the 5*d* and 6*s* sublevels. Its atomic radius is small (1.44 Å) due to lanthanide contraction. The valence electrons in lead are in the 6*p* sublevel giving it a larger atomic radius (1.75 Å).

Oxidation States

Transition metals often form +2 and +3 ions, but they may exist in many oxidation states. The *ns* and $(n-1)d$ electrons are close enough in energy that most (or all) of these electrons can be used in bonding. The highest oxidation state of elements in groups 3B through 7B is equal to the group number. The higher oxidation states cannot exist as hydrated ions because the highly charged metal ion "pulls" electron density away from the water molecules causing them to lose protons. Chromium +3 forms the complex, $Cr(H_2O)_6^{3+}$; chromium +6 forms the oxides CrO_4^{2-} and $Cr_2O_7^{2-}$.

Oxidation Potentials

Metals tend to act as reducing agents. All the transition metals in period 4 except for copper reduce H^+ from aqueous acid to hydrogen gas:

$$Ni(s) + 2H^+(aq) \rightarrow H_2(g) + Ni^{2+}(aq)$$

Reducing strength decreases from left to right across the table, although chromium and zinc are exceptions.

Oxidation states of transition metals

What is the oxidation state for the transition metal in each of the following compounds and/or ions?

a) CrO_4^{2-} c) VO_4^{3-} e) $CrCl_3$ g) VI_2

b) $Cr_2O_7^{2-}$ d) MnO_4^- f) VOCl h) Mn_2O_7

answers:

a) Cr^{6+}, b) Cr^{6+}, c) V^{5+}, d) Mn^{7+}, e) Cr^{3+}, f) V^{3+}, g) V^{2+}, h) Mn^{7+}.

Notice that vanadium, in group 5B can exist in an oxidation state of +5; chromium, in group 6B can exist in an oxidation state of +6, and manganese, in group 7B can exist in an oxidation state of +7.

Colors

Many transition metal compounds have striking colors making them fun to work with in the lab. (See, for example, Figure 23.6 on page 1023 in your text.) Main group chemists, with their colorless, white (or if they're impure, perhaps yellowish-colored) compounds have been known to peer enviously into the hoods of transition metal chemists with their brilliantly colored compounds. The energy difference between partially filled *d* sublevels and unfilled *d* sublevels is the same as the energy of visible light. Ions with partially filled *d* shells are therefore often colored. The color depends on the metal and the groups surrounding the metal.

The many colors of transition metals

Which of the following compounds and/or ions would you expect to be colored, and which would you expect to be colorless?

a) ZnO
b) $Cu(H_2O)_6^{2+}$
c) $TiCl_4$
d) Cr_2O_3
e) $CoSO_4$
f) PCl_3
g) TiO_2
h) $MnSiO_3$
i) Sc_2O_3
j) $CoCr_2O_4$

answers:

Colored compounds (partially filled "*d*" orbitals):

b) Cu^{2+} has the electron configuration $[Ar]3d^9$; $Cu(H_2O)_6^{2+}$ is blue.

d) Cr^{3+} has the electron configuration $[Ar]3d^3$; Chromic oxide is light to dark green. It is used as a pigment in coloring glass and printing fabrics and bank notes.

e) Co^{2+} has the electron configuration $[Ar]3d^7$; Cobaltous sulfate is red to lavender. It is used in Co pigments for decorating porcelain.

h) Mn^{2+} has the electron configuration $[Ar]3d^5$; Manganese silicate is red. It is used as a color for special glass and for producing red glazes on pottery.

j) Co^{2+} has the electron configuration $[Ar]3d^7$; Cobaltous chromate is a brilliant greenish-blue. It is used as a green pigment for ceramics.

Colorless compounds (full or empty "*d*" orbitals):

a) Zn^{2+} has a filled *d* shell (electron configuration $[Ar]3d^{10}$); Zinc oxide is white. It is used as a pigment in white paint, and on the noses of lifeguards to protect them from the sun.

c) Ti^{4+} has no *d* electrons (electron configuration [Ar]); Titanium tetrachloride is a colorless liquid. It is used in the manufacture of iridescent glass and artificial pearls.

f) P^{3+} has no *d* electrons (electron configuration $[Ne]3s^2$); PCl_3 is a colorless fuming liquid.

g) Ti^{4+} has no *d* electrons (electron configuration [Ar]); Titanium dioxide is white. It is used as a whitener in the food industry, and as a white pigment in paints and films.

i) Sc^{3+} has no *d* electrons (electron configuration [Ar]); Scandium oxide is a white powder.

Note: Chromium comes from the Greek word *chroma*, meaning color. "Cobalt blue," the dark blue color seen in blue glass gets its color from mainly cobalt oxide and alumina.

23.2 THE INNER TRANSITION ELEMENTS

Lanthanides

The lanthanides, or rare earths, fill the 4*f* sublevel. All the lanthanides are relatively active metals. They tend to exist in the +3 oxidation state. They react with chlorine to form chloride salts (MCl_3), and with water to form hydroxides ($M(OH)_3$) and evolve hydrogen.

$$2M(s) + 3Cl_2(g) \longrightarrow 2MCl_3(s)$$

$$2M(s) + 6H_2O(l) \longrightarrow 2M(OH)_3(s) + 3H_2(g)$$

Lanthanides are difficult to separate from one another because of their similar properties. Until recently, the only pure sample that was commercially available was cerium, the most abundant lanthanide.

Actinides

The actinides fill the 5*f* sublevel and all are radioactive. The metals are silvery and chemically reactive forming highly colored compounds. Thorium and uranium occur in nature, but the transuranium elements (elements with a molar mass greater than uranium, $Z > 92$) have been produced in controlled nuclear reactions.

23.3 COORDINATION COMPOUNDS

We can think of **coordination compounds** as Lewis acid-base adducts. Remember from Chapter 18 that a Lewis acid is a species that accepts an electron pair from a Lewis base. Boron trifluoride, with an incomplete octet of electrons around boron, acts as a Lewis acid when it accepts an electron pair from ammonia:

$$:HN_3 \quad + \quad BF_3 \longrightarrow H_3N \rightarrow BF_3$$

When ammonia donates an electron pair to a metal ion such as Ni^{2+}, it forms a **complex ion,** and we call the Lewis base a **ligand.** The Ni^{2+} metal ion acts as a Lewis acid, which accepts an electron pair from each ligand:

$$6NH_3 + Ni^{2+} \rightarrow Ni(NH_3)_6^{2+}$$

Ligands may be molecules, like ammonia, or they may be anions such as Cl^-. Some molecules with two separate lone pairs of electrons can occupy two positions on the metal ion. Ethylenediamine (abbreviated "en") is a common **bi-dentate** ("two teeth") ligand:

$$H_2\ddot{N}-\overset{H_2}{C}-\overset{H_2}{C}-\ddot{N}H_2$$

It forms a five-membered ring around the metal ion:

H2 H2
C — C
H2N NH2
M

EDTA (ethylenediaminetetraacetate) can occupy six positions on a metal ion. The ligands stay attached to the central metal in water. **Counter ions** maintain electrical neutrality and separate from the complex ion in water as an ionic salt would:

$$Ni(NH_3)_6Cl_2 \xrightarrow{H_2O} Ni(NH_3)_6^{2+} + 2Cl^-$$

The **coordination number** is the number of ligands bonded directly to the central metal ion. The coordination number of $Ni(NH_3)_6^{2+}$ is 6.

Coordination compounds

For each of the following coordination compounds, give the charge of the complex ion, the charge of the central atom, and list the ligands and counter ions:

a) $[Co(NH_3)_6]Cl_3$

b) $[Co(NH_3)_5Cl]Cl_2$

c) $[Co(NH_3)_4Cl_2]Cl$

d) $[Co(NH_3)_3Cl_3]$

answers:

Ligands are written within the brackets; species written outside the brackets are not bonded directly to the central metal atom and are present as free ions. The NH_3 molecules are neutral and the chloride ions have a charge of –1.

a) charge of complex = +3; charge of Co = +3
ligands: 6 NH_3 molecules; counter ions: 3 Cl^- ions

b) charge of complex = +2; charge of Co = +3
ligands: 5 NH_3 molecules and 1 Cl^- ion; counter ions: 2 Cl^- ions

c) charge of complex = +1; charge of Co = +3
ligands: 4 NH_3 molecules and 2 Cl^- ions; counter ion: 1 Cl^- ion

d) charge of complex = 0; charge of Co = +3
ligands: 3 NH_3 molecules and 3 Cl^- ions; counter ions: none

The "Discovery" of Coordination Compounds

When the cobalt compounds above were discovered, they were written:

$CoCl_3 \cdot 6\ NH_3$ (orange-yellow)

$CoCl_3 \cdot 5\ NH_3$ (violet)

$CoCl_3 \cdot 4\ NH_3$ (green)

$CoCl_3 \cdot 3\ NH_3$ (green)

Ammonia was apparently tightly bound to cobalt since reaction with HCl did not remove the ammonia. Reaction with silver nitrate precipitated all the chlorine from compound 1 as AgCl, 2/3 of the chlorine from compound 2, and only 1/3 of the chlorine from compound 3. When compound 4 was treated with silver nitrate, no precipitate formed.

Alfred Werner proposed in 1893 the idea of coordination compounds. He suggested that all of the compounds above have six ligands attached directly to the cobalt ion (written within square brackets). Only the Cl^- ions outside the coordination sphere (the counter ions) are precipitated by silver nitrate. *(continued on next page)*

He tested his hypothesis by studying the conductivities of water solutions of the four compounds. As predicted, he found that compounds 1–3 behaved as 3:1, 2:1, and 1:1 electrolytes, respectively, and compound 4 behaved as a nonelectrolyte.

Naming Coordination Compounds

The name for $[Co(NH_3)_5Cl]Cl_2$ is pentaamminemonochlorocobalt(III) chloride. OK, it's not short, but it does the job. Break the name up into parts, and it won't seem so cumbersome:

1) First name the cation, then the anion.
   The same rule is used to name sodium chloride. In the example, the cation is not sodium but, pentaamminemonochlorocobalt(III).

2) Name the ligands (in alphabetical order) before the metal in the complex ion.
   The ligands are named before cobalt; ammine is listed alphabetically before chloro.

3) To name the ligands:

   a) Use the name of the molecule for neutral ligands.
      Exceptions are: aquo for H_2O; ammine for NH_3; carbonyl for CO; nitrosyl for NO

      We use ammine for the ammonia ligands

   b) Add "o" to the root name of anions.

      We use chloro for the Cl^- ligand

   c) Prefixes denote the number of a particular ligand.
      - mono-, di-, tri-, tetra-, penta-, and hexa- for simple ligands
      - bis-, tris-, tetrakis- for ligands which already contain a numerical prefix
      - The prefixes do not affect the alphabetical order.

      The ligands are pentaammine (5 ammonia molecules) and monochloro (one Cl^- ion).

4) Use a Roman numeral in parentheses to designate the oxidation state of the central metal ion.
   Cobalt is in the +3 oxidation state, so (III) appears after the name cobalt.

5) If the complex ion is an anion, add the suffix -ate to the name of the metal.
   The complex ion is a cation, so the metal name appears unchanged as cobalt.

Naming coordination compounds

Name the following compounds:

1) $[Zn(H_2O)_3(OH)]NO_3$
2) $K[Pt(NH_3)_3Cl_3]$
3) $[Ni(en)_3]Cl_2$
4) $[Cu(NH_3)_4]SO_4$

answers:

1) Water is a neutral ligand; hydroxide, OH^- and NO_3^- each have –1 charges, so Zn has a +2 oxidation state. The name is:

   Triaquomonohydroxozinc(II) nitrate

2) Potassium ion has a +1 charge; the ammonia ligands are neutral, and the Cl^- ions have –1 charges, so platinum has a +2 oxidation state. The name is:

(continued on next page)

Potassium triamminetrichloroplatinate(II)

3) Ethylenediammine is neutral; the Cl^- ions have –1 charges, so nickel has a +2 oxidation state. The name is:

Tris-ethylenediamminenickel(II) chloride

4) The ammonia ligands are neutral, and the sulfate ion has a –2 charge, so copper has a +2 oxidation state. The name is:

Tetraamminecopper(II) sulfate

Geometry and Isomerism

The geometry of coordination compounds is summarized in the table below.

Coordination number	Geometry
2	Linear
4	Square planar or tetrahedral
6	Octahedral

The complexes of platinum are square planar; the complexes of zinc with four ligands are tetrahedral. **Geometrical isomers** can exist for square planar complexes but not for tetrahedral complexes. Consider the square planar complex $[Pt(NH_3)_2Cl_2]$. Two forms of this compound exist:

Cl, Cl — Pt — NH_3, NH_3

Cis

Cl, H_3N — Pt — NH_3, Cl

Trans

Cis-$Pt(NH_3)_2Cl_2$ ("cisplatin") has identical ligands next to each other. It is a highly effective antitumor agent. *Trans*-$Pt(NH_3)_2Cl_2$ has identical ligands across from each other and is an ineffective antitumor agent.

Recall from Chapter 10 that six groups around a central atom form an octahedral arrangement. All 6 positions are equivalent and can be visualized as a derivative of a square planar complex:

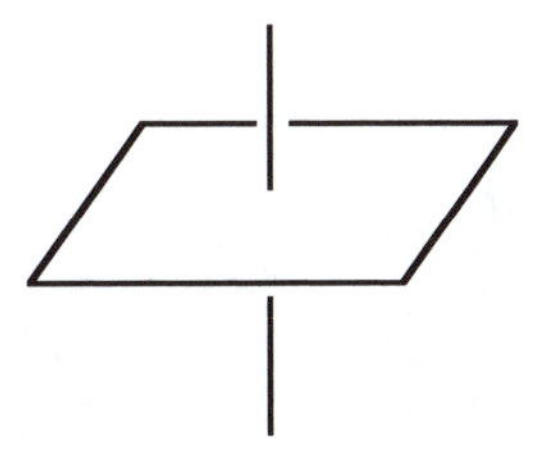

Geometrical isomerism can occur in octahedral complexes. The compound, $[Co(NH_3)_4Cl_2]Cl$, can exist as *cis* and *trans* isomers. The *trans* isomer has chloro ligands opposite each other and is green as described earlier. The *cis* isomer has chloro ligands next to each other and is a violet color:

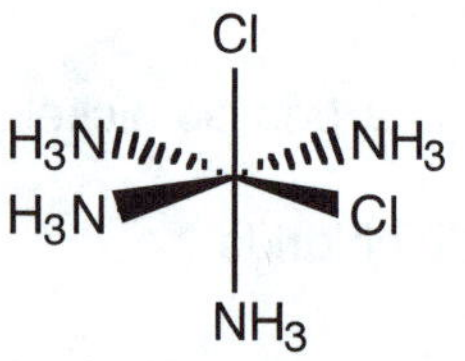

Cl
H_3N NH_3
H_3N Cl
NH_3

Cis

Cl
H_3N NH_3
H_3N NH_3
Cl

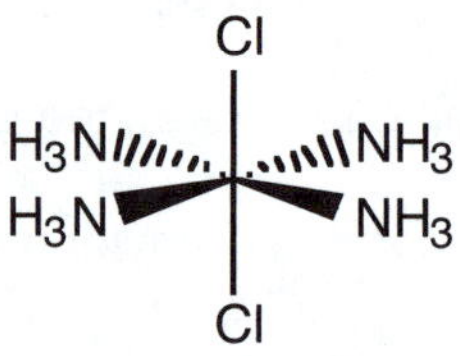

Trans

Alfred Werner, who recognized the possibility of isomers, was able to isolate the *cis* isomer.

The number of isomers

How many isomers are possible for the complex $[Co(NH_3)_3Cl_3]$?

solution:

We can simplify the drawing of isomers by using the general form XA_3B_3 and consider only the "A" ligands. First, if we make two "A" ligands *trans* to each other, all the four remaining positions are equivalent, and we have our first isomer:

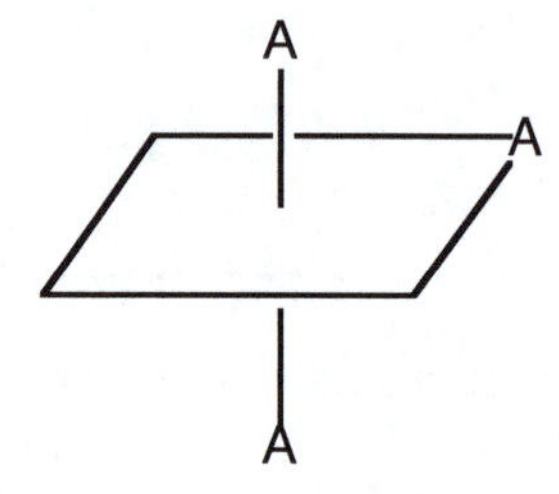

If we move one of the "A" ligands that are *trans* to each other so that all three ligands are *cis*, we arrive with a different isomer:

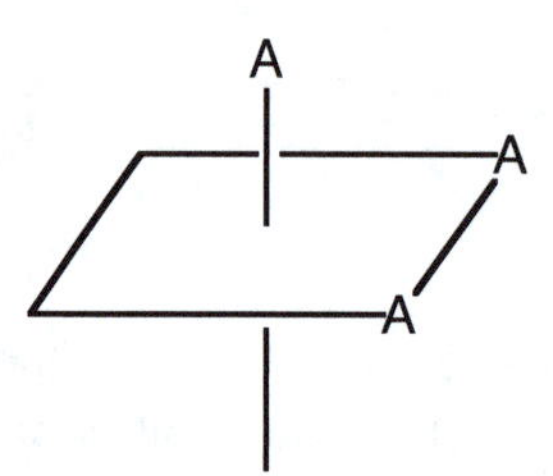

With a little "fiddling," you should be able to convince yourself that these are the only two isomers. Remember that isomers drawn on paper, may look different, but may actually be equivalent. The two compounds drawn below are the same isomer (two groups *trans* with the third group *cis* to each):

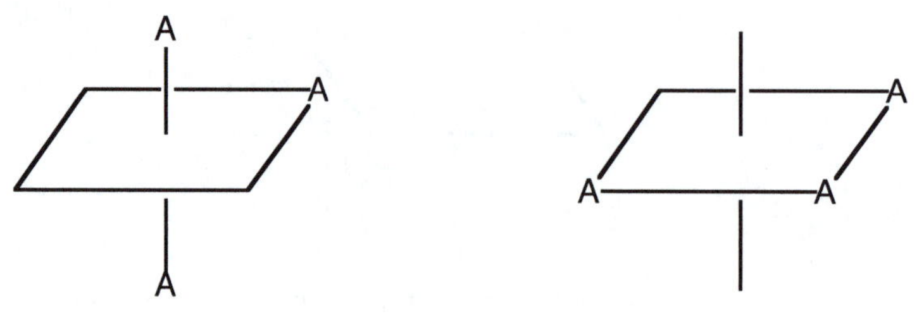

Optical isomers occur when a molecule and its mirror image cannot be superimposed. Optical isomers are identical except for the direction that they rotate the plane of polarized light. If a molecule is superimposable on its mirror image, it is not optically active. This is often difficult to see without making models and trying to superimpose them. If you can find a plane of symmetry in a molecule (one that divides a molecule into two equal parts), the molecule is not optically active. Figure 22.10 in your book shows the *cis* and *trans* isomers of $[Co(en)_2Cl_2]^+$. The *cis* isomer is optically active; the *trans* isomer is not. Can you find the plane of symmetry in *trans*-$[Co(en)_2Cl_2]^+$?

23.4 BONDING IN COMPLEX IONS

The valence bond and molecular orbital theories discussed in Chapter 11 can describe the electronic structures and geometries of complex ions. Another model called **crystal field theory** is more successful at predicting magnetic properties and color of complex ions.

Valence Bond Model

Remember that ligands in complex ions act as Lewis bases and donate electron pairs to metal ions (Lewis acids). In the valence bond model, the electron pairs on the ligands enter hybrid orbitals on the central atom. The hybrid orbitals have the correct geometry to form linear, tetrahedral, square planar, and octahedral complexes.

Hybrid orbital	Coordination number	Shape	Example
sp	2	linear	$[Ag(NH_3)_2]^+$
sp^3	4	tetrahedral	$[Pt(NH_3)_4]^{2+}$
dsp^2	4	square planar	$[Zn(NH_3)_4]^{2+}$
d^2sp^3	6	octahedral	$[Cr(NH_3)_4Cl_2]^+$

You have seen all these hybrid orbitals before in Chapter 11 except for the square planar, dsp^2 orbital. Let's look at an example where the valence bond model successfully predicts the magnetic behavior of a complex ion, $Fe(CN)_6^{4-}$.

The coordination number of $Fe(CN)_6^{4-}$ is 6, so the bonding electrons from the CN^- ligands enter octahedral d^2sp^3 orbitals. How many electrons does the iron atom contribute? The charge on the complex is –4, and each CN^- ligand contributes a –1 charge, so the charge on iron is +2 (electron configuration $[Ar]3d^6$). The 6 electrons from iron remain in the three unhybridized *d* orbitals:

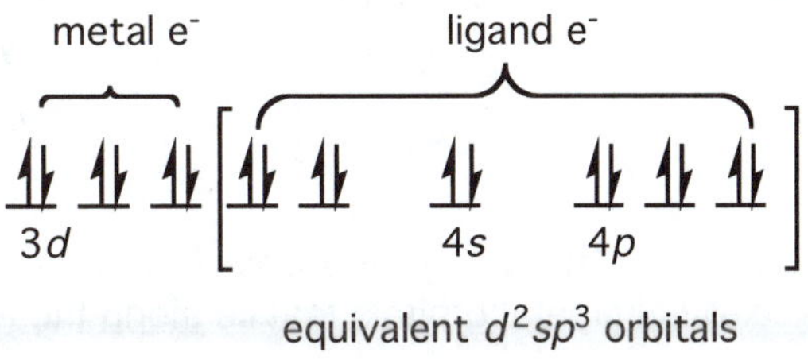

metal e^- ligand e^-

↑↓ ↑↓ ↑↓ | ↑↓ ↑↓ ↑↓ ↑↓ ↑↓ ↑↓

3*d* 4*s* 4*p*

equivalent d^2sp^3 orbitals

Since all electrons are paired, the valence bond theory would predict that this complex is diamagnetic, which experiments confirm.

Orbital diagram for $Fe(H_2O)_6^{2+}$

What is the orbital diagram for the complex ion, $Fe(H_2O)_6^{2+}$?

solution:

The orbital diagram should be identical to $Fe(CN)_6^{4-}$ since $Fe(H_2O)_6^{2+}$ has six ligands bonded to a Fe^{2+} ion. We would predict, based on the valence bond model, that $Fe(H_2O)_6^{2+}$ would have all electrons paired and be diamagnetic.

Now we have a problem. Experiments show $Fe(H_2O)_6^{2+}$ to be paramagnetic with four unpaired electrons. The valence bond theory does not hold up for this complex ion. We need a theory that will predict the magnetic properties of complex ions consistently, and one that can explain the most striking visual feature of complex ions, their color.

Crystal Field Theory

The crystal field model assumes that electrostatic forces between the ligands and metal atom hold complex ions together. In this model, when ligands approach the central metal ion, their electrons do not fill hybrid orbitals on the metal ion. Instead, the electrostatic interaction of ligand electrons with electrons on the metal ion, split the *d* orbitals on the metal ion into different energy levels.

In Chapter 7 we showed the five orientations of the *d* orbitals. Two of the orientations ($d_{x^2-y^2}$ and d_{z^2}) lie along the *x*, *y*, and *z* axes, the other three orientations (d_{xy}, d_{yz}, and d_{xz}) lie between the *x*, *y*, and *z* axes:

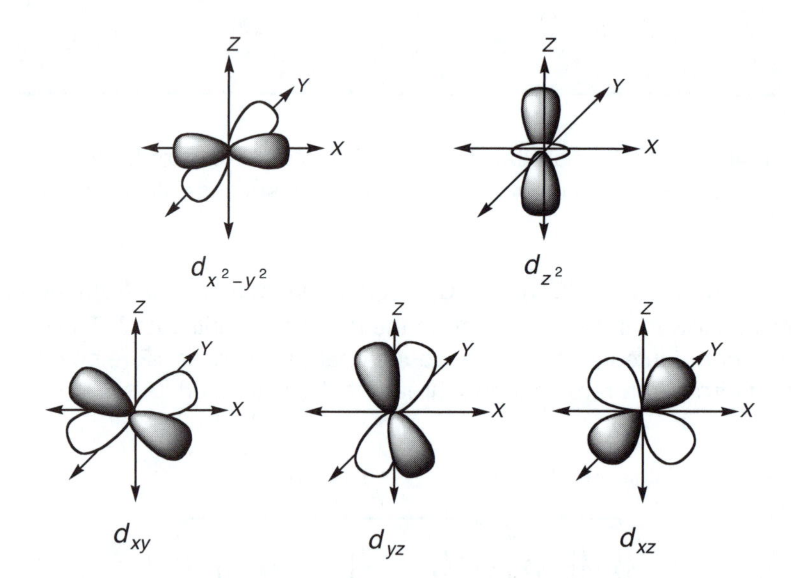

When ligands approach a metal ion along *x*, *y*, and *z* axes, as they do for an octahedral complex, they interact most strongly with electrons in metal orbitals that lie along the axes (the $d_{x^2-y^2}$ and d_{z^2} orbitals). The ligands interact less strongly with electrons in metal orbitals which lie between *x*, *y*, and *z* axes, (d_{xy}, d_{yz}, and d_{xz} orbitals), so these orbitals are lower in energy:

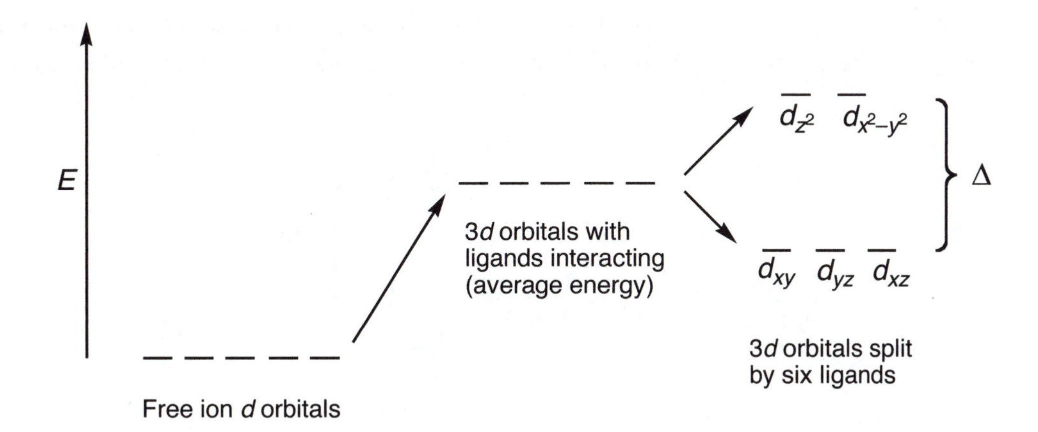

The amount that the *d* orbitals split (called the **crystal field splitting energy,** Δ) depends on how strongly the ligands interact with the metal ion electrons. Some ligands, such as CN^-, interact strongly with metal ion electrons; others, such as water, interact less strongly, and split the *d* orbitals less. Now we have a way to explain the color differences of complex ions. The extent of splitting determines the wavelength of light absorbed, and therefore the color that we see when we look at a complex ion. A large split in the *d* orbitals absorbs a shorter wavelength of light (higher energy); a smaller split absorbs a longer wavelength of light (lower energy).

The order of ligands in their increasing ability to split the *d* orbitals of a metal ion is called the spectrochemical series:

$$Cl^- < F^- < H_2O < SCN^- < NH_3 < \text{en} < NO_2^- < CN^- < CO$$

weak field ligands → strong field ligands

Let's see how the spectrochemical series can explain why $Fe(CN)_6^{4+}$ is diamagnetic and $Fe(H_2O)_6^{2+}$ is paramagnetic with four unpaired electrons. CN^- is a strong field ligand and splits the *d* orbitals farther apart than H_2O, which is a weaker field ligand. When the split in *d* orbitals is large, it takes less energy to pair electrons in the lower energy *d* orbitals than to fill the higher energy orbitals with unpaired electrons:

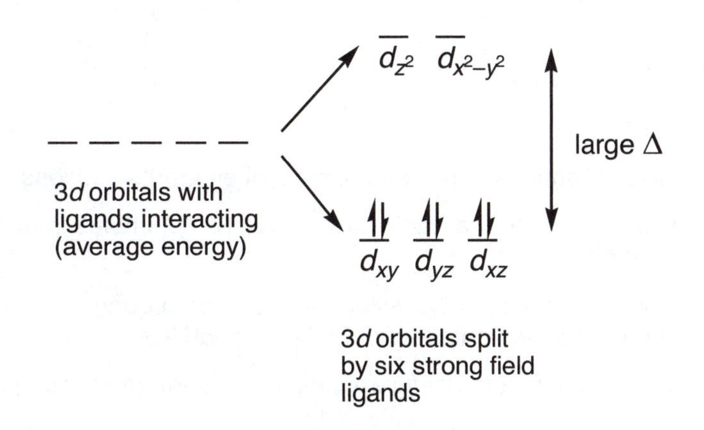

Low-Spin Complex

The six Fe^{2+} electrons are paired in the lower energy *d* orbitals. This type of complex is called a **low-spin complex.** Notice that in this complex, Hund's rule is violated because electrons pair before all *d* orbitals are filled with unpaired electrons.

H_2O is a weaker field ligand than CN^- and splits the *d* orbitals to a smaller extent. In this case, it requires more energy to pair electrons than to fill a higher energy *d* orbital with an unpaired electron:

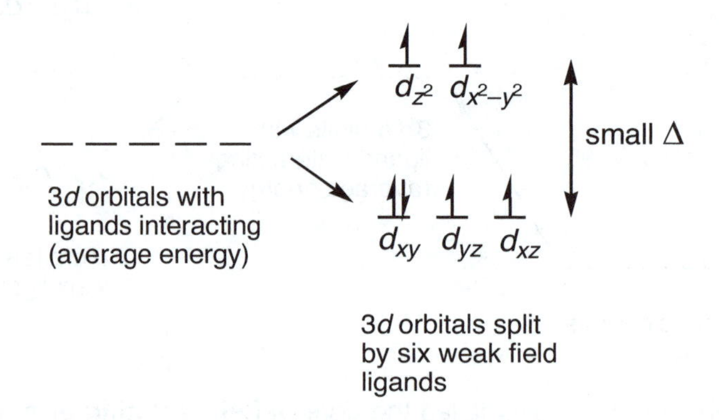

High-Spin Complex

Only two of the six Fe^{2+} electrons are paired, and this theory successfully predicts that $Fe(H_2O)_6^{2+}$ has 4 unpaired electrons. This type of complex is called a **high-spin complex;** Hund's rule is not violated in high-spin complexes.

In tetrahedral complexes, ligands approach the central metal ion <u>between</u> the *x*, *y*, and *z* axes, so the splitting of the *d* orbitals is the reverse of the octahedral case: the d_{xy}, d_{yz}, and d_{xz} orbitals become higher in energy than the $d_{x^2-y^2}$ and d_{z^2} orbitals. The splitting of the *d* orbitals is less than in octahedral complexes, so only high-spin tetrahedral complexes are known.

Square planar complexes split the *d* orbitals into 4 levels. The orbitals that interact most strongly with ligands in a square planar complex are those in the *xy* plane (the $d_{x^2-y^2}$ and d_{xy} orbitals).

Crystal field theory successfully predicts the color and magnetic behavior of complex ions, but it cannot explain why, for example, CO, which is only slightly polar, is a strong field ligand if only electrostatic forces are in effect. **Ligand field theory** includes a covalent contribution in the description of bonding in complex ions in order to explain this result.

CHAPTER CHECK

Make sure you can...

- Write electron configurations for transition metal elements and ions.
- Explain how lanthanide contraction causes transition elements in periods 5 and 6 to be about the same size.
- Determine oxidation states for transition metals in compounds, and identify the highest oxidation state for the elements in groups 3B through 7B.
- Predict, based on *d*-electron configurations, if transition metal compounds are colored or colorless.
- Give the charge of a complex ion and the charge of its central atom; identify all ligands and counter ions.
- Name and write formulas for coordination compounds.
- Determine how many geometric isomers could exist for a coordination compound.

- Use the spectrochemical series to determine if a complex is low or high spin and if it is paramagnetic or diamagnetic.

Chapter 23 Exercises

23.1) According to a dictionary, what does the stem of the word "chromium" mean?

23.2) a. What are the highest oxides of Sc, Ti, V, Cr, and Mn?

b. In terms of the formulas of their highest oxides, Sc, Ti, V, Cr, and Mn are like which elements in the second row of the *p*-block?

c. In what groups are Sc, Ti, V, Cr, and Mn?

23.3) Match the following:

1. Red rouge	a. TiO_2
2. Chrome oxide green	b. Cr_2O_3
3. White paint pigment	c. Fe_3O_4

23.4) Before recognition of the *f*-block, uranium was placed in the same family as what elements?

23.5) OsO_4 corresponds to what compound in the *p*-block?

23.6) Name the following coordination compounds or complex ions:

a. $K_2[CoCl_4]$
b. $[Cr(NH_3)_5I]I_3$
c. $K_3Fe(CN)_6$
d. $Ru(NH_3)_5Cl^{2+}$
e. $[Co(H_2O)_6]I_3$
f. $K_4[PtCl_6]$

23.7) Give formulas for the following:
a. triamminemonobromoplatinum(II) chloride
b. potassium hexafluorocobaltate(III)
c. sodium tri(oxalato)nickelate(II)
d. hexakispyridinecobalt(III) chloride
e. bis(ethylenediamine)dinitroiron(III) sulfate
f. tetramminemonochloromononitrocobalt(III) chloride

23.8) How many isomers are possible for $Co(en)_2(NH_3)(H_2O)$? Are any of the isomers optically active?

23.9) Match the compound on the left with the one on the right that would most closely approach its molar conductivity:

$[Pt(NH_3)_4]Cl_2$	KCl
$[Pt(NH_3)_2Cl_2]$	KNO_3
$K[Pt(NH_3)Cl_3]$	K_2SO_4
$K_2[PtCl_4]$	$CaCl_2$
$[Pt(NH_3)_3Cl]Cl$	CH_3OH

23.10) How many isomers can you draw for the following complexes? (M = metal; X,Y,Z = ligands)

a. MX_5Y
b. MX_4Y_2
c. MX_4YZ
d. MX_3Y_3
e. MX_2YZ (tetrahedral)
f. MX_2YZ (square planar)

23.11) How many octahedral complexes can you write using only a bidentate ligand such as ethylenediamine ("en"), and/or a monodentate ligand, A?

23.12) Consider transition metal ions containing from 0 to 10 *d* electrons. Which ions can form "high-spin" and "low-spin" octahedral complexes?

23.13) Draw an orbital diagram to describe the crystal field splitting for the Ni^{2+} ion in $[Ni(en)_3]Cl_2$. Would you expect Δ to be relatively large or small? Would you expect the compound to be paramagnetic or diamagnetic?

Chapter 23 Answers

23.1) Color. Most compounds of chromium are colored.

23.2)
a. Sc_2O_3, TiO_2, V_2O_5, CrO_3, Mn_2O_7
b. Al, Si, P, S, and Cl
c. 3, 4, 5, 6, 7

23.3) 1-c; 2-b; 3-a.

23.4) Cr, Mo, and W (based on its maximum oxidation state)

23.5) XeO_4

23.6)
a. potassium tetrachlorocobaltate(II) chloride
b. pentamminemonoiodochromium(IV) iodide
c. potassium hexacyanoferrate(III)
d. pentamminemonochlororuthenium(III) ion
e. hexaquocobalt(III) iodide
f. potassium hexachloroplatinate(II)

23.7)
a. $[Pt(NH_3)_3Br]Cl$
b. $K_3[CoF_6]$
c. $Na_4[Ni(C_2O_4)_3]$
d. $[Co(C_5H_5N)_6]Cl_3$
e. $[Fe(en)_2(NO_2)_2]_2SO_4$
f. $[Co(NH_3)_4(NO_2)Cl]Cl$

23.8)

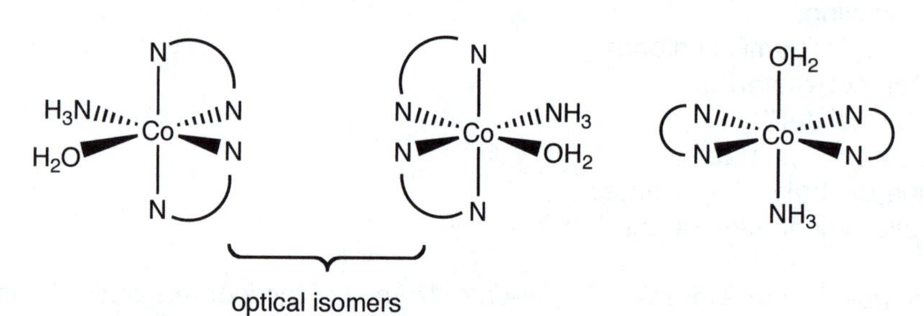

optical isomers

no plane of symmetry

non-superimposable mirror images

23.9)

$[Pt(NH_3)_4]Cl_2$	$CaCl_2$
$[Pt(NH_3)_2Cl_2]$	CH_3OH
$K[Pt(NH_3)Cl_3]$	KNO_3
$K_2[PtCl_4]$	K_2SO_4
$[Pt(NH_3)_3Cl]Cl$	KCl

23.10) a. 1, b. 2, c. 2, d. 2, e. 1, f. 2

23.11) 5 different complexes are possible:

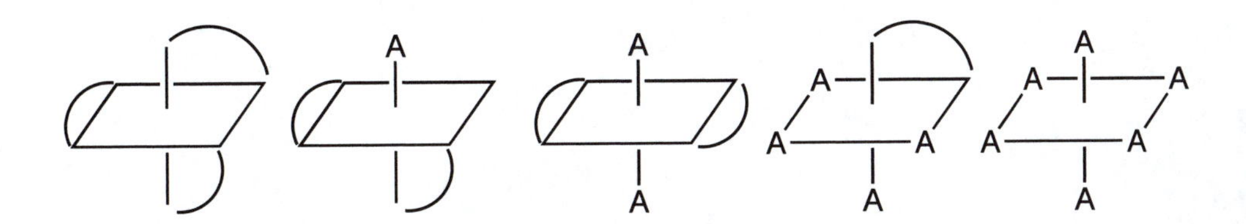

23.12) d^4, d^5, d^6, and d^7

23.13)

E

d_{z^2} $d_{x^2-y^2}$

Δ

3*d* orbitals with ligands interacting (average energy)

d_{xy} d_{yz} d_{xz}

3*d* orbitals split by six ligands

Free ion *d* orbitals

The "en" ligand is a relatively strong field ligand, so Δ is relatively large, and the compound is low spin. It has two unpaired electrons making it paramagnetic.

24

Nuclear Reactions and Their Applications

"The Chemistry of the Nucleus"

For many centuries chemists labored to change lead into precious gold, and eventually found that precious uranium turned to lead without any human effort at all.
–Isaac Asimov

INTRODUCTION

In Chapters 1-23 we studied chemical changes that occur as valence electrons are transferred or shared among atoms to form new compounds. In Chapter 2 we learned that mass is always conserved in chemical transformations; atoms are indestructible particles which are never transformed by chemical reactions, but which merely rearrange (Law of Mass Conservation). Now we consider reactions of the nucleus (nuclear reactions) in which elements *are* transformed, and mass is *not* conserved (the total quantity of mass-energy is conserved). The energy changes of nuclear reactions are so large—maybe a million times greater than those for chemical reactions—that changes in mass ($E = mc^2$) become detectable. Table 24.1 in your textbook provides a side-by-side comparison of chemical and nuclear reactions.

24.1 RADIOACTIVE DECAY

The nucleus is tiny and heavy—it is extremely dense—about 10^{14} g/mL. It contains protons (p^+) with a positive charge, and neutrons (n^o) with no charge. The protons and neutrons account for essentially all the mass of the atom. Many nuclei are unstable **(radioactive).** Over time, they decay as they emit radiation at a characteristic rate.

Terminology

nucleons: The particles that make up the nucleus, the protons and neutrons.

mass number (*A*): The total number of protons and neutrons in an atom.

notation: In Chapter 2, we used the notation for an atom: $^{A}_{Z}X$ where X is the symbol for an element, *A* is the mass number, and *Z* is the atomic number (the number of protons in the nucleus). We extend this notation to elementary particles (protons, electrons, and neutrons) where *Z* is the charge of the particle:

electron: $^{0}_{-1}e$ (electrons have essentially no mass)

proton: $^{1}_{1}p$

neutron: $^{1}_{0}n$ (neutrons have no charge)

nuclide: An atom with a specific nuclear composition. Nuclides are often designated with the element name followed by the mass number:

carbon-12, or carbon-14, for example.

isotopes: Atoms with the same number of protons (the same *Z*), but different numbers of neutrons, and therefore different mass numbers (different *A*).

alpha particle (α): a helium nuclei, $^{4}_{2}He$

beta particle (β or $^{0}_{-1}\beta$): fast electrons

gamma rays (γ): high-energy photons

parent: the decaying nuclide

daughter: the product nuclide

Review exercises

1) How many protons and neutrons are in the nucleus of a carbon-12 atom and a carbon-14 atom?

answer:

All carbon atoms contain 6 protons. The protons and neutrons together make up the mass of the nucleus:

Carbon-12 contains: 12 – 6 = 6 neutrons.

Carbon-14 contains: 14 – 6 = 8 neutrons.

2) When does atomic mass equal mass number?

answer:

Atomic mass equals the mass number when all atoms under consideration are the same isotope. In naturally occurring samples of an element where there is a mixture of isotopes, the atomic mass is the weighted average of the masses of all the isotopes.

Types of Radioactive Decay

1) **Alpha decay:** The loss of an alpha particle $\left({}^{4}_{2}\text{He}\right)$ from a nucleus. All elements heavier than lead (Pb; $Z = 82$), and a few lighter ones exhibit alpha decay. Bismuth (Z=83) is only slightly radioactive.

2) **Beta decay:** The ejection of a beta particle from the nucleus. A neutron converts into a proton and a β particle, which is expelled:

$${}^{1}_{0}\text{n} \rightarrow {}^{1}_{1}\text{p} + {}^{0}_{-1}\beta$$

3) **Positron decay:** The ejection of a positron from the nucleus. A proton converts into a neutron and a positron. Positron decay is the opposite of beta decay. A positron is the antiparticle of an electron:

$${}^{1}_{1}\text{p} \rightarrow {}^{1}_{0}\text{n} + {}^{0}_{1}\beta$$

Positron decay is not observed in natural radioactivity, but is a common mode of decay for radioisotopes produced in the laboratory.

4) **Electron capture:** The nucleus "captures" an electron from the lowest energy level. The net effect is the same as positron decay; a proton is converted into a neutron:

$${}^{1}_{1}\text{p} + {}^{0}_{-1}\text{e} \rightarrow {}^{1}_{0}\text{n}$$

When an electron moves down from a higher energy level to fill the vacancy left by the captured electron, an X-ray photon is generated.

5) **Gamma emission:** The radiation of high-energy gamma photons from an excited nucleus. Gamma emission accompanies most other types of decay. Gamma rays have no mass or charge and so do not change A or Z.

Radioactive decay exercise

Choose words from the list below to complete the following paragraphs:

a) −1	e) alpha	i) excited	m) lower
b) +1	f) beta	j) gamma	n) mass
c) 2	g) charge	k) higher	o) neutron
d) 4	h) electron	l) lead	p) proton

A helium nuclei contains ____________(1) protons and ____________(2) neutrons in its nucleus, giving it a mass number (A) of ____________(3) and a charge (Z) of ____________(4). A helium nuclei is called an ____________(5) particle. An atom that undergoes alpha decay reduces its mass number by ____________(6) and its charge by ____________(7). Every element heavier than ____________(8) undergoes alpha decay.

(continued on next page)

A beta particle is a "fast electron" that is generated when a neutron converts into a _____________(9). Protons and neutrons have the same _____________(10), but a proton has a charge of _____________(11) while a neutron is neutral. Like a valence electron in an atom, a beta particle has no _____________(12) and has a charge of _____________(13). In beta decay, because the product nuclide has one more proton than the parent, the element with the next _____________(14) atomic number forms.

A positron is the "antiparticle" of an _____________(15). Like an electron, a positron has no mass, but instead of a –1 charge, it has a charge of _____________(16). Positron decay is the opposite of _____________(17) decay. In positron decay, a positron is generated when a proton converts into a _____________(18). Because the product nuclide has one less proton than the parent, the element with the next _____________(19) atomic number forms.

Electron capture transforms a proton into a _____________(20) just as in positron decay. However, instead of emitting a positron, a nucleus draws in a surrounding _____________(21), usually one from the lowest energy level. When a higher energy electron fills the orbital vacancy, an X-ray photon is generated.

Many nuclear changes leave the nucleus in an _____________(22) state. These excited nuclei radiate high-energy _____________(23) photons. Gamma rays have no _____________or _____________(24, 25), so gamma emission does not change *A* or *Z*.

answers:

1-c, 2-c, 3-d, 4-c, 5-e, 6-d, 7-c, 8-l, 9-p, 10-n, 11-b, 12-n, 13-a, 14-k, 15-h, 16-b, 17-f, 18-o, 19-m, 20-o, 21-h, 22-i, 23-j, (24,25)-g or n.

Classifying radioactive decay

Determine the type of radioactive decay for each of the following:

1) $^{214}_{84}Po \rightarrow {}^{4}_{2}He + {}^{210}_{82}Pb$

2) $^{63}_{28}Ni \rightarrow {}^{63}_{29}Cu + {}^{0}_{-1}\beta$

3) $^{14}_{6}C \rightarrow {}^{14}_{7}N + {}^{0}_{-1}\beta$

4) $^{11}_{6}C \rightarrow {}^{11}_{5}B + {}^{0}_{1}\beta$

5) $^{55}_{26}Fe + {}^{0}_{-1}e \rightarrow {}^{55}_{25}Mn$ + X ray

6) $^{238}_{92}U \rightarrow {}^{234}_{90}Th + {}^{4}_{2}He + 2\gamma$

7) $^{36}_{17}Cl + {}^{0}_{-1}e \rightarrow {}^{36}_{16}S$

8) $^{232}_{90}Th \rightarrow {}^{228}_{88}Ra + {}^{4}_{2}He$

answers:

1-alpha; 2-beta; 3-beta; 4-positron emission; 5-electron capture; 6-alpha, gamma emission; 7-electron capture; 8-alpha.

Balancing nuclear equations

Balance the following nuclear reactions:

1) ${}^{226}_{88}\text{Ra} \rightarrow {}^{4}_{2}\text{He} + \_\_\_$ 2) ${}^{38}_{19}\text{K} \rightarrow {}^{0}_{-1}\beta + \_\_\_$

solutions:

1) The answer is some element, ${}^{A}_{Z}\text{X}$. First we balance subscripts and superscripts:

$A: \ A + 4 = 226; \ A = 222$

$Z: \ Z + 2 = 88; \ Z = 86$

Now find the element with atomic number 86: ${}^{222}_{86}\text{Rn}$

2) Balance subscripts and superscripts:

$A: \ A + 0 = 38; \ A = 38$

$Z: \ Z - 1 = 19; \ Z = 20$

Now find the element with atomic number 20: ${}^{38}_{20}\text{Ca}$

Nuclear Stability

- All nuclei of atomic number greater than 83 ($Z > 83$) are unstable and decay. Bismuth-209 is the heaviest stable nuclide.

- Light stable nuclei have at least as many neutrons as protons.
  (${}^{1}_{1}\text{He}$ and ${}^{3}_{2}\text{He}$ are exceptions.)
  Heavier stable nuclei have more neutrons than protons.

- Nuclei with an even number of protons and neutrons are more stable than nuclei with odd numbers of protons and/or neutrons.

- There are *magic numbers* of N or Z values which make stable nuclei. Nuclides with N or Z values of 2, 8, 20, 28, 50, 82, and $N = 126$, are exceptionally stable. These numbers are believed to correspond to the numbers of protons or neutrons in filled nucleon shells. Nuclides with double magic numbers tend to be extremely stable.

  Examples of nuclides containing magic and double magic numbers are:

  ${}^{4}_{2}\text{He}$, ${}^{16}_{8}\text{O}$, ${}^{40}_{20}\text{Ca}$, and ${}^{208}_{82}\text{Pb}$

If a nuclide is neutron rich, it will undergo beta decay to convert neutrons into protons. If a nuclide is proton rich, it will undergo positron decay or electron capture to convert protons into neutrons. If a nuclide is heavy ($Z > 83$), it undergoes alpha decay which reduces Z and N values by two units per emission.

Predicting nuclear decay

Predict the mode of decay for the most abundant radioactive isotope of iron, iron-59.

solution:

Neutron rich iron-59 undergoes beta decay converting a neutron into a proton:

$$^{59}_{26}\text{Fe} \rightarrow\ ^{59}_{27}\text{Co} +\ ^{0}_{-1}\beta$$

Iron-59 has a half-life of 45.1 days, and is used as a radiotracer for red blood cells.

24.2 KINETICS OF RADIOACTIVE DECAY

The rate of decay of radioactive nuclei is a first-order process. The number of nuclei that decay over some period of time, t, is proportional to the number of nuclei present:

$$\textbf{Rate} = \frac{-\Delta N}{\Delta t} = kN$$

The integrated form of the rate law is:

$$\ln\left(\frac{N_t}{N_o}\right) = -kt$$

The half-life is the time required for half of the nuclides present to decay. For a first-order process, the half-life, $t_{1/2}$ is:

$$t_{\frac{1}{2}} = \frac{-\ln\left(\frac{\frac{1}{2}N_o}{N_o}\right)}{k} = \frac{\ln(2)}{k}$$

$$t_{\frac{1}{2}} = \frac{0.693}{k}$$

The half-lives of radioactive nuclei vary over a wide range: Uranium-235 has a half-life of 7.1×10^{6} years; lithium-8 has a half-life of 842 milliseconds.

Half-life of $^{222}_{86}\text{Rn}$

$^{222}_{86}\text{Rn}$ has a half-life of 3.82 days.

1) How long does it take for half of a 100 g sample of $^{222}_{86}\text{Rn}$ to decay?
2) How long does it take for half of a 1 g sample of $^{222}_{86}\text{Rn}$ to decay?
3) How much of the 100 g sample of $^{222}_{86}\text{Rn}$ remains after 10 days?

(continued on next page)

solutions:

1) 3.82 days
2) 3.82 days; half-life is independent of the number of nuclei present.
3) We can calculate k from the expression for half-life:

$$k = \frac{0.693}{t_{\frac{1}{2}}}$$

$$k = \frac{0.693}{3.82 \text{ days}} = 0.181 \text{ days}^{-1}$$

Now we can calculate the amount left after 10 days by using the integrated rate law:

$$\ln\left(\frac{N_t}{N_o}\right) = -kt$$

$$N_{10 \text{ days}} = N_o e^{-kt}$$

$$N_{10 \text{ days}} = 100 \text{ g} \times e^{-(0.181 \text{ days}^{-1})(10 \text{ days})}$$

$$N_{10 \text{ days}} = 16.4 \text{ g}$$

Check:

after 1 half-life, 50 g remains

after 2 half-lives, 25 g remains

after 3 half-lives, 12.5 g remains

10 days is about 2.6 half-lives. Our answer, 16.4 g, falls in between the amounts that would remain after 2 and 3 half-lives, so it seems reasonable.

We can count the number of disintegrations of radioactive nuclei per second (dps) to measure the concentration of a radioactive substance. **Radiocarbon dating** is used to determine the age of things that were once living. Cosmic rays in the atmosphere produce nearly constant amounts of carbon-14 from ^{14}N atoms:

$$^{14}_{7}\text{N} + ^{1}_{0}\text{n} \rightarrow ^{14}_{6}\text{C} + ^{1}_{1}\text{p}$$

The carbon-14 formed by this nuclear reaction eventually incorporates into the carbon dioxide in the air giving our atmosphere a constant concentration of carbon-14 atoms (there is about one atom of carbon-14 for every 10^{12} atoms of carbon-12). Living plants, which take in carbon dioxide, have this same $^{14}C/^{12}C$ ratio, as do plant-eating animals and animals who eat plant-eating animals. When an organism dies, the intake of ^{14}C stops, and the $^{14}C/^{12}C$ ratio decreases as ^{14}C decays. The half-life of ^{14}C is 5730 years.

Radiocarbon dating

A sample of linen from a funeral shroud taken from an Egyptian tomb has a carbon-14 content of 0.560 times that of a living plant. Estimate the age of the linen.

solution:

We first calculate the first-order rate constant from the expression for half-life.
(The half-life of ^{14}C is 5730 years.)

$$k = \frac{0.693}{5720 \text{ years}} = 1.21 \times 10^{-4} \text{ yr}^{-1}$$

Use the integrated rate law to solve for time:

$$\ln\left(\frac{N_t}{N_o}\right) = -kt$$

$$\ln\left(\frac{N_t}{N_o}\right) = -\left(1.21 \times 10^{-4} \text{ yr}^{-1}\right) \times t$$

$$N_t = 0.560\,(N_o), \text{ so}$$

$$\ln\left(\frac{0.560\, N_o}{N_o}\right) = \ln(0.560) = -0.580$$

$$-0.580 = -\left(1.21 \times 10^{-4} \text{yr}^{-1}\right) \times t$$

$t = 4790$ years

Note: This method assumes that the ratio of $^{14}C/^{12}C$ in the atmosphere has not changed over the years.

The first atomic explosion

On July 16, 1945, the first atomic bomb was detonated in the desert in New Mexico. What fraction of the strontium-90 ($t_{1/2} = 28.8$ yr) produced by the explosion remains today?

solution:

Calculate the first-order rate constant from the expression for half-life:

$$k = \frac{0.693}{28.8 \text{ yr}} = 2.41 \times 10^{-2} \text{ yr}^{-1}$$

Use the integrated rate law to solve for the fraction of strontium-90 remaining:

(continued on next page)

$$\ln\left(\frac{N_t}{N_o}\right) = -kt$$

$t = 69.0$ yr (on July 16, 2014)

$$\ln\left(\frac{N_t}{N_o}\right) = -\left(2.41 \times 10^{-2}\ \text{yr}^{-1}\right) \times (69.0\ \text{yr})$$

$$\ln\left(\frac{N_t}{N_o}\right) = -1.66$$

$$\frac{N_t}{N_o} = 0.190, \text{ less than 1/5 remains}$$

Check: After 28.8 yr (1 half-life), half of the strontium-90 remained.
After another 28.8 yr (in the year 2003), 1/4 of the strontium-90 remained.
At 69.0 yr, in the year 2014, less than 1/5 remains.

The age of rocks can be determined using "radioactive clocks" such as the radioactive decomposition of uranium-238 to lead-206, and the beta decay of rubidium-87 ($t_{1/2} = 5.7 \times 10^{10}$ years). Rocks dated by these and other radioactive metals gives an approximate age of the earth ($\approx$ 4.5 billion years old).

Nuclear decomposition of uranium-235

Uranium-235 undergoes the following decay series:

$$\alpha, \beta, \alpha, \beta, \alpha, \alpha, \alpha, \alpha, \beta, \alpha, \beta$$

What is the final product of decay of uranium-235? (Hint: See Isaac Asimov's observation at the start of the chapter.)

solution:

To find the final product of this decay series, we'll show the intermediate nuclides with successive decays:

$$^{235}_{92}\text{U} \rightarrow\ ^{4}_{2}\text{He} +\ ^{231}_{90}\text{Th}$$

$$^{231}_{90}\text{Th} \rightarrow\ ^{0}_{-1}\beta +\ ^{231}_{91}\text{Pa}$$

$$^{231}_{91}\text{Pa} \rightarrow\ ^{4}_{2}\text{He} +\ ^{227}_{89}\text{Ac}$$

$$^{227}_{89}\text{Ac} \rightarrow\ ^{0}_{-1}\beta +\ ^{227}_{90}\text{Th}$$

(continued on next page)

$$^{227}_{90}Th \rightarrow \ ^{4}_{2}He + \ ^{223}_{88}Ra$$

$$^{223}_{88}Ra \rightarrow \ ^{4}_{2}He + \ ^{219}_{86}Rn$$

$$^{219}_{86}Rn \rightarrow \ ^{4}_{2}He + \ ^{215}_{84}Po$$

$$^{215}_{84}Po \rightarrow \ ^{4}_{2}He + \ ^{211}_{82}Pb$$

$$^{211}_{82}Pb \rightarrow \ ^{0}_{-1}\beta + \ ^{211}_{83}Bi$$

$$^{211}_{83}Bi \rightarrow \ ^{4}_{2}He + \ ^{207}_{81}Tl$$

$$^{207}_{81}Tl \rightarrow \ ^{0}_{-1}\beta + \ ^{207}_{82}Pb$$

The final product of decay of uranium-235 is lead-207. The overall reaction is:

$$^{235}_{92}U \rightarrow \ ^{207}_{82}Pb + 7\,^{4}_{2}He + 4\,^{0}_{-1}\beta$$

Uranium-235 makes up only about 0.7% of naturally occurring uranium. The more abundant isotope, uranium-238 decays to lead-206, and it is the $^{238}U/^{206}Pb$ ratio that is used to date rocks. As you will see, uranium-235 undergoes fission while uranium-238 does not.

24.3 NUCLEAR TRANSMUTATION

Chemists can convert stable, nonradioactive nuclei to radioactive nuclei by bombarding them with neutrons or positive ions. The first radioactive isotopes made in the laboratory were achieved by bombarding stable isotopes with alpha particles. Bombardment of aluminum foil with alpha particles produced phosphorus-30 and neutrons:

$$^{27}_{13}Al + \ ^{4}_{2}He \rightarrow \ ^{30}_{15}P + \ ^{1}_{0}n$$

A shorthand notation for the above reaction is $^{27}Al(\alpha,n)\ ^{30}P$. Phosphorus-30 is radioactive and decays by positron emission. What are the products of this nuclear reaction? (answer: A positron, the antiparticle of an electron, is emitted as a proton converts into a neutron.)

$$^{30}_{15}P \rightarrow \ ^{30}_{14}Si + \ ^{0}_{1}\beta$$

Particle accelerators, such as the cyclotron (invented in 1930), use electromagnets to accelerate cations to very high velocities so they can overcome coulombic repulsion of the target nucleus and cause nuclear reactions.

Elements with atomic numbers greater than uranium (atomic number 92), the **transuranium** elements have been prepared by bombardment reactions.

Synthesis of transuranium elements

Elements 93-95 can be prepared by neutron bombardment; high-energy positive ions are used to prepare the heavier transuranium elements. Fill in the blanks for the examples below:

1) **Neptunium** and **Plutonium** (atomic numbers 93 and 94)

Uranium-238 is bombarded with neutrons to form uranium-239, which undergoes ___________ decay to form neptunium and plutonium:

$$^{238}_{92}U + ^{1}_{0}n \rightarrow ^{239}_{92}U$$

$$^{239}_{92}U \rightarrow ^{239}_{93}Np + \_\_\_\_$$

$$^{239}_{93}Np \rightarrow ^{239}_{94}Pu + \_\_\_\_$$

answers: beta decay; $^{0}_{-1}\beta$, $^{0}_{-1}\beta$.

2) **Curium** (atomic number 96)

Plutonium-239 is bombarded with alpha particles:

$$^{239}_{94}Pu + \_\_\_\_ \rightarrow ^{242}_{96}Cm + \_\_\_\_$$

answers: $^{4}_{2}He$, $^{1}_{0}n$.

3) **Element 104 (Rf/Ku)***

Californium-249 is bombarded with carbon-12:

$$^{249}_{98}Cf + \_\_\_\_ \rightarrow ^{257}_{104}Rf + \_\_\_\_$$

answer:

$^{12}_{6}C$, $4^{1}_{0}n$.

*Americans have suggested the name Rutherfordium honoring Ernest Rutherford. The Russians prefer the name Kurchatovium after the Russian physicist J.V. Kurchatov.

24.4 EFFECTS OF NUCLEAR RADIATION ON MATTER

Radioactive emissions interact with matter in two ways:

1) Excitation from nonionizing radiation

A relatively low energy particle collides with an atom that absorbs some of the energy and then reemits it. No electrons are lost from the atom. The material may warm up if the absorbed energy causes the atoms to vibrate or rotate more rapidly.

2) Ionization from ionizing radiation

Radiation collides with an atom with enough energy to dislodge an electron. A cation and free electron result, which often collide with another atom and eject a second electron.

The energy absorbed by a tissue is most commonly measured in **rads** (**r**adiation-**a**bsorbed-**d**ose). A rad is equal to 0.01 joule of energy absorbed per kilogram of body tissue:

$$\textbf{1 rad = 0.01 J/kg}$$

To measure tissue damage, we multiply the number of rads by a factor which depends on the effect a certain type of radiation has on a specific tissue (the RBE, relative biological effectiveness). This product is the **rem** (**r**oentgen **e**quivalent for **m**an):

$$\textbf{rems = rads} \times \textbf{RBE}$$

Doses are often expressed in millirems. (The SI unit for dosage equivalent is the **sievert (Sv).** 1 rem = 0.01 Sv.)

When ionizing radiation interacts with molecules, free radicals form. These highly reactive species attack chemical bonds, and in so doing, can damage cells, protective fatty tissue around organs, enzymes, proteins, and nucleic acids. The damage depends both on the penetrating power and the ionizing ability of the radiation. Table 24.7 in your textbook gives examples of typical radiation doses from natural and artificial sources. Notice that the amount of radiation most people are exposed to from nuclear testing and the nuclear power industry is about one-tenth of our exposure to cosmic radiation, and only about 2–3% of the radiation we might be exposed to from the ground and air in a brick house (mostly due to radon).

The penetrating power of emissions is inversely related to the mass and charge of the emission.

Penetrating Power of Radiation

Alpha particles

Alpha particles are large and highly charged; they interact with matter most strongly and penetrate minimally. Light clothing or outer skin stop alpha radiation.

Beta particles and positrons

Beta particles and positrons interact less strongly with matter, but penetrate more deeply than alpha particles. Specialized heavy clothing or a thick piece of metal stop beta particles.

Gamma rays

Gamma rays, with no mass or charge, interact least with matter and penetrate the most. A block of lead several inches thick is required to stop them.

Studies on animals and humans on radiation exposure indicate that either a high, single exposure, or low, chronic exposure is harmful to humans.

24.5 APPLICATIONS OF RADIOISOTOPES

Nonionizing Radiation

Radioisotopes have almost the same chemical properties as nonradioactive isotopes of the same element and thus, a very small amount of a radioisotope mixed with the stable isotope, acts as a **tracer** during a chemical reaction. Radioactive tracers are used to:

- study reaction mechanisms
- study the flow of materials in solid surfaces and the flow of large bodies of water
- determine the elemental composition of an object without destroying the sample (neutron activation analysis, NAA)
- observe organs and body parts
- follow glucose uptake and blood flow changes during brain activity (positron emission tomography, PET)

Ionizing Radiation

Ionizing radiation is used to:

- destroy cancerous tissue
- kill microorganisms that spoil food
- control insect populations by sterilizing captured males

24.6 INTERCONVERSION OF MASS AND ENERGY: FISSION AND FUSION

(Corresponds to sections 24.6 and 24.7 in your textbook)

Binding Energy

The mass of a nucleus is always less than the mass of its constituent protons and neutrons. Consider a carbon-12 atom. One carbon-12 atom weighs exactly 12.00000 amu. The mass of the nucleus is 12.00000 amu minus the mass of 6 electrons:

$$\text{mass of } ^{12}\text{C nucleus} = 12.00000 - 6(0.000549 \text{ amu}) = 11.99671 \text{ amu}$$

The mass of the constituents of a ^{12}C nucleus, 6 neutrons and 6 protons is:

$$\text{mass of 6 protons: } (6 \times 1.00728 \text{ amu}) = 6.04368 \text{ amu}$$

$$\text{mass of 6 neutrons: } (6 \times 1.00867 \text{ amu}) = 6.05202 \text{ amu}$$

$$12.09570 \text{ amu}$$

A ^{12}C nucleus weighs <u>less</u> than the sum of its parts. Mass is not conserved. The missing mass was converted into <u>energy</u> when the nucleus was formed. In order to break apart the nucleus, we would need to provide this energy (called the **binding energy**). We can calculate the binding energy using Einstein's equation:

$$\Delta E = \Delta mc^2$$

The binding energy for a carbon-12 nucleus

How much energy is required to break one mole of carbon-12 nuclei into protons and neutrons?

<u>solution</u>:

The difference in the mass of a carbon-12 nucleus, and the sum of its parts is called the **mass defect,** Δm:

$$\Delta m = m_{\text{final}} - m_{\text{initial}}$$

In this case, since we are breaking the nucleus apart, $m_{\text{final}} > m_{\text{initial}}$ and Δm is positive. If we were calculating the binding energy, $m_{\text{initial}} > m_{\text{final}}$, and Δm would be negative:

$$\Delta m = 12.09570 \text{ amu} - 11.99671 \text{ amu} = 0.09899 \text{ amu}$$

In one mole of carbon-12 nuclei, 0.09899 grams would be "missing":

$$\Delta m = 0.09899 \text{ g/mol} = 9.899 \times 10^{-5} \text{ kg/mol}$$

We use Einstein's equation to solve for ΔE:

$$\Delta E = \Delta m \times c^2$$

$$= 9.899 \times 10^{-5} \text{ kg/mol} \times (2.9979 \times 10^8 \text{ m/s})^2$$

$$= 9.899 \times 10^{-5} \text{ kg/mol} \times (8.9874 \times 10^{16} \text{ m}^2/\text{s}^2)$$

$$= 8.897 \times 10^{12} \text{ kg m}^2/\text{s}^2 \text{ mol} = 8.897 \times 10^{12} \text{ J/mol}$$

$$= 8.897 \times 10^9 \text{ kJ/mol}$$

<u>Note</u>: Energy is required to break apart a nucleus, so ΔE is positive; the binding energy is the energy <u>released</u> when a nucleus forms, so it has a negative ΔE value.

A more convenient energy unit for binding energies is a megaelectron volt, MeV (million electron volts):

$$\mathbf{1 \text{ MeV} = 1.602 \times 10^{-13} \text{ J}}$$

A useful conversion factor converts a <u>mass</u> defect (in amu) directly to its <u>energy</u> equivalent in electron volts:

$$\mathbf{1 \text{ amu} = 931.5 \text{ MeV}}$$

Conversion of mass defect into its energy equivalent

Calculate the binding energy of a carbon-12 atom in MeV given:

a) the binding energy = -8.897×10^{12} J/mol

b) the mass defect, $\Delta m = -0.09899$ amu/nucleus

solution:

a) We use the equality 1 MeV = 1.602×10^{-13} J:

$$\left(-8.897 \times 10^{12} \text{ J/mol}\right) \times \left(\frac{1 \text{ MeV}}{1.602 \times 10^{-13} \text{ J}}\right) = -5.554 \times 10^{25} \text{ MeV/mol}$$

$$-5.554 \times 10^{25} \text{ MeV/mol} \times \frac{1}{6.022 \times 10^{23} \text{ nucleus/mol}} = -92.22 \text{ MeV/nucleus}$$

b) We use the equality 1 amu = 931.5 MeV:

$$\frac{-0.09899 \text{ amu}}{\text{nucleus}} \times \frac{931.5 \text{ MeV}}{\text{amu}} = -92.21 \text{ MeV/nucleus}$$

We can compare the stability of nuclides of different elements by comparing their binding energy per nucleon. For ^{12}C, we have:

$$\frac{-92.21 \text{ MeV}}{\text{nucleus}} \times \frac{\text{nucleus}}{12 \text{ nucleons}} = -7.684 \text{ MeV/nucleus}$$

The most stable nucleus is $^{56}_{26}Fe$. Its binding energy per nucleon is -8.790 MeV. The binding energy per nucleon increases for elements up to $A \approx 60$, and then slowly decreases. Nuclei can gain stability by forming nuclei closer to $A \approx 60$. Heavier nuclei can split into lighter nuclei **(fission);** lighter nuclei can combine to form heavier nuclei **(fusion).**

Uranium-235 can be split into fragments by neutron bombardment. Uranium-238 (the more abundant isotope) does not undergo fission. During World War II, several processes were studied to separate the two isotopes. About 82 million kJ of energy is given off for every gram of U-235 that reacts. In comparison, the heat of combustion of coal is about 33 kJ/g.

Fusion reactions produce 3 to 10 times more energy per gram of starting material as fission processes (see end of Chapter Problem 24.9). Fusion reactions produce stable isotopes rather than hazardous radioactive isotopes formed by fission. The fusion of two deuteriums forms helium:

$$^{2}_{1}H + ^{2}_{1}H \rightarrow ^{4}_{2}He$$

The fusion of only a small percent of deuterium in seawater would supply the annual energy requirements of the world. Unfortunately, nuclear fusion requires extremely high temperatures and so is not yet a practical energy source.

CHAPTER CHECK

Make sure you can...

- Use the $^{A}_{Z}X$ notation to express the mass and charge of a particle.
- Define nucleon, mass number, nuclide, and isotope.
- Define alpha, beta, and positron decay, electron capture, and gamma emission.
- Classify the type of radioactive decay from a nuclear equation.
- Balance nuclear equations.
- Predict nuclear stability based on N/Z ratios and the N and Z values.
- Recognize the magic numbers of N or Z values that make stable nuclei.
- Predict the mode of radioactive decay for a nuclide.
- Calculate the half-life of radioactive nuclei.
- Estimate the age of an object from the half-life of an isotope.
- Write and balance equations for nuclear transmutation.
- Describe the two ways radioactive emissions interact with matter.
- Define a rad and a rem, and calculate radiation dose.
- Relate the mass and charge of an emission to its penetrating power.
- List several uses of nonionizing radiation; list several uses for ionizing radiation.
- Calculate mass defect and its energy equivalent.
- Calculate binding energy per nucleon, and use it to compare nuclide stability.

Chapter 24 Exercises

24.1) Predict the type of nuclear decay for the following nuclei:

a. $^{102}_{47}Ag$ b. $^{226}_{90}Th$ c. $^{24}_{11}Na$ d. $^{83}_{35}Br$

24.2) Complete the following nuclear equations:

a. $^{231}_{91}Pa \rightarrow ^{227}_{89}Ac + ?$

b. $^{166}_{67}Ho \rightarrow ? + ^{0}_{-1}e$

c. $? \rightarrow ^{239}_{93}Np + ^{4}_{2}He$

d. $^{59}_{29}Cu \rightarrow ? + ^{0}_{+1}e$

24.3) Complete the following nuclear equations:

a. $^{131}_{53}I \rightarrow ? + ^{0}_{-1}e$

b. $? \rightarrow ^{222}_{86}Rn + ^{4}_{2}He$

c. $^{67}_{31}Ga + ^{0}_{-1}e \rightarrow ? + \text{X ray}$

d. $? \rightarrow ^{32}_{16}S + ^{0}_{-1}e$

24.4) Complete the following nuclear transformations:

a. $^{239}_{94}Pu + ^{4}_{2}He \rightarrow ? + ^{1}_{0}n$

b. $^{249}_{98}Cf + ? \rightarrow ^{260}_{105}Ha + 4^{1}_{0}n$

c. $^{242}_{96}Cm + 17^{1}_{0}n \rightarrow ? + 8^{0}_{-1}e$

d. $? + ^{54}_{24}Cr \rightarrow ^{262}_{107}Uns + ^{1}_{0}n$

24.5) Uranium-235 undergoes radioactive decay to the final product lead-207. How many alpha and beta particles are emitted during this decay?

24.6) If a radioactive series starts with $^{237}_{93}Np$ and ends with $^{209}_{83}Bi$, how many alpha and beta particles are emitted?

24.7) Technetium-99^m is the radioisotope of choice in diagnostic medical imaging because it is only a gamma ray emitter and has a half-life of only 6.00 hours. If a young patient was administered 40 mCi, how long would it take for 90.0% of the ^{99m}Tc to decay?

24.8) Gadolinium-153 is used in the determination of bone density when diagnosing osteoporosis. If a patient is administered 120 μCi of ^{153}Gd, how much ^{153}Gd remains in her system one year later? The half-life of gadolinium-153 is 242 days.

24.9) Compare the energy obtained from a fusion process and a fission process. Calculate the amount of energy evolved, in kJ per gram, for:

a. fusion reaction: $^{2}_{1}H + ^{2}_{1}H \rightarrow ^{4}_{2}He$ (mass of ^{2}H = 2.0140 amu; mass of ^{4}He = 4.00260 amu)

b. fission reaction: $^{235}_{92}U \rightarrow ^{90}_{38}Sr + ^{144}_{58}Ce + ^{1}_{0}n + 4^{0}_{-1}e$

(masses in amu: ^{235}U = 234.9934; ^{90}Sr = 89.8864; ^{144}Ce = 143.8816; $^{1}_{0}n$ = 1.008665; $^{0}_{-1}e$ = 0.00055)

Chapter 24 Answers

24.1) a. electron capture or positron decay; b. alpha particle decay; c. beta particle decay; d. beta particle decay

24.2) a. $^{231}_{91}Pa \rightarrow ^{227}_{89}Ac + ^{4}_{2}\alpha$

b. $^{166}_{67}Ho \rightarrow ^{166}_{68}Er + ^{0}_{-1}e$

c. $^{243}_{95}Am \rightarrow ^{239}_{93}Np + ^{4}_{2}He$

d. $^{59}_{29}Cu \rightarrow ^{59}_{28}Ni + ^{0}_{+1}e$

24.3) a. $^{131}_{53}I \rightarrow ^{131}_{54}Xe + ^{0}_{-1}e$

b. $^{226}_{88}Ra \rightarrow ^{222}_{86}Rn + ^{4}_{2}He$

c. $^{67}_{31}Ga + ^{0}_{-1}e \rightarrow ^{67}_{30}Zn$ + X ray

d. $^{32}_{15}P \rightarrow ^{32}_{16}S + ^{0}_{-1}e$

24.4) a. $^{239}_{94}Pu + ^{4}_{2}He \rightarrow ^{242}_{96}Cm + ^{1}_{0}n$

b. $^{249}_{98}Cf + ^{15}_{7}N \rightarrow ^{260}_{105}Ha + 4^{1}_{0}n$

c. $^{242}_{96}Cm + 17^{1}_{0}n \rightarrow ^{259}_{104}Ra + 8^{0}_{-1}e$

d. $^{209}_{83}Bi + ^{54}_{24}Cr \rightarrow ^{262}_{107}Uns + ^{1}_{0}n$

24.5) The mass change from ^{235}U to ^{207}Pb is 28 amu. Each alpha particle has a mass of 4 amu, so 28/4 = 7 alpha particles are emitted. Each alpha particle has a charge of +2, so a decrease in 7 × 2 = 14 on the charge results from the emission of the 7 alpha particles. Charge = 92 – 14 = 78. Lead has an atomic number of 82, so 4 beta particles must be emitted to balance the charge to 82:

$$^{235}_{92}U \rightarrow ^{207}_{82}Pb + 7^{4}_{2}He + 4^{0}_{-1}e$$

(The example on pages 431 and 432 shows the individual steps for this decay.)

24.6) The mass change from ^{237}Np to ^{209}Bi is 28 amu. 7 alpha particles are emitted. The decrease in charge from the alpha particles is 14; 4 beta particles are emitted to balance charge:

$$^{237}_{93}Np \rightarrow ^{209}_{83}Bi + 7^{4}_{2}He + 4^{0}_{-1}e$$

24.7) 19.8 hours

24.8) 42.2 μCi remains

24.9) For the fusion process:

Calculate the change in mass per mole of product:

$$\Delta m = 4.00260\ g - 2(2.0140 g) = -0.02540\ g = -2.54 \times 10^{-5}\ kg$$

The change in energy is: $\Delta E = \Delta mc^2$

$$\Delta E = (-2.54 \times 10^{-5}\ \text{kg}) \times (2.9979 \times 10^{8}\ \text{m/s})^2 = -2.28 \times 10^{12}\ \text{J/mol or } -2.28 \times 10^{9}\ \text{kJ/mol}$$

The energy change for a gram of deuterium is:

$$-2.28 \times 10^{9}\ \text{kJ/mol} \times 1\ \text{mol product/2 mol }^{2}\text{H} \times 1\ \text{mol }^{2}\text{H/2.0140 g} = -5.667 \times 10^{8}\ \text{kJ/g}$$

For the fission process:

$$\Delta m = 89.8864\ \text{g} + 143.8816\ \text{g} + 1.008665\ \text{g} + 4(0.00055\ \text{g}) - 234.9934 = -0.2145\ \text{g}$$

$$\Delta E = (-2.145 \times 10^{-4}\ \text{kg}) \times (2.9979 \times 10^{8}\ \text{m/s})^2 = -1.928 \times 10^{13}\ \text{J/mol or } -1.928 \times 10^{10}\ \text{kJ/mol}$$

For one g of ^{235}U: -1.928×10^{10} kJ/mol /235 g = -8.2×10^{7} kJ/g

The fusion process provides about 7 times as much energy per gram of starting material as the fission process. Coal provides about 33 kJ of energy per gram, approximately two and a half million times less than fusion!

NOTES

NOTES

NOTES

NOTES

NOTES

NOTES

NOTES

NOTES

NOTES

NOTES